Human Health Risk Assessment of Capstone Depleted Uranium Aerosols

Attachment 3 of
Depleted Uranium Aerosol Doses and Risks:
Summary of U.S. Assessments

R. A. Guilmette[a]
M. A. Parkhurst[b]
G. Miller[a]
F. F. Hahn[c]
L. E. Roszell[d]
E. G. Daxon[e]
T. T. Little[a]
J. J. Whicker[a]
Y. S. Cheng[c]
R. J. Traub[b]
G. M. Lodde[f]
F. Szrom[d]
D. E. Bihl[b]
K. L. Creek[a]
C. B. McKee, *Project Administrator*[f]

Prepared for the US Army by Battelle under the
Chemical and Biological Defense Information Analysis Center
Task 241, DO 0189, Aberdeen, Maryland

[a]Los Alamos National Laboratory
[b]Battelle, Richland
[c]Lovelace Respiratory Research Institute
[d]U.S. Army Center for Health Promotion and Preventive Medicine
[e]Battelle, San Antonio
[f]Battelle Eastern Science and Technology Center

Battelle Press
Columbus • Richland

DISCLAIMER

For online electronic copies, please refer to www.deploymentlink.osd.mil/, or for hard copies and CDs, please refer to www.battelle.org/bclscrpt/Bookstore/default.cfm.

The Chemical Warfare/Chemical and Biological Defense Information Analysis Center is a Department of Defense Information Analysis Center operated by Battelle Memorial Institute and administered by Defense Information Systems Agency, Defense Technical Information Center (DTIC-AI), Fort Belvoir, VA 22060-6218.

LIBRARY OF CONGRESS CATALOGING-IN-PUBLICATION DATA

Depleted uranium aerosol dose and risk. Attachment 3, Human health risk assessment of Capstone depleted uranium aerosols / R.A. Guilmette . . . [et al.].

p. cm.

"Prepared for the U.S. Army by Battelle under the Chemical and Biological Defense Information Analysis Center Task 241, DO 0189, Aberdeen, Maryland."

"Conducted under the auspices of the U.S. Army Center for Health Promotion and Preventive Medicine"—Acknowledgments p.

ISBN 1-57477-147-7 (alk. paper)

1. Uranium—Health aspects. 2. Uranium—Environmental aspects. 3. United States. Army—Weapons systems. 4. United States. Army—Sanitary affairs. 5. Projectiles. 6. Tank warfare—Health aspects. I. Title: Human health risk assessment of Capstone depleted uranium aerosols. II. Guilmette, Ray A. III. Capstone DU Aerosol Characterization and Risk Assessment Program. IV. Chemical Warfare/Chemical and Biological Defense Information Analysis Center (U.S.) V. U.S. Army Center for Health Promotion and Preventive Medicine.

RA1231.U7D467 2005
363.17'91—dc22

2005042818

Printed in the United States of America

Battelle Press
505 King Avenue
Columbus, Ohio 43201-2693, USA
614-424-6393 or 1-800-451-3543
Fax: 614-424-3819
E-mail: press@battelle.org
Website: www. battelle.org/bookstore

Participating Organizations

Battelle *The* Business *of* Innovation

Lovelace

USACHPPM *Readiness thru Health*

Pacific Northwest National Laboratory
Operated by Battelle for the U.S. Department of Energy

Los Alamos NATIONAL LABORATORY

Chemical and Biological Defense CBIAC Information Analysis Center

Acknowledgments

This Human Health Risk Assessment was conducted under the auspices of the U.S. Army Center for Health Promotion and Preventive Medicine, LTC Mark A. Melanson, PhD, CHP, Health Physics Program Manager. Mr. David P. Alberth, Task Sponsor Representative (Health Risk Assessment for Exposure to Depleted Uranium—CBIAC Task 241) and USACHPPM Master Consultant, provided programmatic guidance and management of the technical review process.

The Human Health Risk Assessment could not have been completed without the oversight, support, and encouragement of Ms. Ellen P. Embrey, the Deputy Assistant Secretary of Defense for Force Health Protection and Readiness (in the Office of the Assistant Secretary of Defense for Health Affairs) and Director of the Deployment Health Support Directorate; and her staff within the Deployment Health Support Directorate.

Many people from many organizations supported this analysis and whose efforts are truly appreciated by the authors. Some of those providing the most direct support included the following:

U.S. Army Center for Health Promotion and Preventive Medicine (USACHPPM): LTC Mark Melanson ensured the availability of his technical staff including Dr. Jerry Falo, certified health physicist, to support the many contributions they made to the overall project. Ms. Robyn B. Lee, USACHPPM statistician, provided consultation.

Battelle Eastern Science and Technology(Best) Center): Ms. Christiana Lenane, executive administrator.

Battelle (Richland)): Publishing staff—Mr. Cary Counts, technical editor, Ms. Jean Cheyney, and Ms. Barbara Wilson, document production.

Lovelace Respiratory Research Laboratory (LRRI): Ms. Sandy Spurlock, librarian.

We wish to thank the independent peer reviewer science work group who contributed richly to project guidance by reviewing and commenting on the DQOs, our modeling approach, and the draft version of this report. This team of peer reviewers for the

DQOs and the draft report, unless otherwise specified, was led by Dr. Roy Reuter and included (in alphabetical order) Dr. Paul Baron, Dr. Wesley Bolch (who reviewed the draft report), Dr. John Doull, Dr. Rogene Henderson (who reviewed the DQOs), Dr. David Hoel, Dr. Anthony James (who reviewed the draft report), Dr. Morton Lippmann, Dr. Paul Strickland, Dr. Arthur Upton, and Dr. Wesley Van Pelt. The reviewers for the modeling methodology were Dr. Wesley Bolch, Dr. Keith Eckerman, Dr. Anthony James, and Dr. Otto Raabe.

And finally, we thank the reviewers in the U.S. DoD community for their assistance in ensuring that the information relative to the DoD health risk categories are appropriately described and that our suggested application of the health risk results to combat situations are realistic.

Working Group Authors

The Capstone Risk Assessment Working Group authors responsible for this effort are listed below along with a short statement of their areas of expertise that were specifically relevant in assessing DU intakes, concentrations, doses, and risks based on use of the Capstone measurement data.

Dr. Raymond A. Guilmette
Health Physicist, Radiation and Inhalation Toxicologist, Dose Assessor, Aerosol Scientist

Ms. Mary Ann Parkhurst
Project Manager, Health Physicist Specializing in Depleted Uranium

Dr. Guthrie Miller
Physicist/Mathematician and Computer Modeler, Dose Assessor, Bayesian Specialist

Dr. Fletcher H. Hahn
Radiation Biologist and Toxicologic Pathologist

Dr. Laurie E. Roszell
Toxicologist, Diplomate of the American Board of Toxicology

Dr. Eric G. Daxon
Retired Colonel, U.S. Army, Certified Health Physicist

Dr. Thomas T. Little
Health Physicist, Dose Assessor Specializing in Uranium

Mr. Jeffrey J. Whicker
Health Physicist, Aerosol Scientist

Dr. Yung-Sung Cheng
Theoretical and Experimental Aerosol Scientist

Dr. Richard J. Traub
Internal Dosimetrist, Biokinetic Modeler

Mr. Gordon M. Lodde
Retired Lieutenant Colonel, U.S. Army, Radiation Specialist

Ms. Frances Szrom
Certified Health Physicist, Specializing in Depleted Uranium

Mr. Donald E. Bihl
Certified Health Physicist, Occupational and Internal Dosimetry

Ms. Kathy L. Creek
Industrial Hygienist, Aerosol Scientist

Dr. Chad B. McKee
Project Administrator, Liaison with USACHPPM

Contents

Acknowledgments . . . iii
Working Group Authors . . . iv
Figures and Tables . . . ix
Executive Summary . . . xv

1.0 Introduction . . . 1
1.1 Depleted Uranium Background Information . . . 2
1.2 Objectives . . . 4
1.3 Process . . . 5
1.4 References . . . 6

2.0 Exposure Scenarios . . . 9
2.1 OSAGWI Exposure Levels . . . 10
2.2 Level I "In-Vehicle" Scenario Modeling Assumptions . . . 12
2.2.1 Individual in the Vehicle during Vehicle Perforation . . . 15
2.2.2 Individuals in the General Area of the Incident and Quickly Responding to the Situation . . . 15
2.3 References . . . 16

3.0 Development of Input Parameter Values . . . 17
3.1 Risk Assessment Parameters . . . 17
3.1.1 Source-Term Analysis . . . 18
3.1.2 Physiological Parameters . . . 20
3.1.3 General Parameters . . . 22
3.2 Capstone Input Data . . . 23
3.2.1 Capstone Test Firing Scenarios . . . 24
3.2.2 Capstone Aerosol Collection and Characterization . . . 26
3.2.3 DU Aerosol Concentrations as a Function of Time . . . 28
3.2.4 Aerosol Particle Size Distribution as a Function of Time . . . 35
3.2.5 DU Aerosol *In Vitro* Solubility/Dissolution . . . 44
3.2.6 Effects of Ventilation Rate on DU Aerosol Concentration . . . 44
3.3 References . . . 52

4.0 Chemical and Radiological Dose Modeling . . . 55
4.1 Human Respiratory Tract Model . . . 55
4.1.1 Structure of the HRTM . . . 56
4.1.2 Physiology of the HRTM . . . 56
4.1.3 Deposition Model . . . 56
4.1.4 Clearance Model . . . 57
4.2 Gastrointestinal Tract Model . . . 60
4.3 Uranium Systemic Biokinetic Model . . . 61
4.4 Dose Calculations for Level I Scenarios Using Capstone Data . . . 63
4.4.1 DU Concentration and Particle Size Distribution . . . 64
4.4.2 Aerosol Lung Dissolution Parameters . . . 70
4.4.3 Biokinetic and Dosimetric Models . . . 76
4.5 Example Calculation of E(50) and $H_T(50)$ . . . 76
4.5.1 Theory . . . 77
4.5.2 Implementation . . . 78
4.6 Software Quality Assurance . . . 81
4.7 References . . . 87

5.0 Chemical Concentrations and Radiological Doses . . . 91
5.1 Radiological Doses . . . 92
5.1.1 Intakes and Committed Effective Doses . . . 92
5.1.2 Organ and Tissue Doses . . . 94
5.1.3 DU Doses from Double Shots . . . 94
5.2 Uranium Concentrations in the Kidney . . . 97
5.2.1 Peak Uranium Concentrations in the Kidney . . . 98
5.2.2 Double Shots and Peak Uranium Concentrations in the Kidney . . . 101
5.3 DU Mass Concentration in the Lung . . . 103
5.4 Modeling Confounders . . . 102
5.4.1 Ventilation Effects on Doses and Concentrations . . . 102
5.4.2 Effect of Mouth Breathing on Dose . . . 103
5.5 Uncertainty in Dose Calculations . . . 105
5.5.1 Measurement Uncertainty . . . 105
5.5.2 Modeling Uncertainty . . . 107
5.6 Summary . . . 119
5.7 References . . . 119

6.0 Human Health Risk Assessment . . . 121
6.1 Background . . . 121
6.2 Uranium Concentrations in Organs and Tissues . . . 122
6.2.1 Biokinetics of Uranium . . . 122
6.2.2 Organ Distribution of Uranium in Human Populations . . . 126
6.3 Chemical Toxicity of Uranium . . . 127
6.3.1 Effects in the Kidney . . . 128
6.3.2 Effects to the Lung . . . 138
6.3.3 Effects to Other Systems . . . 140
6.4 Chemical Health Risks of Inhaled DU Aerosols . . . 142
6.4.1 Abrams Tank Perforated Through Non-DU Armor (Phase I) . . 143
6.4.2 Bradley Vehicle Perforated Through Conventional Armor (Phase II) . . . 144
6.4.3 Abrams Tank Perforated Through DU Armor (Phase III) 145
6.4.4 Fully Operational Abrams Tank Perforated Through DU Armor (Phase IV) . . . 146

6.4.5 Comparison of Risks of Kidney Dysfunction by Phase and Scenario 147
6.4.6 Nonradiological Risks to Lung from Inhaled DU 149
6.5 Radiation Toxicity of Uranium 150
6.5.1 Epidemiology of Uranium Exposure 150
6.5.2 Effects in the Lung 152
6.5.3 Effects in the Bones 152
6.5.4 Other Organs 153
6.6 Radiological Health Risks of Inhaled DU Aerosol 153
6.6.1 Risk Coefficients for Total Cancer Mortality 154
6.6.2 Risk Coefficients for Lung Cancer 154
6.6.3 Risk Coefficients for Cancer in Other Organs 157
6.6.4 Comparison of Whole Body Cancer Risks and Summed Organ-Specific Cancer Risks 158
6.6.5 Cancer Risks by Phase and Scenario 160
6.7 Summary of Health Risks 163
6.7.1 Chemical Health Effects 163
6.7.2 Radiological Health Effects 164
6.8 References 166

7.0 Management of DU Aerosol Risks 173
7.1 Military Risk Management 174
7.2 Military Radiation Exposure Control Philosophy 176
7.2.1 "Peacetime" Occupational Radiation Protection Standards 177
7.2.2 The U.S./NATO Operational Radiation Control System: Operational Exposure Guidance 176
7.3 Risk Assessment Approach and Assumptions 179
7.3.1 Application of Radiation Exposure Status to the Risk Estimates 181
7.3.2 Comparison of DU Best-Estimate Risk with U.S./NATO RES Categories 183
7.3.3. Risks of Adverse Effects to the Kidney 186
7.3.4 Multiple Perforations 187
7.4 Summary of Risks 188
7.5 Recommended Actions Based on Risks 191
7.6 References 194

8.0 Summary, Conclusions, and Recommendations 197
8.1 Adequacy of Mathematical Modeling 199
8.2 Predicted Human Health Effects 200
8.2.1 DU Mass Intakes 201
8.2.2 Chemical Toxicity Effects 201
8.2.3 Radiological Effects 202
8.3 Effect of Ventilation and NBC Overpressure Systems 205
8.4 Applying the Risk Assessment Results 206
8.4.1 Categories of Human Health Effects 206
8.4.2 Multiple Perforations 208
8.4.3 Relationship of Risk Assessment Results to DoD/Army Doctrine and Policy 208
8.5 Human Health Risk Conclusions 209
8.6 Recommendations for Follow-On Actions 211
8.7 References 212

Abbreviations and Acronyms 215

Conversion Factors 218

Glossary 219

Appendix A: Dose and Concentration Tables 225

Appendix B: Uranium Toxicity—Animal Studies 329

Index 341

Figures and Tables

FIGURES

2.1 Schematic Showing Crew Positions in an Abrams Tank 13
2.2 Schematic Showing Crew Positions in a Bradley Fighting Vehicle with Driver in Front and Commander and Gunner in Turret 13
3.1 Schematic of a Sampler Array 28
3.2 Regression Analysis of DU Concentrations of IOM Data Against CI Data 31
3.3 Regression Analysis of DU Concentrations of IOM Data Against Efficiency-Corrected CI Data 31
3.4 Ratios of Measured IOM-to-CI Concentrations by Time Interval 31
3.5 Calibration Data of Andersen Personal Impactor 37
3.6 Experimental Data and Fitted Curve for Stage 1 38
3.7 Experimental Data and Fitted Curve for Stage 2 38
3.8 Experimental Data and Fitted Curve for Stage 3 39
3.9 Experimental Data and Fitted Curve for Stage 4 39
3.10 Experimental Data and Fitted Curve for Stage 5 40
3.11 Experimental Data and Fitted Curve for Stage 6 40
3.12 Experimental Data and Fitted Curve for Stage 7 41
3.13 Experimental Data and Fitted Curve for Stage 8 41
3.14 DU Aerosol Concentration Reduction with Time with and Without Ventilation 50
3.15 DU Concentration Ratio Reduction with Time in a Ventilated Versus Unventilated Abrams Tank and a Bradley Vehicle 52
4.1 Bayesian Posterior Distribution of E(50) for PIII-2, Loader Position, Scenario B 68
4.2 ICRP 66 Lung Dissolution Model 72
5.1 Radiation Doses, E(50), for the Double-Shot Test Series 97

5.2 Ratio of E(50)s for Mouth Versus Nose Breathers 104
5.3 Dependence of Dose Coefficient on Particle Size and Solubility Classification 106
5.4 Notional E(50)s for Different Solubilities Obtained Using the *In Vitro* Solubility Curve Fits 111
5.5 Dose Distribution for Phase I 114
5.6 Dose Distribution for Phase II 115
5.7 Dose Distribution for Phase III 116
5.8 Dose Distribution for Phase IV 117
5.9 Distribution for the Maximum Kidney Uranium Concentration for Phase I 118
6.1 Total Body Retention of DU After a 5-Min Exposure 124
6.2 Retention of DU in Lung After a 5-Min Exposure 124
6.3 Retention of DU in Kidney After a 5-Min Exposure 124
6.4 Retention of DU in Thoracic Lymph Nodes After a 5-Min Exposure ... 125
6.5 Urinary Excretion Rates of DU Resulting from a 5-Min Exposure 125
6.6 Fecal Excretion Rates of DU Resulting from a 5-Min Exposure 125
6.7 Distribution of Acute Kidney Toxicity by REGs 135
6.8 Median Peak Kidney Uranium Concentrations and 10th and 90th Percentiles for All Phases and Scenarios 147
6.9 Normalized Risk from the Intake of U-238, Which Estimates Isotope-Specific Risks 156
7.1 Best-Estimate Risks for Perforation of the Vehicles Shown Compared to the RES Categories 183
7.2 Normalized Accumulated Risk as a Function of Time 186

TABLES

2.1 Incident and Exposure Classification for ODS 12
2.2 Summary of Level I In-Vehicle Exposure Scenarios 14
3.1 Uranium Isotopic Composition of DoD DU Munitions 19
3.2 Respiratory Physiological Parameters for Healthy Adult Males 21
3.3 Respiratory Physiological Parameters that Vary with Physical Activity Level 21
3.4 Wall-Loss Correction Factors 34
3.5 Values of Fitted Curve in Equation (3.6) 42
3.6 Calculated Effective Cutoff Diameters and Size Increments for the Eight-Stage CIs 44
3.7 Ventilation Rates for Intact, Operational Abrams Tanks and Bradley Vehicles 47
3.8 Pre-Shot Ventilation Rates in BHTs for Phases I, II, and III 47
3.9 Fitted Constants for Phase-III DU Aerosol Reduction with Time 49
3.10 Fitted Constants with Increased Ventilation in an Abrams Tank 49

3.11 Constants for a BHT and an Operational Bradley Vehicle 51
4.1 Biokinetic Rate Constants for Uranium 61
4.2 Scenario Breathing Pattern Assumptions 64
4.3 Breathing Parameters for a Normally Nose-Breathing Adult Male 64
4.4 Fractional Collection Efficiency of Monodisperse Aerosols on CI Stages 67
4.5 Results of MCMC Calculations of Means and Standard Deviations for Scenario B Cases 69
4.6 Correspondence Between Particle Size of CI stages with Cyclone Stages 71
4.7 Dissolution Fit Parameters from the Capstone Data 74
4.8 Assignment of Dissolution Parameters to Cyclone Stages 74
4.9 Dissolution Fit Parameters Converted to the Modified ICRP 66 Framework 75
4.10 DU Aerosol Concentrations by CI Stage as a Function of Time 79
4.11 Stage-Specific, Flow-Rate Corrected Intakes by Scenario 80
4.12 Calculated E(50) and $H_T(50)$ from U-238 Only 81
4.13 Compartment and Dose Quantities Considered in BK/AIDE Comparison 82
4.14 Comparison of BK Code Dose Coefficients with ICRP Publication 78 .. 83
4.15 Fractional Decomposition into Intake Parts a and b 84
4.16 Comparison of IMBA and BK Dose Coefficients 84
4.17 Expected Distributions Over CI Stages 85
4.18 Comparison of E(50) Code Results for Various Internal Dosimetry Codes 80
5.1 Intakes and E(50)s by Phase and Scenario 93
5.2 Organ Doses by Phase and Scenario 95
5.3 Comparison of Radiation Doses from the Abrams BHT Double-Shot Tests 95
5.4 Comparison of Radiation Doses from the Bradley BHT Double-Shot Tests 96
5.5 Mean and Median Peak Kidney Uranium Concentrations by Phase and Scenario 98
5.6 Comparison of Intakes and Peak Kidney Uranium Burdens from the Abrams Tank Double-Shot Tests 100
5.7 Comparison of Intakes and Peak Kidney Uranium Burdens from the Bradley Vehicle Double-Shot Tests 100
5.8 Maximum Deposition of DU Aerosols in the Respiratory Tract Regions 101
5.9 Phase IV, Shot 4 Intakes, 50-Year Committed Doses, and Peak Kidney Uranium Concentrations 103
5.10 Deposition Fractions for Particle Sizes Specific to the Marple CI 109

5.11 Descriptive Statistics for the Influence of Solubility Parameters on E(50), (H_T50), and Peak Kidney Uranium Concentration 111
5.12 Means and Standard Deviations Calculated from the Summed Bayesian Posterior Distributions 112
5.13 Phase-I Doses and Assigned Regions by Shot and Position 114
5.14 Phase-II Doses and Assigned Regions by Shot and Position 115
5.15 Phase-III Doses and Assigned Regions by Shot and Position 116
6.1 Uranium in Normal Humans—Environmental Exposures 126
6.2 Uranium in Normal Humans—Occupational Exposures 127
6.3 Acute Human Exposures to Uranium Resulting in Effects in the Kidney 130
6.4 Follow-Up to Acute Human Exposures to Uranium Resulting in Effects in the Kidney 133
6.5 Renal Effect Groups Based on Kidney Uranium Concentration 135
6.6 Risk Characterization—Chemical Risk to the Kidney 136
6.7 Chronic Human Exposures to Uranium Resulting in Effects on the Kidney 138
6.8 Summary of Exposure Scenario Parameters 142
6.9 Predicted Peak Kidney Uranium Concentrations and REGs, Abrams Tank Perforated Through Non-DU Armor 144
6.10 Predicted Peak Kidney Uranium Concentrations and REGs, Bradley Vehicle 145
6.11 Predicted Peak Kidney Uranium Concentrations and REGs, Abrams Tank Perforated Through DU Armor 146
6.12 Predicted Peak Kidney Uranium Concentrations, Fully Operational Abrams Tank Perforated Through DU Armor 147
6.13 Clearance Time for Peak Uranium Kidney Burdens 149
6.14 Lung Burdens of DU or Dust in Exposed Populations 149
6.15 Summary of Mortality for Various Causes in Uranium Workers Compared with the General Population 151
6.16 Lung Cancer Lifetime Mortality Risk Coefficients 155
6.17 Nominal Probability Coefficients for Fatal Cancer 157
6.18 Mean Radiation Doses to Various Organs 159
6.19 Comparison of Mean E(50) and $E(50)_{SOrg}$ for Use in Risk Analysis 160
6.20 Risks for Lifetime Cancer Mortality After DU Aerosol Inhalation 161
6.21 Risks for Increased Lifetime Cancer Mortality from DU Aerosol Inhalation 162
6.22 Risk Characterization—Chemical Risk to the Kidney 163
6.23 Peak Kidney Uranium Concentration, Percentiles and Their Respective REGs 164
6.24 Median Risk Estimates for Lifetime Cancer Mortality from DU Aerosol Inhalation 165
7.1 Radiation Exposure Limits for "Peacetime" Operations 178

7.2 Comparison of RES Categories with Peacetime Standards or Guidelines 180
7.3 Conversion of DoD and NATO RES Categories into Risk of Fatal Cancers Due to Radiation Exposure 182
7.4 Fatal Cancer Risk from DU Inhalation Exposures and the Associated RES Categories 184
7.5 Summary of the 10th, 50th, and 90th Percentile Radiological and Chemical Health Risk Categories 189
7.6 Stochastic Risks, Medical Implication, and Recommended Medical Actions for Each of the RES Categories 192
8.1 Summary of Exposure Scenario Parameters 198
8.2 Summary of Median DU Intakes by Scenario 201
8.3 Summary of Median Peak Kidney Uranium Concentrations by Scenario 202
8.4 Summary of the Median Committed Effective Doses for Most Likely Scenarios 203
8.5 Summary of the Median Committed Equivalent Lung Doses for Most Likely Scenarios 204
8.6 Summary by Scenario of Median Lifetime Risk of Fatal Cancer from DU Aerosol Inhalation 204
8.7 Median Radiological and Chemical Risk Groups by Vehicle and Scenario 207

Executive Summary

Armor-piercing, depleted-uranium (DU) munitions were used on the battlefield for the first time by the U.S. Armed Forces during the 1991 Gulf War (Operation Desert Storm [ODS]). Use of this new munition helped to quickly end that war because of the effectiveness of the high-density munition to reach and perforate distant armored targets. During the course of ODS, misidentification of U.S. forces in distant vehicles led to incidents in which three U.S. Abrams tanks and most of the 14 occupied Bradley vehicles struck by the DU munitions had perforations of the crew compartments. Crewmembers in these vehicles would have been exposed to DU oxide aerosols in addition to any DU fragment injuries resulting from armor perforation.

The U.S. Department of Defense (DoD) determined that the data available at that time were not sufficiently robust to conduct a human health risk assessment for personnel exposed to DU under battlefield conditions. As a result of this determination, DoD committed to obtaining more complete data about aerosols generated by the impact and perforation of armored combat vehicles by DU munitions and to using the data obtained in a revised and updated personnel exposure assessment and human health risk characterization. In the Capstone DU Aerosol Study (Parkhurst et al. 2004), which arose from this commitment, DU aerosols were generated through perforation of armored target vehicles with large-caliber DU (LC-DU) munitions, and the chemical and physical properties of these aerosols generated were characterized.

This human health risk assessment is a companion document to the Capstone study final report and is part of the Capstone DU Aerosol Characterization and Risk Assessment Program. The assessment, sponsored by the U.S. Army

Center for Health Promotion and Preventive Medicine (USACHPPM), uses data documented in the Capstone report as input to risk-assessment models that were used to estimate radiological and chemical health effects to personnel in, on, or near an Abrams tank or a Bradley Fighting Vehicle (hereafter referred to as a Bradley vehicle) at the time it was fired upon. The risk calculations are based on technically sound, well-accepted models used by the international scientific community and include radiological doses to principal organs, the effective dose, and the organ-specific uranium mass concentration as a function of time and time-integrated exposures.

S.1 PURPOSE

In this assessment, the estimated intakes, radiological and chemical doses of DU exposure in, on, and near armored combat vehicles perforated by DU munitions were assessed to determine the probability of health effects from DU to DoD military personnel exposed primarily to DU aerosols immediately after vehicle perforation. The analyses included retrospective scenarios to simulate conditions that occurred during ODS and prospective scenarios of possible future battlefield actions. The aerosol data derived from the Capstone study provided the basis for modeling input parameters by summarizing the bounds and characteristics of typical aerosols generated when armored vehicles are perforated. The test results will be used by DoD to revise human health risk characterizations for ODS exposure scenarios, and to determine if changes in personnel protective measures are warranted to reduce risks to military personnel in the future.

S.2 INVESTIGATIVE TEAM

Independent subject matter experts in the fields of aerosol science, toxicology, health physics, radiological dose assessment, and risk assessment were recruited from various laboratories and research institutes, including Los Alamos National Laboratory (LANL), Pacific Northwest National Laboratory (PNNL) operated by Battelle (Richland, Washington), Lovelace Respiratory Research Institute (LRRI), Battelle (San Antonio, Texas), and Battelle Eastern Science and Technology (BEST) Center (Aberdeen, Maryland) to assist health physicists and toxicologists from USACHPPM with this study. The effort was administered by BEST.

USACHPPM and an U.S. Army steering committee (the Depleted Uranium Research–Integrated Process Team [DUR-IPT]) guided the overall test objectives by establishing Data Quality Objectives (DQOs) for the specific information to be derived from this testing program. Four independent health physicists with expertise in assessing health risk from exposure to radiological materials reviewed the modeling approach. A nine-member peer review panel provided technical feedback on the DQOs and the draft report.

S.3 STUDY OVERVIEW

The overall human health risk assessment was a four-part process. Part 1 involved characterization of the DU aerosols for specific scenarios relevant to Level I exposures as defined by the DoD Office of the Special Assistant for Gulf War Illnesses, Medical Readiness, and Military Deployment, now known as DoD Health Affairs/Deployment Health Support Directorate (OSAGWI 2000). Level I exposures include personnel in, on, or near combat vehicles at the time they were perforated by an LC-DU munition. The Capstone study provided the results of the Part 1 effort (Parkhurst et al. 2004). Parts 2 through 4 of the risk assessment process are discussed in Chapters 5 through 7 of this report. Part 2 documents the use of these aerosol data to estimate radiation and chemical doses for each of the representative scenarios (described in Chapter 5 of this report). In Part 3, the calculated radiological and chemical doses are converted into estimates of human health risk (Chapter 6). Part 4 places these risks into operational perspective so they can be used by combat commanders for operational risk management and by policy makers to assess the adequacy of current policy and doctrine (Chapter 7). A summary of the data and conclusions, together with recommendations for follow-up actions, completes the health risk analysis (Chapter 8). Appendix A consists of tables of the estimated intakes, radiation doses, and chemical concentrations. Appendix B provides a review of animal data related to uranium toxicity.

Inhalation of DU aerosols by crewmembers and first responders was evaluated to assess the associated health risks. These aerosols were created by perforations of armored vehicles with LC-DU munitions. This assessment used aerosol data from the Capstone study as input to standard biokinetic and dosimetric models that were used to estimate radiological doses and kidney uranium concentrations. Information from published literature was then used to establish the relationship between doses and health effects. The radiological and chemical risks were assessed for specific exposure scenarios that defined exposure time and breathing rates.

This human health risk assessment involved the development of exposure scenarios that were used to determine a range of potential risks from exposure to DU in aerosols. These scenarios were based on personnel activities during ODS as reported in interviews conducted by OSAGWI and on assumptions developed to simulate breathing patterns and to bound exposure times. The scenarios include perforations of Abrams M1 series tanks and Bradley vehicles (both M2 and M3 series).

The Capstone study generated extensive data sets from 13 separate experiments in which LC-DU penetrators were fired at and perforated conventional and DU armor on an Abrams tank ballistic hull and turret (BHT) (Phases I and III, respectively), conventional armor on a Bradley BHT (Phase II), and DU armor on a fully operational Abrams tank (Phase IV). The BHTs were intact structures from which instrumentation, flammable materials, and unnecessary structures had been removed. Additionally, for logistical reasons, the BHTs did not contain ventilation or Halon fire-suppression systems, which are standard in operational vehicles. Operational versions of these vehicles are shown in Figures S.1 and S.2.

Several sampling stations with filter cassettes and eight-stage cascade impactors (CIs) collected timed, staged samples that were specifically relevant to assessing the aerosol concentration and particle sizes from the DU aerosols generated by the perforation. Additional samples were collected using a five-stage cyclone and a moving filter sampler to supplement the aerosol particle concentration and size measurements and to provide adequate masses of DU for other analyses (e.g., solubility, chemical form, particle morphology). Priority was placed on characterizing DU aerosols within crew compartments during the first 2 h post shot.

Figure S.1.
Abrams Main Battle Tank

Figure S.2.
Bradley Fighting Vehicle

The aerosol source term for the interior of the armored combat vehicle was well characterized over time for the various shots. The BHTs used for the operational vehicles provided sufficient space to contain the sampler configurations described in this report, and the engineered shielding resulted in a very high survival of the samplers. The Capstone Phase-IV shot provided information on the effects of ventilation and air volume on DU aerosol concentration. The BHTs were treated as models of their corresponding operational vehicles and are discussed in the remainder of this summary as if they were operating vehicles.

Physical and chemical characterization of the DU aerosols was extensive. Aerosol composition, particle size, morphology, and *in vitro* dissolution properties determined from a wide range of samples collected inside the armored vehicles after perforation has led to a better understanding of the properties of these materials, and has provided a better scientific basis for bracketing the range of expected characteristics. The heterogeneous nature of the aerosol particles was identified through scanning electron microscopy, and the ramifications of such heterogeneity on particle size distributions and *in vitro* dissolution were used to interpret the observed variability in the measured results.

Time-dependent aerosol concentrations and particle size distributions were used as source-term input for the modeling effort. In a departure from the traditional practice of using either assumed or fitted unimodal or bimodal lognormal particle size distributions, actual uranium masses from each of the eight size-specific CI stages were used instead. This approach better maintained the quality and variability of the original measurement data and avoided introducing additional uncertainty into the dose assessment from poor data fit. Likewise, *in vitro* solubility data were used to define the rates of DU absorption to blood in a size- and material-specific manner.

The biokinetic/dosimetric models selected for the radiological dose assessment included the following three respiratory tract, ingestion, and biokinetic models, which are the most contemporary published by the International Commission on Radiological Protection (ICRP):

- the inhalation/respiratory tract model described in ICRP Publication No. 66 (1994)
- the ingestion/gastrointestinal tract model described in ICRP Publication No. 30 (1979)
- the uranium systemic biokinetic model in ICRP Publication No. 69 (1995).

The outputs derived from running these models included organ-specific DU mass concentrations and activities as a function of time and time-integrated exposures for each of the scenarios. These results were used to calculate committed effective doses, E(50); committed equivalent doses, $H_T(50)$; and selected dose versus time profiles. The calculated output was benchmarked against ICRP reports and commercially available internal dosimetry codes. Bayesian statistics and the Markov chain Monte Carlo method were employed to calculate probabilistic expansion of means, median, and confidence limits.

Following the modeling of radiological doses to and chemical concentrations in the body's organs, estimates of the health risk as described by the nature and magnitude of the risk were determined. These health risk estimates were based on the radiological dose and chemical concentration for each organ affected.

S.4 RISK ASSESSMENT RESULTS

Chemical and radiation risks are related to the duration of exposure (stay-time) by personnel (including crewmembers and first responders) to DU aerosols inside a perforated vehicle. This relationship is not linear; it depends on 1) the DU aerosol mass concentration, 2) the particle size distribution, both of which decrease over time, and 3) its solubility, which is related to particle size and uranium oxide composition. Five scenarios, designated Scenarios A through E, were developed to evaluate the doses based on exposure duration (Table S.1). Scenarios A through D relate to crewmembers who were in the vehicle when it was perforated and refer to stay-times of 1 min (A), 5 min (B), 1 h (C), and 2 h (D), respectively. Based on the stay-time experiences from ODS, Scenarios A and B are most likely, while stay-times represented by Scenarios C and D represent upper bounds. Scenario E specifically relates to first responders who enter the vehicle shortly after perforation.

Table S.1. Scenario Exposure Times and Durations

Scenario	Time of Exposure	Exposure Duration
	Crewmembers Inside Vehicle	
Most Likely Scenarios		
Scenario A	From impact to exit 1 min post shot	1 min
Scenario B	From impact to exit 5 min post shot	5 min
Upper Bound Scenarios		
Scenario C	From impact to exit in 1 h post shot	1 h
Scenario D	From impact to exit in 2 h post shot	2 h
	First Responders	
Scenario E	Entry 5 min post shot, exit 10 min later	10 min

The number of samples used in the evaluation of DU intake, concentration to organs, radiological doses, and associated risk analysis varied by phase depending on the number of shots within the phase and the number of sampling stations operating at specific time intervals within each shot. Shot lines (trajectories into the crew compartments) were varied in Phases I and II to evaluate differences in DU aerosol produced. As a result, some of these shot tests are replicates and others represent differing conditions. The two Phase-III shot tests are replicates. Only one Phase-IV shot test (LC-DU) was conducted. DU intakes, peak kidney uranium concentrations, and doses are calculated for each shot and sampling position, and these values are listed in Appendix A. Composite results by phase and scenario make up the majority of this assessment's discussions. These results were evaluated using arithmetic means, medians, and 10th and 90th percentiles. The number of shots per phase and the number of cascade impactor (CI) samplers operating during the interval covered by each scenario are listed in Table S.2 by test phase.

The mathematical modeling of organ doses and concentrations from the Capstone study aerosol data relied on integrating standard ICRP models with a new technique to incorporate size-specific stage data from CIs and the direct application of the *in vitro* results. These improvements allowed the quality and variability of the original measurement data to be maintained in the calculations.

Uranium intakes, organ concentrations, and associated doses were modeled for vehicle crewmember positions (i.e., commander, driver, gunner, loader [and scouts in a Bradley vehicle]) using the above exposure scenarios. The modeling results led to predictions about health effects from chemical toxicity and radiological doses for these personnel. These predictions assume one LC-DU round perforates the crew compartment and that no ventilation

Table S.2. Shots per Phase and Operating CI Samplers per Scenario

Phase	Total Shots	Number of CIs Operating per Scenario				
		A	B	C	D	E
Phase I—Abrams Tank through Non-DU Armor, No Ventilation	7	22	22	20	18	22
Phase II—Bradley Fighting Vehicle, No Ventilation	3	11	11	8	8	11
Phase III—Abrams Tank through DU Armor, No Ventilation	2	8	8	8	8	8
Phase IV—Abrams Tank Through DU Armor, EC/NBC[(a)] Ventilation	1	1	1	(b)	(b)	2

(a) Environmental control/nuclear, biological, chemical
(b) Estimated from sample at 25 to 27 min post perforation

systems are operating during or after perforation, except where indicated. The health risks from inhalation of DU oxide aerosols in unventilated vehicles at the levels projected from exposure scenarios (Phases I through IV outlined in the Capstone study) and from this risk assessment (Scenarios A through E) are low and are due primarily to the risk of kidney effects and potential increases in lung cancer.

S.4.1 DU Mass Intakes

The predicted median intakes of DU mass (using a conversion of 0.38 nCi/mg) from the aerosols for the five scenarios are listed for each vehicle configuration in Table 8.2. The intakes range from a low of 10 mg for a 1-min exposure in the ventilated Abrams tank with DU armor to a high of 710 mg for a 5-min exposure and 1000 mg for a 2-h exposure in an unventilated Abrams tank with DU armor. These data were used as input to the models for calculating DU concentrations to tissues and organs and radiological doses.

Table S.3. Summary of Median Intakes by Scenario

Scenario	DU Mass Intakes (mg)			
	Abrams Tank			Bradley Vehicle, Through Conventional Armor, No Ventilation
	Through Conventional Armor, No Ventilation	Through DU Armor, No Ventilation	Through DU Armor, EC/NBC Ventilation	
Crewmembers				
Most Likely Scenario				
A - Exit in 1 min	280	250	10	83
B - Exit in 5 min	590	710	43	220
Upper Bound Scenario				
C - Exit in 1 h	760	970	91	330
D - Exit in 2 h	780	1000	110	380
First Responders				
E - Entry 5 min post shot, exit 10 min later	160	200	27	99

S.4.2 Predicted Chemical Toxicity Effects

Chemical toxicity evaluations of uranium have been conducted over many years and include numerous studies of animals as well as studies of workers in the uranium-processing field. These studies have identified the kidney as the target organs for uranium toxicity. Therefore, peak kidney uranium concentrations were assessed. For the mostly likely exposures in these vehicles, the concentrations predicted are unlikely to cause any adverse health effects, although there is a possibility of transient or protracted indicators of kidney dysfunction for exposures of about 5 min or more in an unventilated Abrams tank perforated through non-DU armor. Table S.4 summarizes the predicted median of peak kidney uranium concentrations for exposure Scenarios A through E.

For reference with the Table S.4 values, the *de facto* occupational guideline is 3 μg U/g kidney (derived from ICRP [2] 1959 and experimental animal studies). Although the peak kidney uranium concentrations were slightly higher after perforation of non-DU armor (because the solubility of the aerosol from the non-DU armor was slightly greater than that from the DU armor), the data suggest that the amount of uranium reaching the kidneys does not vary greatly within these three vehicles in which no ventilation systems were operating. The concentrations predicted for exposures in an Abrams tank with an operating ventilation system are at least an order of magnitude below the ICRP occupational guideline and are well below levels at which health effects might be expected.

A review of renal effects in humans occurring after acute exposures to uranium was used to develop a set of effects groups correlating uranium (U) concentrations in the kidneys with renal effects. These Renal Effects Groups (REGs) were used to predict the potential effects of the modeled exposure concentrations. The predicted kidney concentrations for the five scenarios were compared against the following REGs developed for this assessment (discussed further in Section S.4.3):

- REG 0 consists of kidney concentrations ≤2.2 μg U/g kidney in which the predicted outcome is no kidney dysfunction.

Table S.4. Summary of Median Peak Kidney Uranium Concentrations by Scenario

Scenario	Peak Kidney Uranium Concentrations (μg U/g Kidney)			
	Abrams Tank			Bradley Vehicle, Through Conventional Armor, No Ventilation
	Through Conventional Armor, No Ventilation	Through DU Armor, No Ventilation	Through DU Armor, EC/NBC Ventilation	
Crewmembers				
Most Likely Scenario				
A - Exit in 1 min	3.0	1.1	0.054	1.0
B - Exit in 5 min	6.4	2.6	0.23	2.9
Upper Bound Scenario				
C - Exit in 1 h	8.2	3.5	0.46	3.5
D - Exit in 2 h	8.0[(a)]	3.7	0.56	4.0
First Responders				
E - Entry 5 min post shot, exit 10 min later	1.5	0.67	0.14	1.4

(a) The sampler data used to calculate Scenarios C and D differed slightly, and this was responsible for the lower dose during the longer exposure time.

- REG 1 consists of concentrations between 2.2 and 6.4 µg U/g kidney and, at worst, may cause transient renal dysfunction.
- REG 2 applies to concentrations in the range of >6.4 to 18 µg U/g kidney and may cause protracted (but not permanent) kidney dysfunction.

Using these groups, the health effects predicted for exposure in unventilated vehicles are mostly within REGs 0 and 1. REG 2 was predicted only for the longer exposures when the perforation was through non-DU armor. The kidney uranium concentrations in the Abrams tank with operating ventilation were within REG 0.

S.4.3 Predicted Radiological Effects

Median 50-yr committed effective doses, E(50), for the five scenarios are presented in Table S.5. The E(50)s for the most likely exposures and upper bound exposures in the Abrams/conventional armor and the Bradley vehicles were within 5 rem. The E(50) for a 5-min exposure in the unventilated Abrams tank perforated through DU armor was predicted to be 6 rem, and increases to 8 rem with a 2-h exposure. The upper bound for all other scenarios was ≤5 rem. The U.S. annual occupational radiation dose limits, as promulgated by the Nuclear Regulatory Commission, include a 5 rem total effective dose equivalent (10 CFR 20). Although the E(50) is different from the total effective dose equivalent, the concepts are similar, and for radiation protection purposes, the two quantities can be compared (ICRP [60] 1991). The NRC's planned special exposure limit is 10-rem in a year (two times the

Table S.5. Summary of Median 50-yr Committed Effective Doses by Scenario

	E(50), rem			
	Abrams Tank			Bradley Vehicle, Through Conventional Armor, No Ventilation
Scenario	Through Non-DU Armor, No Ventilation	Through DU Armor, No Ventilation	Through DU Armor, EC/NBC Ventilation	
	Crewmembers			
Most Likely Scenario				
A – Exit in 1 min	2.0	2.2	0.090	0.59
B - Exit in 5 min	3.7	6.0	0.44	1.7
Upper Bound Scenario				
C - Exit in 1 h	4.8	8.3	1.02	2.1
D - Exit in 2 h	5.0	8.7	1.20	2.4
	First Responders			
E - Entry 5 min post shot, exit 10 min later	0.92	1.9	0.41	0.89

annual limit, not to exceed five times the annual limit in a lifetime, 10 CFR 20.1206). Although for some scenarios, the E(50)s exceed the occupational radiation limits, all E(50)s are less than the planned special exposure limit.

Unlike the results of the chemical toxicity analysis in which the Abrams tank perforated through non-DU armor (Phase I) was highest, the radiological doses were highest from the Abrams tank perforated through DU armor (Phase III). The reason for this difference is that the DU aerosols collected in Phase I were slightly more soluble than the aerosols collected from Phase III. Doses to a crewmember in an Abrams tank with its EC/NBC ventilation system operating when perforated through DU armor were about an order of magnitude below the levels in a perforated Abrams tank in which the ventilation system was not operating. For all scenarios modeled, radiation doses at the levels predicted are not expected to cause adverse health effects.

Mean 50-yr committed equivalent doses, $H_T(50)$, to organs were also calculated for the lung, bone surface, kidney, red marrow, and liver. The doses to the lung were higher by at least a factor of 10 than the doses to the bone surfaces, kidney, red marrow, or liver. (Some models calculate higher doses to the lymph nodes and extrathoracic tissues, but these tissues do not have a high susceptibility to radiation-induced cancer, and studies in dogs have demonstrated that the lymph nodes are not target organs for cancer [Hahn et al. 1999]). The predicted *median* $H_T(50)$s calculated for the lung are summarized in Table S.6 for each of the scenarios. The $H_{Lung}(50)$s for the mostly likely exposures in these vehicles were between 0.7 and 44 rem. The

Table S.6. Summary of Median 50-yr Committed Equivalent Doses to the Lung by Scenario

Scenario	$H_{Lung}(50)$, rem			
	Abrams Tank			Bradley Vehicle, Through Conventional Armor, No Ventilation
	Through Non-DU Armor, No Ventilation	Through DU Armor, No Ventilation	Through DU Armor, EC/NBC Ventilation	
Crewmembers				
Most Likely Scenario				
A - Exit in 1 min	14	18	0.66	5.2
B - Exit in 5 min	32	44	3.3	14
Upper Bound Scenario				
C - Exit in 1 h	38	60	7.6	20
D - Exit in 2 h	39	61	8.7	22
First Responders				
E - Entry 5 min post shot, exit 10 min later	8.8	14	3.1	6.7

upper bounds for exposures ranged from 7.6 to 61 rem. The U.S. NRC annual occupational radiation dose limits include a 50-rem committed dose equivalent (10 CFR 20). Although the $H_T(50)$ is different from the committed dose equivalent, the concepts are similar, and for radiation protection purposes, the two quantities can be compared (ICRP [60] 1991). Except for the case in which an Abrams tank was perforated through DU armor and the stay-time was 1 to 2 h, the predicted doses to the organs were less than this annual occupational limit. For all scenarios modeled, organ doses are at levels unlikely to cause adverse health effects.

The long-term radiation health effects of inhaled uranium compounds have been difficult to identify with any certainty in human populations. In fact, natural uranium, which is more radioactive than DU, has not been shown to cause cancer in occupationally exposed workers, although epidemiologic studies of workers in the uranium processing industry suggest that there is a very weak association between exposure to inhaled uranium and cancer. Because of the relative absence of data on increased cancer incidence in exposed populations, no risk coefficients for uranium-induced cancer are available from these populations. However, the assessment of risk from radiation usually employs a separate methodology based on the conservative assumption that any radiation carries some risk. The radiation doses from internally deposited alpha-emitting radionuclides, such as uranium, have been studied in both humans and animals. The human dose data are sufficiently robust to allow the use of risk coefficients derived from internal alpha emitters and the summed tissue/organ-specific risks rather than using the E(50) and the cancer risk coefficients used to convert E(50) to risk.

In situations in which the radionuclide is not uniformly distributed in the body, the calculated whole-body risk may be underestimated when compared to the summation of the risks to each of the tissues/organs (ICRP [60] 1991). The results of this risk assessment indicated that the calculated risks using summed tissue/organ-specific risks were greater than the calculated risks using the E(50) methodology. Consequently, this assessment of radiation effects was based on organ-specific risks, which are driven primarily by the dose to the lungs. The risks for cancer mortality are based on the sum of the individual tissue/organ risks. The lifetime cancer mortality risk from inhalation of DU aerosols was calculated using generic ICRP risk coefficients and the Linear, No-Threshold model of effect that may overestimate risks at the levels predicted in this study and are, therefore, thought to be protective of health. Table S.7 summarizes the median estimated increased lifetime cancer mortality from inhaled, deposited DU aerosols for the five exposure scenarios evaluated.

Table S.7. Summary of Median Lifetime Risk of Fatal Cancer from DU Inhalation by Scenario

Scenario	Lifetime Risk Increase of Fatal Cancer (%)			
	Abrams Tank			Bradley Vehicle, Through Conventional Armor, No Ventilation
	Through Non-DU Armor, No Ventilation	Through DU Armor, No Ventilation	Through DU Armor with EC/NBC On	
Crewmembers				
Most Likely Scenario				
A - Exit in 1 min	0.11	0.12	0.0049	0.034
B - Exit in 5 min	0.20	0.32	0.025	0.099
Upper Bound Scenario				
C - Exit in 1 h	0.27	0.44	0.057	0.12
D - Exit in 2 h	0.28	0.45	0.065	0.14
First Responders				
E - Entry 5 min post shot, exit 10 min later	0.050	0.10	0.023	0.052

The median probability of lifetime cancer mortality from DU aerosol inhalation within the most likely exposure levels ranged from 0.005 to 0.32%. The upper bound ranged from 0.057 to 0.45%. A risk of 0.0045 implies that, on average, the chance over an individual's lifetime of dying of cancer from this exposure is 0.45% greater than the chance of that same individual dying from the natural or background rate of cancer mortality, which is 23.6% for U.S. males (Ries et al. 2003). If a crewmember exited this same tank within 1 min after perforation, the lifetime risk of cancer mortality from inhalation of DU aerosols is predicted to be 0.0012 or 0.12%. Risks to a crewmember in an Abrams tank (perforated through DU armor) in which the EC/NBC ventilation system was operating were about an order of magnitude below the risks in an Abrams tank (perforated through DU armor) without an operating ventilation system.

S.4.4 Applying the Results to Military Risk Management

The total risk to personnel involved in combat ranges from routine operational risks to the battlefield risks of injury and death. Applying occupational standards to combat situations is neither appropriate nor practical. Field commanders need the best possible estimate of risk associated with combat situations so they can appropriately include the predominantly potential long-term risks associated with DU exposure in their mission risk analysis and in their mission risk-management process. The ultimate goals of these processes are to ensure that the benefits of the mission outweigh the risks and, most importantly, to ensure that appropriate actions are taken by the commander to balance the various factors making up total risk, including combat mortality and morbidity as well as long-term potential risks.

The military radiation exposure-control philosophy as stated in U.S. Army Field Manuals provides a framework for this cost-benefit analysis in which the primary cost is defined in terms of casualties or lost lives. Radiation Exposure Status (RES) categories (Mettler et al. 1999) are used to track the total radiation dose received by the combat unit. A RES unit of zero refers to a combat unit receiving no radiation exposure. A RES-1 category indicates a unit has received some radiation exposure, but the dose is ≤75 rad. The primary health endpoint in this dose range is the potential for increased risk of fatal cancer. The RES-1 category is subdivided into these five levels based on external gamma exposures:

- RES-1A: between 0.05 and 0.5 rad
- RES-1B: between 0.5 and 5 rad
- RES-1C: between 5 and 10 rad
- RES-1D: between 10 and 25 rad
- RES-1E: between 25 and 75 rad.

For this risk assessment, project personnel modified the RES risk-based system to make it applicable to internal DU deposition and its related dose. This modification was achieved by converting the RES category doses into levels of increased risk using ICRP/NCRP risk coefficients and then comparing the calculated risks of cancer from inhaled DU aerosols to risks represented by these categories. A dose and dose-rate effectiveness factor of 2 was applied to all absorbed doses less than 20 rad and to any absorbed dose delivered at dose rates less than 10 rad/h. Because there is no similar system developed for chemical toxicity, the REG chemical risk model (discussed above), which is based on the peak kidney uranium concentration, was used to perform a similar function. A summary of the RES categories and REGs for the median radiological risks and chemical effects is presented in Table S.8. Predicted RES categories ranged from 1A to 1B and REGs ranged from 0 to 1 in the most likely exposure scenarios. For upper bound scenarios, the RES categories ranged from 1B to 1D and REGs ranged from 1 to 2. Because the results are conservative, both levels are shown for those medians that were very close to the boundary between RES categories and REGs.

Table S.8. Summary of Median Radiation Exposure Status Categories and Renal Effects Groups

Scenario	Abrams Tank (Through Non-DU Armor)		Abrams Tank (Through DU Armor)		Abrams Tank (Through DU Armor, EC/NBC Ventilation)		Bradley Vehicle (Conventional Armor, No Ventilation)	
	RES	REG	RES	REG	RES	REG	RES	REG
Crewmembers								
Most Likely								
A–Crew, Exit 1 min	1B	1	1B	0	1A	0	1B	0
B–Crew, Exit 5 min	1B	1/2	1C	1	1B	0	1B	1
Upper Bound								
C–Crew, Exit 1 h	1C	2	1C/ID	1	1B	0	1B	1
D–Crew, Exit 2 h	1C	2	1C/1D	1	1B	0	1B	1
First Responders (10-min exposure)								
E–Entrance at 5 min	1B	0	1B	0	1B	0	1B	0

S.5 CONCLUSIONS

The data from the Capstone study provide a sound basis for assessing the inhalation of DU aerosols and associated doses and risks to personnel in combat vehicles at the time of perforation and to those entering immediately after perforation. These Capstone data were robust enough for use in calculating median estimates of Level I exposures to personnel (i.e., both those within vehicles at the time of perforation and first responders) and have reduced many of the uncertainties in estimating DU exposures that existed prior to that study. The data from which the doses and risks were calculated were collected using conditions representative of conservative vehicle configurations—that is, shot lines maximizing DU penetrator erosion and associated DU aerosol production, nonexistent or inoperative ventilation systems, and hatches that were closed at the time of perforation. Additionally, for logistical reasons, a Halon fire-suppression system, which would have reduced aerosol concentrations, was either not present or was not activated during the tests, unlike most of the ODS perforation events.

An evaluation of DU aerosols in a fully equipped, perforated vehicle with an operating ventilation system and additional mathematical modeling of increased air exchange provided data on the effectiveness of ventilation in reducing doses and risks. Although shot-to-shot variations in aerosol generation would be expected in combat situations, the redundancies in the experiment were sufficient to ensure reasonable certainties in the data and in the conclusions drawn from the data.

The doses and risks to human health resulting from the inhalation of DU aerosols in a perforated vehicle are low and would be low when compared with many other combat risks, such as the possibility of combat injuries and of hazardous materials including different heavy metals, chemicals, and other agents being released onto the battlefield and into the environment.

The most important factor for reducing exposure and dose discovered in this analysis was the use of onboard vehicle ventilation. Even though the doses and risks from inhaled DU based on the Capstone data are low, ventilation systems operating during or turned on as soon as possible after a DU perforation can further significantly reduce the intake and its associated exposure received by the crewmembers.

The following are the major conclusions and practical generalizations drawn from this dose and risk assessment.

Dose and Risk Modeling

- The development of exposure-dose models that can use the size-specific stage data from CIs was an important improvement in dose assessment, allowing the quality and variability of the original measurement data to be maintained in the calculations. This technique should be useful for any dose assessment in which size-selective aerosol data are available.
- The transformation of multicomponent exponential functions obtained from *in vitro* dissolution study into the ICRP 66 (1994) solubility model structure allowed the direct application of the *in vitro* results. Although the mechanisms likely responsible for DU dissolution are not compatible with the description of the compartments in the ICRP 66 model, the transformation is mathematically correct and makes the best use of the data.
- Implementation of the Markov chain Monte Carlo method for uncertainty estimation made good use of the extensive aerosol measurement data and allowed propagation of uncertainties in a robust manner. As further analysis of uncertainties in model structures and parameter values become available, they can be incorporated into more extensive uncertainty analysis approaches that will be able to take advantage of the Bayesian statistical approaches.
- Bayesian statistics using the Markov chain Monte Carlo method were employed in calculating probabilistic expressions of means, medians, and confidence limits. This method is believed to improve the estimates of the "true values." Because negative component amplitudes (which

occurred at low concentration levels, caused by subtraction of imperfect field blanks) were disallowed, the resulting positive bias adds an element of conservatism in low concentrations and doses.

- Data from the literature review of human health effects following exposure to uranium were used to develop a set of Renal Effects Groups to predict the potential kidney effects of peak uranium concentrations. This model provided a means of estimating risk of health effects from the chemical effects of uranium toxicity.

Predicted Doses, Risks, and Human Health Effects

- In Abrams tanks with one or more ventilation systems operating, the levels of predicted chemical toxicity and radiological risks for the time periods modeled (up to 2 h) are below routine occupational limits. Personnel are not likely to develop adverse health effects as a result of exposure at these levels.
- In unventilated Abrams tanks, no adverse health effects are predicted for the most likely exposures (5 min or less) with the possible exception of a 5-min exposure in the Abrams tank perforated through conventional armor in which the potential for temporary kidney effects could increase. This potential would increase with stay-time. Past experience during ODS indicated that the crewmembers exited perforated tanks within 5 min.
- The veterans who sustained DU fragment wounds would have also inhaled DU aerosols as a result of armor perforations. Because of the DU fragments, their total intake of DU would be at least as large, and most likely larger than, the modeled Level I intakes. After more than a decade of medical surveillance of these 1991 Gulf War survivors of DU-related incidents, no adverse toxicological effects related to the presence of DU have been identified (McDiarmid et al. 2004). This supports the conclusion of this report that no immediate or long-term adverse health effect, especially to the kidney, would be expected for the most likely exposures presented.
- In unventilated Abrams tanks, most of the intake would occur within the first 5 to 10 min after perforation. This intake would be significantly reduced by activation of the ventilation systems during this timeframe. With an operating ventilation system, the dose and, therefore, the risks would be reduced by about an order of magnitude over the risk incurred in vehicles with no ventilation systems operating during or after perforation. Dose and risk reductions would also be expected in a Bradley vehicle with an operating ventilation system.

- No adverse health effects are expected from exposures in unventilated Bradley vehicles for any of the time periods modeled (up to 2 h).
- The data from the double-shot tests, in which there was a 10- to 15-min delay between shots, suggest that the first shot did not add significantly to the DU aerosol concentrations generated by the second shot. From these data, it appears that the radiation doses and uranium concentrations in the kidney from exposure to aerosols produced in a vehicle twice perforated by LC-DU munitions could be approximated by doubling the doses and kidney concentrations.

Other Considerations

- Because of differences in individual exposures for a given crew in a perforated vehicle, DU bioassays are needed to establish individual dose estimates. This report provides information that the U.S. DoD community could use for deciding when biomonitoring should be implemented.
- Although radiation risks from all scenarios evaluated are predicted to be negligible to low, counseling of affected personnel and their family members is suggested because of the perceived radiological risks associated with exposure to DU.

REFERENCES

10 CFR 20. Code of Federal Regulations, Title 10, *Energy*, Part 20, "Standards for Protection Against Radiation."

Hahn FF, BA Muggenburg, RA Guilmette, and BB Boecker. 1999. "Comparative Stochastic Effects of Inhaled Alpha- and Beta-Emitting Radionuclides in Beagle Dogs." *Radiation Research 152*:S19-S22.

International Commission on Radiological Protection (ICRP). 1959. *Report on the Committee II on Permissible Dose for Internal Radiation.* ICRP Publication 2, Pergamon Press, Oxford, United Kingdom.

International Commission on Radiological Protectection (ICRP). 1979. *Limits for Intakes of Radionuclides by Workers*. ICRP Publication 30, Part 1, Pergamon Press, Oxford, United Kingdom.

International Commission on Radiological Protection (ICRP). 1991. *1990 Recommendations of the International Commission on Radiological Protection.* ICRP Publication 60, Pergamon Press, Oxford, United Kingdom.

International Commission on Radiological Protection (ICRP). 1994. *Human Respiratory Tract Model for Radiological Protection*. ICRP Publication 66, Pergamon Press, Oxford, United Kingdom.

International Commission on Radiological Protection (ICRP). 1995. *Age-dependent Doses to Members of the Public from Intake of Radionuclides: Part 3 Ingestion Dose Coefficients*. ICRP Publication 69, Pergamon Press, Oxford, United Kingdom.

McDiarmid MA, S Engelhardt, M Oliver, P Gucer, PD Wilson, R Kane, M Kabat, B Kaup, L Anderson, D Hoover, L Brown, B Handwerger, R Albertini, D Jacobson-Kram, C Thorne, and K Squibb. 2004. "Health Effects of Depleted Uranium on Exposed Gulf War Veterans: A 10-Year Follow-Up." *Journal of Toxicology and Environmental Health* 67:277-296.

Mettler FA, Jr., Committee Chairman. 1999. *Potential Radiation Exposure in Military Operations: Protecting the Soldier Before, During, and After*. Committee on Battlefield Radiation Exposure Criteria, Institute of Medicine, National Academy Press, Washington, DC.

Parkhurst MA, F Szrom, RA Guilmette, TD Holmes, YS Cheng, JL Kenoyer, JW Collins, TE Sanderson, RW Fliszar, K Gold, JC Beckman, and JA Long. 2004. *Capstone Depleted Uranium Aerosols: Generation and Characterization, Volumes 1 and 2*. PNNL-14168, Prepared for the U.S. Army by Pacific Northwest National Laboratory, Richland, Washington.

Office of the Special Assistant to the Deputy Secretary of Defense for Gulf War Illnesses (OSAGWI). 2000. *Exposure Investigation Report, Depleted Uranium in the Gulf (II)*. Online report available at URL: www.gulflink.osd.mikl in the Environmental Exposure Reports Section.

Ries LAG, MP Eisner, CL Kosary, BF Hankey, BA Miller, L Clegg, A Mariotto, MP Fay, EJ Feuer, and BK Edwards (eds). 2003. *SEER Cancer Statistics Review, 1975-2000*. National Cancer Institute. Bethesda, Maryland. Accessed online in January 2004 at http://seer.cancer.gov/csr/1975_2000,2003.

1.0 Introduction

The use of depleted uranium (DU) in munitions and armor has been a subject of debate and media interest since these munitions were first used by the U.S. Armed Forces during the 1991 Gulf War (Operation Desert Storm [ODS]). Operation Desert Storm and the military operations that followed showed the effectiveness of large-caliber (LC), armor-piercing DU cartridges as well as the intensity of the national and international controversy generated by the use of DU. In light of the questions raised regarding the health risk to U.S. military personnel exposed to DU through inhalation and ingestion of aerosols generated during perforation of Abrams tanks and Bradley Fighting Vehicles (hereafter referred to as Bradley vehicles), the Office of the Special Assistant for Gulf War Illnesses (OSAGWI) directed the U.S. Army Center for Health Promotion and Preventive Medicine (USACHPPM) to conduct a human exposure and health risk characterization for these personnel. The results of their evaluation included an identification of data gaps and future research needs (USACHPPM 2000a).

One of the data gaps identified was the lack of robust data to support a human health risk assessment from exposure to DU aerosols for personnel in, on, or near armored combat vehicles in which DU aerosols were generated in crew compartments. Recognition of this gap led to sponsorship by OSAGWI and the U.S. Army Heavy Metals Office of the Capstone DU Aerosol Study in which DU munitions were fired at ballistic hulls and turrets (BHTs) of Abrams tanks and Bradley vehicles. The DU aerosols generated during the Capstone study were collected and characterized. The equipment and methods used and the results of these characterizations are provided in the report entitled *Capstone Depleted Uranium Aerosol:*

Generation and Characterization, Volumes 1 and 2 (Parkhurst et al. 2004). The Capstone study results provide the primary technical basis for the dose calculations in this human health risk characterization, which addresses categories including personnel in the vehicle at the time of vehicle perforation and individuals entering shortly thereafter to assist in the recovery of personnel and equipment.

The previous USACHPPM assessment (2000a) used data from external measurements and from the burn tests of DU munitions and vehicles uploaded with DU munitions, which were relatively robust in terms of the amounts and particle characteristics. The data from hard target strikes of DU were less consistent. Measuring and characterizing the aerosols from the impact of DU munitions on non-DU and DU armor were the most difficult to conduct because of the violent nature of the experimental environment. The DU aerosols generated from U.S. Army burn and impact tests and the variability in the parameters were summarized by Parkhurst et al. (1995).

Prior to the Capstone study, data from the most comprehensive test of aerosol generated in a test that involved firing DU and non-DU munitions at a DU-armored Abrams tank was reported by Fliszar et al. (1989). The primary objectives of that test were to measure the aerosol plume as a function of distance from the vehicle, estimate the level of soil contamination, and measure the level of DU contamination remaining in the vehicle. A secondary test objective was an attempt to measure the aerosol generated inside the vehicle at the time of crew compartment perforation. The exposure and environmental data collected outside the perforated vehicle were robust and sufficient for use in health and environmental risk assessments. However, measurements inside the vehicle were compromised because of the damage done to the samplers when the armor was penetrated. Data were obtained but were difficult to interpret.

1.1 DEPLETED URANIUM BACKGROUND INFORMATION

Uranium is ubiquitous in the environment and is present in all soils and rocks in the form of a variety of minerals, with granite typically having the highest concentrations. The world average uranium content of soils is about 2 μg U/g soil, and in the United States, the concentration is typically in the range of 1 to 4 μg U/g (NCRP 1976). Trace amounts of uranium are found in all foods. The daily intake of uranium in food in the United States and the United Kingdom ranges from 1 to 4 μg/day (Hamilton 1972; Welford and Baird 1967; ATSDR 1999). Although root uptake by plants is generally quite small (NCRP 1999), some soil-to-vegetation transfer does occur, especially by crops grown with phosphate-based fertilizers. The uranium content of cereals, fruits, and vegetables is greater than that of meat or milk. Drinking water is also a source of uranium and may contribute more than food to the human intake (Cothern and Lappenbusch 1983). The quantity of uranium in drinking water varies widely. Wrenn et al. (1987) estimated that people in the United States ingest between 0.2 to 0.7 μg U/day from water, whereas ATSDR (1999), using results of a U.S. Environmental Protection Agency (EPA) analysis, give an average of 2 μg/d. Air concentrations of uranium also vary widely, but typically are low, ranging from 0.01 to 0.2 ng/m^3. The uranium in air comes principally from suspended soil particles (ATSDR 1999). Exposures from uranium in the air result in a uranium intake estimated at 0.002 to 0.020 μg U/day, which is less than 1% of the uranium intake from food (EPA 2000). Industrial processes, such as mining or milling of uranium ore, airborne effluent releases from coal-fired power plants, and nuclear manufacturing facilities, can increase the uranium concentrations present in the air (WHO 2001).

Three isotopes of uranium, all radioactive, occur together in natural uranium: U-238, U-234, and U-235. The chemistry of these isotopes, which is identical, determines the reactions of the isotopes within the environment as well their transport and reactions within the body. The U-238 isotope is the longest lived with a half-life of 4.5 billon years—a half-life so long that U-238 might be considered nominally stable. This isotope constitutes more than 99% of the mass of natural uranium and approximately half of its radioactivity. Uranium-234, a decay product of U-238, is responsible for nearly all the remainder of the radioactivity of natural uranium. A small amount (0.7% by weight) of U-235 is present in natural uranium.

All the isotopes of uranium emit primarily alpha particles. Natural uranium also emits beta particles and some gamma radiation from uranium decay products, also known as progeny. Separation of uranium from its progeny by the enrichment and milling processes removes most of the decay progeny present in the natural radioactive decay chain. Within six months, secular equilibrium is reestablished with the U-235 and U-238 isotopes, and the decay products emit beta and gamma radiation. As a result, DU emits alpha, beta, and gamma radiation, but only the alpha particles are relevant in studies of DU inhalation.

Depleted uranium is derived as a by-product of the enrichment process in which about 70% of the U-235 in natural uranium is separated from U-238. The exact percentage depends on the degree of enrichment. This process also reduces the percentage of U-234 in DU. The remaining uranium is DU, which contains about 0.2% U-235 by mass and emits about 60% of the radioactivity of natural uranium. DU emits about 60% of the radioactivity of natural uranium primarily because of the reduction in the U-234. DU is defined as any uranium that has less than 0.711% by mass of U-235 (10 CFR 40.4; AEPI 1995). U. S. military specifications designate that DU for use by the U.S. Department of Defense (DoD) contain less than 0.3 percent U-235, 0.0006 percent U-234, and 0.003 percent U-236 by weight (AEPI 1995). U-236 is not a naturally occurring isotope of uranium but was introduced during the DU production process. In reality, the military uses DU that contains approximately 0.2% U-235 (AEPI 1995). The specific activity of DoD DU is 0.38 nCi/mg (14 Bq/mg U) compared with 0.69 nCi/mg (25 Bq/mg U) for natural uranium. This material contains traces of transuranic elements that contribute less than 1% to the dose (USACHPPM 2000a).

The interaction of the DU penetrator with armor that it perforates creates DU oxides and aggregates with iron, aluminum, or other armor components (most of which would be trace amounts). Projecting potential health effects for Soldiers who inhale these DU oxides from armor perforations is the focus of this report. The contributing effect, if any, of inhaling the non-DU particles is outside the scope of this work.

1.2 OBJECTIVES

The overall objective of this Human Health Risk Assessment (HHRA) is to use the data generated in the Capstone study to develop health risk assessments that the DoD can make use of in answering three primary questions as it reviews its internal policies (USACHPPM 2000b):

Primary Question 1. Are the health risks to personnel in, on, or near an armored vehicle at the time the crew compartment is perforated by DU munitions high enough to warrant changes in the medical policy, in the medical treatment and medical monitoring, and in personnel protective measures for Soldiers potentially exposed to DU?

Primary Question 2. Following the initial perforation of an armored vehicle crew compartment, are the health risks to personnel who enter the armored vehicle still with residual aerosolized DU and also resuspended DU in the air high enough to warrant changes in the medical policy, in the medical treatment and medical monitoring, and in personnel protective measures for Soldiers potentially exposed to DU?

Primary Question 3. Are the health risks to personnel who entered an armored vehicle well after the vehicle was perforated by DU munitions and who re-suspended DU into the air high enough to warrant changes in the medical policy in the medical treatment and medical monitoring, and in personnel protective measures for Soldiers potentially exposed to DU?

Specific objectives for the HHRA as related to these questions are as follows:

- Use the Capstone data to estimate DU intakes for credible military scenarios and provide a measure of uncertainty.
- Use current models recognized by the national and international community to estimate organ concentrations and doses.
- Use organ concentrations and radiological doses to assess the chemical and radiological risk using appropriate risk models.
- Use risk management techniques to determine if changes in policy or procedures are required.
- Use good risk communication to provide the estimated risks of DU exposure so that appropriate decisions can be made and actions can be taken.

1.3 PROCESS

Beginning with the Capstone study and continuing through this HHRA, a methodology that involved rigorous peer review at critical stages of the process was adopted. For the Capstone study, an independent peer review panel was formed to review the data quality objectives (DQOs) that established the ultimate objectives and the quality required. The peer review team also reviewed the experimental design, the experimental setup, the conduct of the first experiment, and the draft report. The DQOs for the Capstone test are provided in USACHPPM (2000b).

The U.S. Army selected a scientist from Pacific Northwest National Laboratory (PNNL) as the independent Principal Investigator to direct the Capstone DU Aerosol Study and lead the Capstone test team, which included scientists and engineers from the U.S. Army, national laboratories, and private research institutes.

The same process for external review was used for this HHRA. A peer review panel reviewed the DQOs (USACHPPM 2003) for this project and the draft report. Additionally, four dose modeling experts reviewed the dose assessment methodology to ensure that the proposed modeling process and model types were appropriate for this work. The Army selected a management team from Battelle to lead the risk assessment team, which included subject matter experts from national laboratories, research institutes, and the U.S. Army.

The HHRA process included:

- Developing exposure scenarios for the Level I exposure group defined as those Soldiers in, on, or near combat vehicles at the time these vehicles were struck by DU munitions or those Soldiers who entered immediately afterward, primarily to recover personnel or equipment (Chapter 2).
- Developing intake parameters such as source-term and physiological data for use in modeling. The Capstone aerosol database was the basis for the source-term calculations. The particle size fractionation of the Capstone data was used directly instead of fitted unimodal or bimodal lognormal distributions to describe the particle size distribution (Chapter 3).
- Selecting appropriate chemical and radiological dose models for the respiratory tract, ingestion, and uranium biokinetic behavior (Chapter 4).
- Calculating committed effective doses and committed equivalent organ doses (Chapter 5).
- Evaluating known chemical and radiological levels that cause clinical effects (Chapter 6).
- Characterizing the nature and magnitude of human health risks from DU aerosol intake (Chapter 7).
- Summarizing the data and making recommendations about its use (Chapter 8).

This report describes the risk assessment process in more detail, the results of the chemical and radiological analysis of exposure to DU aerosols, and a characterization of the associated risks to human health.

1.4 REFERENCES

10 CFR 40. Code of Federal Regulations, Title 10, *Energy*, Part 40, "Domestic Licensing of Source Material."

Army Environmental Policy Institute (AEPI). 1995. *Health and Environmental Consequences of Depleted Uranium in the U.S. Army: Technical Report.* US Army Environmental Policy Institute, Atlanta, Georgia.

Agency for Toxic Substances and Disease Registry (ATSDR). 1999. *Toxicological Profile for Uranium*. US Department of Health & Human Services, Report TP 90-29, Atlanta, Georgia.

Cothern CR and WL Lappenbusch. 1983. *Occurrence of Uranium in Drinking Water in the U.S. Health Physics*, 45:89-95.

Fliszar RW, EF Wilsey, and EW Bloore. 1989. *Radiological Contamination from Impacted Abrams Heavy Armor*. Technical Report BRL-TR-3068, Aberdeen Proving Ground, Ballistic Research Laboratory, Aberdeen, Maryland.

Hamilton EI. 1972. *The Concentration of Uranium in Man and His Diet. Health Physics* 22:149-153.

National Council on Radiation Protection and Measurements (NCRP). 1976. *Environmental Radiation Measurements.* NCRP Report No. 50, Washington, DC.

National Council on Radiation Protection and Measurements (NCRP). 1999. *Recommended Screening Limits for Contaminated Surface Soil and Review of Factors Relevant to Site-Specific Studies*. NCRP Report No. 129, Bethesda, Maryland.

Parkhurst MA, JR Johnson, J Mishima, and JL Pierce. 1995. *Evaluation of DU Aerosol Data: Its Adequacy for Inhalation Modeling*. PNL-10903, Prepared for the US Army by Pacific Northwest National Laboratory, Richland, Washington.

Parkhurst MA, F Szrom, RA Guilmette, TD Holmes, YS Cheng, JL Kenoyer, JW Collins, TE Sanderson, RW Fliszar, K Gold, JC Beckman, and JA Long. 2004. *Capstone Depleted Uranium Aerosols: Generation and Characterization, Volumes 1 and 2.* PNNL-14168, Prepared for the US Army by Pacific Northwest National Laboratory, Richland, Washington.

US Army Center for Health Promotion and Preventive Medicine (USACHPPM). 2000a. *Depleted Uranium—Human Exposure Assessment and Health Risk Characterization in Support of the Environmental Exposure Report "Depleted Uranium in the Gulf" of the Office of the Special Assistant to the Secretary of Defense for Gulf War Illnesses, Medical Readiness and Military Deployments (OSAGWI), OSAGWI Levels I, II and III Scenarios, 15 September 2000.* Health Risk Assessment Consultation No. 26-MF-7555-00D, Aberdeen Proving Ground, Maryland. Online report available at URL: www.gulflink.osd.mil in the Environmental Exposure Reports Section.

US Army Center for Health Promotion and Preventive Medicine (USACHPPM). 2000b. MCHB-TS-OHP, Statement of Data Quality Objectives for Testing to Support Human Health Risk Assessments of Potential Depleted Uranium Exposures during the 1991 Gulf War, Final Version, 14 July 2000.

US Army Center for Health Promotion and Preventive Medicine (USACHPPM). 2003. MCHB-TS-OHP, Statement of Data Quality Objectives for the Depleted Uranium Exposure Assessment and Health Risk Characterization Updated with Capstone Test Data, Final Version, 2 April 2003.

US Environmental Protection Agency (EPA). 2000. "Radionuclides (Uranium, Radium and Radon)." Technology Transfer Network Air Toxics Website, Office of Air Quality, Planning and Standards. Online report available at URL: http:/www.epa.gov/ttn/atw/hlthef/radionuc.html.

Welford GA and R Baird. 1967. *Uranium Levels in Human Diet and Biological Materials. Health Physics* 13:1321-1324.

World Health Organization (WHO). 2001. *Depleted Uranium: Sources, Exposure and Health Effects.* Department of the Protection of the Human Environment, World Health Organization, Geneva, Switzerland. Online report available at URL: www.who.int/environmental_information/radiation/depleted_uranium.htm.

Wrenn ME, PW Durbin, DL Willis, and NP Singh (1987). "The Potential Toxicity of Uranium in Water." *Journal of American Water Works Association*, 79: 177-184.

2.0 Exposure Scenarios

The Capstone DU Aerosol Study characterized the DU aerosol concentrations and particle sizes inside the vehicle as a function of time (Parkhurst et al. 2004). The approach used to estimate the time in the vehicle and the other physiological parameters required to calculate radiation dose and kidney concentrations are described in this chapter. The intent was to develop credible exposure scenarios that accounted for each of these parameters.

The exposure scenarios developed for the Human Health Risk Assessment (HHRA) were based on an OSAGWI framework for categorizing DU exposures during ODS. This framework was based on interviews conducted by OSAGWI with veterans of ODS, after-action reports, and discussions with other military experts (OSAGWI 1998, 2000; USACHPPM 2000). A joint team of DoD and the Department of Veterans Affairs personnel reviewed the interviews, consulted with subject matter experts, and developed a three-level grouping of exposures for ODS that was based on the potential for internalizing DU (Section 2.1). This risk assessment addresses inhalation for personnel who were in a vehicle at the time of perforation by a DU munition or first responders who entered the vehicles shortly after perforation—both groups of which were designated by OSAGWI as Level I personnel (further described in Section 2.1). The exposure scenarios selected were those that 1) were generally representative of the description of events provided by ODS veterans who fit the criteria for Level I or 2) represented a credible upper bound for crewmember stay-times. The assumptions used to develop scenarios that bound the potential exposures to personnel exposed at this level follow the discussion of the exposure levels.

Because it was not possible to evaluate every conceivable scenario, operational conditions were selected for modeling that would reflect relatively long stay-times and the associated potential for resulting in the highest doses or kidney uranium concentrations. Certain conditions that would increase aerosol removal, such as operation of ventilation systems and opening of hatches, were not modeled but were qualitatively discussed. Additionally, evaluation of other confounders, such as the operation of the fire-suppression system after perforation, was not addressed.

The impact of shot lines other than those presented in the Capstone study was not considered. The Capstone study shot lines were based on either composites of actual events during ODS or shot lines that did not occur but would have increased aerosol concentrations. For example, DU armor was not perforated during ODS, but it was tested as one of the Capstone shot lines. Fluke penetrations (e.g., through creases between hatches, through open hatches, or at the junction of the turret and hull) were not modeled and would typically generate less aerosol than generated in the shot lines used in the Capstone study.

The scenarios described in Section 2.1 will likely result in estimates of the radiological dose and kidney concentrations for personnel in vehicles perforated by DU that are higher than those that actually occurred during ODS or would occur in future conflicts.

2.1 OSAGWI EXPOSURE LEVELS

These three levels of exposure in ODS were defined by OSAGWI (2000) as follows:

- ***Level I*** includes military personnel in, on, or near combat vehicles at the time of impact and perforation by DU munitions, or personnel who entered vehicles immediately after they were struck (and perforated) by DU munitions. These personnel could have been exposed to DU by fragments resulting from impact, inhalation of DU aerosols, ingestion of DU residues, or settling of DU particles on open wounds, burns, or other breaks in an individual's skin or embedded fragments—or by any combination of these possibilities. This level also includes personnel occupying a vehicle in which DU armor is perforated by non-DU munitions.
- ***Level II*** includes military personnel and a small number of DoD civilian employees whose job function required them to work in and around vehicles containing DU fragments and particles. These individuals

were not in the vehicle at the time of impact and did not immediately enter the vehicle after it was struck. This group performed a variety of tasks, such as battle damage assessment, repairs, explosive ordnance disposal, and intelligence gathering. They typically entered vehicles well after the initial suspended aerosol had dissipated or settled onto interior surfaces. They may have inhaled DU residues that were re-suspended by their activities, ingested DU through hand-to-mouth transfer, or spread contamination on their clothing. DoD personnel who were involved in cleaning up DU residues generated during other events, such as the July 11, 1991, explosion and fires at the Camp Doha North Compound, are also included in this group.

- ***Level III*** is an "all others" group whose exposures were brief or incidental. This group includes personnel who entered DU-contaminated Iraqi equipment, were downwind from burning Iraqi or U.S. equipment struck by DU rounds, or were downwind from burning DU ammunition (e.g., personnel at Camp Doha during the July 11, 1991, explosions and fire). While these individuals could have inhaled airborne DU particles, the possibility of receiving an intake high enough to cause health effects is unlikely.

The primary difference between a Level I exposure and a Level II or a Level III exposure is the time delay between the armor perforation and the initiation of personnel activities that may lead to DU intakes. For Level I exposures, personnel were in, on, or near the vehicle at the time the vehicle was perforated, or they entered soon enough after peforation that they were exposed to both DU that remained in the air from the perforation event and from resuspended DU caused by the disturbance of deposited DU particles as a result of their entrance and activities. At Level I, personnel could also internalize DU through wound contamination and embedded fragments (OSAGWI 2000). (Evaluation of the health risks from embedded fragments and wound contamination is beyond the scope of this report.) For Level II and Level III exposures, the time delay between perforation and vehicle entry would be hours to days, and the exposure would be primarily from resuspended DU and, potentially, from hand-to-mouth transfer of the DU remaining in the vehicle. Table 2.1 lists each of the three exposure levels developed for ODS with examples of the types of exposures that were included in each (USACHPPM 2000). The table also specifies if personal protective equipment (PPE) was worn for each of the subcategories listed.

Table 2.1. Incident and Exposure Classification for ODS (OSAGWI 2000)

Exposure Classifications: Scenarios	Numbers of Persons	Personal Protection Worn
Level 1		
Soldiers in or on a U.S. vehicle when a DU munition penetrated it.	104	None
Soldiers who entered U.S. vehicles to rescue occupants immediately after friendly fire DU impacts.	≈30-60[(a)]	None
Level II		
Explosives Ordnance Disposal (EOD) and unit personnel who removed equipment and munitions from U.S. vehicles struck by DU munitions.	≈10-20[(a)]	None
Unit personnel who performed maintenance on or recovered items from U.S. vehicles struck by DU munitions.	≈60-80[(a)]	None
Logistics Assistance Representatives (LARs) who inspected U.S. vehicles struck by DU munitions.	≈6-12[(a)]	Some wore PPE[(b)]
Battle Damage Assessment Team (BDAT) members who examined U.S. combat vehicles damaged and destroyed by DU munitions.	16	Most wore PPE[(b)]
144th Service and Supply Co. personnel who processed damaged equipment, including some struck by DU munitions.	29	None
Radiation control (RADCON) team members.	11	Most wore PPE[(b)]
Personnel exposed to DU during cleanup operations at Camp Doha's North Compound.	≈600	None
Level III		
Personnel exposed to smoke from burning DU rounds at Camp Doha.	Hundreds	None
Personnel exposed to smoke from burning Abrams tanks.	Unknown	None
Personnel who entered DU-contaminated equipment.	Unknown	None
Personnel exposed to smoke from Iraq's DU-struck equipment.	Unknown	None

(a) Number is not final; remains under investigation.
(b) Personal Protective Equipment (PPE) includes respirator, coveralls, boots and gloves. Reports of respiratory protection ranged from the military M25 and M17A2 respirators to industrial dust mask, surgical-type paper mask, etc.

2.2 LEVEL I "IN-VEHICLE" SCENARIO MODELING ASSUMPTIONS

As stated above, the Level I exposure category includes personnel in, on, or near the vehicle at the time it was perforated. The Capstone study provided data to support an evaluation of exposures to personnel on or near the vehicle in addition to inside an Abrams tank and a Bradley vehicle, but modeling doses on or near the vehicles is beyond the scope of this report. The focus of this modeling effort is to provide upper-bound estimates of potential DU aerosol exposures and intakes to personnel **inside** a perforated Abrams tank (Figure 2.1) or a Bradley vehicle (Figure 2.2).

Potential DU exposures and intakes to personnel outside the vehicle are lower because DU aerosol concentration is lower outside the vehicle, and the aerosol would diffuse faster in the open environment (USACHPPM 2000; Fliszar et al. 1989).

The modeling of the Level I exposures inside the vehicle is a matter of varying the aerosol source characteristics within the vehicle (DU concentration,

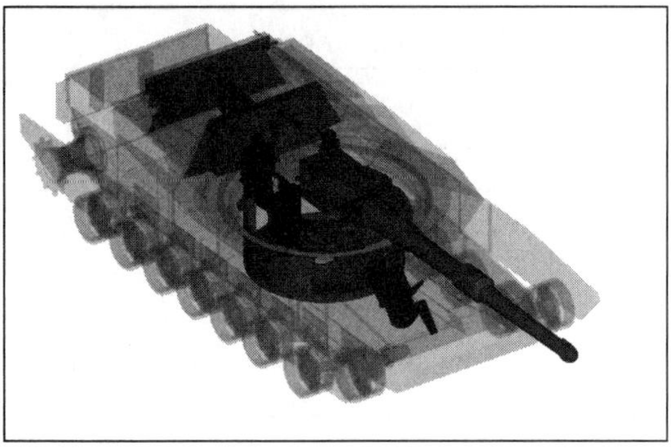

Figure 2.1. Schematic Showing Crew Positions in an Abrams Tank

Figure 2.2. Schematic Showing Crew Positions in a Bradley Fighting Vehicle with Driver in Front and Commander and Gunner in Turret (not shown are scouts or passengers in compartment behind the turret)

particle size, solubility), the timing, the duration of exposure, and the breathing rates associated with the physical activities being performed. These factors influence the magnitude of the intake and the consequent doses. Five Level I exposure scenarios were modeled and are listed in Table 2.2. These include 1) four for crewmembers present in the vehicle at the time of perforation and 2) one for first responders who enter the vehicle after perforation. The intake via inhalation for these scenarios is based on measured DU aerosol concentrations inside the vehicle during the time of exposure. Ingestion was not modeled as a part of this effort because the doses from inhalation were known to predominate (USACHPPM 2000).

The ODS veterans' statements (OSAGWI 2000) provided little information specifically related to exposure times for the various scenarios, but to the extent possible, the information was used in developing duration estimates. Some of the scenario exposure durations were extended to further bound probable exposure times within the vehicle that may have been greater than those experienced in ODS, but are considered probable in future conflicts. Evacuation tests with uninjured crewmembers and with crewmembers extricating an unconscious driver showed that an exit time of less than a minute was reasonable (Henry and Baetz 1988).

Table 2.2. Summary of Level I In-Vehicle Exposure Scenarios

Scenario	Time of Exposure	Exposure Duration	Breathing Rate
Crew Inside Vehicle			
A	From impact to exit 1 min post shot	1 min	3 m^3/h
B	From impact to exit 5 min post shot	5 min	3 m^3/h
C	From impact to exit up to 1 h post shot	1 h	3 m^3/h for first 15 min, 1.5 m^3/h thereafter
D	From impact to exit up to 2 h post shot	2 h	3 m^3/h for first 15 min, 1.5 m^3/h thereafter
First Responder			
E	Entry 5 min post shot, exit 10 min later	10 min	3 m^3/h

Four scenarios (subsequently referred to as Scenarios A through D) were developed to model DU aerosol concentrations from the time of perforation through exit times ranging from 1 min to 2 h. In a fifth scenario (Scenario E), entrance into the vehicle was assumed to occur 5 min after perforation and the length of stay was limited to 10 min.

Breathing rates are taken to be those specified in the human respiratory tract model (ICRP 1994) for light and heavy exercise breathing. Heavy exercise refers to two-thirds of the maximum workload capacity and assumes deep rather than shallow breathing regardless of breathing frequency, and is typical of firemen, construction workers, athletes, etc. (ICRP 1994). ICRP considered this high workload to be unsustainable for more than 2 h per day. Light exercise assumes one-third of the highest workload capacity, corresponding to working in laboratories and workshops, painting, woodworking, etc. (ICRP 1994).

Assumptions used for the timing of exposure (start and end times relative to the perforating event), duration of exposure, and breathing rates for Level I exposure scenarios are discussed in the following text and listed in Table 2.2.

With the exception of Phase IV, the scenarios assumed ventilation systems were not operating for the entire exposure time. For the sake of simplicity, it is assumed that the hatches were closed at the time of perforation. In actual operations, at least minimal ventilation would be expected, and any number of hatch positions would be possible. In the Capstone study, the hatches were initially closed, but pressure from perforation caused partial or full opening of a hatch during most shots. As shown in the Capstone study, operation of the ventilation and/or the environmental control/nuclear, biological, chemical (EC/NBC) protection systems reduce the aerosol concentrations significantly. Each scenario assumed that the fire suppression system was not activated. The activation of the fire-suppression system will reduce the aerosol concentrations.

2.2.1 Individual in the Vehicle during Vehicle Perforation

Four exposure times were selected: 1 min, 5 min, 1 h, and 2 h. The first two exposure times assumed that the crewmembers would be able to readily exit the vehicle. The second two times assumed continued exposure in a still-functioning vehicle. Longer stay-times would increase exposure, but the DU aerosol concentrations after 2 h are orders of magnitude smaller, and therefore add little to the intake. The scenarios assume no respirator use. Respirators reduce possible exposure by at least an order of magnitude. Most tests were conducted using vehicle structures that did not have operable ventilation systems. The assumptions used in creating the four scenarios for personnel in the vehicle during vehicle perforation are as follows:

Scenario A: These individuals would be in the vehicle during perforation and would exit the vehicle by 1 min after impact. Their breathing rate is taken to be that for heavy exercise (3 m^3/h) for the full duration of exposure.

Scenario B: These individuals would be in the vehicle during perforation and would exit the vehicle by 5 min after impact. The breathing rate is taken to be that for heavy exercise (3 m^3/h) for the duration.

Scenario C: The vehicle is assumed to function after perforation, and the crewmembers would continue their mission. These individuals would not exit the vehicle for up to 1 h. Breathing rates are taken to be that for heavy exercise (3 m^3/h) for the first 15 min followed by that for light exercise (1.5 m^3/h) for an additional 45 min.

Scenario D: The vehicle is assumed to function after perforation, and the crewmembers would continue their mission. These individuals would not exit the vehicle for up to 2 h. Breathing rates are taken to be that for heavy exercise (3 m^3/h) for the first 15 min followed by that for light exercise (1.5 m^3/h) for an additional 105 min.

2.2.2 Individuals in the General Area of the Incident and Quickly Responding to the Situation (Specifically First Responders)

First responders are assumed to enter the struck vehicle soon after vehicle perforation usually to recover personnel or equipment. Their exposures depend primarily on how soon they enter after the perforation has occurred and secondarily on how long they remain inside the vehicle. It is assumed that entry personnel are not wearing respiratory protection.

Scenario E: These individuals are NOT in or on the combat vehicle during perforation but enter the vehicle 5 min post shot to help evacuate the vehicle occupants. The estimated length of stay for these responders is 10 min, and they would be breathing at rates consistent with heavy exercise (3 m^3/h) during that time.

2.3 REFERENCES

Fliszar RW, EF Wilsey, and EW Bloore. 1989. *Radiological Contamination from Impacted Abrams Heavy Armor*. Technical Report BRL-TR-3068, Aberdeen Proving Ground, Ballistic Research Laboratory, Aberdeen, Maryland.

Henry MA and WA Baetz, Jr. 1988. *Initial Production Test (IPT) of Automotive and Weapons Phase of M1A1 Tank System, Final Report, Volume I.* US Army Combat Systems Test Activity, Aberdeen Proving Ground, Maryland.

International Commission on Radiological Protection (ICRP). 1994. *Human Respiratory Tract Model for Radiological Protection.* ICRP Publication 66, Pergamon Press, Oxford, United Kingdom.

Office of the Special Assistant to the Deputy Secretary of Defense for Gulf War Illnesses (OSAGWI). 1998. *Exposure Investigation Report, Depleted Uranium in the Gulf.* Online report available at URL: www.gulflink.osd.mikl in the Environmental Exposure Reports Scction.

Office of the Special Assistant to the Deputy Secretary of Defense for Gulf War Illnesses (OSAGWI). 2000. *Exposure Investigation Report, Depleted Uranium in the Gulf (II).* Online report available at URL: www.gulflink.osd.mikl in the Environmental Exposure Reports Section.

Parkhurst MA, F Szrom, RA Guilmette, TD Holmes, YS Cheng, JL Kenoyer, JW Collins, TE Sanderson, RW Fliszar, K Gold, JC Beckman, and JA Long. 2004. *Capstone Depleted Uranium Aerosols: Generation and Characterization, Volumes 1 and 2.* PNNL-14168, Prepared for the US Army by Pacific Northwest National Laboratory, Richland, Washington.

US Army Center for Health Promotion and Preventive Medicine (USACHPPM). 2000. Depleted Uranium—Human Exposure Assessment and Health Risk Characterization in Support of the Environmental Exposure Report "Depleted Uranium in the Gulf" of the Office of the Special Assistant to the Secretary of Defense for Gulf War Illnesses, Medical Readiness and Military Deployments (OSAGWI), OSAGWI Levels I, II and III Scenarios, 15 September 2000. Health Risk Assessment Consultation No. 26-MF-7555-00D, Aberdeen Proving Ground, Maryland. Online report available at URL: www.gulflink.osd.mikl in the Environmental Exposure Reports Section.

US Army Center for Health Promotion and Preventive Medicine (USACHPPM). 2003. Memorandum for Record, MCH-TS-OHP, Subject: *Transuraniucs (TRUs) and Technicium-99 (Tc-99) Contaminants in Depleted Uranium Armor, 14 February 2003.*

3.0 Development of Input Parameter Values

General and task-specific input parameters required for risk assessment modeling are discussed in this chapter, and field data from the Capstone DU Aerosol Study relevant to this application are presented. This chapter briefly discusses the general input parameters including source-term data requirements; default and specific physiological parameters; and ventilation options likely to influence the source term; and Capstone-specific data and parameters.

The Capstone study conducted 12 separate firings with LC-DU cartridges into Abrams tank and Bradley vehicle BHTs. Additionally, one shot was fired into an operational Abrams tank. Aerosol samples were collected in these target vehicles as a function of elapsed time post shot, and the samples were analyzed for their uranium content, particle size distribution, and other chemical and physical characteristics. These data were reviewed for relevance to the risk assessment and appropriateness for use in the modeling.

3.1 RISK ASSESSMENT PARAMETERS

The parameters considered for the health risk characterization involve the following:

- *Source-term analysis.* This is the quantity of DU aerosol available for exposure, and is directly related to the amount of penetrator erosion caused by perforation of armor and other vehicle components. The uranium concentrations calculated from the cascade impactors (CIs) in the Capstone report were used in this assessment (discussed in Section 3.2).

- *Pathway analysis.* Pathways of DU aerosol intake include inhalation and ingestion (hand-to-mouth). Dermal and subcutaneous tissue contamination pathways are also possible. The dose-modeling portion of this characterization addresses DU intake by inhalation only.
- *Exposure, intake, and dose assessment.* Information from personnel interviews conducted by OSAGWI was used to develop generalized exposure levels (Levels I, II, and III). This information was used to generate specific Level I exposure scenarios as described in Chapter 2. The Capstone database provided input about DU aerosol concentrations for calculating intake during the selected time of exposure. The dose assessment relates DU aerosol activities and concentrations to estimates of radiological doses and DU mass concentrations in organs (discussed in Chapter 5).
- *Human health risk characterization.* After the estimates of radiological dose and DU mass concentrations in organs have been obtained, the human health risk characterization can be conducted. This characterization is done in terms of the nature and likelihood of adverse human health effects resulting from intake of DU aerosol (discussed in Chapter 6).
- *Modeling confounders and uncertainty analysis.* The uncertainty in the measurements on which the analysis is based is addressed in the uncertainty analysis. The influence of uncertainty that key parameters might have on the magnitude of the result is evaluated in the uncertainty analysis (discussed in Chapter 5).
- *Military Risk Management.* Practical aspects of balancing combat risks are evaluated with the introduction of radiation exposure status (RES) categories and toxicological effects groups (discussed in Chapter 6) to provide best estimates of risks and their relevance within the context of combat operations (Chapter 7).

3.1.1 Source-Term Analysis

USACHPPM (2000) identified the parameters needed for a source-term analysis of DU aerosols. The data or assumptions for these parameters are described below:

- *Type of Munition (as it relates to DU mass).* This risk assessment used the DU concentrations measured in the Capstone aerosols to estimate source term as opposed to the use of mass balance based on total weight of the LC-DU penetrator and an estimation of the portion aerosolized.
- *Isotopic and Elemental Composition of DU and Residue.* USACHPPM reported that the isotopic composition of DoD DU munitions was

assayed and found to contain 99.8% U-238, 0.2% U-235, 0.0006% U-234, and 0.003% U-236 by mass (Table 3.1). The isotope activities are reported in microcuries per gram (μCi/g) and becquerels per microgram (Bq/μg). Several of the Capstone aerosol samples underwent isotopic analysis, which supported these results. DU may have trace quantities of impurities other than U-236 from fuel reprocessing, but these impurities add less than 1% to the radiological dose and are considered inconsequential from a radiological and chemical toxicity standpoint (USACHPPM 2003; WHO 2001).

Table 3.1. Uranium Isotopic Composition of DoD DU Munitions

Isotope	Percent by Mass	Isotope Activity in μCi/g (Bq/μg) of DU[(a)]
U-234	0.0006	0.37E-02 (1.39E-03)
U-235	0.2	4.3E-03 (1.60E-04)
U-236	0.003	2.0E-04 (7.18E-06)
U-238	99.8	3.4E-01 (1.24E-02)
(a) Total DU-Specific Activity: 0.38 μCi/g (14 Bq/mg).		

- *DU Armor (whether it was involved in perforation incident).* In each Capstone experimental firing, the DU munition perforated either standard armor or DU (heavy) armor depending on the intent of the shot. Complete information is provided in Section 3.2.1.
- *Airborne Release Fraction.* The total quantity of DU aerosol actually generated from impact with armor cannot be measured directly because of losses of absorption into or spallation of the DU onto the target. During the Capstone tests, the intact portion of each penetrator was captured in the armor catch plate behind the target vehicle, from which it could not be recovered and weighed. Additionally, aside from two small fragments recovered inside the vehicle, non-aerosolized DU fragments created by the penetrator-armor interaction and landing outside the vehicle were not weighed. However, based on Capstone aerosol data, it is estimated that a maximum of 7% of the LC-DU penetrator was aerosolized inside the heavily armored Abrams tank and a maximum of 1% in the lighter armored Bradley vehicle.
- *Respirable Fraction.* The aerosol generated in each test was sampled using CIs so that particle size distributions could be evaluated. The portion of the aerosol within the nominal respirable range could be determined from these results.
- *Chemical Form (and its associated solubility in lung or extracellular fluid).* The oxidation phases of the uranium oxide powder were

characterized in the experimental tests. The powders consisted of relatively insoluble forms and were highest in U_3O_8 and U_4O_9. Lung fluid solubility tests were performed on these samples to establish their dissolution behavior.

- *Particle Size.* Particle size distributions were evaluated in the experimental tests using CIs. The actual collected size fractions, each represented by a characteristic (median) aerodynamic diameter, were used in the dose modeling.

Additionally, the time of exposure after the shot occurred and the duration of the exposure strongly influence the estimated intake and the associated dose. Experimental data were collected during the first 2 h post shot for each experimental test. Following two of the tests, additional aerosols were sampled 3.5 to 4 h post shot to evaluate resuspension during sampler recovery operations.

Experimental DU aerosol data that were used in modeling the radiological and chemical doses include the DU aerosol concentration as a function of time, particle size distribution (also as a function of time), and *in vitro* solubility.

3.1.2 Physiological Parameters

In addition to the source term, additional parameters are used in modeling internal doses. Some of these parameters are related to human physiology and are standardized in the respiratory tract, gastrointestinal tract, and systemic biokinetic models discussed in Chapter 4. Other physiological parameters including breathing rate and type (route) of breathing also affect dose. These parameters are discussed in more detail below:

- *Reference Man (a standardization of physical characteristics).* The physical characteristics programmed into these models are based on the "Reference Man" characteristics, which are provided in several ICRP reports (ICRP [23] 1975, [66] 1994, [70] 1995, and [89] 2003).

  The characteristics relevant to the respiratory tract are based on ICRP 66, those related to bone are based on ICRP 70, and those related to the remainder of the organs are from ICRP 23 and ICRP 89. Only adult males were part of the Level I exposure cohort. The respiratory physiological parameter default values from the human respiratory tract model that do not vary with physical activity level were used in the dose assessment and are listed in Table 3.2. Although these values vary by lung size, they are not dependent on breathing capacity.

Table 3.2. Respiratory Physiological Parameters for Healthy Adult Males (ICRP [66] 1994)

Physiological Parameter	Value
Functional residual capacity	3301 cm^3
Extrathoracic dead space	50 cm^3
Bronchial (BB) dead space	49 cm^3
Bronchiolar (bb) dead space	47 cm^3
Height	176 cm
Tracheal diameter	1.65 cm
First bronchial diameter	0.165 cm

- *Breathing Rate (or ventilation rate).* Breathing rates are related to type and amount of physical exertion. ICRP 66 default values for a healthy adult male (Reference Man) were used assuming light and heavy exercise conditions. Physiological parameter values that vary with physical activity level and that were used in the dose assessment are listed in Table 3.3.
- *Type (or route) of Breathing.* The path of inhalation entry, whether through the nose or mouth, affects the deposition patterns of the aerosol in the respiratory tract. Most people breathe normally through their noses, particularly at lower physical activity levels. When the requirement for inspired air increases due to exertion level, then nasal airway resistance makes it difficult to move the requisite air volumes to the lung, and the breathing pattern changes by partitioning air flow between the nasal and oral airways. The default nasal airway inspiration fractions for a healthy adult male (Reference Man) for light and heavy exercise as well as sitting are given in Table 3.3. For those individuals whose normal breathing pattern favors oral breathing (usually because of anatomical, vasomotor, or pathological obstruction of the nasal airways), the nasal inspiration fractions are also listed. Nasal/oral partitioning of air flow affects the particle-size-dependent deposition

Table 3.3. Respiratory Physiological Parameters that Vary with Physical Activity Level (ICRP [66] 1994)

Parameter	Light Exercise	Heavy Exercise	Sitting
Ventilation Rate (m^3/h)	1.50	3.00	0.54
Respiratory frequency (per min)	20	26	12
Tidal volume (cm^3)	1250	1923	750
Volumetric flow rate (cm^3/sec)	833	1667	300
Fraction through nose (nose breather/mouth breather)	1.0/0.4	0.5/0.3	1.0/0.7

patterns within the various region of the respiratory tract. In general, more aerosol reaches the lung during oral breathing because the nasal airways are more effective filters for the inspired aerosol particles. The default case used in these calculations was for a normally nose-breathing individual. However, a sensitivity analysis was done to estimate the effect of oral breathing on intake and dose. This analysis is described in Section 5.5.1.

3.1.3 General Parameters

Additional factors that are likely to influence the concentration of the aerosol within the interior of the vehicle will also influence the intake of DU aerosols. These factors dilute the aerosol within the interior of the vehicle. Four of these factors are related to the vehicle's equipment. The fifth factor, aerosol coagulation/aggregation of the aerosol particles, will cause the aerosol to deposit quickly. The sixth factor refers to a mitigating measure to reduce intake of existing aerosol.

- *Ventilation Systems.* If the Abrams tank EC/NBC system or the Bradley vehicle ventilation systems continue to operate after the perforation, the aerosol concentration inside the vehicle would be reduced quickly, thus reducing exposure to personnel (See Section 3.2.6 for a description of these systems).
- *Opening of the Hatches or Loss of the Vehicle's Structural Integrity.* If one or more hatches is open or is forced open by the shot, passive venting through this opening would occur, thus reducing the aerosol concentration within the vehicle. If more extensive structural losses occur in addition to the entry and exit holes, this venting may proceed even more quickly.
- *Operation of the Fire Suppression System.* If the shot or related actions caused the fire suppression system to operate, the aerosol concentrations would be expected to drop significantly by becoming aggregated with the fire suppression aerosol and being deposited more rapidly than it otherwise would. The aerosol size distribution would also be affected.
- *Vehicle Deposition Surfaces.* Compared with the BHTs used in the Capstone tests, functional vehicles are equipped with various instruments and other materials that decrease the volume of air within the vehicle. Perforation of the equipped vehicle would be expected to create the same mass of DU (unless massive components are hit that further erode the penetrator) over a smaller volume, possibly leading to

a higher concentration. However, the additional surface area would increase deposition, leading to a faster reduction of concentration with time.

- *Coagulation and Aggregation of Particles.* The very high initial mass and particle number concentrations of DU aerosols measured after each shot lead to a high probability that the particles originally formed as a result of target-penetrator interaction will aggregate within the vehicle interior soon after penetration. This aggregation results in growth of the aerosol particle size, and increases the rate of loss to surfaces by sedimentation.
- *Use of PPE.* Respiratory protection in the form of a respirator reduces the amount of aerosol entering the respiratory system. The degree of reduction will be related to the efficiency of the respirator.

3.2 CAPSTONE INPUT DATA

The Capstone study was designed to fill the data gaps that existed for Level I exposures for actual firing incidents (retrospective scenarios) and, to a limited extent, for possible future engagements (prospective scenarios). The emphasis was placed on potential exposures inside the vehicle, although aerosol and wipe samples were collected on and near the vehicle to assist with evaluations of exposures to personnel in these positions. The study included field tests conducted at the Aberdeen Test Center's (ATC) DU Containment Facility (Superbox) located at Aberdeen Proving Ground. The field tests were designed to be surrogates for:

- the types of shots into crew/passenger compartments that occurred during ODS as a result of DU perforation through standard tank armor and Bradley vehicles
- shots that maximized aerosol generation by perforating the tank armor and the massive breech
- perforation of DU armor using munitions specifically loaded for that purpose.

These firing scenarios, which are surrogates for battle situations, provided representative concentrations for nonventilated vehicles with the hatches closed at the time of perforation (at least one of which opened at least briefly following nearly every shot). The time and duration of exposure and the types of exposure (e.g., aerosols versus deposited material) that occurred during ODS formed the time-element basis for modeling the Level I retrospective and prospective scenarios in the human health risk assessment scenarios. Deposition samples within and exterior to the vehicles were collected, but these data are not included in this analysis.

3.2.1 Capstone Test Firing Scenarios

The test shots performed during the four phases of the Capstone study, which physically simulated the types of shots that breached Abrams tanks and Bradley vehicles during ODS or could occur in future actions, are discussed in this section. A partial listing is provided of the types of aerosols collected and characterized that were specifically relevant as input to the human health risk modeling effort. The phases and shots are abbreviated as Phase (PI, PII, PIII, and PIV) and Shot (1 to 7) so that Phase I, Shot 1 is represented as PI-1, and the double shots are abbreviated as PI-3/4 and PII-1/2.

3.2.1.1 Physical Simulation of Scenarios

The Capstone study involved a four-phase series of shot tests in which test vehicles were impacted with tank DU munitions to simulate specific battlefield conditions. Scenarios were developed to determine an upper bound on aerosol concentrations and to quantify other aerosol characteristics from retrospective scenarios and from possible prospective scenarios. The field tests, which incorporated various firing angles, were divided into phases so the investigators could separately focus on circumstances similar to 1) ODS actions (Phases I and II) and 2) possible future actions (Phases III and IV). Aerosol samplers were arranged in sampling arrays located near crew positions. In the Abrams, these arrays were placed at each of the crew positions, noted as commander, driver, gunner, and loader. In the Bradley vehicle, the crew positions for which arrays were emplaced are the commander, driver, left scout, and right scout. Brief descriptions of the test objectives and simulated scenarios follow.

- **PHASE I—**
  Abrams Tank Perforated Through Conventional (Non-DU) Armor

  Some Abrams tanks used in ODS were equipped with conventional (non-DU) armor packages and some were equipped with heavy (DU) armor packages. The six tanks that were struck by DU rounds during ODS did not involve perforation of heavy armor. Consequently, the majority of the Capstone study simulations of these ODS incidents used an Abrams BHT with conventional armor. In Phase I, seven shots were fired at the Abrams BHT through conventional armor. In six of the seven shots, the shot lines were more severe (perforated more armor) than the shots fired during actual ODS incidents. The first four Phase-I shots were crossing shots in which the round was fired at and perforated the side armor of the turret and exited through the opposite side. Two of the four crossing shots (PI-3/4) were fired 13 min apart and are referred to in this report as a double shot. Two of the remaining three shots purposely were fired into the gun breech to maximize (presumably) the amount of aerosol generated. In the seventh test, the shot was fired through the hull within the turret basket to simulate the most severe Abrams tank perforation that occurred during ODS.

- **PHASE II—**
  Bradley Fighting Vehicle Perforated through Conventional Armor

  The Bradley vehicle is a lightly armored, maneuverable vehicle that is equipped with a medium caliber gun. LC-DU munitions easily penetrate the Bradley vehicle's light armor, and they generally traverse the vehicle with little penetrator erosion. During ODS, 15 Bradley vehicles were perforated by DU munitions, and in some cases, there were multiple perforations. Most of the strikes that crossed crew compartments were fired into the Bradley vehicle's passenger compartment rather than through its turret.

  In Phase II of the Capstone study, two crossing shots through the passenger compartment and one shot into the gun feeder in the turret (similar to a gun breech shot in an Abrams tank) of a Bradley BHT were fired to simulate the Bradley vehicle incidents that occurred during ODS. The two crossing shots (PII-1/2) were fired 14 min apart and are referred to as a double shot.

- **PHASE III—**
  Abrams Tank Perforated Through DU Armor

  Although the Abrams heavy (DU) armor was not perforated during ODS, aerosol generation and potential aerosol levels that would result from perforation of DU armor were also evaluated because this could occur in future conflicts. In Phase III, two shots were fired into the DU armor of a second ballistic turret on the Phase-I hull using LC-DU munitions specifically loaded to perforate the armor.

- **PHASE IV—**
  Operational Abrams Tank Perforated Through DU Armor

  In a series of tests conducted under a separate program, LC munitions were fired at a fully operational M1A2 Abrams Main Battle Tank equipped with heavy armor and uploaded with aluminum surrogate penetrators rather than DU munitions. Unlike the Phase I through III tests, the Abrams Live-Fire Tests were performance-based and were not designed specifically to evaluate DU aerosols produced during the tests. In Phase IV, one DU munition was fired through the DU armor. Unlike Phases I through III in which ventilation systems were absent, the vehicle's EC/NBC ventilation system operated during and after the Phase-IV shot. This vehicle was fully equipped and contained mannequins at each crew position. The Phase-IV shot is referred to in the Capstone report as PIV-4 because of its designation in the test series. Although there were other shots fired at this vehicle, only the PIV-4 shot contained a DU penetrator.

3.2.2 Capstone Aerosol Collection and Characterization

The aerosol collection system used in the Capstone tests was designed to 1) function in a high-temperature, high-pressure environment; 2) survive damage from fragments; and 3) provide sample collection redundancy, confidence in timing, adequate sampling flow rates, and sufficient sample collection for chemical analysis of selected samples. Aerosols were collected inside the vehicle as a function of time, position, and shot line. Additionally, samples of aerosols exterior to the vehicle and material deposited inside and outside the vehicle were collected.

Analysis of an extensive assortment of samples was conducted to address the primary factors likely to affect Gulf War/ODS exposure assessments. The four analyses most relevant to the risk assessment are discussed in more detail below.

- *Interior DU concentration over time.* The results from Institute of Occupational Medicine (IOM) cassette filters and CIs provided the DU aerosol concentration as a function of time in the post-shot environment. Settling-rate curves were generated based on IOM data. The DU aerosol concentration provided the source term for each time interval. A five-stage cyclone with a backup filter collected time-integrated aerosol over 2 h post shot.
- *Interior DU particle size distribution over time.* The results from CI sampler data provided both the DU aerosol concentration and the distribution of particle size as a function of time in the post-shot environment. The particle size distribution generally determines where the particles (and in what amounts) will deposit in the respiratory tract, which are key considerations when estimating dose.
- *Lung fluid solubility and dissolution rates by particle size.* The dissolution rates of DU aerosols determine the rate at which the DU may enter body fluids, which in turn establishes the distribution of chemical and radiological dose. Solubility changes as a function of particle size, both

because of the change in available surface area and because chemical composition (primarily the oxidation phase in this case) may vary with particle size. The size-separated cyclone residues were used for lung fluid solubility analysis and for additional chemical and physical characterization of the aerosol.

- *Ventilation systems effects on interior DU concentration by particle size over time.* The heating, ventilation, and air conditioning (HVAC) and EC/NBC systems, as well as the fire-suppression system and turret blower, will reduce the DU airborne concentration as well as change the particle size distribution. To avoid the additional complications, target vehicles (BHTs) used in Phases I, II, and III of the Capstone study were not equipped with such systems. In addition, to provide data for possible scaling to functional vehicles, air exchanges in vehicles with operating ventilation systems were measured. Many of the ventilation and fire-suppression systems were operational (though their operating status may not be known) in the armored vehicles that were perforated during ODS. The ventilation systems would be expected to increase the removal rate of DU aerosols from the vehicle interior, thus reducing DU concentration and associated doses. Operation of the fire-suppression system may result in the formation of compounds that not only will influence the settling rate of the aerosol particles, but also may affect the solubility and dissolution rate of DU once it is internalized.

Input data used in the human health risk assessment modeling effort were taken from the Capstone study database as listed in detail in Appendix A of the report or as summarized in the main report text (Parkhurst et al. 2004).

The factors listed below also were evaluated in the Capstone study and may have some bearing on the interpretation of results of the risk assessment. However, these parameters are either secondary to or confirmatory to the elements discussed above (e.g., the influence of uranium oxide composition on solubility); therefore, they were either not assessed separately in this report or were outside the scope of the risk assessment.

- Chemical forms by particle size over time
- Particle morphology by particle size over time
- Total elemental composition by particle size
- Resuspension rates from interior and exterior surfaces
- Transferable DU to hands from interior and exterior surface contamination
- Exterior aerosol concentrations, including particle size distribution.

3.2.3 DU Aerosol Concentrations as a Function of Time

The Capstone study used a computer-controlled sampling system to collect aerosols at three or four crew positions at various time points post shot. The primary samplers used in the Capstone study for evaluating DU concentration as a function of time were the 25-mm (SKC Gulf Coast, Inc., Houston, Texas) stainless-steel IOM filter cassettes (referred to as IOMs or in the data tables as FS [filter samples]). These data were not used in this report for reasons discussed below (Section 3.2.3.1). Particle size distributions as a function of time were measured using eight-stage Marple CIs (Andersen Instruments, Inc., Smyrna, Georgia). Samples from the Marple CIs were also used to evaluate the total airborne DU concentration. Inertial CIs are widely used for the size-selective collection of aerosol particles. The Marple CI is capable of measuring the aerosol's aerodynamic particle size distribution covering a range of aerodynamic particle diameters from 20 to 0.5 μm.

A description of the instrumentation is provided in Section 3.5 of Capstone Vol. 1 (Parkhurst et al. 2004). Briefly, each of four sampling stations within each vehicle (except Phase IV) were equipped with a sampling array, which contained nine sampling positions with a filter cassette/CI pair (Figure 3.1), with sampling ports positioned horizontally. Aside from the samplers in the center, the CIs form the exterior circle and the IOMs form the interior circle. Eight of the sampling pairs ran sequentially. The ninth pair was used as a

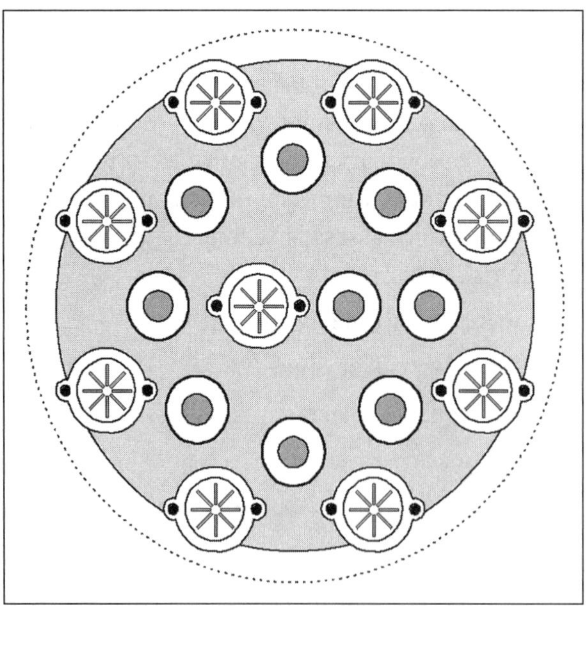

Figure 3.1. Schematic of a Sampler Array

field blank. In Phase IV, five CIs each were attached to the mannequin at the loader's position and five were attached at the driver's position.

A five-stage cyclone collected aerosol beginning 5-sec post shot through 2-h post shot. Aerosols collected by the cyclone were used to evaluate certain chemical and physical properties of the aerosol. A moving filter collected aerosol within the first 5 sec post shot and periodically thereafter as confirmation for the IOM data. Although it provides concentration information specifically during the first few seconds of five of the shots, concentrations collected on this filter were similar to the concentrations collected in the first interval using IOM and CI. Because the moving filter sampler did not operate during every shot and does not provide particle size data, it was not as useful as the IOMs and CIs for evaluating concentrations over time. Therefore, the moving filter samples were not used in this dose assessment. All other time-sequenced sampling except for the moving filter was conducted beginning no earlier than 5 sec post shot to avoid possible damage to the filter or sampler from the initial blast pressure.

As demonstrated within the Capstone IOM data set, DU aerosol concentration peaks almost immediately after armored vehicle perforation, and the concentration decreases rapidly thereafter. The mass concentrations for the various shots could be fitted well with four-component negative exponential functions (see Capstone report, Section 6.1, Parkhurst et al. 2004). With the Capstone data, by 30 min post shot, most uranium concentrations had declined at least two orders of magnitude below their peak concentration, and in several cases, the decline was even greater. Beyond about 30 min, the reduction in DU aerosol concentration as a function of time was slower.

The DU aerosol concentrations measured by the IOMs and CIs provide the source term for each time interval. The selection of Capstone DU aerosol concentration data as input to this modeling effort is discussed in the following sections.

3.2.3.1 Selection of Sampling Monitors—Cascade Impactors Versus IOMs

This risk assessment was conducted using the CI dataset for total DU concentration rather than the IOM dataset because the IOMs collected a larger fraction of particles ≥20-μm AED than the CIs. Applying the CI particle size distribution to the IOM concentration would bias the dose. The total mass concentration of DU in the aerosol samples was measured by both the IOMs and the CIs, which were run temporally and spatially in pairs to measure the concentration of aerosols from three to eight separate time intervals after vehicle perforation. Each IOM sampler collects particles of a wide range of respirable sizes on a single filter whereas a CI collects particles on eight staged substrates and a backup filter (Rubow et al. 1987; Li et al. 2000), referred to as a bottom filter in the Capstone report. A practical advantage of using the IOM data as the basis for DU concentration rather than the CI data is that there was less potential for sample loss from the IOMs (one IOM sample vs. nine Marple substrate and filter samples) and less possibility of underestimating the concentrations. However, the IOMs collected a larger range of particle sizes than the CIs, particularly particles larger than the 21-μm aerodynamic diameter (AD) cutoff diameter of the first CI stage. Based on the published collection characteristics for the IOM and CI samplers, it was concluded that the higher mass values of the IOMs were due to greater collection efficiency of particles ≥20-μm AD compared to the CIs (Rubow et al. 1987; Li et al. 2000). Therefore, it was not considered appropriate to use the mass concentration data from the IOMs together with the particle size distribution data from the CIs because applying the latter distribution to the IOM concentrations would overestimate the amount of respirable DU aerosol. To maintain consistency, the CI concentration data, obtained after correcting for stage collection efficiency, were used with the particle size data.

To compare the concentrations measured by the IOMs and CIs, the respective DU masses collected on the IOM filters and the eight CI substrates and the backup filter (sum of all DU mass collected on the filters for each CI stage) were regressed. Concentrations were calculated by dividing the total masses of DU collected by the sample volumes. Initially, this comparison was performed using the CI stage data uncorrected for internal sampler wall losses. A least-squares regression of the IOM concentrations against the paired summed concentrations measured by the CI (all phases and shots) showed a significant difference (Figure 3.2). The IOM concentrations were on average about 2.3 times greater than the CI concentrations.

This bias between the samplers led to a reanalysis of these data after accounting for the particle collection efficiency as outlined in the manual for the CI (discussed in Section 3.2.3.3). Using the wall-loss correction decreased the bias to about 1.7 (Figure 3.3).

Additional analysis showed that the differences in concentration was significant only during the early time periods post shot (i.e., times less than 10 min). In the "box and whisker" plot of the ratios (Figure 3.4), the ratios of the measured IOM concentrations divided by the summed CI concentrations are categorized by time interval. This plot shows ratios of the median, the 25 to 75%

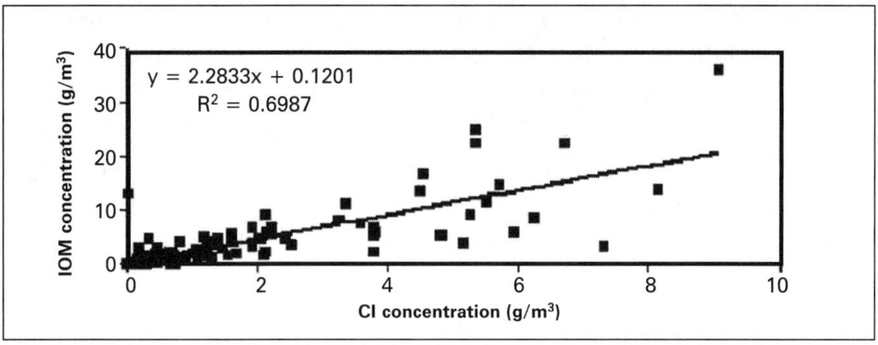

Figure 3.2. Regression Analysis of DU Concentrations of IOM Data Against CI Data

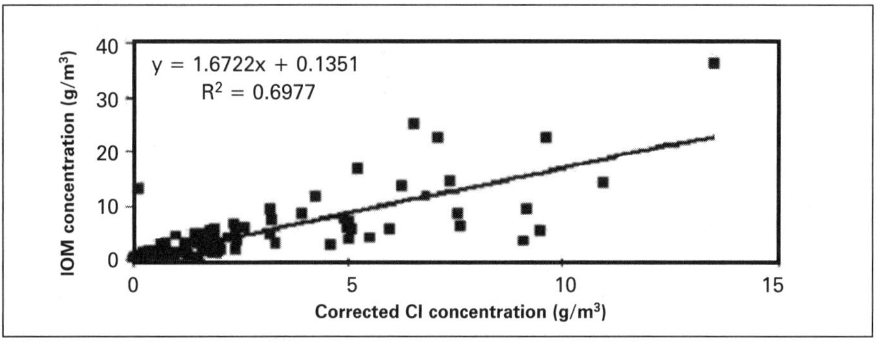

Figure 3.3. Regression Analysis of DU Concentrations of IOM Data Against Efficiency-Corrected CI Data

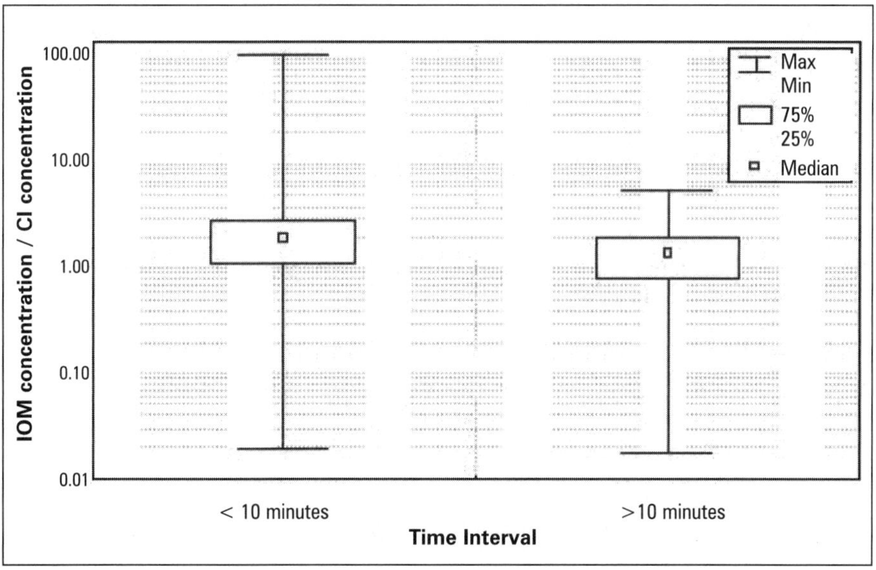

Figure 3.4. Ratios of Measured IOM-to-CI Concentrations by Time Interval

range, and the minimum and maximum range in responses. After 10 min post shot, no significant differences between the IOM concentrations and the summed CI measured concentrations were found. This finding suggests that the difference between the IOM and CI measurements may be caused primarily by different sampling efficiencies between the two sampler types for larger particles (e.g., >5 μm), a situation preferentially noted in the sampling that occurred shortly after vehicle perforation (Parkhurst et al. 2004).

3.2.3.2 Field Blank Adjustment of Data

In the Capstone test, each data sampling station (array) included nine CIs and nine IOMs, which operated in pairs. In most cases, air was drawn through eight pairs of samplers to measure the concentrations and particle sizes of the aerosol as a function of time. The ninth pair did not have air drawn through the inlet and were used as field blanks in the Capstone study. Even though air was not drawn through the blanks, the initial blast or pressure pulse produced by the DU perforation forced particulates into the filters or top substrates of all samplers. Subsequent diffusion (turbulent and Brownian) of aerosol particulates is also assumed to have caused deposition onto the blanks and sample filter surfaces. Therefore, the total amount of particulates on each non-blank sampler is the sum of the particulates from the initial blast, the particulates captured during the sampling interval, and probably, to a lesser extent, aerosol deposited during times in which the air sampler was not in operation. Given that no air was pulled through the field blank samplers, the mass deposited on these blanks was assumed to represent the mass of the particulates due to the initial blast and from any additional convection or diffusion-driven deposition. Such deposition is likely to have been random among samplers, but using field blanks to measure this deposition is believed to be representative. Each DU concentration data set listed in the Capstone report is based on blank-adjusted concentrations.

In this dose and risk assessment, a different approach was taken to adjust for non-flow-related deposition in the samplers. In Phases I through III, the first set of samplers that began operating within a few seconds after the blast were not blank corrected because it was reasoned that an individual in the vehicle would be subjected to the initial blast, which could have driven an amount of DU particles into the respiratory tract and contributed to their overall intake. That is, a conservative assumption was made that the initial impact and subsequent pressure wave caused additional intake of DU into the lung beyond the amount taken into the respiratory tract through active inhalation.

Using uncorrected samples for samples 2 through 8, in which the background had not been subtracted, would have led to an overestimation of the amount of aerosol present in the air at the times these samples were collected. To avoid this, samplers in time periods 2 through 8 were blank corrected. When possible, the field blank from the same sampling array was used. However, if the sample was damaged, collected a large DU particle, or otherwise was not available, the average of other blanks from the same shot or a similar surrogate was used for the blank correction.

The CI results from Phase IV (PIV-4, in which a DU munition was fired through DU armor) were not blank corrected because the uranium masses collected were so low that the greater concern was underestimating uranium concentrations by not using the total uranium mass present.

3.2.3.3 Adjustment of Data for CI Wall Losses

The mass of DU collected on the CI substrates and backup filters was derived from beta counting and comparison with radiochemical analysis of selected samples using the regression procedure outlined in the Capstone report (Parkhurst et al. 2004). Net masses on the substrates and filters were determined as the arithmetic difference of the gross DU mass and the mass collected on the blank sample.

CI samplers inherently lose a fraction of the particles that enter a given stage. These losses are caused by a combination of imperfect aspiration of particles into the sampler inlet and unwanted deposition of particles on the internal walls of the sampler prior to encountering the collection substrates. Adjustment of CI data for wall losses is not a traditional practice in aerosol science because stage-specific collection efficiencies are seldom available. These efficiencies vary as a function of flow rate, and applying them invariably complicates data analysis. However, wall-loss adjustments were used in this analysis to increase the accuracy of the results. The sampling collection efficiencies for the Marple CI stages (Andersen Model 290) has been characterized, and published correction factors are listed in the sampler manual (Table 3.4) to adjust for these losses in each stage (Rubow et al. 1987).

As shown in the table, the correction factors are greatest for stages 1 and 2 and are moderate for stages 3 and 4. Wall loss is minimal on stages 5 through 8.

As recommended by the manufacturer, for each individual impactor stage (i), the DU mass was adjusted for wall losses (Mass_{adj}) by dividing the collected DU mass on the sample filter ($\text{Mass}_{\text{filter}}$) by the fractional sampling efficiency (ε) as shown in Equation 3.1.

$$\text{DU Mass}_{\text{adj}}(\text{i}) = \text{DU Mass}_{\text{filter}}(\text{i})/\varepsilon(\text{i}) \tag{3.1}$$

Table 3.4. Wall-Loss Correction Factors

Impactor Stage No.	Sampling Efficiencies	Correction Factor
1	0.52	× 1.92
2	0.61	× 1.64
3	0.78	× 1.28
4	0.89	× 1.12
5	0.95	× 1.05
6	0.96	× 1.04
7	0.97	× 1.03
8	0.99	× 1.01
Backup Filter	1.00	× 1.00

Equation 3.2 shows the formula used for calculating the net amount of DU on each filter. The masses on the field blanks were not corrected for sampling efficiency because ε(i) is not known for samples collected with no airflow, or for particles driven onto the substrate primarily from the initial pressure pulse on impact. Therefore, the material collected on the sample filters solely from impact and diffusion is assumed to have been collected with 100% efficiency. This assumption creates a bias for the $Mass_{net}(i)$ that is high and results in somewhat higher calculated DU concentrations in air.

$$\text{DU Mass}_{net}(i) = \text{DU Mass}_{adj}(i) - \text{DU Mass}_{FB}(i) \tag{3.2}$$

The DU air concentration (DU_{conc}) was then calculated as the net mass divided by the measured volume of the air drawn through the CI sampler (Equation 3.3). The volume of sampler air was taken to be equal to the product of the volumetric flow rate (AR) and the sampling time (T).

$$DU_{conc}(i) = \text{DU Mass}_{net}(i)/(AR * T) \tag{3.3}$$

The uncertainties in the measured DU concentrations were calculated by propagating the errors of each of the individual factors (Knoll 2000). Error propagation used in the DU concentration results reported in the Capstone study is based on statistical variability in the counting, background subtraction, and flow rates, uncertainty in regression model parameters (used to compare DU concentrations based on beta counting with those from chemical analysis of the same samples), and uncertainty in the ingrowth correction applied to account for the state of disequilibrium of the beta-emitting progeny Th-234 and Pa-234m in the collected aerosols (see sections A.2.1.5 through A.2.1.7, Capstone Report, Parkhurst et al. 2004). In this analysis, the CI sampling efficiency (ε) was included in the calculations of $DU_{conc}(i)$. Therefore the DU concentration uncertainty was recalculated as shown in

Equations 3.4 through 3.6. The standard deviation of sampling efficiency σ_E as a function of wall loss was assumed to be 10% and the standard deviation of the airflow rate (σ_{AR}) was estimated as 3%.

$$\sigma_{\text{DU\_Mass-adj(i)}} = \left|DU\ Mass_{adj(i)}\right| \sqrt{\left(\frac{\sigma_{DUMass\_stage(i)}}{DU\ Mass_{stage}(i)}\right)^2 + \left(\frac{\sigma_\varepsilon}{\varepsilon}\right)^2} \quad (3.4)$$

$$\sigma_{\text{DU\_Mass-net(i)}} = \sqrt{(\sigma_{DU\_Mass-adj(i)})^2 + (\sigma_{Field-Blank})^2} \quad (3.5)$$

$$\sigma_{\text{DUconc}} = \left|DU_{conc}\right| \sqrt{\left(\frac{\sigma_{DU\_Mass-net(i)}}{DU\_Mass-net(i)}\right)^2 + \left(\frac{\sigma_{AR}}{AR}\right)^2} \quad (3.6)$$

3.2.4 Aerosol Particle Size Distribution as a Function of Time

DU aerosol particle size distributions were evaluated as a function of time, and were expected to change with time as the larger particles deposited more rapidly than the smaller particles. Particle size distribution is an important parameter in the calculation of dose from inhaled radionuclides. In the ICRP 66 model (ICRP 1994), the input aerosol size can be either a monodisperse particle size with a geometric standard deviation (GSD) of 1 or a polydisperse size with user-defined values for mass median aerodynamic diameter (MMAD) and GSD.

In the Capstone study, each distribution was analyzed to determine its activity median aerodynamic diameter (AMAD) using SigmaPlot 2000 (SPSS Inc., Chicago, Illinois). In most cases, better fits were obtained using a bimodal distribution rather than unimodal distribution. However, partly because the particulate mass collected on the first stage included relatively large particles, these distributions were seldom adequate to describe the distributions and were often skewed toward the higher particle sizes. Although the AMADs tended to decrease with time, this pattern was not entirely consistent.

The difficulties in interpreting the AMAD data led to the conclusion that the aerosols were heterogeneous and multimodal, and the decision was made to use the individual stage data entirely in the dose and risk assessment calculations. This choice obviated possible influences of imposing assumed lognormal distributions on data that were poorly fitted and maintained the intrinsic variability of the datasets.

The computer modeling developed to apply Capstone DU particle size distribution data as input to the risk assessment modeling effort is discussed below.

3.2.4.1 DU Aerosol Concentration as a Function of Particle Size

Both the CIs and the cyclone used to sample aerosols were multi-stage samplers that segregated particles based on size. In the eight-stage CI with a nominal flow rate of 2 Lpm, the air is pulled first through a 21-µm ECD (effective cutoff diameter) stage, then sequentially through 15-, 10-, 6-, 3.5-, 2.0-, 0.9-, and finally 0.5-µm ECD stages. Particles <0.5 µm that pass through the final stage are collected on the backup filter. Traditionally, the CI data are used to determine a particle size distribution by fitting either a unimodal or multimodal lognormal distribution to the data. However, the Capstone data clearly showed that many of the CI data sets were not well described using any formulation of the lognormal distribution. So, instead of using the Capstone unimodal or bimodal AMAD results to simulate a continuous distribution for the aerosol in terms of the particle size, the uranium concentration from each stage of each CI was used directly in calculating intakes and associated radiation doses. This method takes advantage of the raw data (adjusted for field background and sampling efficiency) without imposing fit structure, good or bad, to the data. To do this, nine parallel dose models were formulated, one corresponding to each stage, which has its characteristic median particle size.

3.2.4.2 Particle Size Distribution Modeling

The CI measurements provided amounts of DU collected in each impactor stage and back-up filter. Generally, there were three analysis techniques used to estimate particle size distributions from the impactor data: 1) cumulative distribution of aerosol mass distribution as a function of the ECD (with 50% collection efficiency in each stage), 2) a histogram method where ECDs of adjacent stages are used to define the size-increment boundaries, and 3) a data inversion technique. The first two techniques make the assumption that the impactor collection efficiency characteristics are well defined for every impactor stage, which is not completely accurate. The third technique is based on the following equation (Equation 3.7) that assumes that the mass of DU collected on impactor stage i ($M_{measured}$) can be expressed as:

$$M_{measured}\,(i^{th}\text{ stage}) = \sum_{j=1}^{j=J} P_i(x_j)\,E_i(x_j)\,M_{calculated}(x_j) \qquad (3.7)$$

where $M_{calculated}$ is the DU mass for particle size increment x_j, $E_i(x_j)$ is the collection/impaction efficiency at stage i for DU aerosol in the particle size increment x_j, and $P_i(x_j)$ is penetration of aerosol in the particle size increment

x_j in the impactor stage i. The penetration functions can be expressed as (Equation array 3.8):

$$\begin{aligned} P_1 &= 1 \\ P_2(x_j) &= 1 - E_1(x_j) \\ P_3(x_j) &= [1 - E_1(x_j)][1 - E_2(x_j)] \end{aligned} \tag{3.8}$$

The impaction efficiency, $E_i(x_j)$, as a function of AD, (d_{ae}), was experimentally determined for each stage of the Andersen personal impactor (Rubow et al. 1987) as shown in Figure 3.5. The efficiency curve was measured for a flow rate of 2 Lpm.

The aerosol size distribution data documented in the Capstone report (Parkhurst et al. 2004) were obtained in a flow rate ranging from 1.7 to 2.7 Lpm. The collection efficiency of the CI changes as the flow rate changes; therefore, it is necessary to estimate the collection efficiency of each impactor stage as a function of flow rate and particle size. Based on the impactor theory (Marple 1970; Rader and Marple 1985), the collection efficiency is a function of Stokes number (Stk= $[d_{ae}]^2U/9\eta W$) and Reynolds number (Re= $\rho WU/\eta$), where U is the average velocity at nozzle, W is the diameter of a round nozzle or the width of the slit for a rectangular nozzle, d_{ae} is the AD of the particle, η is the air viscosity, and ρ is the air density. Figures 3.6 to 3.13 show the calibration curves plotted in collection efficiency versus the square root of Stokes number ($\sqrt{Stk}$). The collection

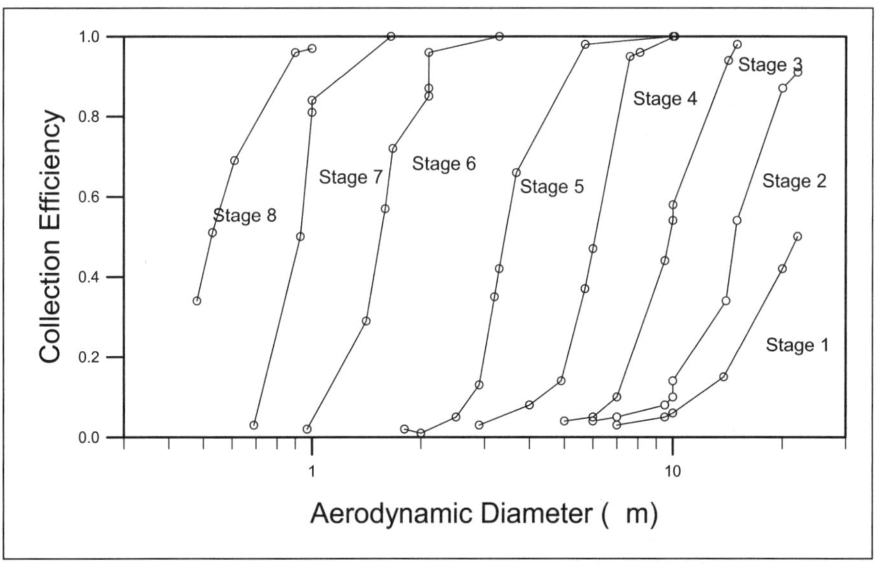

Figure 3.5. Calibration Data of Andersen Personal Impactor (Rubow et al. 1987)

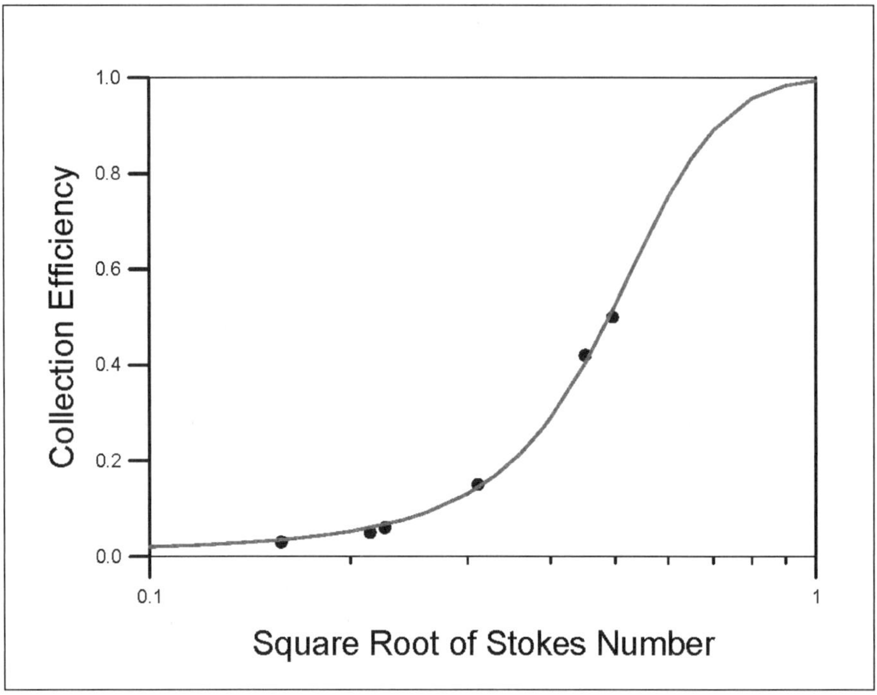

Figure 3.6. Experimental Data and Fitted Curve for Stage 1

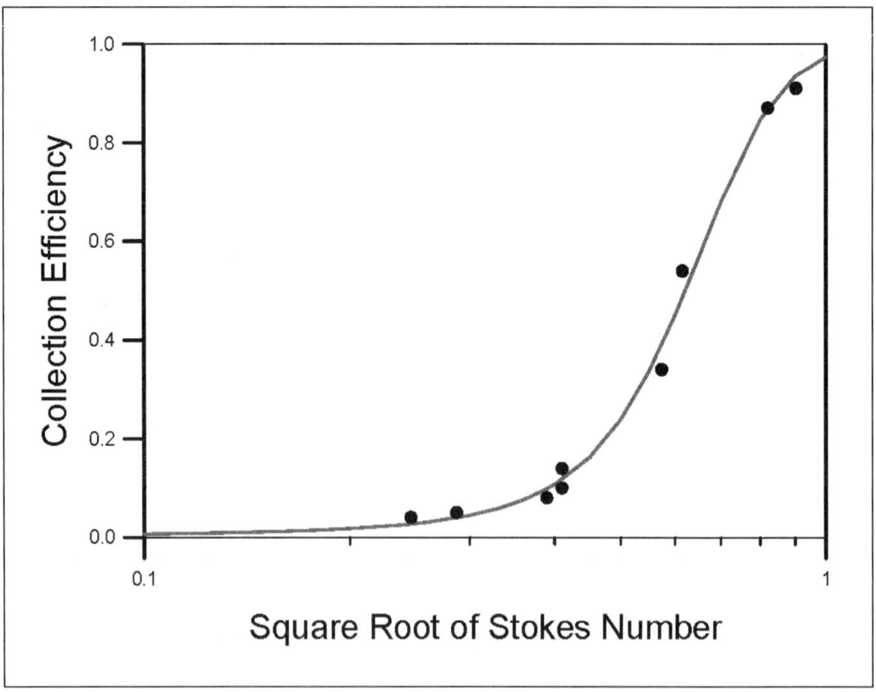

Figure 3.7. Experimental Data and Fitted Curve for Stage 2

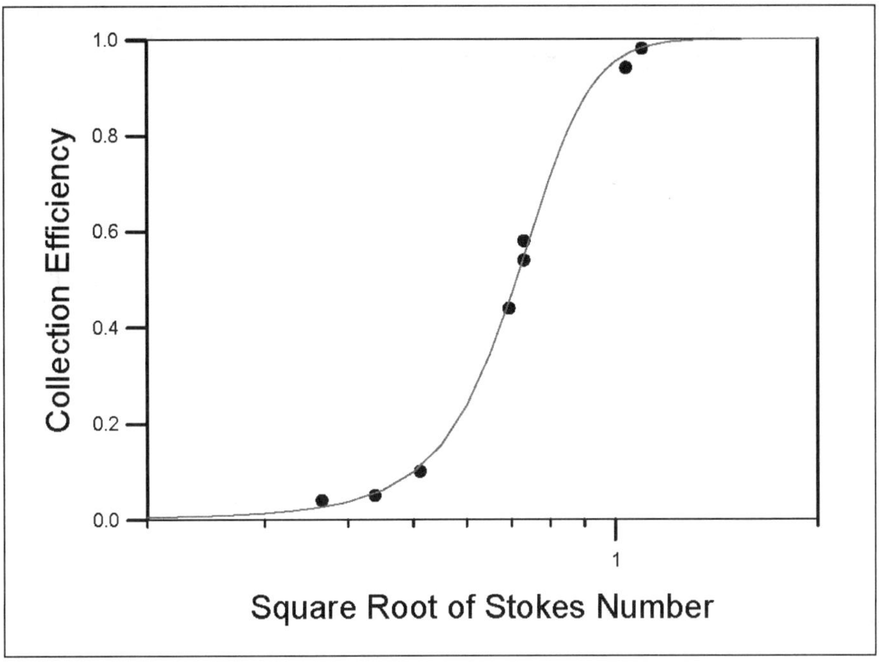

Figure 3.8. Experimental Data and Fitted Curve for Stage 3

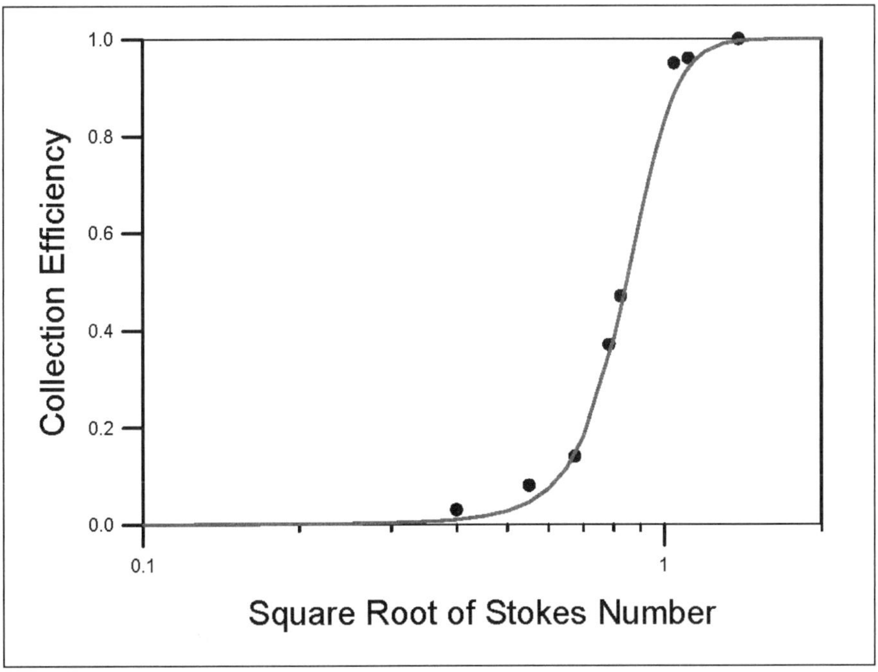

Figure 3.9. Experimental Data and Fitted Curve for Stage 4

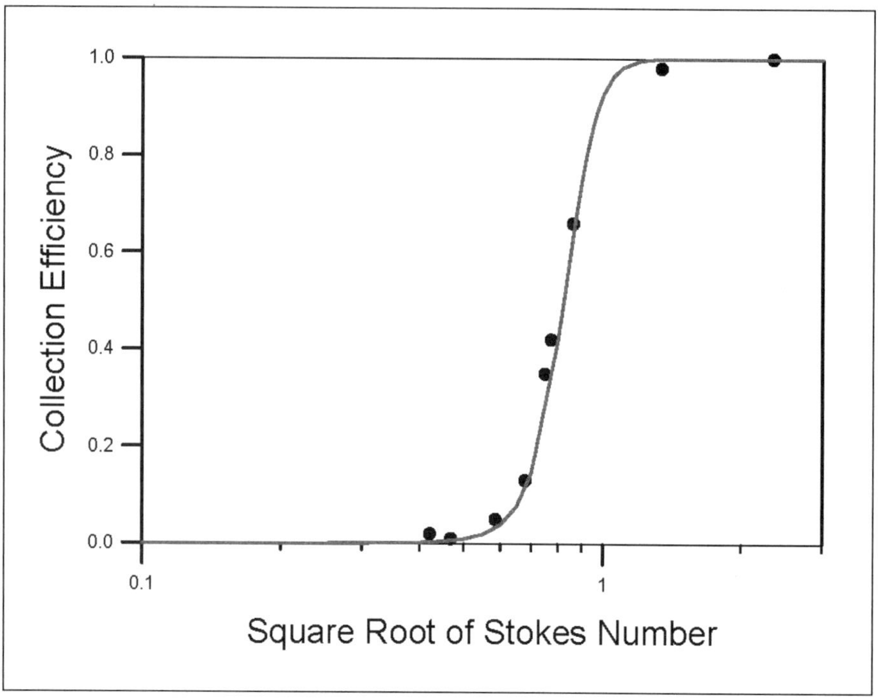

Figure 3.10. Experimental Data and Fitted Curve for Stage 5

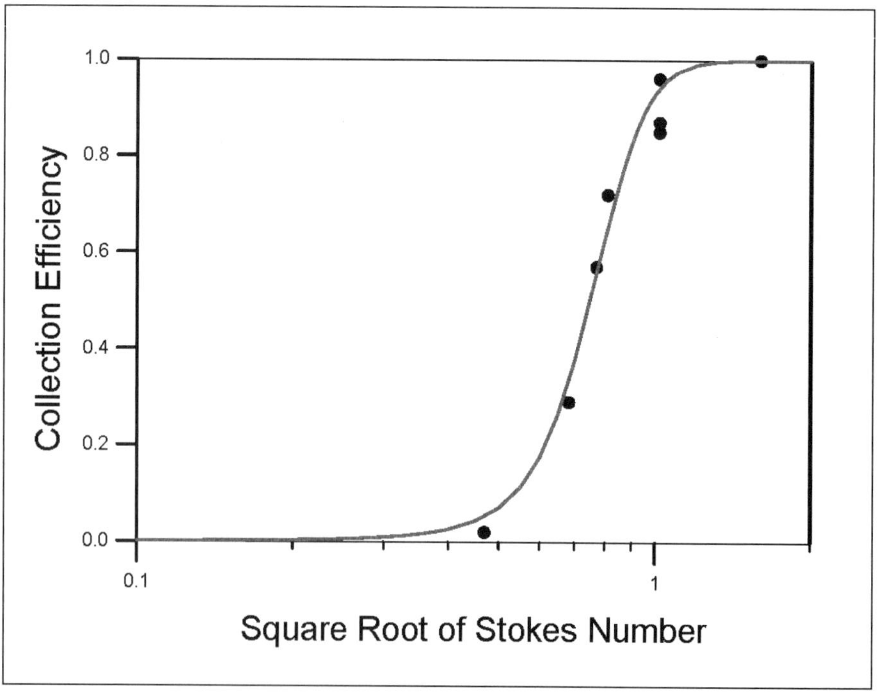

Figure 3.11. Experimental Data and Fitted Curve for Stage 6

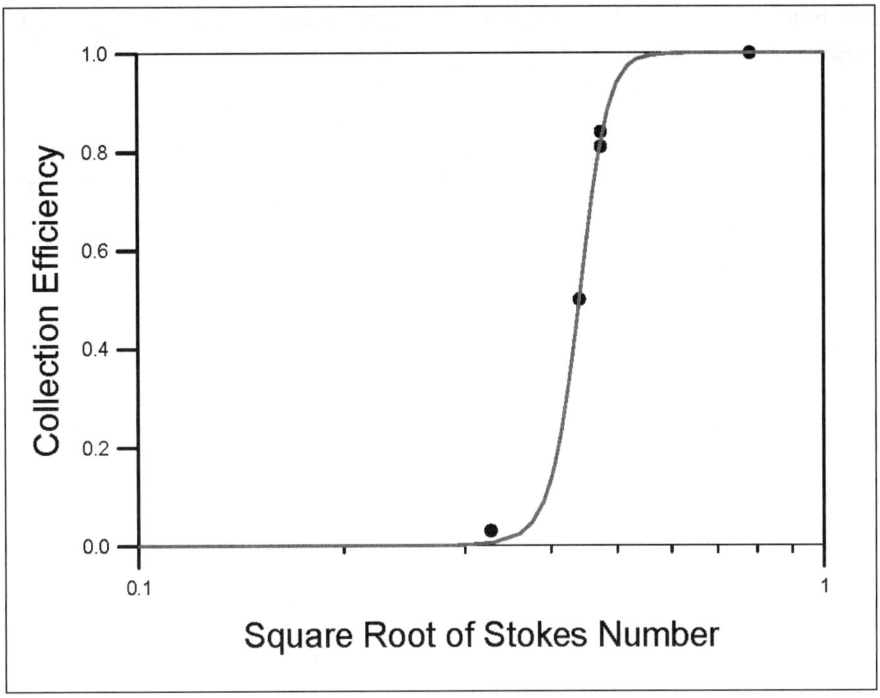

Figure 3.12. Experimental Data and Fitted Curve for Stage 7

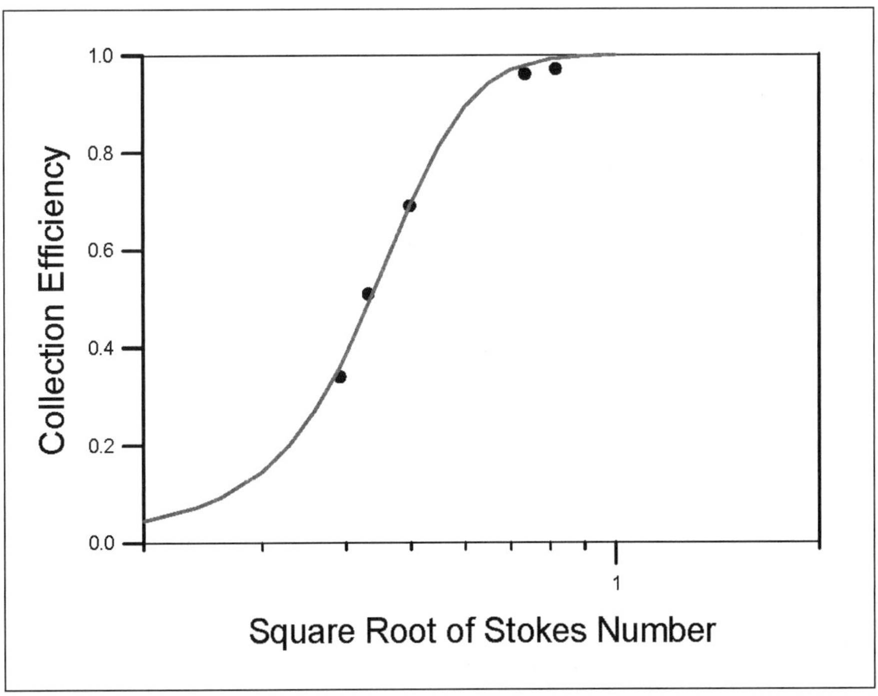

Figure 3.13. Experimental Data and Fitted Curve for Stage 8

efficiency data (E) as a function of (√Stk) in each impactor stage were expressed in Equation 3.9:

$$E(x) = \frac{1}{1+\exp\left[-\frac{(S-S_o)}{b}\right]} \tag{3.9}$$

where S = (√Stk), S_o is the 50% value of (√Stk), and b is a constant relating to the shape of the efficiency curve. Values of x_o and b for each impactor stage were obtained by the nonlinear regression routine of SigmaPlot and are listed in Table 3.5. The coefficients of determination (r^2) are also listed in Table 3.4, indicating good agreement between the fitted equation and experimental data (Figures 3.6 to 3.13).

Table 3.5. Values of Fitted Curve in Equation (3.6)

Stage	x_0	b	r^2
1	0.4896	0.1001	0.995884
2	0.6211	0.1045	0.991263
3	0.7123	0.0959	0.997284
4	0.8476	0.0974	0.996826
5	0.8232	0.0695	0.997304
6	0.7488	0.0965	0.969548
7	0.4406	0.0217	0.99758
8	0.4363	0.0771	0.992304

Once the collection equations arc cstablished, these equations can be used to calculate the collection efficiency and 50% cutoff diameter at different flow rates under the assumption that these equations yield valid results in the range of flow rates used in the Capstone experiments. The collection efficiency in each stage (E_i, where i = 1 to 8) can be calculated as a function of aerodynamic size (d_{ae}) and flow rate (Q in cm^3/sec) through the impactor (Equation array 3.10):

$$\begin{aligned} E_i(S) &= \frac{1}{1+\exp\left[-\frac{(S-S_o)}{b}\right]} \\ S &= \sqrt{\frac{d_{ae}^2\,Q}{9n_i\,\eta W^2\,L}} \quad \text{for stage } i=1 \text{ to } 6 \\ S &= \sqrt{\frac{4d_{ae}^2\,Q}{9\pi n_i\,\eta W^3}} \quad \text{for stage } i=7 \text{ and } 8, \end{aligned} \tag{3.10}$$

where d_{ae} is the AD;

Q is the flow rate through the impactor;

W is the width of slit (rectangular) nozzles for impactor stages 1 through 6 and the diameter of the round nozzles for stages 7 and 8;

L is the length of slit (rectangular) nozzles for impactor stages 1 to 6;

n_i is the number of nozzles in stage i; and

η is the air viscosity.

Table 3.5 lists the parameters of the best-fit curves to the impactor calibration data shown in Figures 3.5 to 3.12. Equation 3.7 is a good expression of the calibration data as shown in the figures and values of r^2. Using the best fit parameters in Table 3.5, the collection efficiency of an aerosol in each impactor stage can be calculated as a function of particle size and flow rate. For example, the 50% cutoff diameter in each stage can be calculated from S_o, the value of the square root of Stokes number corresponding to 50% collection efficiency on each impactor stage.

From values of the square root of Stokes number corresponding to 50% collection efficiency for impactor stage i (S_{oi}), the 50% cutoff AD (ECD_i) for each stage at flow rate Q can be calculated as (Equations 3.11):

$$ECD_i = \frac{9\eta x_{oi}^2 W^2 L}{Q} \quad \text{for stage } i = 1 \text{ to } 6$$

(3.11)

$$ECD_i = \frac{9\pi\eta x_{oi}^2 W^3}{4Q} \quad \text{for stage } i = 7 \text{ and } 8$$

Table 3.6 shows examples of ECD_i for a flow rate of 2 Lpm. ECDi were used to define the nine size increments. For each size bin, the aerosol is assumed to be a monodisperse particle with diameter x_i equal to the square root of the product of ECD_i and ECD_{i-1} ($x_i = [ECD_i\ ECD_{i-1}]^{0.5}$).

If these nine size increments are used to describe the particle size distribution from personal impactor measurement, then Equation (3.1) becomes (Equation 3.12):

$$M_{measured} = R\ M_{calculated} \qquad (3.12)$$

where $M_{measured}$ and $M_{calculated}$ are vectors with nine elements and R is a response matrix with $R_{ij} = P_i(x_j)E_i(x_j)$. The particle size distribution in terms of concentration of DU in each size increment can be calculated by the method of data inversion as (Equation 3.13):

$$M_{calculated} = R^{-1}\ M_{measured} \qquad (3.13)$$

where R^{-1} is the inversion of matrix R.

Table 3.6. Calculated Effective Cutoff Diameters and Size Increments for the Eight-Stage CIs

Stage	ECD (m)	x (m)
--	50	--
1	21.7	33.0
2	15.1	18.1
3	9.7	12.1
4	6.2	7.7
5	3.5	4.7
6	1.6	2.3
7	0.93	1.2
8	0.53	0.70
Filter	0.10	0.23

3.2.5 DU Aerosol *In Vitro* Solubility/Dissolution

To evaluate aerosol solubility as a function of particle size, the Capstone study evaluated the *in vitro* solubility of 15 cyclone samples and 4 backup filters, which sampled aerosols over the entire 2-h sampling period. Additionally, five IOM samples were analyzed to evaluate solubility as a function of time. Because of resource limitations, cyclone samples were analyzed only for PI-3/4, PI-7, PII-1/2, and PIII-2 for stages 2 through 5 and the backup filter. Particles from cyclone Stage 1 are generally too large to be respirable, so only one Stage 1 sample (PI-7) was analyzed. Calculating doses from Capstone shots for which samples were not analyzed for solubility involved 1) cross-mapping the particle size stages from the cyclone onto the more numerous CI stages, and 2) grouping the *in vitro* dissolution results as indicated by the data to produce averaged particle-size-specific dissolution patterns. The data support these groupings for most shots, although expert judgment was used in grouping some of the samples. Additional information about the process is provided in Section 4.4.2.

3.2.6 Effects of Ventilation Rate on DU Aerosol Concentration

During the Capstone study (Phases I, II, and III), aerosol measurements were taken in a combat vehicle and in BHT, which were vehicle structures without instrumentation, ventilation systems, NBC protective systems, wiring, or flammable components. Systems normally found inside a combat-ready vehicle, such as communications and control equipment, were not present in the BHTs. This affects the aerosol results in two ways: 1) no ventilation systems or NBC protective systems (the EC/NBC overpressure system in the Abrams tanks and the crew NBC system in the Bradley vehicles) to remove aerosols are available, 2) the interior air volume is increased compared to a fully equipped vehicle, and 3) the surface area available for removal of aerosol from the air by deposition of particles is less in the BHT versus the fully equipped vehicles. A description of the ventilation measurement procedures is provided in Appendix G, Volume II of the Capstone report (Parkhurst et al. 2004).

3.2.6.1 Description of Air Filtration Systems

Abrams Tanks

In an operational vehicle, the Abrams M1A1 and M1A2 tanks are equipped with a crew and compartment EC/NBC ventilation system. The systems are the same in these two vehicles.[a] The system protects the crew in two ways. First, it provides filtered air directly to the ventilated face pieces worn by each member of the crew. These face pieces are masks equipped with an inlet for the compressed, filtered air from the Abrams EC/NBC ventilation system. Second, the system uses filtered, compressed air to generate a positive pressure in the crew compartment (General Dynamics 1987).

The EC/NBC ventilation system uses air bled directly off the compressor of the Abrams turbine engine. The compressor draws air from outside the vehicle and filters the air prior to entering the compressor. This filtration system is designed to remove dirt and debris from the air and does not filter for NBC contaminants. The air is cooled and routed through two parallel 2800 Lpm (100 cfm) NBC filters. After filtration, a fraction of the filtered air is routed to the hoses that connect to the face pieces, and the remainder of the air is routed directly into the crew compartment to create the overpressure inside the vehicle. The overpressure is designed to ensure that NBC contaminants do not enter the vehicle though the normal air gaps in the vehicle or through breeches in the structure.

The EC/NBC ventilation can be activated at any of the crew positions and is sometimes used to clear the air of the low level of combustion products that enter the crew compartment when the coaxial machine gun or the main gun is fired. If the main EC/NBC system is not functioning, a backup NBC system, if activated by the tank commander supplies filtered air to the crew face pieces only. This backup system does not provide an overpressure inside of the vehicle but does provide filtered air to the crew. Additionally, there are ventilation systems in the Abrams that recycle inside air but provide no filtration for the air (denoted as "fan on" in Table 3.7 and in the Capstone report, Chapter 4 and Appendix G, Parkhurst et al. 2004).

---

[a]Personal communication, Mr. W. A. Baetz, Test Director, U.S. Army Aberdeen Test Center, October 8, 2003.

In actual battle situations, the EC/NBC ventilation system will likely be in constant operation only if there is a threat of NBC agents being used but will be periodically used to remove the combustion products from the firing of the weapons systems in the vehicle. The nominal flow rate of the EC/NBC system is 5600 Lpm (200 cfm), which is equivalent to a ventilation rate of 47.2 air exchange per hour assuming the total net air volume in an equipped and occupied Abrams tank is 7.2 m^3. The high ventilation rate provided by the EC/NBC system would affect the aerosol concentration inside the vehicle mainly by removing particles from within the vehicle to the outside.

Bradley Fighting Vehicles

The Bradley vehicle is equipped with two ventilation fans, one located behind the driver and one in the crew compartment.[a] These systems draw outside air though a particulate filter and into the passenger compartment and the driver's compartment. The systems can be activated by the driver or by anyone in the passenger compartment.

The Bradley vehicle has an NBC system that filters inside air and provides pressurized air to the crews' protective masks through ventilation hoses. The system does not provide positive pressure for the crew compartment. A ventilation system mounted in the turret vents the gasses from firing the machine gun to the outside environment.

3.2.6.2 Comparison of Ventilation Rates

The Capstone study evaluated the air exchange rates in the Abrams and Bradley BHTs used in this study and compared them with fully functioning and fully equipped Abrams M1 and M1A1 tanks and a Bradley vehicle. Air exchange measurements were taken in these vehicles with these systems operating and not operating and hatches open and not open with the expectation that results could be scaled for other needs. The test configurations for each vehicle are listed in Table 3.7 for each tested vehicle. The M1A1 and M1A2 tank EC/NBC systems were not tested. The aerosol concentration and particle size distributions were measured during Phase IV in which an EC/NBC system was operating.

Ventilation rates of the BHTs before perforation and with the hatches closed (and no fans operating) yielded the volume exchange data presented in Table 3.8. The rates from the Abrams tank PI-1 through PI-3 shots varied little and

[a]Personal communication, Mr. D. M. Blankenbiller, U.S. Army Test and Evaluation Command, November 2003.

Table 3.7. Ventilation Rates for Intact, Operational Abrams Tanks and Bradley Vehicles (volumes exchanged per hour)

Fan OFF Hatches Closed	Exchange Rate	Fan ON,[(a)] Hatches Closed	Exchange Rate	Fan OFF, Hatches Open	Exchange Rate
Abrams M1 Tank					
1	6.02	1	7.51	1	19.7
2	3.90	2	7.38	2	18.2
3	4.18	--	--	--	--
Abrams M1A1 Tank					
1	1.17	1	6.59	1	2.96
2	1.09	2	6.58	2	2.86
Abrams M1A2 Tank					
1	0.27	1	NA	1	4.83
2	0.23	2	NA	2	5.63
Bradley Vehicle					
1	7.66	1	45.27	1	7.14
2	11.5	2	34.57	2	9.21

(a) In the Abrams tanks, "fan on" refers to the system that recycles inside air but provides no filtration for the air; in the Bradley vehicle, "fan on" refers to ventilation fans in the driver's and crew compartments, which filter outside air entering the vehicle.
NA–not available

Table 3.8. Pre-Shot Ventilation Rates in BHTs for Phases I, II, and III

Shot	Exchange Rate (volumes exchanged per hour)
Phase I (Abrams tank)	
1	2.98
2	3.49
3/4	2.70
5	NA
6	NA
7	7.15
Phase II (Bradley vehicle)	
1/2	NA
3	33.9
Phase III (Abrams tank with DU armor)	
1	5.55
2	6.28

NA—not available

ranged from 2.7 to 3.5 exchanges per hour. The ventilation rate from the PI-7 shot was similar to those tested during Abrams Phase III, ranging from 5.55 to 7.15 exchanges per hour. The single Bradley vehicle Phase-II result was considerably higher at 33.9 exchanges per hour. This result is much larger than the results from the operational vehicle except for the readings in the operational vehicle when the fan was on and the hatches were closed.

With one or more additional holes in the vehicle caused by the shot, the post-shot ventilation rates would be higher than the pre-shot values. Additionally, the hatches that remained partially or fully opened from the perforation blast would allow a somewhat faster removal of suspended particles. Measurements inside the BHTs were taken immediately after perforation and for several hours post shot, but these results were not interpretable due to unknown confounders (Capstone report, Appendix G, Parkhurst et al. 2004).

In Phase IV, the target vehicle was fully equipped, and although the onboard ventilation system was not operating, the EC/NBC overpressure system was operating during and after perforation. No pre- or post-test ventilation measurements were performed with this vehicle.

3.2.6.3 Projected Effects of Ventilation on DU Aerosol Removal

The effect that a higher ventilation rate has on aerosol concentration by removing the aerosol from the vehicle was examined using the following simplified model to describe the aerosol concentration as a function of time after the shot (Equation 3.14):

$$\frac{dC}{dt} = -\lambda C \tag{3.14}$$

where C is the aerosol concentration and λ is the rate constant for particle removing processes. In this case, two mechanisms, ventilation and particle deposition on surfaces, were considered (Equation 3.15):

$$\lambda\ (\text{min}^{-1}) = v + \beta \tag{3.15}$$

where v is the minute ventilation rate and β is the surface deposition rate constant, a function of particle size. In Equation 3.14, effects such as particle coagulation were neglected. It has the following solution (Equation 3.16):

$$C = C_o \exp(-\lambda t) \tag{3.16}$$

In reality, the measured aerosol has a particle size ranging from 0.5 to about 20 μm (excluding particles above the sampler cutoff diameter), and a single equation such as Equation 3.16 is not an accurate description of the aerosol concentration as a function of time. Rather the aerosol should be treated as a collection of several size increments, each of which has an aerosol decay function. The total concentration then can be expressed as sum of the size increments or components (Equation 3.17):

$$C = \sum C_i \exp(-\lambda_i t) \tag{3.17}$$

Abrams Tank Tests

As an example, the mean uranium aerosol concentration decay data obtained using the IOM filter samplers for Phase III, Shots 1 and 2, which were single-shot tests, can be expressed by a four-component exponential decay function (Equation 3.18), the constants of which are listed in Table 3.9.

$$C\ (\mu g\ m^{-3}) = C_1 \exp(-\lambda_1 t) + C_2 \exp(-\lambda_2 t) + C_3 \exp(-\lambda_3 t) + C_4 \exp(-\lambda_4 t), \quad (3.18)$$

Table 3.9. Fitted Constants for Phase-III DU Aerosol Reduction with Time

Shot	C_1	λ_1	C_2	λ_2	C_3	λ_3	C_4	λ_4
1	8.125E+06	1.520E+01	2.207E+06	1.799E-01	1.623E+05	5.400E-02	3.761E+03	8.179E-10
2	1.045E+07	3.429E+00	7.715E+06	8.007E-01	2.353E+06	3.697E-01	3.769E+04	6.175E-03

Assuming the same initial concentrations (C_i) when the ventilation rate is increased, the ventilation rate constant increases from measured values of 5.55 and 6.28 per hour to 47.2 per hour. Assuming that the surface deposition rate constants stay unchanged, the DU aerosol concentration can be expressed as equation (5) with the constants shown in Table 3.10.

Table 3.10. Fitted Constants with Increased Ventilation in an Abrams Tank

Shot	C_1	λ_1	C_2	λ_2	C_3	λ_3	C_4	λ_4
1	8.125E+06	1.590E+01	2.207E+06	8.735E-01	1.623E+05	7.476E-01	3.761E+03	6.936E-01
2	1.045E+07	4.111E+00	7.715E+06	1.482E+00	2.353E+06	1.051E+00	3.769E+04	6.876E-01

Figure 3.14 shows the DU aerosol concentration as a function of time for PIII-1 for the measured samples in the unventilated BHT (dotted and solid lines for minimum, mean, and maximum) and the projected reduction for an increased ventilation rate (of 47 exchanges per hour; shown by dashed line). Actual measurements of the Phase-IV data are provided and are consistent with the model predictions during the first few minutes but reflect the temporal pattern of the Phase III concentration data at longer times.

Modeling an Abrams tank using the Capstone data to evaluate the reduction in aerosol concentrations to be expected from a ventilation rate of 5600 Lpm (200 cfm) as provided by the EC/NBC system, the aerosol concentration was determined by simplifying the aerosol dynamics in the chamber with assumptions such as 1) the initial aerosol concentration and particle size remained constant, 2) the wall deposition rate constant did not change, and 3) there was no particle coagulation, which would alter the particle size

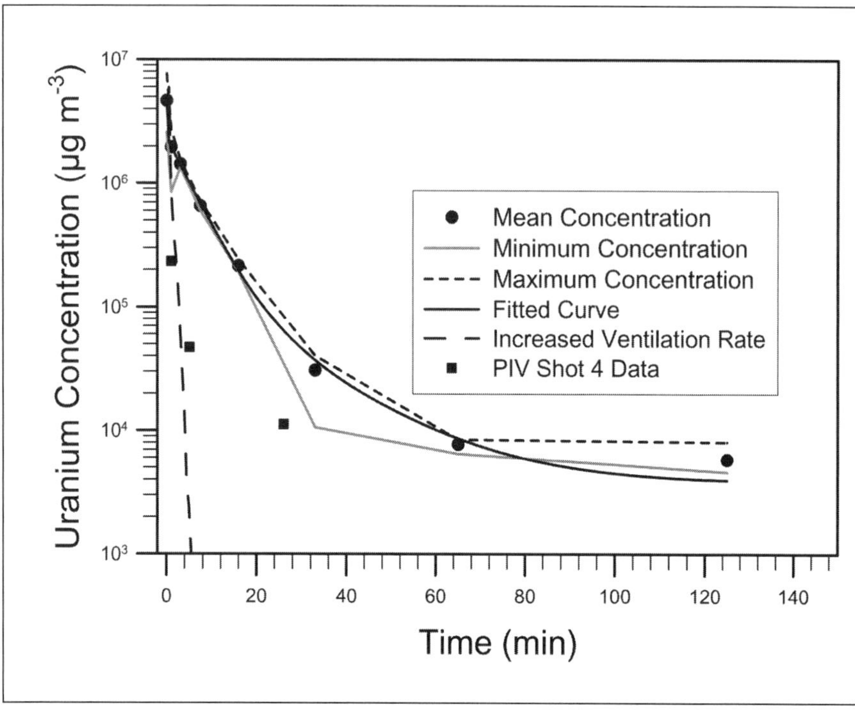

Figure 3.14. DU Aerosol Concentration Reduction with Time with (PIV-4) and Without (PIII-1) Ventilation

distribution. These calculations led to the following conclusions regarding DU concentration reduction:

- The concentration decreased by a factor of about 2 in about 1 min.
- The concentration decreased by about an order of magnitude in 3 min.
- The concentration decreased by about two orders of magnitude in 6.5 min.

Bradley Fighting Vehicle

For the Bradley vehicle, using the Phase II, Shot 3 data, the DU aerosol concentrations in the BHT and in an operational vehicle with an operating ventilation system could be expressed using Equation 3.18 with the constants shown in Table 3.11. The measured pre-shot ventilation rate in the PII-3 for the tested (BHT) vehicle was 33.9 exchanges per hour. The measured ventilation rate (average of two measurements) in an operational vehicle with the fan on and hatch closed was measured as 39.9 volume exchanges per hour. The effect of ventilation on the aerosol concentration inside an operational

Table 3.11. Constants for a BHT and an Operational Bradley Vehicle

Vehicle Type	C_1	λ_1	C_2	λ_2	C_3	λ_3	C_4	λ_4
BHT	6.816E+05	3.265E+00	5.677E+05	1.492E+00	9.887E+04	1.303E+00	6.988E+04	9.093E-03
Operational	6.816E+05	3.366E+00	5.677E+05	1.592E+00	9.887E+04	1.403E+00	6.988E+04	1.094E-01

Bradley vehicle was estimated using as the same assumptions described in the previous section. Under these assumptions the uranium aerosol concentration can be expressed as Equation 3.18 using the constants listed in Table 3.11.

The DU aerosol concentration reduction was modeled in a Bradley vehicle using the Capstone data and increasing the ventilation rate to 39.9 exchanges per hour. The modeling included such assumptions as 1) the initial aerosol concentration and particle size stay constant, 2) the wall deposition rate constant did not change, and 3) there was no particle coagulation, which would alter the particle size distribution. These calculations led to the following conclusions concerning the DU aerosol concentration reduction with the slight increase in ventilation:

- The concentration decreased by a factor of about 2 in about 7 min.
- The concentration decreased by about an order of magnitude in 23 min.

Comparison of DU Concentration Reduction

With increased ventilation rates in operational Abrams tanks and Bradley vehicles, the DU aerosol concentration would be significantly reduced from data measured in unventilated BHTs in the Capstone study. To examine the influence of the ventilation rate in a BHT versus the operational vehicles with the ventilation systems as tested, and making the assumption that the particle size distribution does not change between the tested vehicles and operational vehicles, the dose ratio can be calculated for operational vehicles versus the BHTs that were perforated in the Capstone study. Figure 3.15 shows the concentration ratio (reduction in DU aerosol concentration in a ventilated vehicle) as a function of time post shot for the Abrams and Bradley operational vehicles as compared to the test vehicles. The DU concentration reduction for the Abrams tanks is much more substantial than for the Bradley vehicle because the ventilation increase for an Abrams tank with an operating ventilation system was much greater than the ventilation increase in the Bradley vehicle.

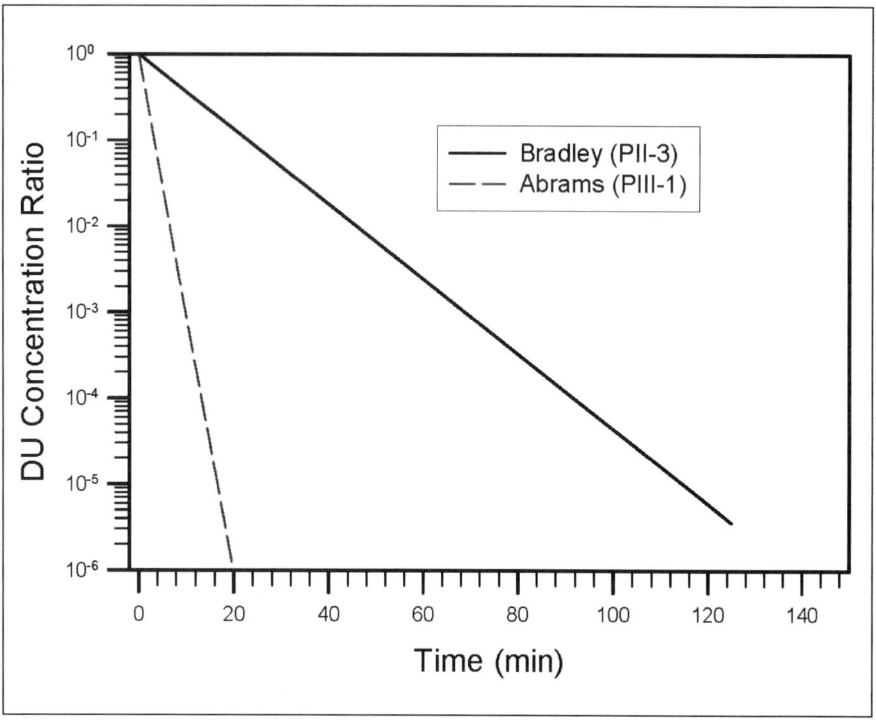

Figure 3.15. DU Concentration Ratio Reduction with Time in a Ventilated Versus Unventilated Abrams Tank and a Bradley Vehicle

3.3 REFERENCES

General Dynamics. 1987. *M1A1 Tank: Characteristics and Description Book.* Systems Technical Support Services for the Abrams Tank Systems, Contract DAAE07-86-C-R032, Document Number NP-86-13098-001, September 1987.

International Commission on Radiological Protection (ICRP). 1975. *Report of the Task Group on Reference Man.* ICRP Publication 23, Pergamon Press, Oxford, United Kingdom.

International Commission on Radiological Protection (ICRP). 1994. *Human Respiratory Tract Model for Radiological Protection.* ICRP Publication 66, Pergamon Press, Oxford, United Kingdom.

International Commission on Radiological Protection (ICRP). 1995. *Basic Anatomical and Physiological Data for Use in Radiological Protection: the Skeleton.* ICRP Publication 70, Pergamon Press, Oxford, United Kingdom.

International Commission on Radiological Protection (ICRP). 2003. *Basic Anatomical and Physiological Data for Use in Radiological Protection: Reference Values.* ICRP Publication 89, Pergamon Press, Oxford, United Kingdom.

Knoll GF. 2000. *Radiation Detection and Measurement.* Third Edition, John Wiley & Sons, New York.

Li, S-N., DA Lundgren, D Rovell-Rixx. 2000. "Evaluation of Six Inhalable Aerosol Samplers." *American Industrial Hygienists Association Journal*, 61:506-516.

Marple VA. 1970. *A Fundamental Study of Inertial Impactors*. PhD Thesis, University of Minnesota, Minneapolis, Minnesota.

Parkhurst MA, F Szrom, RA Guilmette, TD Holmes, YS Cheng, JL Kenoyer, JW Collins, TE Sanderson, RW Fliszar, K Gold, JC Beckman, and JA Long. 2004. *Capstone Depleted Uranium Aerosols: Generation and Characterization, Volumes 1 and 2*. PNNL-14168, Prepared for the U.S. Army by Pacific Northwest National Laboratory, Richland, Washington.

Rader DJ, and VA Marple. 1985. "Effect of Ultra-Stokesian Drag and Particle Interception on Impaction Characteristics." *Aerosol Science and Technology*, 4:141-156.

Rubow KL, VA Marple, J Olin, and MA McCawley. 1987. "A Personal Cascade Impactor: Design, Evaluation and Calibration." *American Industrial Hygienists Association Journal*, 48:532-538.

US Army Center for Health Promotion and Preventive Medicine (USACHPPM). 2003. Memorandum for Record, MCHB-TS-OHP, 14 February 2003, Subject: Transuranics (TRUs) and Technetium-99 (TC-99) Contaminants in Depleted Uranium (DU) – Health Assessment.

World Health Organization (WHO). 2001. Department of the Protection of the Human Environment, World Health Organization, Geneva, Switzerland. Online report available at URL: www.who.int/environmental_information/radiation/depleted_uranium.htm.

4.0 Chemical and Radiological Dose Modeling

Determining the concentration of uranium in organs in the body as a function of time and the radiation dose from the uranium requires the use of models that describe the intake and deposition of the uranium into the respiratory tract, clearance from the respiratory tract to the gastrointestinal tract, removal from the gastrointestinal tract either to the blood or by elimination from the body, clearance from the respiratory tract to the blood and lymphatic system, distribution through blood and lymphatic system to other organs of the body concurrent with elimination through the urinary tract, and finally retention in the blood-fed organs and feedback from those organs to the blood. The International Commission on Radiological Protection (ICRP) has developed the models needed to mathematically describe the processes listed above. These models represent scientific consensus based on years of studies on animals and from human data where possible, and they represent the latest in a series of progressive refinements from models dating from at least 1959 (ICRP [2] 1959). For uranium, the total of the processes described above are addressed by three different models that are integrated in the computer programming. These models are the Human Respiratory Tract Model (HRTM), the Gastrointestinal (GI) Tract Model, and the Uranium Systemic Biokinetic Model. Discussions of these models is followed by a description of the dose calculations for Level I exposures using Capstone data as input.

4.1 HUMAN RESPIRATORY TRACT MODEL

The HRTM was introduced in ICRP Publication 66 (1994a) and was summarized in subsequent ICRP Publications 68, 71, and 78 (1994b, 1995c, 1997, respectively). The model is also summarized in USACHPPM (2000)

Appendix J. The model consists of five sub-models that deal with 1) the anatomy of the respiratory tract (RT), 2) the physiology of breathing, 3) deposition of material in the various parts of the tract, 4) the clearance of material from the RT, and 5) the dosimetry for various types of lung tissue from the radioactive particles deposited in the RT.

4.1.1 Structure of the HRTM

The model subdivides RT into five distinct anatomic compartments: 1) the anterior portion of the nose (anterior nasal passage [ET_1, where ET stands for extrathoracic]); 2) the back of the nose and mouth and the throat (posterior nasal and oral passages, the pharynx, and larynx [ET_2]); 3) the trachea, the split of the air pathway into the two lobes of the lung, and the first seven subsequent divisions of the two pathways into the lung (trachea, main bronchi, and generations 2-8 bronchi [BB]); 4) the next seven divisions (bronchioles and terminal bronchioles [bb]), and 5) the final divisions of the airway and the alveolar sacs where gas exchange with blood occurs (alveolar-interstitial [AI] region). Compartments 1 and 2 comprise the ET airways. Compartments 3, 4, and 5 comprise the thoracic region. Separate lymphatic tissues are associated each with the ET region and the thoracic region, and both are included in the RT model.

4.1.2 Physiology of the HRTM

Physiological data include the breathing rate and the amount of the air space in the lung that is typically used during breathing. For this study ICRP 66 (1994a) default values were used for total lung capacity, functional residual capacity, vital capacity, and dead space. The ICRP 66 breathing rates for heavy exercise (3.0 m^3/h) and light exercise (1.5 m^3/h) for a healthy adult male were used as well.

4.1.3 Deposition Model

The deposition model describes deposition of particles as a fraction of the intake within each of the five regions of the respiratory tract as a function of particle sizes, ranging from 0.6 nm to 100 μm. Each region is modeled as a filter, and all five regions act as a series of filters for both the inhalation and exhalation phases of breathing. The efficiency of each filter is determined by both aerodynamic processes, basically gravitational settling and inertial impaction, and the thermodynamic process of diffusion acting competitively. The thermodynamic process has its greatest effect on submicrometer particles. The important parameters for the deposition model are the particle size distribution, which includes a measure of the median size either by the activity median aerodynamic diameter (AMAD) or activity median thermodynamic diameter (AMTD), the dispersity of the distribution such as the geometric standard deviation for lognormally distributed aerosols, the particle density, and the particle shape factor, which compensates for non-spherical particles. The ICRP 66 default values for aerosols found in a typical work environment are 5 µm, 2.5, and 1.5 for AMAD, geometric standard deviation, and shape factor, respectively; for the Capstone study, the AMADs and geometric standard deviations were determined based on the sample measurements of the Marple CIs (Capstone report, Appendix B, Parkhurst et al. 2004). In this report, an alternative treatment of the aerosol data was used to mitigate the difficulties encountered in the Capstone study due to inadequate fitting of the CI data (see Section 4.4).

Deposition is not considered to be dependent on the element, radionuclide, or chemical form.

The breathing rate in conjunction with time-varying air concentrations of uranium in various particle size bins were used to obtain the total intake of uranium and deposition in the five RT regions (each particle-size bin was treated as a separate intake; see Section 4.4.1).

4.1.4 Clearance Model

Particulate material is removed from the RT by two methods that act independently and competitively on the material in each deposition region: 1) particle transport (mechanical clearance) and 2) absorption into blood. Diagrams depicting these models are contained in several ICRP publications, most recently in ICRP 78 (1997) and in the USACHPPM (2000) health risk report.

In the anterior nose (ET_1), only mechanical clearance applies, and material is removed directly out of the body through the front of the nose at a fast clearance rate (1 d^{-1}).

In the other RT regions, particle transport includes clearance to the GI tract and to the lymphatic system. Rates of removal by particle transport are independent of the element, radionuclide, chemical form, or the parameters that determined the original deposition, for instance, particle size (except for the fraction cleared slowly by mucous in the bronchial (BB_2) and bronchiolar (bb_2) regions, which does depend on particle size). In the ET_2 region, clearance is directly to the GI tract (swallowed) or to the lymphatic system that drains that region; the rates of clearance are vastly different, however, with extremely fast clearance rate to the GI tract (100 d^{-1}) and very slow clearance to the lymphatic system (0.001 d^{-1}). This means that almost no material actually clears to the lymphatic system from ET_2.

In the thoracic region, particles cleared to the GI tract move proximally within the conducting airways by cellular and mucociliary action to the back of the throat and are swallowed. The bronchial and bronchiolar regions each have two rates of mechanical clearance upward to the region above it, and the AI region has three rates of clearance to the region above it, representing some particles that clear rapidly to the bronchioles and other particles that clear more slowly. Particle clearance to the thoracic lymphatic system from "sequestered" compartments in the bronchial and bronchiolar regions (BB_{seq} and bb_{seq}) and the slowest cleared compartment of the AI region (AI_3) occurs at the rate of 0.01 d^{-1} from the bronchial/bronchiolar compartments and at the rate 0.00002 d^{-1} from AI_3.

Concurrent with the mechanical clearance of particles is dissolution of the particles and absorption to blood. This mechanism depends on the physical and chemical form of the deposited material, and the rates of dissolution are modeled to occur at the same rates for all RT regions (except ET_1). Dissolution is a time-dependent process but is mathematically modeled by assuming that a fraction of the deposited material dissolves relatively rapidly, and the rest dissolves more slowly. In the HRTM, material deposited in the RT is assigned to compartments labeled ***particles in initial state*** in which a portion dissolves and is transferred to blood at a constant rate, s_p. Concurrently, some of the material is transferred (at a rate s_{pt}) to a compartment labeled ***particles in transformed state*** within which it is transferred to blood at a different rate, s_t (ICRP [78] 1997).

Different chemical forms of a given radioelement can have different dissolution rates. The HRTM provides default values for three broad classes of solubility, called absorption Types F, M, and S representing fast, moderate, and slow rates of dissolution/absorption. ICRP 66 (1994a) summarizes the behavior of these default materials as follows:

> **Type F**: 100% absorbed with a half-time of 10 min. There is rapid absorption of almost all material deposited in the bronchi, bronchioles, and AI regions, and 50% of material deposited in ET_2; the remaining 50% is cleared to the GI tract by particle transport.
>
> **Type M**: 10% is absorbed with a half-time of 10 min and 90% with a half-time of 140 d. There is rapid absorption of about 10% of the deposit in the bronchi and bronchioles, and 5% of the material deposited in ET_2.

About 70% of the deposit in the AI region eventually (in approximately two years) reaches body fluids.

Type S: 0.1% absorbed with a half-time of 10 min and 99.9% with a half-time of 7000 d. There is little absorption from ET, bronchi, or bronchioles, and about 10% of the deposit in the AI region eventually (in approximately 100 years) reaches body fluids.

The ICRP absorption model also allows for another compartment called the ***bound state***. The ***bound state*** is not used as part of the default F, M, or S absorption types, but can be used when there is evidence that a fraction of the material retained in the lung for a very long time is due to binding of dissolved radionuclide to local lung tissue constituents. When the ***bound state*** is modeled in this way, a fraction of the material in the initial state goes to the bound state at the same rate as dissolution, $f_b s_p$, and a fraction of the material in the transformed state also goes to the bound state, $f_b s_t$, from which it is cleared to the blood with a new rate, s_b. Because the bound radionuclide is envisaged to be locally bound biochemically, mechanical clearance does not apply, and clearance from the bound state occurs only to blood.

Uranium compounds can be categorized into any of the three absorption types depending on chemical form. However, the ICRP also recommends that material-specific rates of absorption should be used for compounds for which reliable human or animal experimental data exist. In this study, dissolution rates obtained from the Capstone study *in vitro* dissolution experiments are used to define specific dissolution rate constants and associated fractions for the mixtures of uranium forms encountered in the aerosols measured in the vehicle interior.

Typical dissolution characteristics for many radionuclide aerosols can be described by two-component negative exponential equations. This behavior can be suitably modeled using the ***particles in initial state*** and ***particles in transformed state*** compartments of the HRTM. However, the dissolution characteristics of the Capstone aerosols were fitted better by three-component negative exponential functions, in which the slowest dissolving component had half times of years to tens of years. To accommodate the three-component dissolution behavior, the ***bound state*** compartment of the HRTM was invoked but with one important difference. Because the *in vitro* dissolution behavior of the DU aerosols was considered to be a chemical property of the particles, it was concluded that both mechanical clearance and dissolution should be applied to the ***bound state*** compartment for these DU aerosols. This differs from the HRTM definition, in which mechanical clearance does not occur. Thus, to explicitly differentiate between the usage of this compartment in this report and that of the HRTM, the ***bound state*** compartment is renamed as the ***very slow dissolution (VSD)*** compartment, signifying the fraction of DU aerosol that dissolves very slowly.

4.2 GASTROINTESTINAL TRACT MODEL

The GI tract model used by the ICRP was first introduced in Publication 30 (ICRP 1979) and is based on the model of Eve (1966). A diagram depicting the model is provided in ICRP Publication 78 (1997) and in Appendix J of the USACHPPM health risk report (USACHPPM 2000). It consists of four compartments representing the stomach, small intestine, upper large intestine, and lower large intestine. Material enters the stomach and clears to the small intestine, from which a fraction, f_1, is absorbed to blood and the remainder clears to the upper large intestine. The material in the upper large intestine clears to the lower large intestine and from the lower large intestine and is excreted in feces. There is no feedback between the compartments. The clearance rates from one compartment to the other, in units of d^{-1}, are 24, 6, 1.8, and 1 for stomach to small intestine, small intestine to upper large intestine, upper large intestine to lower large intestine, and lower large intestine to excretion, respectively. The clearance rate to blood depends on the f_1 factor according to

$$\lambda_B = \frac{f_1 \lambda_{si}}{1 - f_1} \tag{4.1}$$

where λ_{si} is the clearance rate from the small intestine to the upper large intestine. The f_1 factor depends on the solubility of the material in GI tract fluids, and is generally related to the absorption type used in the RT model. Uranium is generally poorly absorbed from the GI tract. The default f_1 values for uranium follow:

- 0.002 for highly insoluble compounds and most tetravalent compounds (e.g., UO_2, U_3O_8, or compounds considered Type S)
- 0.02 for soluble compounds or moderately soluble compounds (e.g., UO_3, or compounds considered Type F or M in ICRP [78] 1997).

For many elements there is input to the upper large intestine from material that has gotten into the blood, either from the blood itself or via bile from the liver. This material is modeled as part of the biokinetic model, but then is acted upon by the GI tract model like any other material in the intestine. Uranium is not transferred from the liver to the GI tract and only weakly transferred from the blood to the GI tract (ICRP [69] 1995a; ICRP [78] 1997).

4.3 URANIUM SYSTEMIC BIOKINETIC MODEL

Systemic biokinetic models describe the fate of elements after uptake to the blood from organs (e.g., lung, small intestine) or through a wound or abraded skin. The general form of the biokinetic model for uranium is called the alkaline earth model. Its structure applies to radium and strontium as well, but the transfer rates between compartments of the model are specific for uranium. The model is complex and is generally described below. The diagram depicting the model is provided most recently in ICRP Publication 78, Section 3.3.1 (1997), and in Appendix J of the USACHPPM health risk report (CHPPM 2000). The rate constants specific for uranium are found in ICRP Publication 69, Section 5, and are presented in Table 4.1. A more detailed description of the basis for the rate constants is found in ICRP Publication 69, Section 5 (ICRP [69] 1995a).

Table 4.1. Biokinetic Rate Constants for Uranium

Transfer Compartment	Adult Transfer Rate (d^{-1})
Plasma to ST0	1.050E+01
Plasma to RBC	2.450E-01
Plasma to urinary bladder	1.543E+01
Plasma to urinary path	2.940E+00
Plasma to other kidney tissue	1.220E-02
Plasma to ULI contents	1.220E-01
Plasma to liver 1	3.670E-01
Plasma to ST1	1.630E+00
Plasma to ST2	7.350E-02
Plasma to trabecular surfaces	2.040E+00
Plasma to cortical surfaces	1.630E+00
ST0 to plasma	8.320E+00
RBC to plasma	3.470E-01
Other kidney tissue to plasma	3.800E-04
Liver 1 to plasma	9.200E-02
Liver 2 to plasma	1.900E-04
ST1 to plasma	3.470E-02
ST2 to plasma	1.900E-05
Bone surfaces to plasma	6.930E-02
Nonexch. trab. vol. to plasma	4.930E-04
Nonexch. cort. vol. to plasma	8.210E-05
Urinary path to urinary bladder	9.900E-02
Liver 1 to liver 2	6.930E-03
Bone surfaces to exch. volume	6.930E-02
Exch. Bone vol. to bone surfaces	1.730E-02
Exch. Bone vol. to nonexch. volume	5.780E-03

Abbreviations: ST is soft tissue, ULI is upper large intestine, RBC is red blood cells.

The biokinetic model for uranium (as with the newer ICRP models) is a recycling model, which means that uranium in the blood (more properly the blood plasma) flows into organs and then back again to blood. Red blood cells are treated like an organ, so there is a small flow of uranium into red blood cells (<1%) and quite rapidly back again to the plasma. Other major compartments of the model are the skeleton, liver, kidneys, and a generic soft tissue compartment. The generic soft tissue compartment is modeled as all other systemic organs and tissues, excluding the skeleton, liver, and kidneys, and it receives all of the distribution from the blood not accounted for by the other organs or excretion (about 37%). There are three subcompartments in the soft tissue compartment, referred to in Table 4.1 as ST0, ST1, and ST2. Most of the uranium in soft tissue is retained only briefly, but a very small amount (<1%) is retained essentially indefinitely.

The skeleton is modeled as two compartments: one representing trabecular bone and one representing cortical bone (ICRP [70] 1995b). There are additional subcompartments representing movement of uranium from the bone surfaces to bone volume and back. The skeleton receives about 15% of the uranium leaving the blood but most of that amount is returned fairly rapidly (retention half-times of 30 d or less); a very small amount of the uranium is retained in the skeleton for time periods (years) consistent with normal turnover rates. The bone marrow is also included in the skeleton compartment to allow calculation of dose specific to red marrow.

The liver is modeled as two compartments: one that receives the uranium from the blood but returns most of it with a 7-d half-time, and one that receives uranium from the other liver compartment and retains it with a long half-time of about 10 years. About 1.5% of the uranium leaving the blood goes to the liver.

The kidney is modeled as two compartments: 1) kidney tissue and 2) a urinary path that flows one way only to the bladder and excretion. The kidney tissue itself receives only about 0.05% of the uranium leaving the blood but retains it with a moderately long half-time of about 5 years.

Excretion represents the largest pathway of removal of uranium from the blood. Of the uranium in the blood, 63% is modeled to be transported directly to the bladder, and another 12% passes through the kidney (renal tubules), is retained briefly (days), and then passes on to the bladder. The one unusual aspect of the uranium biokinetic model is that no transfer of systemic uranium from the liver to the GI tract occurs and only a very small percent of uranium in the blood transfers to the GI tract (0.5%). Consequently, systemic uranium is eliminated virtually entirely through urine.

4.4 DOSE CALCULATIONS FOR LEVEL I SCENARIOS USING CAPSTONE DATA

This section discusses the methods used to calculate radiological and toxicological doses based on the Capstone DU aerosol data and Level I scenario parameters. The primary measurement data were obtained from two sources: 1) the eight-stage CIs (located at four different positions inside the vehicles) that collected particles for median particle sizes from 0.2 to 31 μm AMAD at sequenced times after perforations and 2) the five-stage cyclone that collected aerosol samples differentiated by particle size, but integrated in time over a 2-h collection period. Scenario assumptions used in the calculations included time duration and breathing rate based on expected level of physical exertion during various time periods. The biokinetic and dosimetric models used in the calculations were those described in the ICRP Publications 66, 69, and 78 (1994a, 1995a, and 1997) and are discussed in the previous sections.

The calculated dose quantities (committed effective dose, organ doses, concentration in kidneys) depend on the amount and composition of the DU aerosol inhaled, other specific parameters (e.g., breathing rate, nose/mouth breathing characteristics, and solubility of the DU aerosol in lung/extracellular fluids), and physiological parameters. For the latter, the ICRP reference adult male was assumed. The aerosol concentration as a function of time and particle size, which is discussed in the next section, was obtained from the CI measurements, including a quantitative determination of measurement uncertainties. The solubility characteristics of the aerosol in simulated lung fluids, which are discussed in Section 4.4.2, were also measured, using powder samples collected by the cyclone. Other factors used in the calculations included an average DU aerosol density of 9.0 gm/cm^3 and a shape factor of 1.5 (USACHPPM 2000), which is not uncommon for irregularly shaped particles and is a default from ICRP 66 (1994a). The uranium isotopic composition of DU is listed in Table 3.1. About 98% of the radioactivity of DoD DU is from U-238 and U-234, and for simplicity, only U-234 and U-238 isotopes were used in the modeling calculations and in the experiments conducted to verify the model. The dose contributions from U-235, U-236, and all of the short-lived progeny radioisotopes that will grow in during a human lifetime (Th-234, Pa-234m, Pa-234, and Th-231) were incorporated in the dose calculations used in subsequent chapters by applying a correction factor (1.020 derived by the relative dose contributions) to each of the doses calculated from U-234 and U-238. Assumptions about the duration of exposure and breathing patterns (light or heavy) for the five scenarios are listed in Table 4.2 in which all times are indexed to the time of perforation.

Table 4.2. Scenario Breathing Pattern Assumptions[a]

Scenario	Description
A	0-1 min heavy exercise
B	0-5 min heavy exercise
C	0-15 min heavy exercise, followed by 15-60 min light exercise
D	0-15 min heavy exercise, followed by 15-120 min light exercise
E	5-15 min heavy exercise
(a) From ICRP 66 (1994a)	

The default breathing rates listed in ICRP 66 for a healthy adult male are 3.0 m^3/h for heavy exercise and 1.5 m^3/h for light exercise (ICRP 1994a). Calculations for these scenarios were carried out assuming normally nose-breathing individuals only. (The effect of mouth breathing is discussed in Section 5.4.2) Table 4.3 lists the parameters used in the assumptions of ventilation rate, respiratory frequency, and fraction breathed through the nose (for an adult male who normally breathes through the nose; denoted by "normally" in the table title) versus through the mouth in relation to heavy and light exercise.

Table 4.3. Breathing Parameters for a Normally Nose-Breathing Adult Male[a]

Parameter	Heavy Exercise	Light Exercise
Ventilation rate	3 m^3/h	1.5 m^3/h
Respiratory frequency	26/min	20/min
Fraction breathed through nose	0.5	1.0
(a) From ICRP 66 (1994a)		

4.4.1 DU Concentration and Particle Size Distribution

The measurement data used to calculate DU aerosol concentrations inside the vehicles were obtained from the eight-stage CIs substrates (referred to in this section as nine stages, which includes the backup filter; see Parkhurst et al. (2004) (Appendix A, Tables A.19 through A.32). As discussed in Section 3.2.4, the actual stage data were used directly rather than making use of the unimodal or bimodal lognormal fits presented in the Capstone report, Appendix B. Many of these fits poorly describe the data.

The CI aerosol concentration data are denoted as the multicomponent quantity y consisting of DU concentrations measured from the nine stages at different times (midpoints of the sampling intervals). The time dependence of y was obtained by linear interpolation/extrapolation from the measurements at discrete times. To avoid the occasional negative concentration that occurred when extrapolating to time zero (t_0), the aerosol concentration at t_0 was assumed to equal the first measured air concentration. For scenario modeling that extended beyond the last measured aerosol concentration (t_{last}), the final concentration was set to equal t_{last}.

In classical statistics, uncertainties are usually described in terms of the distribution of measurement results expected for a given true value. In Bayesian statistics, measurement uncertainties are described in terms of this same distribution, but are now considered a function of the true value of the quantity and termed the likelihood function. The likelihood function is a function of the true value because the measurement result is known and fixed, whereas the true value is unknown. The data y obtained from analysis of the CI samples is assumed to have the following Gaussian likelihood function:

$$P(y\,|\,\psi)=\prod_j \frac{\exp\left(-\frac{(y_j-\psi_j)^2}{2\sigma_j^{\,2}}\right)}{\sqrt{2\pi}\sigma_j} \tag{4.2}$$

where ψ_j is the "true" value of the jth DU concentration, and σ_j is the uncertainty standard deviation of y_j. To clarify Equation (4.2), the data consist of CI stage measurements from the nine stages at various midpoint times, so the index j runs over some multiple of nine depending on how many time points contribute to the exposure period of interest. The likelihood function is a product of independent Gaussian distributions for different j values. The notation $P(y\,|\,\psi)dy$ stands for the probability that y is in the interval dy given the value of ψ.

In terms of the notation used in Chapter 3, ψ, the true value of the measured DU mass, is equivalent to $M_{measured}$ in the equations (3.5 through 3.10). Measurement uncertainty is the reason for the distinction between the actual measured value y and $M_{measured}$.

The form of the likelihood function is not well known in this case. In similar situations with a background subtraction, the Gaussian distribution is found empirically to apply. The log-normal distribution is not suitable because of the possibility of zero or negative measured values. In actual fact, since there are a large number of measurements, the form of the likelihood function is not critical, and Equation 4.2 can be replaced by a uniform distribution with the same standard deviation (half width = ($\sqrt{3}\sigma_j$) with very little change in the results shown below. This is a consequence of the central limit theorem of statistics (Feller 1957).

The uncertainty σ_j in Equation 4.2 results mostly from counting statistics, but other components were considered as well, as discussed in Section 5.5.2. In the few cases for which uncertainty data were not available, a default value of 40% of the measurement value (but not less than 25 μg/m^3) was assumed based on a random sampling of cases.

The intake depends on the time integral of the aerosol concentration times the breathing rate

$$\zeta = C \int B(t)\psi(t)dt \tag{4.3}$$

where ζ is the intake (Bq), $\psi(t)$ is the concentration (μg/m^3), $B(t)$ is the breathing (ventilation) rate (m^3/min), C is the specific activity (Bq/μg), and t is the time (min). The time dependence of $\psi(t)$ is obtained from the discrete values ψ_j in Equation 4.2 using linear interpolation or constant extrapolation.

The actual aerosol is polydisperse; that is, it is composed in a continuous manner of monodisperse components each having a different AMAD. The term monodisperse refers to an aerosol comprised entirely of a single-size particle. Rather than using an exact continuous decomposition, the aerosol was approximated with nine discrete components corresponding to the median diameters collected by the nine CI stages. The intake amounts of these monodisperse aerosol components are denoted by ζ_i (related to $M_{calculated}$ in Chapter 3). In terms of ζ_i, the stage-specific intake ζ_j is given by

$$\zeta_j = \sum R_{j,i}\,\zeta_i \tag{4.4}$$

where $R_{j,i}$ is the CI response matrix shown in Table 4.4 to four decimals. The role of the response matrix is to account for the fact that particles of any particular size will not all deposit on a single CI stage, but rather some fraction will deposit on earlier or later stages. For example, the first entry in the matrix signifies that 95.38% of a 31-μm AMAD monodisperse aerosol (i.e., an aerosol comprised entirely of 31-μm AD particles) would deposit on CI stage 1 and 4.61% on stage 2.

The AMAD values, which were rounded for modeling simplicity, were initially based on a sampler flow rate of 2.1 Lpm. The data were later revised to more closely model the average Capstone CI flow rate of 2.6 Lpm. The values in Table 4.4 are based on the 2.6 Lpm average. The columns of this matrix show how the nine assumed monodisperse aerosol components would be measured on the nine CI stages.

The total intake of DU is a sum of intakes from the nine aerosol components. In terms of the true amounts measured by the CI, the discretized aerosol components are given by

$$\zeta_j = \sum R^{-1}_{i,j}\,\zeta_j \tag{4.5}$$

where R^{-1} is the inverse of the response matrix.

Table 4.4. Fractional Collection Efficiency of Monodisperse Aerosols on CI Stages

CI Stage	Monodisperse Aerosol AMAD (μm)								
	31	16	11	7	4	2	1	0.6	0.2
1	0.9538	0.3083	0.1036	0.0412	0.0210	0.0126	0.0098	0.0087	0.0079
2	0.0461	0.5288	0.2111	0.0497	0.0157	0.0064	0.0041	0.0034	0.0029
3	0.0000	0.1626	0.5911	0.1628	0.0195	0.0034	0.0015	0.0010	0.0007
4	0.0000	0.0003	0.0942	0.6746	0.1005	0.0044	0.0009	0.0004	0.0002
5	0.0000	0.0000	0.0000	0.0716	0.8254	0.0177	0.0004	0.0001	0.0000
6	0.0000	0.0000	0.0000	0.0000	0.0179	0.9374	0.1463	0.0142	0.0014
7	0.0000	0.0000	0.0000	0.0000	0.0000	0.0182	0.8347	0.0071	0.0000
8	0.0000	0.0000	0.0000	0.0000	0.0000	0.0000	0.0023	0.8289	0.0448
Filter	0.0000	0.0000	0.0000	0.0000	0.0000	0.0000	0.0000	0.1361	0.9421

Dose quantities depend not only on the total intake but also on additional parameters of the problem (e.g., breathing rate, nose/mouth breathing characteristics, and solubility of the DU aerosol in lung/extracellular fluids) denoted by the multicomponent quantity λ. To calculate the dose *D* and its uncertainty from Bayes' Theorem, the probability distribution of *D* given (conditioned on) the measurements was applied using the following formula:

$$P(D\,|\,y)=\frac{\int d\lambda d\psi D(\lambda,\xi(\psi))P(\lambda)P(y|\psi)P(\psi)}{\int d\lambda d\psi P(\lambda)P(y|\psi)P(\psi)} \tag{4.6}$$

where $P(\lambda)$ and $P(\psi)$ are the probability distribution of λ and the prior probability distribution of ψ. For the latter, a uniform distribution in terms of ψ on positive values of ζ was assumed; that is, all values of concentration are taken to be equally likely *a priori* as long as they result in positive component amplitudes.

Since the distribution of additional parameters is not well known, the posterior distribution (the answer that results from a Bayesian statistical analysis) of *D* given above was calculated assuming that the parameters have a single value rather than a distribution, and then the effect of various discrete choices of parameter values was investigated. The evaluation of the posterior distribution was done using Markov chain Monte Carlo (MCMC) (Gilks et al. 1996). A typical distribution of E(50) is Gaussian and relatively narrow (relative standard deviations from 10 to 20%) as shown in Figure 4.1 and Table 4.5. For simplicity, both Figure 4.1 and Table 4.5, which were run with only U-238 and a flow rate of 2.1 Lpm, are intended to illustrate the model used. All of the doses used in the risk assessment include the contributions from all uranium isotopes and the progeny that will grow in during the near term.

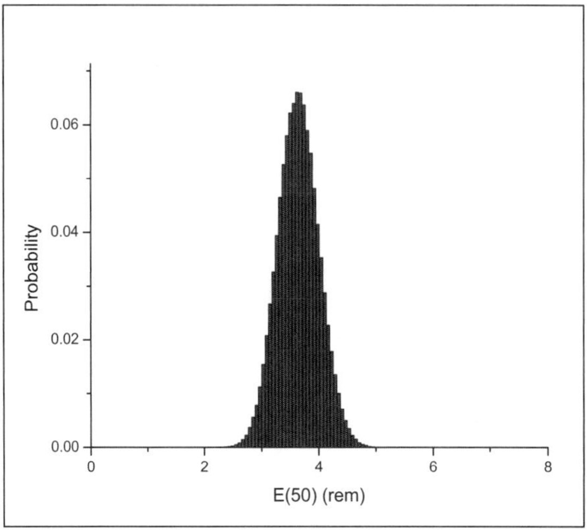

Figure 4.1. Bayesian Posterior Distribution of E(50) for PIII-2, Loader Position, Scenario B

The Bayesian approach taken to estimate the radiation doses to individuals exposed to the aerosols measured by this project has two advantages over other methods that may have been employed. The first advantage is that the method calculates a distribution of doses, thus allowing the means and standard deviations of the doses received to be determined rather than just a single estimate of dose. Second, the Bayesian approach allows the use of a predefined set of reasonable air concentrations (the Bayesian prior) from which the Monte Carlo method is allowed to sample. For the doses reported here, the Bayesian prior was that only positive aerosol components amplitudes were reasonable. The use of a Bayesian prior that disallows negative component amplitudes is subject to some discussion when the distribution of population results is near zero (Strom 1986). Using such a prior will result in higher dose estimates than would be achieved by a prior that included negative air concentrations and is thus conservative. Because the application of the Bayesian approach to human health risk assessments is not widespread, results were also calculated using a more conventional statistical approach. This approach also avoided negative intakes and doses by changing both negative concentrations and negative aerosol component amplitudes to zero.

The relative narrowness of the posterior distribution is a consequence of the large number of measurements. For the above example, four time points and thus 36 independent CI measurements contribute to the result. Means and relative standard deviations for all Scenario B cases (U-238 only) are given in Table 4.5. The phases and shots are abbreviated as Phase (PI, PII, PIII) and Shot (1 to 7) so that Phase I, Shot 1 is represented as PI-1. The letters

Table 4.5. Results of MCMC Calculations of Means and Standard Deviations for Scenario B Cases

Phase-Shot-Position	E(50)		Peak Kidney Conc.		Intake	
	Avg (rem)	Rel SD	Avg (μg/g)	Rel SD	Avg (nCi)	Rel SD
PI-1-C	2.719	0.092	5.10	0.084	131	0.064
PI-1-D	0.714	0.09	1.52	0.089	45.2	0.078
PI-1-G	2.80	0.123	4.95	0.091	133	0.08
PI-1-L	2.97	0.092	5.83	0.094	139	0.062
PI-2-C	2.65	0.078	4.85	0.082	118	0.059
PI-2-D	2.41	0.142	4.86	0.126	121	0.093
PI-2-G	2.59	0.093	5.03	0.087	131	0.061
PI-2-L	2.42	0.07	4.57	0.067	116	0.052
PI-3-C	8.78	0.175	14.7	0.151	400	0.129
PI-3-D	5.69	0.202	7.69	0.122	246	0.13
PI-3-G	8.78	0.153	14.8	0.144	401	0.096
PI-3-L	8.78	0.175	14.7	0.151	400	0.129
PI-4-C	4.66	0.175	7.32	0.125	238	0.105
PI-4-D	2.81	0.102	5.36	0.119	158	0.087
PI-4-G	5.02	0.136	9.14	0.114	310	0.093
PI-4-L	4.48	0.090	9.88	0.104	408	0.116
PI-5-C	3.32	0.088	6.91	0.081	171	0.063
PI-5-D	3.42	0.117	8.01	0.113	194	0.076
PI-5-L	2.66	0.084	5.87	0.084	156	0.074
PI-6-C	3.40	0.095	7.66	0.094	201	0.069
PI-6-D	1.62	0.138	3.60	0.143	77.0	0.086
PI-6-L[(a)]	0.017	0.163	0.036	0.169	1.10	0.126
PI-7-C	2.92	0.122	6.79	0.101	215	0.077
PI-7-D	5.78	0.097	11.8	0.093	309	0.064
PI-7-G	3.79	0.105	7.75	0.102	195	0.060
P1-7-L[(a)]	0.017	0.175	0.043	0.187	1.20	0.118
PII-1-C	2.32	0.150	4.65	0.140	106	0.098
PII-1-D	2.52	0.157	4.85	0.149	108	0.108
PII-1-G	1.87	0.142	3.71	0.124	90.5	0.110
PII-1-L	1.42	0.101	2.78	0.079	74.4	0.073
PII-2-C	4.34	0.180	8.45	0.158	190	0.121
PII-2-D	1.64	0.156	3.22	0.127	80.6	0.097
PII-2-G	1.23	0.116	2.45	0.088	63.4	0.080
PII-2-L	1.39	0.119	2.82	0.082	76.2	0.075
PII-3-C	0.229	0.136	0.515	0.123	15.4	0.080
PII-3-D	0.173	0.109	0.331	0.101	11.6	0.091
PII-3-G	0.312	0.130	0.674	0.127	14.5	0.098
PII-3-L	0.133	0.164	0.389	0.118	15.6	0.099
PIII-1-C	5.04	0.115	2.45	0.074	168	0.070
PIII-1-D	4.75	0.122	2.76	0.128	152	0.098
PIII-1-G	4.89	0.127	2.03	0.084	135	0.076
PIII-1-L	4.03	0.122	1.70	0.083	102	0.078
PIII-2-C	4.35	0.106	3.53	0.082	285	0.090

Table 4.5. *(continued)*

Phase-Shot-Position	E(50)		Peak Kidney Conc.		Intake	
	Avg (rem)	Rel SD	Avg (µg/g)	Rel SD	Avg (nCi)	Rel SD
PIII-2-D	4.35	0.124	2.54	0.078	149	0.078
PIII-2-G	10.1	0.104	7.41	0.101	614	0.124
PIII-2-L	3.65	0.098	1.80	0.069	112	0.065
PIV-4-D	0.379	0.174	0.232	0.153	15.1	0.152

(a) Sample was collected for resuspension studies and was used in this example only, not in actual dose calculations.

following the phase and shot numbers represent the commander (C), driver (D), gunner (G), and loader (L) positions. (In the Bradley vehicle [Phase II], the gunner designation refers to the right scout position and the loader designation refers to the left scout position.)

The problem of determining the distribution of particle sizes from impactor measurements has been extensively discussed (Kandlidar and Ramachandran 1999; Cooper and Wu 1990; Cooper 2001; O'Shaughnessy and Raabe 2003). Difficulties arise when the condition number of the response matrix is large (the condition number of A is the norm of A times the norm of A^{-1}, where the norm is the maximum absolute value of the elements of the matrix [see Cooper and Wu 1990]). In this case, the condition number of the response matrix is about two (not large), and in any case the effects of ill conditioning are taken into account in the calculation of the posterior distribution and act to increase the width of the posterior distribution.

The Bayesian prior, even though uniform for positive component amplitudes, is essential in ensuring that the aerosol component amplitudes take on only positive values. This has an important effect on the posterior distribution for small or negative measured values.

4.4.2 Aerosol Lung Dissolution Parameters

The *in vitro* dissolution analysis process and the resulting dissolution parameters of aerosol samples collected by a multi-stage cyclone were documented in Appendix E of the Capstone report Volume II (Parkhurst et al. 2004). Of the total 54 cyclone stage samples collected (6 stages from 9 shots), 19 were analyzed for *in vitro* dissolution. These results were mapped by particle size to the CI stages, and composite surrogate dissolution rates were applied to samples that were not individually analyzed. The dissolution model used to evaluate particles moving into the blood is presented.

4.4.2.1 Using Cyclone *In Vitro* Dissolution Data to Map to CI Stages

Bulk aerosol samples collected by the five-stage cyclone with a backup filter (referred to here as the sixth stage) were used in the evaluation of the solubility of the DU aerosols in lung/extracellular fluid. Unlike the CIs, which collected samples in various sampling time windows, the cyclone collected aerosol for a single 2-h period beginning 5 sec after each perforation. This presents two challenges in applying the solubility results from the cyclone stages to the CI data. The first is in correlating the solubility data from six cyclone stages with the concentrations from nine CI stages. The second is the uncertainty of assigning the results from the 2-h, time-integrated sample to the time-resolved CI samples. This second issue can be addressed by comparing the cyclone data with the *in vitro* results obtained from five IOMs taken at different times post shot (PI-2). These results (Capstone report, Appendix E, Parkhurst et al. [2004]) are within the range of solubilities measured using the cyclone samples, and do not show uniform time-dependent changes in solubility.

To address the first issue, the stages in the two samplers were "cross-mapped" based on the particle size median diameters (geometric mean of the cutoff diameters of the stage and the stage below). The correspondences between the CI and the cyclone stages are listed by stage and median particle size diameters in Table 4.6. The median diameters are a weak function of flow rates, and in this case, the diameters were calculated for the Capstone CI average of 2.6 Lpm. For modeling simplicity, all AMAD values were rounded as shown in Table 4.6. The backup filter is abbreviated as BU.

Table 4.6. Correspondence Between Particle Size of CI stages with Cyclone Stages

CI Stage	Median AMAD (μm)	Cyclone Stage
1	31	1
2	16	1
3	11	1
4	7	2
5	4	3
6	2	4
7	1	5
8	0.6	BU
9	0.2	BU

4.4.2.2 *In Vitro* Dissolution Kinetic Parameters

The dissolution model used in this analysis, as shown in Figure 4.2, is adapted from ICRP Publication 66 (ICRP 1994a) in which N_{vsd} represents very slowly dissolving material, N_p refers to particles in the initial state, and N_t refers to particles in the transformed state.

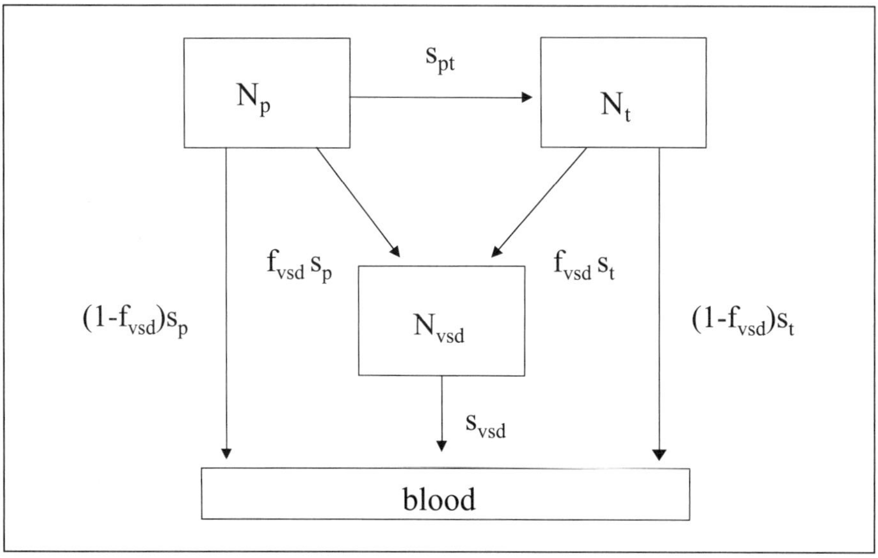

Figure 4.2. ICRP 66 Lung Dissolution Model

This model assumes five parameters: s_{pt}, s_p, s_t, s_{vsd}, and f_{vsd} in which

s_{pt} = transformed rate of dissolution
s_p = initial rate of dissolution
s_t = final dissolution rate
s_{vsd} = rate at which f_{vsd} material goes into blood
f_{vsd} = fraction of material in a very slowly dissolving state.

Dissolution studies of aerosol samples from different stages of the cyclone using simulated lung fluid produced three-exponential fits having a rapid component, a slow component, and a very slow component. The rates and fractions of each component are denoted by s_r, s_s, s_{vs}, f_s, and f_{vs}, where f_s is the slow fraction and f_{vs} is the very slow fraction. The rapid fraction is given by $f_r = 1 - f_s - f_{vs}$. The following inversion formulas (the Equation 4.7 array) were used to convert the measured parameters to the parameters of Figure 4.2.

$$
\begin{aligned}
s_t &= s_s \\
s_{vsd} &= s_{vs} \\
s_{pt} &= f_s(s_r - s_s)\left[\frac{1}{1 - f_{vsd}\dfrac{s_s}{s_s - s_{vs}}}\right] \\
s_p &= s_r - s_{pt} \\
f_{vsd} &= f_{vs}(s_r - s_s)\left[\frac{1}{\dfrac{s_r(s_p - s_s)}{s_r - s_{vs}} + \dfrac{s_s s_{pt}}{s_s - s_{vs}}}\right]
\end{aligned}
\tag{4.7}
$$

The last three formulas above are nonlinear and were solved by iteration starting with $f_{vsd} = 0$.

An alternative to the introduction of the *vsd* compartment, in order to represent three dissolution components, is to have two intake components. The first component can be used to represent two of the dissolution components, and the second intake component can be used to represent the third.

The dissolution parameters obtained from the lung fluid dissolution study (Capstone Report Vol. II, Appendix E, Parkhurst et al. 2004) are shown in Table 4.7. The samples are listed by their phase (PI through PIII), shot number, and cyclone stage. Most of the parameters sets shown in Table 4.7 correspond to a certain phase, shot, and cyclone (abbreviated as CY) monitor stage. Dissolution of cyclone samples was analyzed for four of the 10 shot events (two of which included double shots). Additionally, only one Stage-1 sample (PI-7) was analyzed, and there was insufficient sample material collected from Stages 1 through 3 on PIII-2 for this analysis. To estimate dissolution for the shots for which no samples were evaluated, surrogate dissolution functions were obtained using data from grouped samples that were judged representative of the missing stages. These surrogate composites are also listed in the table. The other columns include an index label that is referred to in the discussion that follows.

The selections of dissolution data for each of the six cyclone stages for assignment to the CI stages are listed in Table 4.8.

The six character sequences in Table 4.8 represent the assignment of dissolution parameter sets (shown in Table 4.7) to the six cyclone stages. These

Table 4.7. Dissolution Fit Parameters from the Capstone Data

Sample ID	Label	f_s	f_{vs}	s_r (d^{-1})	s_s (d^{-1})	s_{vs} (d^{-1})
PI-3/4-CY-2	a	0.03	0.842	17.1	0.14	0.00073
PI-3/4-CY-3	b	0.135	0.642	21.7	0.15	0.0032
PI-3/4-CY-4	c	0.21	0.587	31.5	0.043	0.0021
PI-3/4-CY-5	d	0.082	0.707	63	1.9	0.0084
PI-3/4-CY-BU	e	0.99	0	25.7	0.00086	NA
PI-7-CY-1	f	0.102	0.649	5.1	0.084	0.00067
PI-7-CY-2	g	0.05	0.848	21.8	0.11	0.00062
PI-7-CY-3	h	0.129	0.69	11.4	0.21	0.0025
PI-7-CY-4	i	0.109	0.69	16.2	0.19	0.0026
PI-7-CY-5	j	0.12	0.685	29.2	0.083	0.0024
PI-7-CY-BU	k	0.064	0.868	14.5	0.19	0.0010
PII-1/2-CY-2	l	0.047	0.779	7.8	0.091	0.0022
PII-1/2-CY-3	m	0.076	0.892	15.8	1.1	0.0035
PII-1/2-CY-4	n	0.168	0.576	7.1	0.039	0.0024
PII-1/2-CY-5	o	0.112	0.628	12.2	0.055	0.0034
PII-1/2-CY-BU	p	0.088	0.772	7.7	0.039	0.0010
PIII-2-CY-4	q	0.075	0.898	3.8	0.044	0
PIII-2-CY-5	r	0.8	0.15	2.8	0.0029	4E-17
PIII-2-CY-BU	s	0.045	0.847	4.8	0.061	0.00041
Composite CY-2	t	0.04	0.845	19	0.12	0.00068
Composite CY-3	u	0.09	0.746	19.9	0.79	0.0052
Composite CY-4	v	0.174	0.604	11.9	0.06	0.00104
Composite CY-5	w	0.06	0.726	27.1	0.96	0.00065
Composite CY-BU	x	0.047	0.872	7.4	0.17	0.00094

Table 4.8. Assignment of Dissolution Parameters to Cyclone Stages

Phase	Shot	Dissolution Parameters Stage 1,2,3,4,5,BU
I	1	ttuvwx
I	2	ttuvwx
I	3	tabcde
I	4	tabcde
I	5	ttuvwx
I	6	ttuvwx
I	7	fghijk
II	1	llmnop
II	2	llmnop
II	3	lluvwx
III	1	qqqqrs
III	2	qqqqrs
IV	1	qqqqrs

were then mapped to the respective CI stages as described in Table 4.6. The "ttuvwx" labels listed for PI-1 in Table 4.8 refer to the labels t, u, v, w, and x in Table 4.7, which are assigned to composites of cyclone stages 2 through 6 (backup filter), respectively. The composite labels and their associated rates were assigned to samples in which dissolution rates were not measured. For example, the cyclone samples were not measured for PI-1, so composite surrogate dissolution parameters were applied as shown in Table 4.8. For PIII-1 and PIII-2, the results labeled "q," which were measured for PIII-2 cyclone stage 4, were used for cyclone stages 1 through 3 as well. Because the perforation in Phase IV was similar to the Phase-III shots (i.e., through DU armor), results from samples collected from the Phase-III cyclone were applied to Phase IV.

The dissolution fit parameters for Table 4.7, converted to the modified ICRP 66 lung model framework (from Equations 4.7), are shown in Table 4.9.

Table 4.9. Dissolution Fit Parameters Converted to the Modified ICRP 66 Framework

Label	s_t (d^{-1})	S_{vsd} (d^{-1})	s_{pt} (d^{-1})	s_p (d^{-1})	f_b
a	0.14	0.00073	3.29	13.8	0.841
b	0.15	0.0032	8.32	13.4	0.637
c	0.043	0.0021	16.6	15.0	0.572
d	1.9	0.0084	17.2	45.8	0.706
e	0.00086	0	25.4	0.258	0.000
f	0.084	0.00067	1.47	3.63	0.647
g	0.11	0.00062	7.29	14.5	0.846
h	0.21	0.0025	4.73	6.67	0.686
i	0.19	0.0026	5.74	10.5	0.687
j	0.083	0.0024	11.5	17.7	0.677
k	0.19	0.001	7.06	7.44	0.866
l	0.091	0.0022	1.76	6.04	0.774
m	1.1	0.0035	10.4	5.38	0.890
n	0.039	0.0024	2.95	4.15	0.561
o	0.055	0.0034	3.95	8.25	0.615
p	0.039	0.001	3.12	4.58	0.764
q	0.044	0	2.76	1.04	0.898
r	0.0029	4E-17	2.63	0.167	0.150
s	0.061	0.00041	1.43	3.37	0.845
t	0.12	0.00068	4.99	14.0	0.844
u	0.79	0.0052	6.85	13.1	0.744
v	0.06	0.00104	5.28	6.62	0.599
w	0.96	0.00065	5.73	21.4	0.726
x	0.17	0.00094	2.72	4.68	0.870

4.4.3 Biokinetic and Dosimetric Models

The compartmental biokinetic model for uranium recommended by the ICRP is summarized in ICRP 78 (1997) and Section 4.3; compartment definitions and transfer rates are also listed in Appendix J of USACHPPM (2000). The quantity f_1, which governs absorption from the small intestine to blood, is taken to be the same as for Type M uranium, 0.02.

The resulting differential equations were solved using the code BK, developed at Los Alamos National Laboratory, which uses the Gear method solver LSODE (Livermore Solver for Ordinary Differential Equations) (Gear 1971; Hindmarsh 1981). Initial lung depositions were calculated using the LUDEP module DEP20 kindly provided by Alan Birchall and James Marsh from the National Radiological Protection Board. The doses in target tissues were calculated using the specific effective energy (SEE) matrices calculated using the SEECAL code developed at Oak Ridge National Laboratory (supplied courtesy of Dr. Keith Eckerman). ICRP 68 (1994b) weighting factors and algorithms were used to determine effective doses with the following modifications: 1) the definition of the gonadal $H_T(50)$ is that of the testes only rather than the ICRP-suggested maximum of the $H_T(50)$ to the testes or ovaries, 2) the remainder organs, the thymus and uterus, were eliminated from the calculations, and 3) the tissue-weighting factors were fixed (no splitting rule) with the ET airways split from the remainder group of tissues and assigned a constant weighting of 0.025.

4.5 EXAMPLE CALCULATION OF E(50) AND $H_T(50)$

The following text provides a conceptual framework that will allow the reader to obtain the doses in appendix Table A.2 from the data in Tables A.4 through A.7 using an internal dosimetry code of their choice. The algorithm described here, along with tabulated intermediate data obtained at selected breakpoints, provide convenient benchmarks for independent evaluations. The intake and dose values shown here were calculated using the System for Internal Dosimetry (SID) (Traub 2004), an internal dosimetry toolbox that was implemented in the MATLAB/Simulink programming environment. (MATLAB and Simulink are software programs available from The MathWorks, Inc., Natick, Massachusetts.)

The following information is presented in a manner that approximates the derivations provided by the ICRP in their various publications. The order in which the calculations proceed in an actual internal dosimetry code may differ markedly from the derivations. The modifications are generally done to conserve computer memory and to enhance the speed of the calculations. For example, the decay constant for U-234 is so small that virtually no uranium decay occurs during the 50 years following an intake; therefore, it is convenient to base the calculations on the mass of uranium inhaled and to account for the different isotopes and decay progeny by applying conversion factors at the end of the calculations.

4.5.1 Theory

The committed effective dose E(50) is defined by the ICRP by Equation 4.8

$$E(50) = \sum_{T} w_T H_T(50) \tag{4.8}$$

where w_T is the tissue weighting factor and $H_T(50)$ is the committed equivalent dose for tissue or organ T. ICRP assigns explicit w_T values to a number of, but not all, organs. In the calculations done here, all tissues were assigned explicit, fixed values of w_T (no remainder group splitting) so that E(50) would be additive over intakes.

The committed (50 yr) equivalent dose to an organ or target tissue, T, is given by Equation 4.9

$$H_T(50) = \sum_{S} U_S(50)\mathrm{SEE}(T \leftarrow S) \tag{4.9}$$

where $U_S(50)$ is the number of nuclear transformations (Bq s) in 50 yr in source region S following an acute intake. SEE(T←S) is the Specific Effective Energy, equivalent dose in T per transformation in S expressed as (Sv/(Bq-s) (ICRP 1991). The SEE values used for the doses reported in Table A.2 were obtained from SEECAL 2.0.[a]

The purpose of the computer code is to calculate U_S based on the biokinetic models for the respiratory tract, systemic organs, and gastrointestinal tract that were described previously. The total intake of a particulate material, I_0 in Bq, is expressed as:

$$I_0 = C_A t B \tag{4.10}$$

where C_A is concentration of activity in air in Bq/m^3, t is exposure duration in h, and B is breathing rate m^3/h (ICRP 1994a).

[a]SEECAL is available from the Radiation Safety Information and Computation Center, Oak Ridge, TN as code package CCC-620.

If the air concentration and breathing rate vary with time, it is necessary to compute the total intake, I_o in Bq, using equation 4.11:

$$I_0 = \sum_i \bar{C}_{Ai} \times \tau_i \times B_i \tag{4.11}$$

where C_A is the average concentration of activity (Bq/m^3) in air during the ith time interval expressed as

$$\frac{C_A(t_i) - C_A(t_{i-1})}{2};$$

τ_i is exposure duration in h during the ith time interval; and B_i is the average breathing rate, m^3/h, during the ith time interval.

Not all of the material that comprises an intake is deposited in the lungs; some is exhaled. The fraction of the intake that is deposited in the lung and the distribution of that deposition in the various compartments of the lung need to be calculated. The equations for the deposition calculations are described in ICRP 66 (1994a).

The fraction of an intake that is deposited in each of the five RT regions, described in Section 4.1, is a complex function of several variables that include the breathing rate, B, of the individual, the aerodynamic diameter, d_{ae}, of the particle, the density of the particle, and its shape factor.

4.5.2 Implementation

The aerosol to which the personnel in Level I scenarios are exposed consists of varying concentrations of DU in the aerosols that was assumed to consist of nine discrete aerodynamic diameters (ADs). The nine particle sizes and their relationship to the stages of a CI are given in Table 4.6. A one-to-one relationship is established between a particle's d_{ae} and the stage of the CI on which it was collected, after correcting for the CI response function.

The measured DU aerosol concentrations used in the calculations were obtained from Appendix Tables A.4 through A.7 and are assumed to occur at the midpoints of the measurement intervals. Note that the time intervals and their associated midpoints vary among the phases and shots. To estimate the intakes for Scenarios A through E, two DU aerosol concentrations must be inferred from the existing data—the aerosol concentration at the very beginning of the first measurement period and the air concentration at the end of the last time of Scenario E, 120 min after sampling commenced. The aerosol concentration at time = 0 min (at the beginning of the first sampling interval, which was actually 5 sec after perforation) was set equal to the first measured air concentration. The aerosol concentration at the end of Scenario E was set equal to the last measured value. If any of the air concentrations were negative, as might be the case after background subtraction, the air concentration was set equal to zero. Table 4.10 shows the DU aerosol concentrations for the PI-1 driver's position. The aerosol concentrations listed by stage under the column headings 0.25 through 95.5 min are measured air concentrations taken directly from Table A.4.

Appendix Tables A.4 through A.7 list the median particle diameter for each stage of the CI. These particle diameters were not used directly for the dose calculations. Rather, the particle diameters listed in Table 4.6, which were based on averaging and rounding these diameters, were used for all dose estimates.

The deposition characteristics of a given d_{ae} vary for breathing patterns, which were modeled for light and heavy exercise. The breathing pattern assumptions for each scenario are listed in Table 4.2, and the ventilation rate and frequency values for light and heavy exercise are provided in Table 4.3. Scenarios A, B, and E assume heavy exercise. Scenarios C and D assume a combination of light and heavy exercise. To account for the differences in breathing rates in these two scenarios from the difference in exercise levels, separate deposition calculations were performed. For each scenario shown in Table 4.2, intakes for each CI stage were calculated by performing the numerical integration from Equation 4.11. The DU aerosol concentrations

Table 4.10. DU Aerosol Concentrations by CI Stage as a Function of Time (PI-1 Driver's Position)

	Concentration of Uranium at Midpoint Intervals (min) after Initiation of CI Sampling, µg/m^3									
Stage	0[(a)]	0.25[(b)]	1.0[(b)]	2.5[(b)]	5.5[(b)]	11.5[(b)]	23.5[(b)]	47.5[(b)]	95.5[(b)]	127.5[(c)]
1	8.66E+05	8.66E+05	2.19E+05	1.31E+05	1.84E+04	5.28E+03	6.50E+03	3.31E+03	4.90E+03	4.90E+03
2	3.34E+04	3.34E+04	4.59E+04	1.81E+04	2.38E+03	5.96E+02	0	9.19E+00	4.31E+02	4.31E+02
3	6.03E+03	6.03E+03	6.08E+04	3.56E+04	1.10E+04	3.79E+03	1.18E+03	3.54E+02	1.87E+02	1.87E+02
4	1.85E+04	1.85E+04	8.97E+04	2.59E+04	1.20E+04	3.82E+03	9.25E+02	3.10E+02	3.57E+02	3.57E+02
5	1.98E+04	1.98E+04	4.87E+04	6.01E+04	2.70E+04	5.13E+03	2.50E+03	1.51E+03	1.19E+03	1.19E+03
6	1.72E+04	1.72E+04	8.09E+04	2.61E+04	1.67E+04	4.40E+03	2.56E+03	7.70E+02	8.05E+02	8.05E+02
7	1.54E+04	1.54E+04	5.22E+04	4.28E+04	2.05E+04	5.89E+03	2.33E+03	1.52E+03	1.05E+03	1.05E+03
8	1.19E+04	1.19E+04	7.78E+04	9.59E+04	2.53E+04	7.74E+03	2.95E+03	1.88E+03	1.38E+03	1.38E+03
9	1.41E+04	1.41E+04	2.31E+05	4.99E+04	9.84E+03	1.00E+04	9.98E+03	8.03E+03	7.08E+03	7.08E+03

(a) Obtained from the aerosol concentration at 0.25 min.
(b) Average measured aerosol concentration for the specified time interval.
(c) Obtained from the aerosol concentration at 95.5 min.

were obtained by linear interpolation of the DU concentrations for each stage. The result of these calculations was an array of intakes, one associated with each stage of the CI, that correspond to the values of ζ_j described in Section 4.4.1. The stage-specific intakes were multiplied by the inverse of matrix R (from Table 4.4) to obtain the aerosol component amplitudes, also referred to as corrected stage-specific intakes. See Equation 4.5 for the calculation of ξ_i. The corrected intakes, in µg U, are shown in Table 4.11.

Table 4.11. Stage-Specific, Flow-Rate Corrected Intakes by Scenario

	Heavy Exercise Scenarios					Light Exercise Scenarios (µg U)				
Stage	A	B	C	D	E	A	B	C	D	E
1	3.23E+04	5.50E+04	6.00E+04	6.00E+04	5.01E+03	(a)	(a)	5.57E+03	1.25E+04	(a)
2	0	2.38E+02	0	0	0	(a)	(a)	0	0	(a)
3	1.42E+03	1.03E+04	1.50E+04	1.50E+04	4.66E+03	(a)	(a)	1.78E+03	2.14E+03	(a)
4	2.90E+03	1.01E+04	1.32E+04	1.32E+04	3.09E+03	(a)	(a)	6.25E+02	9.99E+02	(a)
5	1.56E+03	1.28E+04	2.02E+04	2.02E+04	7.41E+03	(a)	(a)	2.81E+03	5.05E+03	(a)
6	1.86E+03	7.20E+03	1.05E+04	1.05E+04	3.34E+03	(a)	(a)	1.50E+03	2.37E+03	(a)
7	1.69E+03	1.07E+04	1.71E+04	1.71E+04	6.40E+03	(a)	(a)	2.79E+03	4.83E+03	(a)
8	1.95E+03	1.91E+04	2.73E+04	2.73E+04	8.16E+03	(a)	(a)	3.07E+03	5.15E+03	(a)
9	4.79E+03	1.79E+04	2.21E+04	2.21E+04	4.19E+03	(a)	(a)	1.02E+04	2.15E+04	(a)

(a) Scenarios A, B, and E do not include light exercise.

Calculation of U_S is performed using computer implementations of the biokinetic models described in Sections 4.1 through 4.4. The solubility parameters for the respiratory tract clearance model vary according to the phase and shot of the experiment and the particle size. Sets of dissolution parameters are given in Table 4.9, and the conditions for using each dissolution set are given in Table 4.8.

The U_S values, calculated with the internal dosimetry computer code, are multiplied by the SEE values to obtain $H_T(50)$ for target tissues, and then E(50) is computed (see Equation 4.9). For this report, the algorithm to calculate E(50) was a modified version of the algorithm published in ICRP 68 (1994b). Briefly, the modifications are 1) the gonadal dose equivalent was the testes dose equivalent rather than the maximum of the testes or ovaries dose equivalent, 2) the thymus and uterus were deleted from the remainder tissues, and 3) the dose to the ET airways was given a weight of 0.025, always, and the mass-weighted average equivalent dose to the remaining remainder tissues was given a weight of 0.025. Table 4.12 shows the stage-specific E(50) and $H_T(50)$s for Scenario B for U-238 only.

To calculate the E(50) and $H_T(50)$ values shown in Appendix A from the air concentration data in Appendix A, it was necessary to account for the progeny of U-238 and the dose resulting from exposure to U-234. To do this, calculations were performed similar to those used to obtain the doses for U-238 shown in Table 4.12, but using the mass fraction SEE values for U-234. The final dose is calculated as the sum of the U-238 plus U-234 doses multiplied by 1.02. The factor 1.02 accounts for the influence of U-238 short-lived progeny.

The algorithm described in this section provides one method that can be used to obtain the E(50) and $H_T(50)$ values shown in Tables A.4 through A.7 in Appendix A. Intermediate results of the internal dose calculations at selected breakpoints help the reader to independently calculate the radiation doses shown in this report. Alternate algorithms may also be employed.

Table 4.12. Calculated E(50) and $H_T(50)$ from U-238 Only (PI-1 Driver's Position, Scenario B)

		$H_T(50)$ for Selected Organs and Tissues (rem)						
Stage	*E*(50)	**Lung**	**Bone Surfaces**	**ET**	**Kidneys**	**LN-TH**	**Red Marrow**	**Liver**
1	0.065	0.006	0.037	2.473	0.014	0.127	0.004	0.005
2	0.000	0.000	0.000	0.013	0.000	0.004	0.000	0.000
3	0.030	0.109	0.013	0.630	0.005	0.534	0.001	0.002
4	0.111	0.795	0.022	0.579	0.008	1.470	0.002	0.003
5	0.174	1.385	0.093	0.130	0.034	0.209	0.010	0.013
6	0.066	0.503	0.046	0.118	0.017	0.416	0.005	0.006
7	0.055	0.416	0.035	0.110	0.013	0.629	0.004	0.005
8	0.085	0.671	0.043	0.082	0.016	0.592	0.004	0.006
9	0.220	1.770	0.115	0.058	0.042	1.485	0.012	0.016
Sum[a]	0.805	5.656	0.405	4.192	0.148	5.465	0.042	0.055
Total[b]	0.779							

(a) The sum of the stage-specifiic values.
(b) The E(50) for the intake, calculated using the unmodified ICRP-68 algorithm.

4.6 SOFTWARE QUALITY ASSURANCE

Data entry reviews and code validations and verifications (V&V) were carried out as part of the software quality assurance. A comprehensive comparison between the BK code and a completely independent code, AIDE (Bertelli and Lipsztein 1987; Wren et al. 1994; Bertelli et al. 1998; Bertelli et al. 2002), was performed for uranium isotopes U-234 and U-238, using Type S and Type M, standard 5-μm AMAD aerosol, and ICRP default breathing assumptions. The quantities shown in Table 4.13 were compared

Table 4.13. Compartment and Dose Quantities Considered in BK/AIDE Comparison

Compartment Quantities	Equivalent Doses to Tissues		
Urine excretion	Lung	Colon	Breast
Lung burden	LN-Th	Small Intestine	Esophagus
Fecal excretion	ET	Stomach	Skin
Kidney burden	Bone Surface	Muscle	Pancreas
LN-Th burden	Kidneys	Adrenals	Thyroid
Total body burden	Liver	Brain	Bladder
	Red Marrow	Spleen	Gonads

between the two codes for 300 separate times between 1 d and 26,000 d. The agreement between the output of the two codes was within 3% for all of these quantities.

Table 4.14 shows a comparison of the dose coefficients obtained with the BK code versus those given in ICRP 78 for selected radionuclides. (Because the ICRP E(50) values are listed in Sv/Bq, this convention is maintained in this table rather than converting the numbers to rem/nCi.) The table lists radionuclide evaluated, whether calculations included short-lived progeny, the dissolution type, the final dissolution rate (f_1), the E(50) per intake for the two codes, and the ratio between their results. Agreement between these codes ranged from 0.965 to 1.035. For the uranium isotopes, the agreement ranged from 1.004 to 1.035. Most of the discrepancies arise because ICRP 78 values are listed with only two significant figures.

As a check of the lung fluid dissolution model, ICRP standard Type M dissolution behavior (which involves only the ***particles in initial state*** and ***particles in transformed state*** components and not the "vsd" box shown in Figure 4.2) was calculated using three different but equivalent methods. Standard Type M has only two dissolution components; these were described in the expanded three-compartment formalism in three ways by letting the "rapid," "slow," and "very slow" components successively be zero. The same results were obtained in each case.

As a further check of the dissolution model, calculations for two of the actual aerosol components used in this study were performed to compare with the IMBA uranium code (Birchall et al. 1998). The IMBA uranium code is limited to ICRP default breathing assumptions for a standard occupational worker (31.3% sitting and 68.8% light exercise with a mean ventilation rate of 1.2 m^3/h). The two aerosol components were 4-μm AMAD (CI stage 5) and 2-μm AMAD (CI stage 6) (see Table 4.6). The solubility parameters for these CI stages are obtained from cyclone stages 3 and 4 as shown in Table 4.6.

Table 4.14. Comparison of BK Code Dose Coefficients with ICRP Publication 78

Nuclide	Progeny (y/n)	Type	f_1	BK E(50)/intake (Sv/Bq)	ICRP 78 E(50)/intake (Sv/Bq)	Ratio
Cs-137 (5 µ AMAD)	y	F	1	6.74E-09	6.70E-09	1.005
Co-60	n	M	0.1	7.13E-09	7.10E-09	1.004
		S	0.05	1.66E-08	1.70E-08	0.975
Am-241	n	M	5.00E-04	2.72E-05	2.70E-05	1.006
Pu-238	n	M	5.00E-04	3.01E-05	3.00E-05	1.002
		S	1.00E-05	1.06E-05	1.10E-05	0.965
Pu-239	n	M	5.00E-04	3.26E-05	3.20E-05	1.018
		S	1.00E-05	8.37E-06	8.30E-06	1.008
Pu-240	n	M	5.00E-04	3.26E-05	3.20E-05	1.018
		S	1.00E-05	8.39E-06	8.30E-06	1.010
U-234	n	F	2.00E-02	6.54E-07	6.40E-07	1.022
		M	2.00E-02	2.11E-06	2.10E-06	1.005
		S	2.00E-03	6.83E-06	6.80E-06	1.004
U-235	y	F	2.00E-02	6.07E-07	6.00E-07	1.012
		M	2.00E-02	1.82E-06	1.80E-06	1.012
		S	2.00E-03	6.11E-06	6.10E-06	1.002
U-238	y	F	2.00E-02	5.85E-07	5.80E-07	1.008
		M	2.00E-02	1.66E-06	1.60E-06	1.035
		S	2.00E-03	5.73E-06	5.70E-06	1.005
Sr-90	y	F	3.00E-01	3.09E-08	3.00E-08	1.029
		S	1.00E-02	7.86E-08	7.70E-08	1.021
Fe-59	n	F	1.00E-01	3.00E-09	3.00E-09	1.001
		M	1.00E-01	3.15E-09	3.20E-09	0.984
Ru-106	y	F	5.00E-02	9.79E-09	9.80E-09	0.999
		M	5.00E-02	1.67E-08	1.70E-08	0.981
		S	5.00E-02	3.44E-08	3.50E-08	0.983
Ra-226	y	M	2.00E-01	2.21E-06	2.20E-06	1.005
Ra-228	y	M	2.00E-01	1.72E-06	1.70E-06	1.014

Assuming the solubility parameters for PI-1, Table 4.8 shows that cyclone stages 3 and 4 have solubility types "u" and "v." The dissolution parameters for these types are given in Table 4.7.

The two-intake method was used with the IMBA code to represent the three dissolution components. Intakes are divided into two parts, a and b. Part a describes the rapid and slow dissolution components, and part b describes the very slow dissolution component. Table 4.15 shows the fractional decomposition into parts a and b.

For example, for the 4-µm (type "u") aerosol component, a 1 Bq intake is divided for the sake of the IMBA code into two parts: a) 0.254 Bq having dissolution parameters $s_r = 19.9\ d^{-1}$, $s_s = 0.79\ d^{-1}$, $f_s = 0.354$, and b) 0.746 Bq with a single dissolution component having $s = 0.0052\ d^{-1}$.

Table 4.15. Fractional Decomposition into Intake Parts a and b

Dissolution Type	$s_r(d^{-1})$	$s_s(d^{-1})$	$s_{VS}(d^{-1})$	f_r	f_s	f_{VS}	Fraction a or b
u	19.9	0.79	0.0052	0.164	0.09	0.746	--
u: Part a	19.9	0.79	--	0.646	0.354	--	0.254
u: Part b	--	--	0.0052	--	--	1	0.746
v	11.9	0.06	0.00104	0.222	0.174	0.604	--
v: Part a	11.9	0.06	--	0.561	0.439	--	0.396
v: Part b	--	--	0.00104	--	--	1	0.604

Table 4.16. Comparison of IMBA and BK Dose Coefficients

AMAD (μm)	Dissolution type		Fraction a or b	IMBA Dose Coeff (rem/nCi)	
				U-238	U-234
4	u	part a	0.254	1.38E-03	1.57E-03
		part b	0.746	1.05E-02	1.37E-02
		average	--	8.18E-03	1.06E-02
		BK	--	8.14E-03	1.07E-02
2	v	part a	0.396	3.32E-03	4.03E-03
3	v	part b	0.604	2.10E-02	2.58E-02
		average	--	1.40E-02	1.72E-02
		BK	--	1.39E-02	1.73E-02

The dose coefficients calculated using IMBA and BK are compared in Table 4.16. The agreement is better than 1%. The other parameters assumed for these calculations were as follows: f_1=0.02, σ_g =1.0, density=3.0 g/cm^3, shape factor=1.5.

Three other calculations were performed to compare the codes used in this study with the IMBA uranium code, again assuming ICRP default breathing assumptions. For these comparisons, standard Type M, f_1=0.02 material was assumed. The three test cases were the following:

1. Monodisperse aerosol with an aerodynamic diameter of 2 μm, σ_g=1.0
2. Monodisperse aerosol with an aerodynamic diameter of 4 μm, σ_g=1.0
3. Polydisperse aerosol with AMAD=5 μm, σ_g=2.5.

For each of these cases, the DU aerosol concentration data were simulated assuming 1 g/m^3 total DU concentration at a midpoint time of 5 min and 0 g/m^3 at a midpoint time of 20 min and assuming a nominal CI flow rate of 2.6 Lpm. Since linear interpolation/extrapolation is used, the mean air concentration during the assumed exposure time period of 0 to 10 min is 1 g/m^3.

The simulated CI stage data were obtained by multiplying the total air concentration by the factors in Table 4.17 that represent the expected distribution over the CI stages for these three cases. (For the monodisperse 2- and 4-μm cases, these are columns of the response matrix.)

Table 4.17. Expected Distributions Over CI Stages

Stage	2 μm	4 μm	5 μm AMAD
1	0.0126	0.0210	0.0720
2	0.0064	0.0157	0.0711
3	0.0034	0.0195	0.1363
4	0.0044	0.1005	0.1870
5	0.0177	0.8254	0.2345
6	0.9374	0.0179	0.2218
7	0.0182	0.0000	0.0534
8	0.0000	0.0000	0.0190
Backup Filter	0.0000	0.0000	0.0049

The intake is the time integral of the linearly interpolated/extrapolated air concentration data over the assumed exposure period of 0 to 10 min, times the specific activity numbers (nCi μg^{-1}) listed in Table 3.1 (page 19), times the mean ventilation rate of 1.2 m^3/h (ICRP [66] 1994a) breathing rate for a light worker, defined as a worker sitting or engaging in light exercise. Although the four isotopes of uranium are listed in Table 3.1, only U-234 and U-238 were used in the calculations.

A comparison of the results for E(50) of the three V&V cases are shown in Table 4.18. The designation P0-S0 (phase 0, shot 0) refers to the QA checks. This table contains the E(50) results of three codes: the BK code, the IMBA Expert Version 3.0.64 (Available from A.C. James and Associates, Richland, Washington), and the System for Internal Dosimetry (SID) that was implemented in the Matlab/Simulink programming environment (Matlab and Simulink are software programs available from The MathWorks, Inc., Natick, Massachusetts). While Table 4.18 shows only the E(50) values, $H_T(50)$ values were also calculated by each of the three codes. The agreement between the three codes for the E(50) values are representative of the agreement found for the $H_T(50)$ values. As shown in Table 4.18, the SID code is capable of reproducing the doses calculated by the BK and IMBA codes. Thus, the SID code provides a computational platform that allows the same code to reproduce the calculations of both the BK and IMBA codes, thus providing the equivalent of a transfer standard. The SID code also provided a mechanism to show the how various assumptions would impact the final dose values calculated by the BK code.

Table 4.18. Comparison of E(50) Code Results for Various Internal Dosimetry Codes (in rem)

V&V Case[a]	Sol. Type	LANL BK[b]	SID[c] BK emulation	SID[c] IMBA emulation	IMBA[d]
PO-S0 2 μm Monodisperse	M	1.066	1.065	0.987	0.979
	S	3.138	3.139	2.67	2.67
	'v'	1.240	1.235	NC[e]	NC
PO-S0 4 μm Monodisperse	M	0.853	0.854	0.754	0.753
	S	2.193	2.193	2.17	2.17
	'v'	0.709	0.710	NC	NC
PO-S0 5 μm Polydisperse	M	0.583	0.583	0.470	0.470
	S	1.722	1.722	1.62	1.62

(a) The intake for all cases was 2480.11 Bq U-238 and 277.87 Bq U-234.
(b) The BK code accounts for the progeny of U-238 and doses from U-234 and U-236 by multiplying the sum of the U-238 E(50) plus the U-234 E(50) by 1.02. The tissue-weighting factor for the ET regions is always 0.025.
(c) The SID code accounts for the progeny of U-238 in the same way as the code it emulates.
(d) The IMBA code accounts for the progeny of U-238 by merging the SEE values of the parent and progeny radionuclides. The tissue-weighting factor for the ET regions depends on the dose equivalent to the ET regions relative to the dose equivalent to the named organs (see ICRP [68] 1994b). The doses from U-234 and U-236 were not included in these calculations.
(e) NC – Not calculated. IMBA cannot be modified by the user to account for the "very slowly dissolving" compartment in the RT model.

The BK and IMBA codes differ in two substantial ways. First, the two codes differ in how they account for the dose from radioactive progeny of U-238. The BK code accounts for the dose from the progeny by multiplying the sum of the U-238 dose and the U-234 dose by 1.02, a factor that was found to approximate and always overestimate the dose contribution from the progeny. The IMBA code, on the other hand, accounts for the progeny by either merging the SEE values of the progeny with the SEE values of U-238, or by explicitly calculating the dose from U-238 and its progeny. For the dose calculations reported here, the first method (i.e., the merged SEE values) was used. The second way in which the BK and IMBA codes differ is in how they handle the $H_T(50)$ from the ET regions when calculating E(50). The BK code always assigns a tissue weighting factor of 0.025 to the $H_T(50)$ of the ET regions, whereas the IMBA code has implemented the algorithm described in ICRP-68 (ICRP 1994b) in which case the tissue-weighting factor assigned to the ET region depends on the magnitude of the ET region $H_T(50)$ dose.

The differences between the BK and IMBA code for the 2 μm and 4 μm monodosperse aerosol cases are indicative of how the assumptions used by the two codes influence the estimate of E(50). The BK code used the aerosol distribution shown in Table 4.17, which after correction with the R matrix (see Equations 4.3 and 4.4 and Table 4.4), results in a true monodisperse aerosol. The IMBA code does not accept the stage-specific concentrations shown in Table 4.17 but can be set to calculate doses for monodisperse aerosols. Thus, the two codes are calculating doses for the same aerosol distribution. In the case of the 5-μm AMAD aerosol, the BK code assumed that the aerosol was distributed on a nine-stage CI having the stage-specific distribution shown in Table 4.17. The IMBA code was used to calculate the dose from a true lognormal distributed aerosol (with a σ_g of 2.5). Thus, the differences between the two codes provides some information as to how the method of decomposing the aerosol distribution into nine monodisperse aerosol components will influence the estimate of E(50). As the number of aerosol components is increased, convergence of the two results would be expected. It is noted that this validation checks the performance of the nine-component decomposition of an idealized lognormal distribution. Using real CI stage data directly, a greater uncertainty in intake and dose estimation would result from force-fitting lognormal distributions than using nine-component decomposition.

The SID code was used to check the dose calculations reported in Appendix A, Table A.5. The doses for four example cases, as calculated by SID, are shown in Appendix Table A.6. These doses are very similar to those calculated by the BK code and provide another verification that the doses calculated by the BK code are reproducible by independently developed computer codes for internal dosimetry.

4.7 REFERENCES

Bertelli L and JL Lipsztein. 1987. "A Mathematical Simulation for the Study of Radionuclide Kinetics in the Human Body." *Radiation Protection Dosimetry* 18:209-214.

Bertelli L, A Puerta, ME Wrenn, and JL Lipsztein. 1998. "Bioassay Interpretation and Dosimetry Using Specific Absorption Parameters For UO_2 and U_3O_8." *Radiation Protection Dosimetry* 79(1-4):111-113.

Bertelli L, RA Guilmette, and NP Singh. 2002. "Sensitivity of Doses and Associated Uncertainties in Bioassay Estimates to ICRP Publication 66 Respiratory Tract Dosimetric Model With Respect to the Parameters: Ventilation Rates, Time Budget, Total and Regional Deposition, Oral Versus Nasal Breathing and Clearance Rates of Compartments." *Journal of Radioanalytical and Nuclear Chemistry* 252(2):327-338.

Birchall A, NS Jarvis, MS Pease, AE Riddell, and WP Battersby. 1998. "The IMBA Suite: Integrated Modules for Bioassay Analysis." *Radiation Protection Dosimetry* 79:107-110.

Cooper DW. 2001. "Methods of Size Distribution Data Analysis and Presentation." *Aerosol Measurement: Principles, Techniques, and Applications.* Second Edition, eds. PA Baron, and K Willeke, Wiley-Interscience.

Cooper DW and JJ Wu. 1990. "The Inversion Matrix and Error Estimation in Data Inversion: Application to Diffusion Battery Measurements." *Journal of Aerosol Science* 21:217-226.

Eve IS. 1966. "A Review of the Physiology of the Gastrointestinal Tract in Relation to Radiation Doses from Radioactive Material." *Health Physics* 12:131-161.

Feller W. 1957. *An Introduction to Probability Theory and Its Applications*, 2nd ed. Wiley & Sons, New York.

Gear CW. 1971. *Numerical Initial Value Problems in Ordinary Differential Equations*. Prentice Hall, Englewood Cliffs, New Jersey.

Gilks WR, S Richardson, and DJ Spiegelhalter. 1996. *Markov Chain Monte Carlo in Practice*. Chapman and Hall, New York.

Hindmarsh AC. 1981. LSODE (Livermore Solver for Ordinary Differential Equations). Lawrence Livermore National Laboratory, Livermore, California. Accessible through www.llnl.gov/casc/odepack/.

International Commission on Radiological Protection (ICRP). 1959. *Report of Committee II on Permissible Dose for Internal Radiation*. ICRP Publication 2, Pergamon Press, Oxford, United Kingdom.

International Commission on Radiological Protectection (ICRP). 1979. *Limits for Intakes of Radionuclides by Workers*. ICRP Publication 30, Part 1, Pergamon Press, Oxford, United Kingdom.

International Commission on Radiological Protection (ICRP). 1994a. *Human Respiratory Tract Model for Radiological Protection*. ICRP Publication 66, Pergamon Press, Oxford, United Kingdom.

International Commission on Radiological Protection (ICRP). 1994b. *Dose Coefficients for Intakes of Radionuclides by Workers*. ICRP Publication 68, Pergamon Press, Oxford, United Kingdom.

International Commission on Radiological Protection (ICRP). 1995a. Age-dependent Doses to Members of the Public from Intake of Radionuclides: Part 3 Ingestion Dose Coefficients. ICRP Publication 69, Pergamon Press, Oxford, United Kingdom.

International Commission on Radiological Protection (ICRP). 1995b. Basic Anatomical and Physiological Data for Use in Radiological Protection: The Skeleton. ICRP Publication 70, Pergamon Press, Oxford, United Kingdom.

International Commission on Radiological Protection (ICRP). 1995c. *Age-dependent Doses to Members of the Public from Intake of Radionuclides: Part 4 Inhalation Dose Coefficients*. ICRP Publication 71, Pergamon Press, Oxford, United Kingdom.

International Commission on Radiological Protection (ICRP). 1997. *Individual Monitoring for Internal Exposure of Workers*. ICRP Publication 78, Pergamon Press, Oxford, United Kingdom.

Kandlidar M and G Ramachandran. 1999. "Inverse Methods for Analysing Aerosol Spectrometer Measurements: A Critical Review." *Journal of Aerosol Science* 30:413-437.

O'Shaughnessy PT, and OG Raabe. 2003. "A Comparison of Cascade Impactor Data Reduction Methods." *Aerosol Science and Technology* 37:187-200.

Parkhurst MA, F Szrom, RA Guilmette, TD Holmes, YS Cheng, JL Kenoyer, JW Collins, TE Sanderson, RW Fliszar, K Gold, JC Beckman, and JA Long. 2004. *Capstone Depleted Uranium Aerosols: Generation and Characterization, Volumes 1 and 2.* PNNL-14168, Prepared for the U.S. Army by Pacific Northwest National Laboratory, Richland, Washington.

Strom DJ. 1986. "The Bayesian Approach to Estimation." *Health Physics* 50(4):553.

Traub RJ. 2004. "A Toolbox for Internal Dosimetry." (in preparation)

US Army Center for Health Promotion and Preventive Medicine (USACHPPM). 2000. *Depleted Uranium-Human Exposure Assessment and Health Risk Characterization in Support of the Environmental Exposure Report "Depleted Uranium in the Gulf" of the Office of the Special Assistant to the Secretary of Defense for Gulf War Illnesses, Medical Readiness and Military Deployments (OSAGWI), OSAGWI Levels I, II and III Scenarios, 15 September 2000.* Health Risk Assessment Consultation No. 26-MF-7555-00D, Aberdeen Proving Ground, Maryland. Online report available at URL: www.gulflink.osd.mikl in the Environmental Exposure Reports Section.

Wrenn ME, L Bertelli, PW Durbin, NP Singh, JL Lipsztein, and KF Eckerman. 1994. "A Comprehensive Metabolic Model for Uranium Metabolism and Dosimetry Based on Human and Animal Data." *Radiation Protection Dosimetry* 53:255-258.

5.0 Chemical Concentrations and Radiological Doses

Depleted uranium intakes, chemical concentrations, and radiation doses to selected organs were calculated for each phase, shot, and sampling position for each scenario. Although the quantity of aerosol collected differed from shot to shot among the sampler arrays, no single sampling location was consistently high or low. As a result, each of the sampling locations can be considered to provide an estimate of the dose to the hypothetical occupants of the vehicle, and results for each shot can be appropriately represented by the mean or the median of the doses measured in each sampling location. Each shot within each phase is treated as a sampling of the universe of shot lines that results in the perforation of the vehicle. Using this construct, each of the shots within a test phase, representing a specific vehicle and configuration, can be combined to provide statistical estimates of DU intakes and associated doses and peak kidney uranium concentrations from exposure within these vehicles from perforation in combat by a LC-DU munition.

Best estimates of intakes, radiological doses, and chemical concentrations were calculated for perforations of 1) an Abrams BHT through non-DU armor (Phase I), 2) a Bradley BHT (Phase II), 3) an Abrams BHT through DU armor (Phase III), and 4) the DU armor of an operational Abrams tank in which the EC/NBC ventilation system was operating (Phase IV). The dose distributions could not be readily quantified as either normal, lognormal, or any other standard distribution (See Section 5.5.2.4). The median based on Bayesian Monte Carlo statistics is used to describe the radiological dose and the chemical concentration. Variation of the data about the median is shown by using the 10th and 90th percentiles and by reporting the minimum and maximum radiological doses and chemical concentrations. In most cases the mean and standard deviations are also presented for comparison. The data for each shot and sampling position on which these statistics are calculated are tabulated in Appendix A. This chapter summarizes:

- The radiological whole body effective doses and equivalent doses to various organs.
- The DU peak mass concentrations in the kidney.
- The projected reduction in doses and concentrations resulting from operation of the vehicle ventilation system.
- Modeling confounders.
- Uncertainty in dose calculations.

5.1 RADIOLOGICAL DOSES

The results from Phases I, II, III, and IV are summarized in Section 5.1.1 in terms of intakes and committed effective doses and in Section 5.1.2 for organ committed equivalent doses. A separate section on DU mass concentration in the lung is discussed in Section 5.3. Doses from the single Phase-IV shot are further discussed and summarized in Section 5.4.1. The phases and shots are abbreviated as Phase (PI, PII, PIII, and PIV) and Shot (1 to 7) so that Phase I, Shot 1 is represented as PI-1, and the double shots are abbreviated as PI-3/4 and PII-1/2. The single DU shot fired in Phase IV at a fully operational tank was identified as Shot 4 in the test series, and this convention (PIV-4) is used in this report.

The nomenclature used in discussion of doses include the symbol "E(50)" to represent the 50-yr committed effective dose (also referred to as CED) and the symbol "$H_T(50)$" to represent the 50-yr committed equivalent organ dose (also referred to as CDE). Doses are listed in traditional units. Conversion factors between traditional and SI units are provided on page 218.

5.1.1 Intakes and Committed Effective Doses

The 10th, 50th (median), and 90th percentile E(50) were calculated for each scenario, phase, shot, and sampling position. The arithmetic means and standard deviations for these values were also calculated and are presented for comparison. In Phases I and II, each of which included one double-shot test, the double shots are treated as two single shots. An analysis of these shots showed that each shot could be treated independently because the first shot did not significantly affect the results of the second shot (see Section 5.1.3). No ventilation systems were present in the Phase-I through Phase-III BHTs, and the calculated intakes and doses assume that no respiratory protection was used. For Phase IV, the EC/NBC ventilation system was operating prior to and following vehicle perforation. The Phase-IV results are based on one operating sampler at each of four time intervals. Therefore, there are no standard deviations for these measurements, indicated as "not applicable" (NA). The intakes in nCi using 0.381 nCi/mg DU and doses for the four phases are presented in Table 5.1. The shots used in the calculation of each phase are listed.

The ratio of the median and the arithmetic mean (last column in Table 5.1) shows that there is good agreement between the two metrics, suggesting that the data are approximately normally distributed. The 10th and the 90th percentiles provide an indication of the spread of the data.

The data illustrate the range of intakes and doses with varying sampling positions and shot configurations for each scenario. These data are not a random sample of actual exposure events and may not be representative of all intakes and doses from actual events. However, because the test conditions tended to

Table 5.1. Intakes and E(50)s by Phase and Scenario

		Intake, nCi		E(50), rem		E(50), rem – Percentiles			Median/
Phase	Shots	Mean	SD	Mean	SD	10th	Median	90th	Mean
Scenario A (exposure from 0 to 1 min, heavy exercise)									
Phase I	1 to 7	105	46.5	2.00	0.820	1.33	1.97	2.78	0.985
Phase II	1 to 3	31.7	25.0	0.676	0.645	0.155	0.589	1.01	0.872
Phase III	1 and 2	96.1	68.2	2.60	1.23	1.49	2.16	5.26	0.829
Phase IV	4	3.70	NA	0.0900	NA	NA	0.0900	NA	NA
Scenario B (exposure from 0 to 5 min, heavy exercise)									
Phase I	1 to 7	225	112	4.42	2.10	3.03	3.66	7.14	0.829
Phase II	1 to 3	82.8	61.6	1.81	1.53	0.335	1.73	3.09	0.956
Phase III	1 and 2	268	223	6.71	2.62	4.52	6.00	12.7	0.894
Phase IV	4	16.5	NA	0.436	NA	NA	0.436	NA	NA
Scenario C (exposure from 0 to 15 min, heavy exercise, and 15 to 60 min, light exercise)									
Phase I	1,2,4,5,6,7	291	159	5.32	1.99	3.15	4.85	8.56	0.910
Phase II	2 and 3	127	91.1	2.60	2.02	1.00	2.08	6.84	0.801
Phase III	1 and 2	370	309	9.08	3.18	5.92	8.33	16.4	0.917
Phase IV	4	34.7	NA	1.02	NA	NA	1.02	NA	NA
Scenario D (exposure from 0 to 15 min, heavy exercise, and 15 to 120 min, light exercise)									
Phase I	1,2,4,5,6,7	297	174	5.40	1.99	2.89	4.95	9.04	0.918
Phase II	2 and 3	143	89.2	2.88	1.99	1.13	2.44	7.06	0.847
Phase III	1 and 2	398	324	9.36	3.25	6.33	8.68	16.8	0.927
Phase IV	4	43.1	NA	1.20	NA	NA	1.20	NA	NA
Scenario E (5 to 15 min, heavy exercise)									
Phase I	1 to 7	59.6	42.6	1.20	0.803	0.451	0.916	2.24	0.765
Phase II	1 to 3	37.9	21.3	0.856	0.493	0.263	0.890	1.33	1.04
Phase III	1 and 2	77.2	75.2	2.01	0.695	1.05	1.94	3.08	0.967
Phase IV	4	10.4	NA	0.412	NA	NA	0.412	NA	NA

maximize the generation and retention of DU aerosol in the crew compartment, the ranges are more likely to overestimate rather than underestimate combat doses.

5.1.2 Organ and Tissue Doses

In addition to E(50), the $H_T(50)$s were calculated for key organs and tissues for each scenario and experimental phase for which data were collected. These doses are summarized in Table 5.2. To facilitate comparison of organ doses, the arithmetic means and standard deviations are presented instead of medians and percentiles. However, median doses (listed in Chapter 6) are used to calculate risks (Chapter 7).

As shown in Table 5.2, the lung receives the most dose for each of the phases and in each of the scenarios. The doses to most other organs, including those of greatest interest (bone surfaces, red marrow, kidney, and liver), are well below the dose to the lung. The lung dose ranges from a factor of 10 to a factor of 60 greater than the dose to the bone surfaces. The lung dose is 100 to 600 times greater than the red-marrow dose. Intermediate ratios are obtained for the kidney and liver. Extrathoracic airways and lymph nodes (listed in Appendix A) receive doses similar to the lung doses, but these tissues are not particularly sensitive to chemical effects nor are they target organs for cancer.

5.1.3 DU Doses from Double Shots

The effects of double shots (in which the same vehicle was hit twice within 15 min) on DU aerosol concentration and dose were tested in both the Abrams tank (PI-3/4) and the Bradley vehicle (PII-1/2). The DU concentrations and doses from these shots were evaluated to develop a correction factor that could be used to estimate the chemical and radiological dose for a double shot from single-shot test data. In both cases, the shot lines were essentially replicates of each other. Intuitively, one would expect that the second shot would double the dose; however, several confounders may affect the chemical and radiological doses in a nonlinear fashion. These include resuspension, unanticipated changes in particle sizes, or enhanced ventilation rates due to damage from the first shot.

The E(50) for each of the two Phase-I and Phase-II double shots were compared for Scenarios A, B, and E, the scenarios for which data were available for both shots. Arithmetic means and standard deviations were used for simplicity. Sampling ended at 10 min following the first shot, so Scenarios C and D were not calculated. Scenario E used an extrapolation based on three data points to estimate the DU concentration between 10 and 15 min.

Table 5.2. Organ and Tissue Doses by Phase and Scenario (HT[50], mean and standard deviation, rem)

Phase	Lung		Extrathoracic Airways		Bone Surface		Red Marrow		Kidney		Liver	
	Mean	SD	Mean	SD	Mean	SD	Mean	SD	Mean	SD	Mean	SD
Scenario A (exposure from 0 to 1 min, heavy exercise)												
Phase I	14.3	6.13	8.87	4.81	0.987	0.419	0.102	0.0432	0.361	0.153	0.135	0.0574
Phase II	5.21	5.06	1.22	0.738	0.382	0.369	0.040	0.0381	0.140	0.135	0.0524	0.0506
Phase III	17.5	6.98	19.3	15.5	0.381	0.186	0.039	0.0192	0.140	0.068	0.0523	0.0255
Phase IV	0.662	NA	0.373	NA	0.0226	NA	0.00233	NA	0.00829	NA	0.00310	NA
Scenario B (exposure from 0 to 5 min, heavy exercise)												
Phase I	32.1	15.7	18.1	10.7	2.21	1.09	0.229	0.112	0.810	0.398	0.303	0.149
Phase II	14.0	11.9	3.04	1.87	1.03	0.880	0.107	0.0909	0.378	0.322	0.141	0.121
Phase III	44.1	11.3	54.6	50.9	0.913	0.412	0.0943	0.0426	0.335	0.151	0.125	0.0565
Phase IV	3.32	NA	1.28	NA	0.10	NA	0.0103	NA	0.0367	NA	0.0137	NA
Scenario C (exposure from 0 to 15 min, heavy exercise, and 15 to 60 min, light exercise)												
Phase I	38.3	13.8	23.2	17.1	2.73	1.08	0.282	0.112	1.00	0.396	0.374	0.148
Phase II	20.0	15.7	4.39	2.66	1.52	1.22	0.157	0.126	0.557	0.446	0.209	0.167
Phase III	60.0	12.6	72.8	70.3	1.25	0.503	0.129	0.052	0.459	0.184	0.172	0.0689
Phase IV	7.58	NA	4.12	NA	0.181	NA	0.0187	NA	0.0664	NA	0.0248	NA
Scenario D (exposure from 0 to 15 min, heavy exercise, and 15 to 120 min, light exercise)												
Phase I	39.0	13.7	23.0	18.7	2.79	1.12	0.289	0.115	1.023	0.409	0.383	0.153
Phase II	22.2	15.5	4.97	2.53	1.70	1.22	0.175	0.126	0.622	0.446	0.233	0.167
Phase III	61.3	12.6	77.5	72.4	1.31	0.515	0.135	0.0532	0.479	0.189	0.179	0.0706
Phase IV	8.65	NA	5.88	NA	0.210	NA	0.0217	NA	0.0771	NA	0.0288	NA
Scenario E (5 to 15 min, heavy exercise)												
Phase I	8.81	5.90	4.24	4.16	0.65	0.45	0.0668	0.0466	0.237	0.165	0.0887	0.0617
Phase II	6.67	3.84	1.20	0.718	0.49	0.28	0.0501	0.0290	0.178	0.103	0.0665	0.0384
Phase III	13.6	4.12	14.4	17.6	0.26	0.11	0.0271	0.0111	0.0962	0.0392	0.0360	0.0147
Phase IV	3.14	NA	1.305	NA	0.052	NA	0.00532	NA	0.0189	NA	0.00707	NA

The Abrams shots were fired through the conventional side armor. From the resulting data (Table 5.3), it is apparent that for all three scenarios the DU aerosols from the first of the two shots (based on two operating sampler arrays [driver and gunner]) resulted in higher intakes and doses than those from the second shot (in which four sampler arrays operated). In fact, the PI-3 test produced the highest Phase-I dose, about twice the Phase-I mean. For each PI-3 scenario, the maximum dose occurred at the gunner's position. The doses from the PI-4 shot are also higher than the Phase-I mean, but were

Table 5.3. Comparison of Radiation Doses from the Abrams BHT (through non-DU armor) Double-Shot Tests

Scenario	First Shot (PI-3)			Second Shot (PI-4)			Phase-I Mean (of 7 Shots)		
	Intake (nCi)	E(50) (rem)	E(50) SD	Intake (nCi)	E(50) (rem)	E(50) SD	Intake (nCi)	E(50) (rem)	E(50) SD
A	145	3.35	1.23	140	2.17	0.612	105	2.00	0.820
B	381	8.98	2.60	327	5.29	1.22	225	4.42	2.10
E	134	3.15	0.0891	103	1.70	0.363	60	1.20	0.803

within one standard deviation of the overall phase mean. For each PI-4 scenario, the maximum dose occurred at the loader's position and may have been partially due to premature release of the armored shield covering this sampling array, which exposed this array to elevated pressure and more DU fragments. The reason for the high doses calculated from the PI-3 aerosols was because of a higher concentration of small particles. The results of the PI-3 and PI-4 data suggest that the PI-3 shot contributed little to the PI-4 doses. The two new perforation holes in the armor from the PI-3 shot may have allowed faster clearance of aerosols formed from the PI-4 shot.

Similar results were obtained for the Bradley vehicle double shots (Table 5.4). The Bradley series (Phase II) had only three total shots, two of which (shots 1 and 2) comprised the double shot. These shots were fired through the passenger compartment. Shot 3, a single shot through the turret, was expected to create the highest concentrations but actually resulted in the lowest concentrations and doses of the three shots. In this case, for Scenarios A and B, the intakes and doses for PII-1 and PII-2 were similar with the higher value occurring with PII-2 shot at the commander's position, a location very close to the penetrator's exit hole. However, these differences were not statistically significant. For Scenario E, the PII-1 intakes and doses were higher than the PII-2 intakes and doses. However, the difference was not significant. In the PII-2 test, the maximum doses occurred at the commander's position for Scenarios A and B, but in the PII-1 test, the highest dose occurred at the driver's position for Scenario E. As with the Abrams data, the second shot concentrations were well within one standard deviation of the phase mean values, and little, if any, influence from PII-1 is evident in the doses of PII-2.

Figure 5.1 graphically presents the intake results of the double shots for the Abrams BHT (through non-DU armor) and the Bradley BHT. The figure shows all of the doses at all of the crewmember positions for the first and second shots of the double-shot series, arranged from the lowest dose to the highest dose for Scenario B. For the Abrams BHT, some of the doses calculated for the first shot were actually higher than the doses calculated for the

Table 5.4. Comparison of Radiation Doses from the Bradley BHT Double-Shot Tests

Scenario	First Shot (PII-1)			Second Shot (PII-2)			Phase-II Mean (of 3 Shots)		
	Intake (nCi)	E(50) (rem)	E(50) SD	Intake (nCi)	E(50) (rem)	E(50) SD	Intake (nCi)	E(50) (rem)	E(50) SD
A	37.5	0.802	0.272	48.6	1.09	0.864	31.7	0.676	0.645
B	113	2.56	0.707	121	2.66	1.79	82.8	1.81	1.53
E	57.6	1.42	0.193	48.0	0.985	0.167	37.9	0.856	0.493

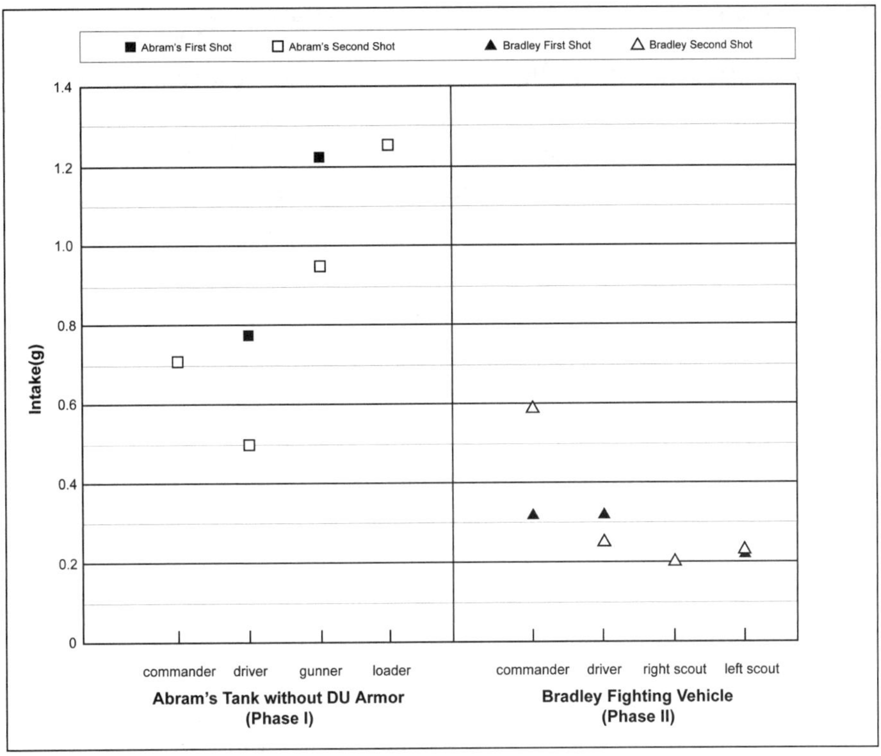

Figure 5.1. Radiation Doses, E(50), for the Double-Shot Test Series (plotted from lowest dose to highest dose for the Abrams [Phase I] and Bradley [Phase II] double-shot tests, Scenario B)

second shot. For the Bradley BHT double-shot series, the first shot doses are intermingled with the second shot doses.

From the results presented in Table 5.3 and 5.4 and Figure 5.1 in which the shots were 10 to 15 min apart, it appears that the shots were essentially independent, i.e., the first shot did not add significantly to the DU aerosol concentration of the second shot.

5.2 URANIUM CONCENTRATIONS IN THE KIDNEY

To evaluate chemical risk, peak kidney uranium concentrations (one day after intake) were calculated for each of the phases and scenarios. These concentrations are summarized in Section 5.2.1 in which all shots are evaluated as single shots. As with the radiological doses, an analysis of the kidney uranium concentrations for the double shots (Section 5.2.2) showed that the first shot did not have a significant impact on the results for the second. The individual results are presented in Appendix A.

5.2.1 Peak Uranium Concentrations in the Kidney

The 10th, 50th (median), and 90th percentile peak kidney uranium concentrations (μg DU/g kidney) were calculated for each scenario, phase, shot, and position. The arithmetic means and standard deviations for these values were also calculated and are presented for comparison. In Phases I and II, which included one double shot each, the double shots were treated as two single shots. No ventilation systems were present in the Phase-I through Phase-IIII BHTs, and the intakes and doses assume that no respiratory protection was used. For Phase IV, the EC/NBC ventilation system was operating during and after vehicle perforation. The peak kidney concentrations are presented in Table 5.5.

As with the radiological doses, the data presented in Table 5.5 are specific to the particular shots performed and the vehicle configurations used, which are

Table 5.5. Mean and Median Peak Kidney Uranium Concentrations by Phase and Scenario

Phase	Intake (mg DU)	Mean Peak U Conc. (μg U/g kidney)		Percentile Peak U Conc. (μg U/g kidney)			Ratio of Median to Mean
		Mean	SD	10th	Median	90th	
Scenario A (exposure from 0 to 1 min, heavy exercise)							
Phase I	280	3.03	1.20	1.94	2.99	4.20	0.990
Phase II	83	1.11	1.02	0.271	0.99	1.63	0.887
Phase III	250	1.31	0.753	0.566	1.08	2.91	0.826
Phase IV	10	0.0542	NA	NA	0.0542	NA	NA
Scenario B (exposure from 0 to 5 min, heavy exercise)							
Phase I	590	6.91	2.88	4.64	6.44	10.0	0.933
Phase II	220	2.99	2.45	0.611	2.88	4.96	0.964
Phase III	710	3.27	2.08	1.72	2.58	7.98	0.789
Phase IV	43	0.225	NA	NA	0.225	NA	NA
Scenario C (exposure from 0 to 15 min, heavy exercise, and 15 to 60 min, light exercise)							
Phase I	760	8.88	3.48	5.34	8.17	14.0	0.921
Phase II	330	4.33	3.43	1.43	3.53	11.5	0.815
Phase III	970	4.38	2.73	2.44	3.55	10.8	0.811
Phase IV	91	0.458	NA	NA	0.458	NA	NA
Scenario D (exposure from 0 to 15 min, heavy exercise, and 15 to 120 min, light exercise)							
Phase I	780	9.1	3.72	5.12	8.03	16.1	0.885
Phase II	380	4.79	3.43	1.83	4.02	11.9	0.839
Phase III	1000	4.60	2.80	2.70	3.66	11.2	0.794
Phase IV	110	0.556	NA	NA	0.556	NA	NA
Scenario E (5-15 min, heavy exercise)							
Phase I	160	2.15	1.51	0.815	1.54	4.77	0.717
Phase II	99	1.37	0.814	0.364	1.41	1.98	1.03
Phase III	200	0.87	0.616	0.344	0.672	2.32	0.770
Phase IV	27	0.140	NA	NA	0.140	NA	1.00

not a random sample of actual exposure events. The data given illustrate the range of intakes and doses to be expected as position and shot configuration vary for each scenario. The medians represent the best estimates of radiological doses and chemical concentrations for each of the phases.

5.2.2 Double Shots and Peak Uranium Concentrations in the Kidney

The peak kidney uranium concentrations from the double shots were analyzed to evaluate the dependence of the aerosol characteristics on successive shots, and to determine whether an approach could be developed from these data to predict aerosol concentrations and associated kidney uranium concentrations for multiple shots. The effect on peak kidney uranium concentration from the double shots, in which the same vehicle was hit twice within 15 min, was tested in both the Abrams tank (PI-3/4) and the Bradley vehicle (PII-1/2). For both vehicles, the shot lines were essentially replicates. Intuitively, one would expect that the second shot would double the dose; however, several confounders may affect the similarity in chemical and radiological doses. These include resuspension, unpredictable changes in particle sizes and solubility, or different ventilation rates due to damage from the first shot.

As with the radiological dose, the peak kidney uranium concentrations of the Phase-I and Phase-II double shots were compared for Scenarios A, B, and E, the scenarios for which data were available for both shots. Once again, arithmetic means and standard deviations were used for simplicity. Sampling ended at 10 min following the first shot, so Scenarios C and D were not calculated. Scenario E used an extrapolation based on three data points to estimate the DU concentration between 10 and 15 min.

Intakes and peak kidney uranium burdens for double shots were compared for significant differences using an unpaired *t*-test for normally distributed data and a Mann-Whitney Rank Sum Test for non-normally distributed data or data with unequal variances.

Table 5.6 shows the mean intakes for PI-3 and PI-4 ranged from 353 to 1000 mg and 272 to 858 mg, respectively, for the three most likely scenarios. Predicted mean peak kidney uranium burdens for PI-3 ranged from 3.64 to 11.4 μg U/g kidney. Predicted mean peak kidney uranium burdens for PI-4 ranged from 3.20 to 8.02 μg U/g kidney. There were no significant differences between PI-3 and PI-4 in the mean intakes or kidney uranium concentrations for any of the scenarios.

Table 5.6. Comparison of Intakes and Peak Kidney Uranium Burdens from the Abrams Tank Double-Shot Tests

Scenario	First Shot (PI-3)		Second Shot (PI-4)		First Shot (PI-3)		Second Shot (PI-4)	
	Intake (mg)	SD	Intake (mg)	SD	Peak (μg U/g kidney)	SD	Peak (μg U/g kidney)	SD
A	380	143	367	148	3.64	2.38	3.21	1.02
B	1000	323	858	323	11.4	4.97	8.02	2.00
E	353	49	272	125	5.69	0.143	3.20	1.07

The Bradley double-shot tests (PII-1 and PII-2) mean intakes and the mean kidney uranium burdens and standard deviations for Scenarios A, B, and E are shown in Table 5.7. Intakes ranged from 73 to 319 mg and 75 to 555 mg, respectively. Predicted kidney uranium burdens for PII-1 ranged from 0.83 to 7.65 μg U/g kidney, and PII-2 kidney uranium burdens ranged from 0.96 to 11.6 μg U/g kidney. The kidney uranium burdens from these shots were not significantly different.

As with the radiological doses, some of the first-shot concentrations in the Abrams BHT were actually higher than those measured for the second shot. For the Bradley vehicle double-shot test series, the first shot concentrations are intermingled with the second shot concentrations. The lowest concentrations for the Bradley vehicle were actually measured for the second shot of the double-shot series.

From the limited data presented in Table 5.6 and 5.7 in which the shots were between 10 and 15 min apart, it was concluded that the shots were essentially independent, i.e., the first shot did not add significantly to the concentration of the second shot. The peak kidney uranium concentrations from exposure to rapidly fired shots were not evaluated because the impact of the higher aerosol concentration on particle sizes and aerosol concentration as a function of time could not be estimated. It is believed that the higher concentrations would result in larger particles being formed and a rapid reduction in

Table 5.7. Comparison of Intakes and Peak Kidney Uranium Burdens from the Bradley Vehicle Double-Shot Tests

Scenario	PII-1		PII-2		PII-1		PII-2	
	Intake (mg)	SD	Intake (mg)	SD	Peak (μg U/g kidney)	SD	Peak (μg U/g kidney)	SD
A	98.6	19.4	128	83.5	1.32	0.41	1.76	1.36
B	296	58.2	317	179	4.18	1.14	4.36	2.88
E	151	23.1	126	26.0	2.23	0.452	1.65	0.292

concentration. These postulated mechanisms would tend to decrease aerosol concentrations and associated peak kidney uranium burdens.

5.3 DU MASS CONCENTRATION IN THE LUNG

The dosimetry of DU aerosols depositing in the respiratory tract (RT) has focused on the radiation dose received by that organ system, based on the likelihood that the prevalent risk to the lung would be due to that dose. It is useful, however, to calculate the mass concentration of DU within the lung and associated extrathoracic (ET) regions so that impressions of the mass loading during the first minutes after a DU penetration event can be obtained. For this purpose, the highest calculated intake was used (PIII-2, gunner position, Scenario B). For example, the intake calculated from this exposure was 879 nCi (final calculated value of 778 nCi, Appendix A, Table A.2), which corresponds to a DU mass of 2310 mg (final calculated value of 2040 mg) using the conversion of 0.38 nCi/mg DU. (The initial results were based on conventional statistics and a 2.1 Lpm sampler flow rate. The final calculations used Bayesian statistics and were based on the Capstone sampler mean of 2.6 Lpm. In most cases, the Bayesian statistics produced higher values. This was one of few cases in which the Bayesian result was notably lower, albeit still the highest.)

To calculate the deposition resulting from the 5-min exposure, the stage-specific intake data obtained from the BK model was multiplied by the particle-size-dependent deposition efficiencies for the nine RT compartments in the HRTM (ET_1-surface, ET_2-surface, ET_2-sequestered, BB-gel, BB-sol, BB-sequestered, bb-gel, bb-sol, bb-sequestered, AI_1, AI_2, AI_3). These deposited amounts of DU, in activity, were converted to masses in mg, and then summed over the four respiratory tract regions. The results are summarized in Table 5.8.

The amount of DU deposited in the RT depends on the particle size distribution for this shot and the heavy-exercise breathing rate being maintained for

Table 5.8. Maximum Deposition of DU Aerosols in the Respiratory Tract Regions (PIII-2, gunner's position)

RT Region	Deposition (nCi)	Deposition (mg)	Fraction RT Deposition (%)	Fraction Total Intake (%)
ET	408	1070	90.7	46.4
BB	17.6	46.2	3.9	2.0
bb	4.6	12.0	1.0	0.52
AI	19.3	50.8	4.3	2.2

5 min. Because of the predominance of large-sized aerosol particles during this period, the maximum deposition occurred in the ET regions of the RT, representing 91% of the total DU mass deposited. About 4% of the total deposition was calculated for the bronchial region (BB), 4% for the alveolar region (AI), and 1% for the bronchiolar region (bb).

According to the calculations, 5 min of breathing in an atmosphere that initially contained about 5000 mg/m^3 DU resulted in an alveolar deposition of about 51 mg DU. Based on an assumption that the DU particles were uniformly deposited throughout the 1100-g alveolar region, the highest alveolar concentration predicted from the Capstone data is 46 μg DU/g lung.

Because DU comprised about 50% of the total aerosol mass in this example calculation (Capstone range was 18% DU [minimum in Bradley] to 72% DU [maximum in Abrams]), a concentration of inhalable particles about twice the DU concentration was assumed for this purpose. The 5-min RT deposition would be about 2 g at the heavy-exercise-breathing rate. However, this amount of aerosol might induce the vagal-mediated laryngeal reflex (The Royal Society 2001), which would restrict breathing and consequently reduce intake and dose.

5.4 MODELING CONFOUNDERS

Modeling confounders include factors such as those that change the DU aerosol concentration and those that affect the intake of the aerosol. The calculated doses apply best to circumstances similar to the Capstone test circumstances in which stripped BHT target vehicles were used and no ventilation or respiratory protection was used. Reduction in the DU aerosol concentration in the crew compartment would be reduced if the ventilation systems were operating. Donning respiratory protection equipment would also reduce intake by the crew. Hatches that were open at the time of impact or that were forced open by the shot would probably increase the air exchange rate and reduce the aerosol concentration within the crew compartment(s), although to a lesser degree than the operation of ventilation systems. Aerosol deposition patterns within the respiratory tract would be affected by mouth breathing because the higher filtration efficiency of the nasal airways is lost.

5.4.1 Ventilation Effects on Doses and Concentrations

Aerosol concentration and particle size data for an Abrams tank that had an operating EC/NBC ventilation system were obtained in PIV-4. The EC/NBC ventilation system was operating at a nominal flow rate of 5600 Lpm (200 cfm) during and after perforation. Neither the internal air circulation system nor the secondary NBC system was operating during this test. The hatches were closed prior to perforation and remained closed after perforation. In this shot, the DU armor of an operational M1A2 Abrams tank was perforated by a LC-DU munition. The vehicle was uploaded with standard instrumentation, mannequins at each crew position, and an abundance of sampling equipment to meet the Live-Fire Test objectives. As explained in the Capstone report, Marple CIs were mounted on mannequins seated in the driver and loader positions. Because the Live Fire Test objectives precluded shielding the sampler control and vacuum lines, fragments severed the lines at the loader's position during the perforation event. The four samplers at the driver position were used (the fifth sampler was a field blank) to calculate the intakes and doses shown for Phase IV in Tables 5.1, 5.2, and 5.6 and summarized in Table 5.9.

Table 5.9. Phase-IV, Shot 4 Intakes, 50-Year Committed Doses (rem), and Peak Kidney Uranium Concentrations

Scenario	Intake (nCi)	Intake (mg)	E(50)	Lung $H_T(50)$	Bone Surface $H_T(50)$	Red Marrow $H_T(50)$	Kidney $H_T(50)$	Kidney Concentration at 24 h (µg U/g)
A	3.70	9.71	0.0900	0.662	0.023	0.002	0.008	0.0542
B	16.5	43.3	0.436	3.32	0.100	0.010	0.037	0.225
C	34.7	91.1	1.02	7.58	0.181	0.019	0.066	0.458
D[(a)]	43.1	113	1.20	8.65	0.210	0.022	0.077	0.556
E	10.4	27.3	0.412	3.14	0.052	0.005	0.019	0.140

(a) Extrapolated from sample collected between 25 and 27 min.

The E(50)s, which ranged from 0.0900 to 1.20 rem, were significantly lower than the E(50)s obtained with the Abrams BHTs (Phases I and III). The difference is due primarily to the significantly lower DU mass concentrations measured in the operational vehicle. In general, the DU concentrations in this Phase-IV shot were a factor of 5 to 24 less than those measured in the Phase-III shots, in which the penetrator-target interactions were comparable (i.e., DU on DU). The primary reason for this significant decrease in aerosol concentrations in Phase IV is due to the working EC/NBC ventilation system in the operational Abrams tank.

5.4.2 Effect of Mouth Breathing on Dose

The deposition pattern in the RT airways, and hence also the delivered dose, is sensitive to the breathing pattern of an individual. In addition to parameters such as breathing frequency, tidal volume, and airway size, the partitioning of airflow between the nasal and oral head airways can affect the deposition pattern within the body (ICRP [66] 1994). The nasal airways are an effective filter, particularly for particles larger than about 1 μm. By reducing deposition in the deep lung, where the contribution to dose is the highest, the E(50) is also effectively reduced. The dose results presented in this report use the default "nose-breather" parameters from the ICRP HRTM: during light exercise, the individual is assumed to be breathing entirely through the nose, and during heavy exercise, half of the air is taken in through the mouth. The actual fraction of air breathed through the nose can vary among individuals. For example, someone experiencing nasal congestion from a cold, allergies, or vasomotor rhinitis may breathe only a small fraction of air through the nose.

To evaluate the effect of breathing pattern on dose, E(50)s were calculated for all exposure scenarios assuming total mouth breathing during both light and heavy exercise. This represents an extreme case and, therefore, is an upper bound of the increase in dose caused by differences in breathing patterns. The distribution of the ratio of E(50)s for 100% mouth breathing compared with default nose breathing is shown in Figure 5.2. Doses were typically about 50% higher under the assumption of total mouth breathing. In a worked example,

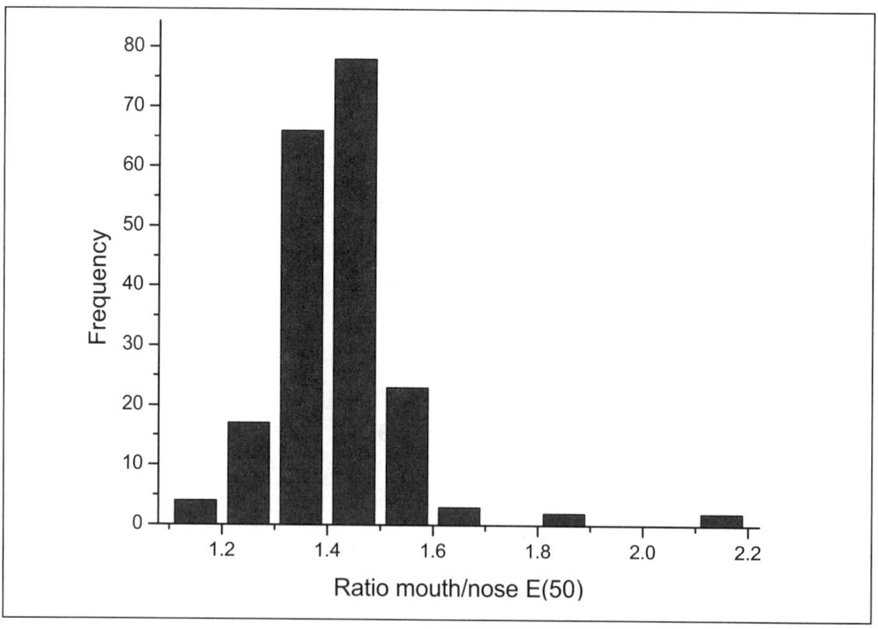

Figure 5.2. Ratios of E(50)s for Mouth Versus Nose Breathers

the highest dose used in this calculation was 24.8 rem, a preliminary E(50) for Scenario D, PIII-2, gunner position (final calculated value of 16.8 rem), compared with 16.9 rem E(50) for the same case assuming standard nose breathing. The exceptional cases shown in Figure 5.2, in which the ratios were greater than 1.8, are associated with samples taken in relation to a resuspension study (discussed in Parkhurst et al. 2004, Section 5.4) and with the PIV-4 sample. In each of these cases, there was a shift in the particle size distribution toward smaller particle sizes compared to the distributions found in the other shots. The smaller particle sizes amplify the differences between mouth and nose breathing because 1) the differences in deposition efficiencies for mouth and nose breathing at the smaller particle size are greater, and 2) the dose coefficients for the smaller particles are also greater than for larger particles. These latter cases are exceptional in some respects, but they illustrate one of the uncertainties associated with variable breathing patterns.

5.5 UNCERTAINTY IN DOSE CALCULATIONS

In addition to the previously mentioned confounders, two principal types of uncertainties are associated with the calculation of organ doses and committed effective doses for inhaled DU aerosols: 1) uncertainty due to variability in the measurement data, and 2) uncertainty in the biokinetic and dosimetric models used to calculate doses. These two types of uncertainties are discussed below.

5.5.1 Measurement Uncertainty

The uranium concentration values were derived from beta-radioactivity counting of the CI collection substrates. The QA applied to the interpretation of these original measurement data, and the propagated uncertainty calculations used for each measured value are described in detail in Appendix A, Section A.2 of the Capstone report. These uncertainty calculations took into account counting statistics, uncertainty in regression model parameters (used to compare DU concentrations based on beta counting with those from chemical analysis of the same samples), and uncertainty in the ingrowth correction applied to account for the state of disequilibrium of the beta-emitting progeny Th-234 and Pa-234m in the collected aerosols (Parkhurst et al. 2004, Section A.2.1.5).

Because there is a strong size dependence of collection efficiency (see Section 3.2.4), the CI data used in this report were modified to incorporate the stage-specific collection efficiencies. In the absence of measured values, the uncertainties in the stage-collection efficiencies were taken to be 10% of the values for the Marple CI published in the technical literature (Rubow et al.

1987). This uncertainty was then included, in quadrature, into the one-sigma error propagation equations used originally in the Capstone study. The adjusted equations are shown in Section 3.2.3.3.

The uncertainties in measured DU values are for the most part dependent on the particle size, with the larger particle sizes having somewhat larger relative standard deviations. However, there is considerable variability in standard deviations among the measured samples. For this dose assessment, a value for the relative standard deviation of each stage of 40% (from a random sample of about 30 of the Capstone CI values with their associated standard deviations) was taken to represent the mean standard deviation for the population of samples. This approach is likely to be conservative. The contribution to dose is also dependent on particle size (Figure 5.3). Comparing the dose coefficients for 0.1- and 20-µm AMAD particle sizes, the ratios vary by about a factor of 50 for Type M uranium and 25 for Type S uranium. Maximum dose coefficients are seen for the smaller particle sizes, within the range sampled by the Marple CI (0.1 to 50 µm AMAD). Therefore, if the largest standard deviations pertain to the largest particle sizes, their influence on dose variability will be dampened because of the smaller dose coefficients. Conversely,

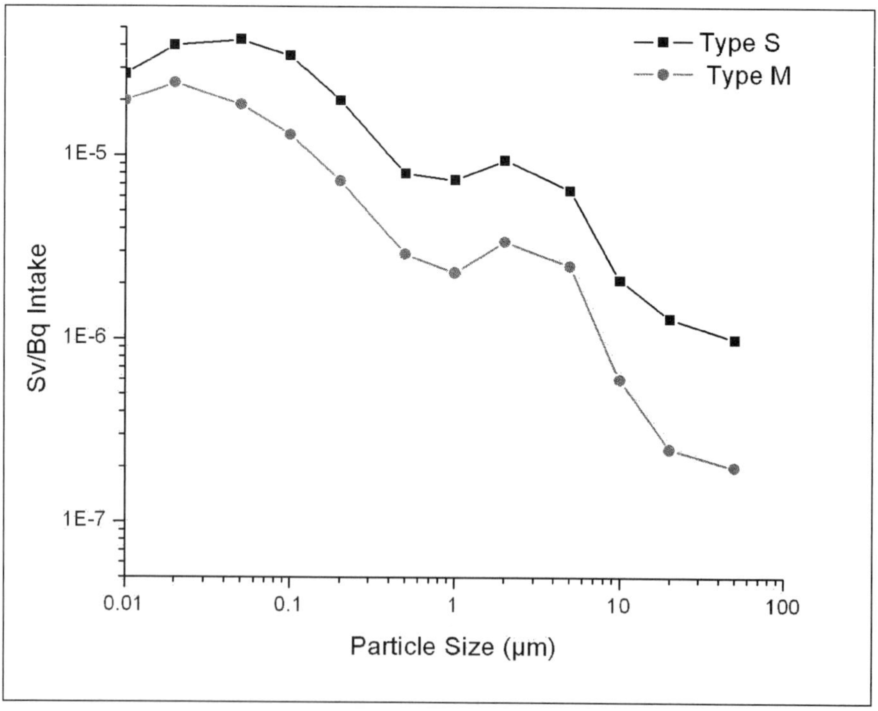

Figure 5.3. Dependence of Dose Coefficient on Particle Size (monodisperse, AMAD) and Solubility Classification

however, overestimating the standard deviations for the smaller particles is conservative, as most of the dose arises from these smaller particles. Typically the standard deviations for sizes <2μm were seen to be about 20%.

5.5.2 Modeling Uncertainty

The models used in this report, particularly the HRTM (ICRP [66] 1994) and the uranium systemic biokinetic model (ICRP 69) are complex mathematical systems, which are not amenable to simple uncertainty estimation. For example, 69 parameters have been associated with the HRTM, 26 of which are associated with particle deposition, 23 with particle clearance, and 20 with the dosimetry calculations. Although recent work by Farfan et al. (2003), Bolch et al. (2003), and Huston et al. (2003) provide comprehensive investigations of model uncertainties using classical statistics, ways to incorporate these new approaches into comprehensive uncertainty analyses, particularly in a Bayesian context, have not been developed. Therefore, the results of Bolch and his co-workers were recognized but were not incorporated into the present uncertainty analysis scheme.

5.5.2.1 Particle Deposition Uncertainty

Bolch et al. (2001) developed a computational code, the Lung Dose Uncertainty Calculation (LUDUC), that solves the ICRP 66 HRTM and provides multiple sets of values for all selected input variables using a Latin hypercube sampling of the assigned probability density functions for each input variable. The literature review of Huston (1995) was used to parameterize the probability density functions. Normal, lognormal, or uniform best fits were developed for the primary and secondary input parameters to the model. The primary deposition-related parameters used in the ICRP 66 model include AD; thermodynamic diameter; diffusion coefficient; volumetric flow rate; nasal volumetric flow rate; oral volumetric flow rate; tracheal, bronchiolar, and alveolar scaling factors; residence times in BB, bb, and AI regions; tidal volume; volumes of ET, BB, and bb regions; and functional residual capacity. Other parameters that were used in modeling particle deposition were age, body height, body mass, body mass index (BMI), ventilation rate, oxygen consumption rate, ventilatory equivalent ratio, basal metabolic rate, vital capacity, oxygen consumption factor, and basal multiplier. The latter set of parameters was used to develop the age- or size-dependent variabilities employed in the uncertainty analyses.

The deposition model parameters used in LUDUC and their associated distribution parameters are listed in Table 6 of Bolch et al. (2001). The product of this multifactorial uncertainty analysis is a set of particle-size-specific deposition fractions for the ET_1, ET_2, BB, bb, and AI regions of the HRTM. All of the distributions are lognormal with geometric means and standard deviations summarized for selected particle sizes in Table 7 of Bolch et al. (2001). For this analysis, the data from Table 7 were used to interpolate deposition fractions (geometric means and standard deviations) for particle sizes specific to the nine stages of the Marple CI used throughout the Capstone study. The means were interpolated as geometric means, whereas the geometric standard deviations were interpolated as arithmetic means. These deposition fractions are summarized in Table 5.10.

5.5.2.2 Particle Clearance Uncertainty

The HRTM treats the clearance of inhaled deposited particles as two competitive processes: 1) mechanical clearance of particles, either bare or inside lung phagocytes, via mucociliary clearance or translocation to interstitium and draining lymph nodes, and 2) *in vivo* dissolution of deposited particles followed by absorption to blood or binding to local tissue constituents within the respiratory tract. Bolch et al. (2003) addressed the uncertainties in particle clearance due to these competing mechanisms using methods similar to those used in their previous study of the uncertainties due to deposition (Bolch et al. 2001). Recognizing that the dissolution/absorption parameters were material-specific, Bolch and his co-workers chose to evaluate the model parameter uncertainties for plutonium and uranium oxides. Their results, which are presented as particle-size-specific integrated disintegrations within the ICRP HRTM compartments (ET, BB, bb, AI, LN), showed relatively weak particle size variations of the central values of integrated deposition for the range 0.1 to 5 μm. The ranges of uncertainties, expressed as the ratio of 95 to 5% confidence levels, were also reasonably stable for the same particle size range, with the ratio being between about 7 and 20 (i.e., an order of magnitude for uranium [Bolch et al. 2003, Figure 9]). The authors correctly point out that the calculated, propagated uncertainty values incorporate variability due to both region-specific and particle-size-specific factors. They also recognize that there are correlations among the variables that must be taken into account in future analyses. Thus, it is clear that these published studies point the way to future approaches to more comprehensive uncertainty analyses that will incorporate not only measurement uncertainties, but also uncertainties in both model structures (not yet addressed analytically for the RT) and model parameters. However, more research is required to determine the best ways to incorporate these new approaches into comprehensive uncertainty analyses, so a decision was made not to

Table 5.10. Deposition Fractions for Particle Sizes Specific to the Marple CI (derived from Bolch et al. [2001], Table 7)

RT Region	Marple Stage	Marple Size (μm)	$Mean_{geom}$	SD_{geom}
ET1	1	31	0.272	1.24
ET1	2	16	0.324	1.22
ET1	3	11	0.373	1.22
ET1	4	7	0.399	1.22
ET1	5	4	0.344	1.32
ET1	6	2	0.263	1.41
ET1	7	1	0.143	1.64
ET1	8	0.6	0.066	1.80
ET1	9	0.2	0.015	1.42
ET2	1	31	0.291	1.20
ET2	2	16	0.353	1.15
ET2	3	11	0.411	1.15
ET2	4	7	0.457	1.14
ET2	5	4	0.435	1.35
ET2	6	2	0.377	1.43
ET2	7	1	0.178	1.76
ET2	8	0.6	0.067	1.80
ET2	9	0.2	0.013	1.29
BB	1	31	0.001	2.21
BB	2	16	0.004	2.20
BB	3	11	0.009	2.20
BB	4	7	0.020	2.14
BB	5	4	0.025	2.04
BB	6	2	0.024	1.74
BB	7	1	0.010	1.71
BB	8	0.6	0.005	1.47
BB	9	0.2	0.004	1.38
bb	1	31	NA[(a)]	NA[(a)]
bb	2	16	0.000	5.59
bb	3	11	0.001	5.59
bb	4	7	0.004	3.00
bb	5	4	0.012	2.00
bb	6	2	0.016	1.73
bb	7	1	0.008	1.46
bb	8	0.6	0.017	1.33
bb	9	0.2	0.022	1.31
AI	1	31	NA[(a)]	NA[(a)]
AI	2	16	0.001	5.31
AI	3	11	0.002	5.31
AI	4	7	0.005	1.95
AI	5	4	0.042	2.20
AI	6	2	0.117	1.61
AI	7	1	0.110	1.41
AI	8	0.6	0.092	1.32
AI	9	0.2	0.102	1.31

(a) Deposition fractions too small to calculate a central estimate.

incorporate the results of Bolch and his co-workers into the present uncertainty analysis scheme.

5.5.2.3 Uncertainty Due to Solubility

As pointed out in Bolch et al. (2003), the uncertainties associated with particle clearance are a combination of the uncertainties due to particle-size-specific deposition, mechanical clearance, and solubility *in vivo*. The final uncertainty is clearly dependent on the physical and chemical properties of the inhaled aerosol. ICRP recommends that, when material-specific knowledge is available, it should be preferred over the default absorption Type F, M, S parameter values.

In the Capstone study, a total of 19 particle-size-segregated samples from four different shots from the first three test phases were analyzed using *in vitro* dissolution methods. To evaluate the uncertainty associated with solubility for the heterogeneous aerosols produced as a result of DU penetrator interaction with armor, the experimentally determined *in vitro* dissolution results obtained from all 19 samples were used to test the influence of solubility on E(50). In a worked example, a standardized scenario was used for all dissolution parameter sets. This scenario used the parameters and preliminary aerosol data associated with the largest intake, which occurred in Phase III, Shot 2, Scenario B, gunner position. To amplify the effect of the different *in vitro* dissolution patterns, a single set of dissolution parameters from the PIII-2 gunner's position was used for all nine of the particle size fractions. This approach is conservative in terms of estimating the uncertainty because incorporation of many of the various dissolution behaviors for the different CI stages would tend to average out the variability for a given calculation.

The E(50)s, calculated as described above, are summarized graphically in Figure 5.4. Summary statistics on these E(50)s as well as those for the H_Ts from selected tissues are shown in Table 5.11. The data demonstrate that differences in solubility of the various DU aerosols impact both the E(50)s and organ committed equivalent doses (H_T)s. The relative standard deviations (coefficient of variation: CV) range from about 29% for the kidney to 51% for bone surfaces. The maximum kidney concentration showed the largest variation, 69%. The variability in CVs is to be expected because the relative amounts of DU in the various tissues are different depending on the intrinsic solubility of a particular aerosol. The mean variation in $H_T(50)$ for the three listed tissues is 39%, which is similar to the variation in E(50), 36%. It would

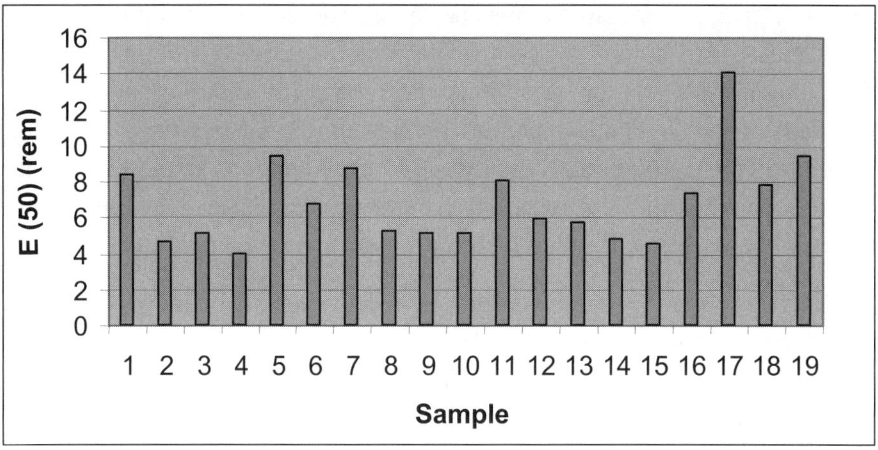

Figure 5.4. Notional E(50)s for Different Solubilities Obtained Using the *In Vitro* Solubility Curve Fits

Table 5.11. Descriptive Statistics for the Influence of Solubility Parameters on E(50), (H_T50), and Peak Kidney Uranium Concentration

Type of Dose	Mean	Median	Std Dev	Coeff Var (%)	Range	Minimum	Maximum
E(50)(rem)	6.95	6.08	2.46	35.5	10.1	4.00	14.1
Lung H_T(50) (rem)	104	84.1	53.1	51.1	210	40.1	250
Bone Surface H_T(50) (rem)	13.1	12.9	5.05	38.4	19.1	7.92	27.0
Kidney H_T(50) (rem)	1.12	1.07	0.32	28.7	1.28	0.50	1.78
Max Kidney (μg U/g kidney)	22.1	17.8	15.2	68.7	65.3	7.70	73.0

therefore appear reasonable that 36% is a representative CV for uncertainty due to differences in solubility *in vivo*.

A sensitivity analysis of the influence of solubility on dose was performed using all modeled variables. The test case was applied using exposure parameters from PIII-2-gunner and modeling a single scenario—Scenario B (Figure 5.4). Doses shown are notional, not actual doses. For doses used in this evaluation, see Appendix A.

5.5.2.4 Summed Bayesian Posterior Distributions and Percentiles

As discussed in Section 4.4.1, measurement uncertainties were evaluated in terms of the likelihood function using Bayesian statistics. Posterior distributions were calculated using the MCMC. An example of one of the distributions generated from this calculation is shown in Figure 4.1. It represents the probability distribution of E(50) for a particular shot, position, and intake scenario, given the measurements and the measurement uncertainties. Then, all of these distributions for a particular phase (vehicle type) are summed. The summed distribution represents the probability distribution of E(50) when all of the shots and positions making up the dataset for a particular phase are considered equally likely. In other words, this is the probability distribution when the type of vehicle and the intake scenario are known, but the exact geometry of the shot or subject placement are not known.

The Bayesian analysis calculates the dose for a single shot and position as a distribution in contrast to the single numbers usually calculated (where the true values of all measured quantities are simply taken to be equal to measured values). The arithmetic means and standard deviations of the dataset for a particular phase calculated from single-number doses are close to those calculated using the summed MCMC distributions. This is illustrated in Table 5.12, which shows the means and standard deviations of the distributions shown in Figures 5.5 through 5.8, as well as the same quantities calculated using measured, single-number dose results. In this table, the "Number Shots/Positions" heading refers to the total number of sampling arrays on which the data are based. Table 5.12 shows that shot/position variation are much greater than measurement uncertainty.

Rather than considering means and standard deviations for these distributions, which are not normal or any other standard distribution, it is more useful to consider percentiles, which are derived from cumulative probability distributions. The cumulative probability distributions allow the reading off of percentiles. For example, 10th and 90th percentiles are obtained by projecting horizontal lines at heights of 0.1 and 0.9 to the cumulative probability curve and then projecting vertically down to find the E(50) intersection points. The 10th percentile is the E(50) point for which the probability is 10%

Table 5.12. Means and Standard Deviations Calculated from the Summed Bayesian Posterior Distributions (Scenario B)

Phase	Number Shots/Positions	Parameter (Units)	Mean ± SD using MCMC Distribution	Mean ± SD using Measured Values
I	22	E(50) (rem)	4.42 ± 2.10	4.45 ± 2.13
I	22	Kidney (μg U/g)	6.91 ± 2.88	5.96 ± 2.49
II	11	E(50) (rem)	1.81 ± 1.53	1.81 ± 1.56
III	8	E(50) (rem)	6.71 ± 2.62	6.87 ± 2.87
IV	1	E(50) (rem)	0.436	0.384

that the true result will be less than this value, and similarly for the 90th percentile. The 90th percentile is the E(50) point for which the probability is 10% that the true result is greater than this value. The 50th percentile, or median, is the E(50) point for which it is equally likely that the true result is either above or below. Because it is permissible to neglect measurement uncertainty since the shot/position uncertainty is much greater, percentiles can be calculated using the distribution of single-number results from different shots and positions.

The median was selected as the best estimate of central tendency for both the radiological doses and the peak kidney uranium concentrations. The 10th and the 90th percentiles are used to provide an indication of the spread of the data.

A worked example illustrates the variation of the data. Figures 5.5 through 5.9 each consist of two graphs for each of the four phases. The upper graph illustrates the cumulative probability distributions from which the 10th, 50th, and 90th percentiles are derived. The lower part of these figures show the summed Bayesian posterior distributions of E(50) for Scenario B. The graphs are based on *preliminary calculations* in which a sampler flow rate of 2.1 Lpm was used and the doses were based solely on U-238. The associated doses that are listed in Tables 5.13 through 5.15 are *final data calculations*, including the contributions from all uranium isotopes and their progeny, and are based on a sampler flow rate of 2.6 Lpm. As a result, the data in the table are slightly different from the corresponding doses in the graphs.

Table 5.13 lists the dose associated with each of the positions and the seven Phase-I shots that correspond (numbers do not match exactly) to the "peaks" in Figure 5.5. For purposes of discussion, the dose distribution was divided into five regions based on doses. The dose regions in the graph were assigned as follows:

Region 1 for doses less than 1 rem
Region 2 from 1 rem to 2 rem
Region 3 from 2 to 4 rem
Region 4 from 4 rem to 7 rem
Region 5 from 7 rem to 12 rem.
Regions 1, 2, and 5 are from single data points.

The doses ranged from 0.934 to 10.8 rem. The driver's position in PI-1 had the lowest dose; the next lowest was a factor of two higher. In the PI-3 test, the gunner's position received the highest dose, which was a factor of approximately 1.5 higher than the next lowest dose. The remainder of the doses fell in a broad range between these two peaks.

Table 5.13. Phase-I Doses and Assigned Regions by Shot and Position

Phase	Shot	Position	E(50) (rem)	Region
I	1	driver	0.934	1
I	6	driver	2.04	2
I	2	loader	3.03	3
I	2	driver	3.04	3
I	2	gunner	3.22	3
I	2	commander	3.33	3
I	5	loader	3.34	3
I	1	commander	3.38	3
I	1	gunner	3.45	3
I	4	driver	3.54	3
I	7	commander	3.64	3
I	1	loader	3.69	3
I	5	commander	4.22	3
I	5	driver	4.33	3
I	6	commander	4.53	4
I	7	gunner	4.77	4
I	4	loader	5.54	4
I	4	commander	5.76	4
I	4	gunner	6.33	4
I	3	driver	7.14	4
I	7	driver	7.24	4
I	3	gunner	10.8	5

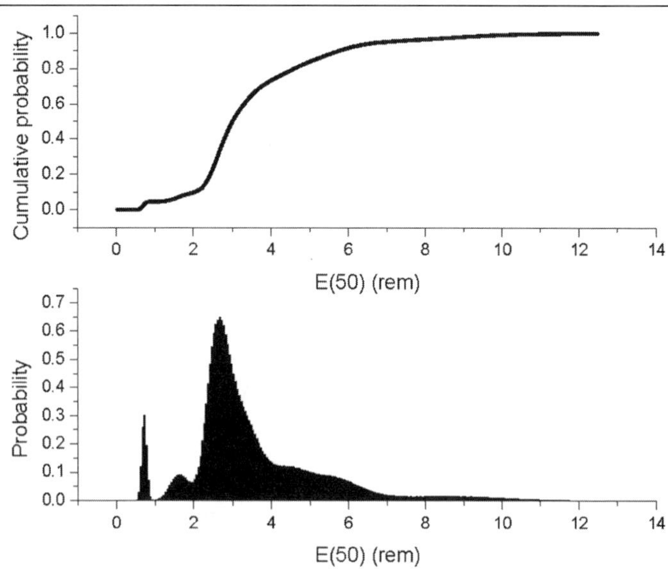

Figure 5.5. Dose Distribution (cumulative and differential) for Phase I (Abrams BHT through non-DU armor, Scenario B, no ventilation)

Table 5.14 lists the doses by positions for the three Phase-II shots corresponding to the results (numbers do not match exactly) depicted in Figure 5.6. For purposes of discussion, the dose distribution was divided into four regions. The dose regions were assigned as follows:

Region 1 for doses less than 0.5 rem

Region 2 from 0.5 to 2 rem

Region 3 from 2 to 3 rem

Region 4 from 3 rem to 6 rem.

Table 5.14. Phase-II Doses and Assigned Regions by Shot and Position

Phase	Shot	Position	E(50) (rem)	Region
II	3	driver	0.317	1
II	3	commander	0.335	1
II	3	gunner	0.403	1
II	3	loader	0.497	1
II	2	gunner	1.56	2
II	2	loader	1.73	2
II	1	loader	1.76	2
II	2	driver	2.02	2
II	1	commander	2.84	3
II	1	driver	3.09	3
II	2	commander	5.32	4

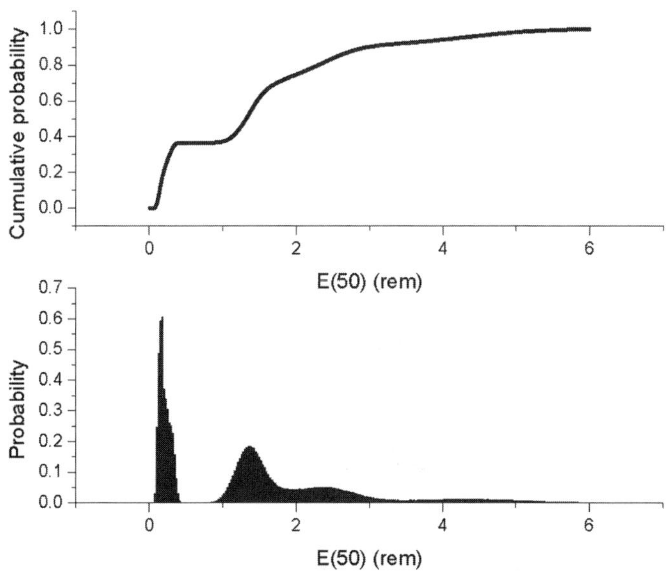

Figure 5.6. Dose Distribution (cumulative and differential) for Phase II (Bradley BHT, Scenario B, no ventilation)

Doses ranged from 0.317 to 5.32 rem. Shot 3 accounts for all of Region 1. The commander's position on PII-2 is the only value in Region 4, and it is a factor of 1.7 higher than the next lowest dose. For the remainder of the exposures, the doses range from 1.56 to 3.09 rem.

Table 5.15 lists the doses by positions for the two Phase-III shots corresponding to the results (numbers do not match exactly) depicted in Figure 5.7. For purposes of discussion, the dose distribution was divided into two regions. The dose regions were assigned as follows:

Region 1 for doses less than 7 rem

Region 2 from 7 rem and higher.

Table 5.15. Phase-III Doses and Assigned Regions by Shot and Position

Phase	Shot	Position	E(50) (rem)	Region
III	2	loader	4.52	1
III	1	loader	4.93	1
III	2	commander	7.88	1
III	2	driver	5.41	1
III	1	gunner	6.15	1
III	1	driver	5.85	1
III	1	commander	6.26	1
III	2	gunner	12.7	2

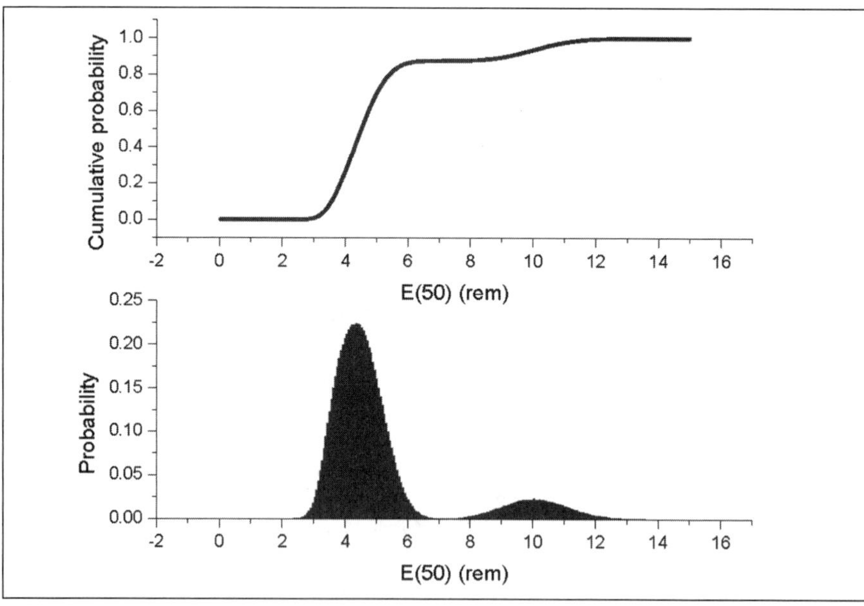

Figure 5.7. Dose Distribution (cumulative and differential) for Phase III (Abrams BHT through DU armor, Scenario B, no ventilation)

The second region is entirely due to the PIII-2, gunner's position dose. With the exception of the PIII-2, gunner's position, the results from both shots at all positions exhibit little variation, ranging from 4.52 to 6.26 rem. The result of 12.7 rem at the PIII-2 gunner's position was approximately a factor of two higher than the second highest result, apparently due, at least in part, to a smaller aerosol size distribution and its associated higher dose. This result is especially interesting in that PIII-1 aerosols were created about the same time as a hydraulic fluid flash fire occurred and coated the exposed vehicle surfaces (and the particles) with hydraulic fluid.

Figure 5.8 is based on the single DU munition fired at the operating Abrams tank equipped with DU armor and in which the EC/NBC ventilation system was operating.

Figure 5.9 is the same type of plot showing the distribution of the maximum (1-d) kidney uranium concentration rather than E(50), for Phase I (Abrams), Scenario B.

5.6 SUMMARY

The intakes of DU aerosols and their corresponding radiological doses and organ concentrations were calculated for five exposure scenarios and four

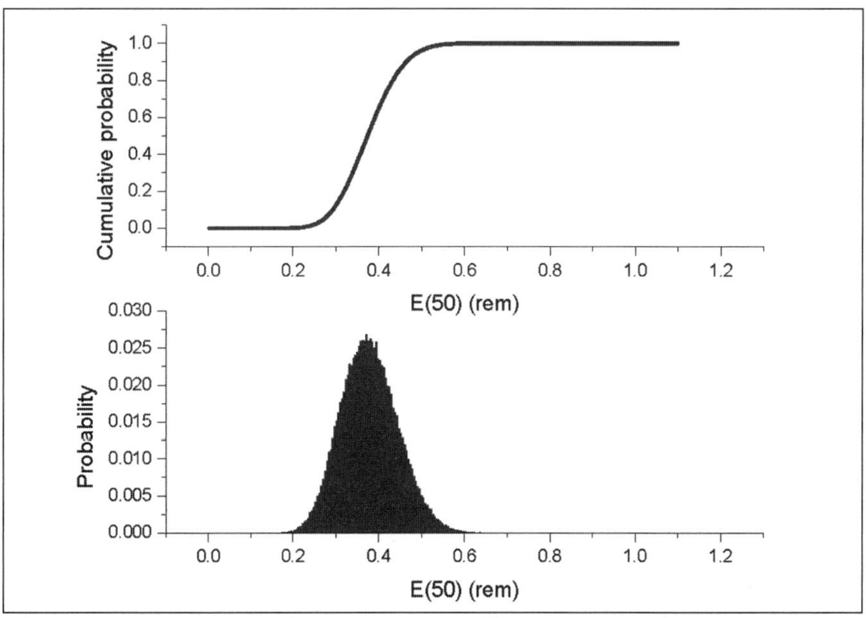

Figure 5.8. Dose Distribution (cumulative and differential) for Phase IV (Abrams tank through DU armor, Scenario B with ventilation)

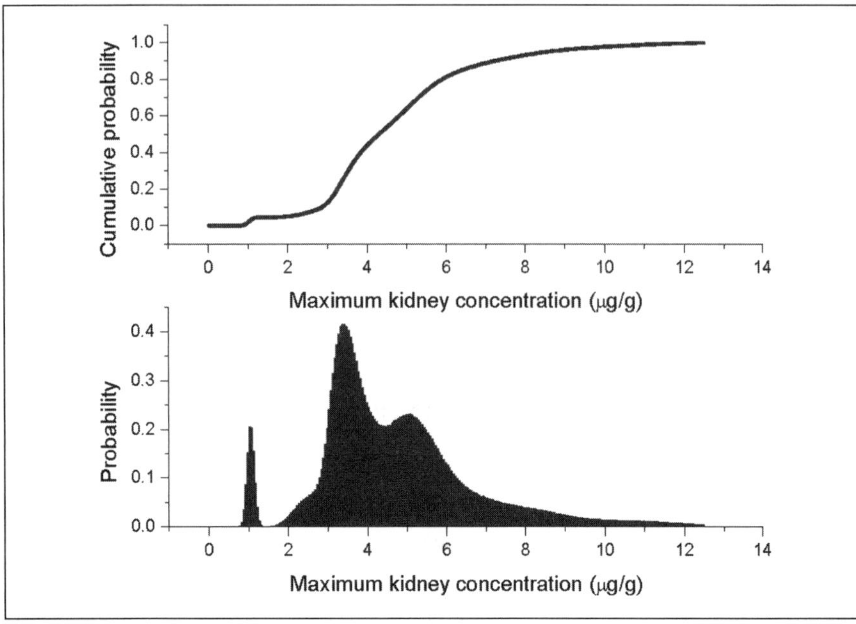

Figure 5.9. Distribution (cumulative and differential) for the Maximum (1-d) Kidney Uranium Concentration for Phase I (Abrams BHT non-DU armor)

phases representing vehicles perforated by DU munitions during combat. These vehicles included an Abram BHT perforated through standard non-DU armor with no vehicle ventilation (Phase I), a Bradley BHT with no vehicle ventilation (Phase II), an Abrams BHT perforated through DU armor with no vehicle ventilation (Phase III), and an Abrams tank perforated through DU armor with the EC/NBC overpressure system operating (Phase IV). The result of each data sampling point for each shot line was treated as an independent data point in the calculation of a single best estimate of the radiological dose and DU concentrations to the crewmember for each phase and scenario.

The median was selected as the best estimate of central tendency for both the radiological doses and the peak kidney uranium concentrations. The 10th and the 90th percentiles are used to provide an indication of the spread of the data.

An analysis of the DU aerosols from the double-shot data, in which the shots were fired about 15 min apart, showed that the results of the second shot were not dependent upon the first shot. In both cases tested, DU concentrations and doses from the first and second shots were similar. Neither residual aerosol nor resuspension of previously deposited aerosol significantly affected the concentrations and doses of the second shot. As a result, in the evaluation in this report of projected risk, double-shot effects are estimated by doubling the single-shot estimate.

Modeling confounders related to vehicle ventilation and dominant route of personnel breathing (nose versus mouth) were reviewed. Increased ventilation significantly reduced the DU aerosol's concentrations as well as the corresponding doses. Doses associated with oral breathing were generally 50% higher than for normally nose-breathing individuals. Modeling uncertainties were evaluated, and led to a new approach for evaluating doses based on particle sizes rather than a measure of the distribution. Bayesian statistics were used to generate probability distributions of doses and organ concentrations for all phases using Scenario B.

5.7 REFERENCES

Bolch WE, Farfan EB, Huh C, Huston TE. 2001. "Influences of Parameter Uncertainties Within the ICRP 66 Respiratory Tract Model: Particle Deposition." *Health Physics* 81:378-394.

Bolch WE, Huston TE, Farfan EB, Vernetson WG. 2003. "Influences of Parameter Uncertainties Within the ICRP 66 Respiratory Tract Model: Particle Clearance." *Health Physics* 84:421-435.

Farfan EB, Huston TE, Bolche E, Vernetson WG, and Bolch WE. 2003. "Influences of Parameter Uncertainties Within the ICRP-66 Respiratory Tract Model: Regional Tissue Doses for $^{239}PuO_2$ and $^{238}UO_2/^{238}U_3O_8$." *Health Physics* 84:436-450.

Huston TE. 1995. *Quantifying Uncertainties In Lung Dosimetry with Application to Plutonium Oxide Aerosols.* Ph.D. Dissertation, University of Florida, Gainesville, Florida.

Huston TE, Farfan EB, Bolche E, and Bolch WE. 2003. "Influences of Parameter Uncertainties Within the ICRP-66 Respiratory Tract Model: A Parameter Sensitivity Analysis." *Health Physics* 85:553-566.

International Commission on Radiological Protection (ICRP). 1994. *Human Respiratory Tract Model for Radiological Protection.* ICRP Publication 66, Pergamon Press, Oxford, United Kingdom.

Parkhurst MA, F Szrom, RA Guilmette, TD Holmes, YS Cheng, JL Kenoyer, JW Collins, TE Sanderson, RW Fliszar, K Gold, JC Beckman, and JA Long. 2004. *Capstone Depleted Uranium Aerosols: Generation and Characterization, Volumes 1 and 2.* PNNL-14168, Prepared for the US Army by Pacific Northwest National Laboratory, Richland, Washington.

Rubow KL, Marple VA, Olin J, McCawley MA. 1987. "Operator Manual Series 290 Marple Personal Cascade Impactors, Thermo Anderson; A personal Cascade Impactor: Design, Evaluation and Calibration." *American Industrial Hygienists Journal* 48(6):532-538.

The Royal Society. 2001. *The Health Hazards of Depleted Uranium Munitions Part I.* Policy Document 6/01, London, United Kingdom. Online report available at www.royalsoc.ac.uk in the Science Policy section.

6.0 Human Health Risk Assessment

This chapter reviews the chemical toxicity and radiological risks associated with uranium beginning with human exposure data presented in the literature. Organ concentrations and radiological doses associated with the Capstone Level I scenarios (from Chapter 5) are used to estimate the human health risks from inhaled depleted uranium. The health risks are low and are most closely associated with the risk of kidney injury and lung cancer.

6.1 BACKGROUND

The chemical toxicity of uranium compounds is well known when compared to the toxicity of most other compounds. In 1824, a treatise described uranium salts as "feeble poisons" when given by mouth to animals (cited in Hodge et al. 1973). In the late 1800s, uranium salts were studied as homeopathic therapeutic agents in humans, primarily for diabetes. In the early 1900s, the renal toxicity of uranium became apparent in humans, and its use as a therapeutic agent ceased.

Radiological toxicity studies were initiated during World War II as nuclear weapons were developed. After the war, the U.S. Atomic Energy Commission supported continuing studies of the toxicity of uranium and other compounds related to nuclear weapons. The results of these toxicity studies are documented in volumes of the National Nuclear Energy Series (Tanenbaum 1951; Voegtlin and Hodge 1949, 1953). A comprehensive review of the hazards of uranium and protection criteria was published in 1973 (Hodge et al. 1973). More recently, reviews of the toxicity (ATSDR 1999) and carcinogenicity

(BEIR IV 1988) of uranium have been published, as well as several reviews of the toxicity of DU used in munitions (Harley et al. 1999; Fulco et al. 2000; WHO 2001; The Royal Society 2001, 2002).

This chapter summarizes the radiological and chemical data that are available concerning the radiological and chemical health effects of uranium. The chapter presents human data and relies on animal data to support and to fill gaps in the human data. The analysis reviews both the deterministic and stochastic effects of uranium. The radioactivity of DU is so low that radiation-induced deterministic effects are not an issue when it is either inhaled or ingested. However, the chemical effects of inhaled or ingested DU are of concern, so our analysis will focus on using available human data to define chemical doses that result in clinical endpoints that will either affect an individual's ability to function, result in permanent injury, or require treatment.

Occupational safety levels for uranium are based on the kidney as the target organ. These safety levels have been defined and are set low enough to ensure that morbidity and mortality do not occur for acute, one-time exposures and for chronic, protracted exposures. The current standards are the subject of national and international debate, but this debate is not centered on the levels required to cause overt clinical or performance effects. Rather, it is based on the levels at which there are measurable changes in human kidney biomarkers or renal damage in animal models. These levels will be used as reference points for occupational follow-up recommendations.

6.2 URANIUM CONCENTRATIONS IN ORGANS AND TISSUES

To cause toxic effects, uranium, like other heavy metals, must be present at a certain level. The concentrations of uranium or any toxic substance present in various organs and tissues after deposition are not constant; rather, they change over time in accordance with varying clearance rates. The ultimate toxicity is determined by the response of these organs and tissues to the deposited uranium.

6.2.1 Biokinetics of Uranium

Inhalation and ingestion are the primary routes for uranium to enter the body. After absorption from the intestines or the lung, uranium enters the blood and is deposited in the soft tissues and bone or is excreted by the kidney. More than 90% of the uranium remaining in systemic tissues and organs (but not including sites of deposition like the lung, skin, GI tract, wound sites) at one day is excreted with half-times ranging from 2 to 6 days and the remainder with half-times ranging from 30 to 340 days (Durbin 1984). After a few days, most of the remaining uranium in the body is found in the kidney and skeleton (ICRP [69] 1995a). A substantial fraction of the metal ions filtered by the kidney is retained in the renal tubules before it is passed into the urinary bladder. It is during this retention that damage and killing of the renal tubular cells may occur at high renal concentrations, and this damage may result in illness or death.

Uranium deposited in the skeleton shows chemical similarities to calcium and deposits on all bone surfaces, especially in areas of active bone growth or remodeling. With time, there is a slow redistribution throughout the volume of the bone. Data for uranium in humans suggest that 80 to 90% of a skeletal deposit of injected uranium is removed from the body within about 1.5 years (Luessenhop et al. 1958).

The biokinetics of inhaled DU have been computed using the previously described ICRP models (Chapter 4) and data from the Capstone study. The time-dependent results for selected organs are shown in Figures 6.1 to 6.4 for a 5-min intake scenario (from the highest calculated intake, which occurred with Phase III, Shot 2, gunner's position, Scenario B). The retention and distribution of DU *in vivo* along with its excretion patterns (Figures 6.5 and 6.6) relate to the initial deposition patterns in the RT and the *in vivo* solubility of the inhaled DU aerosol. For the example selected, the particle size distribution consisted primarily of large particles that deposit preferentially in the upper RT, especially the ET airways (see Section 5.3). About 2% of the intake was deposited in the alveolar region of the lung. Therefore, the initial rapid loss of DU from the total body during the first week after exposure is due to mechanical clearance of particles deposited in the upper RT and conducting airways, and consequent excretion in feces (Figure 6.6). During the same time period, DU is also excreted in the urine, albeit at about 0.1 the amount found in feces. This urinary excretion accounts for an approximately 60% fraction of DU that reaches the blood during the first few days. The remainder of the systemic burden of DU is distributed mainly between the mineralized bone compartments and kidney. After about 10 days, the total body retention is due principally to the longer-term retention of DU in the lung, and this pattern continues for essentially the remainder of the calculated period. Both urine and fecal excretion rates continue to decrease monotonically after the first 2 days, due mainly to decreased rates of dissolution for the fraction excreted in urine and mechanical clearance from the RT for the fraction excreted in feces.

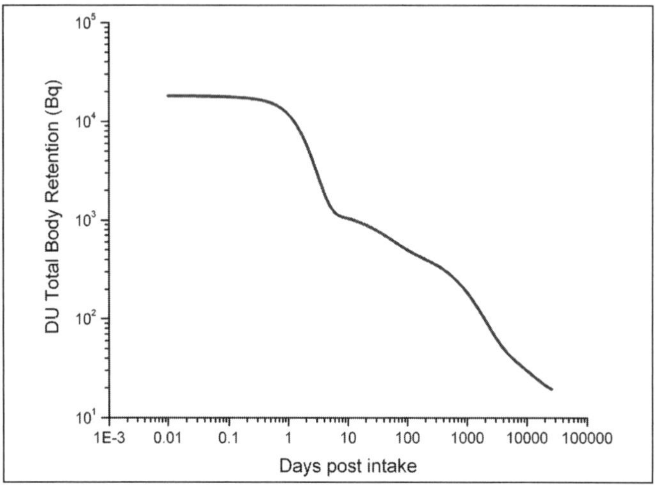

Figure 6.1. Total Body Retention of DU After a 5-Min Exposure

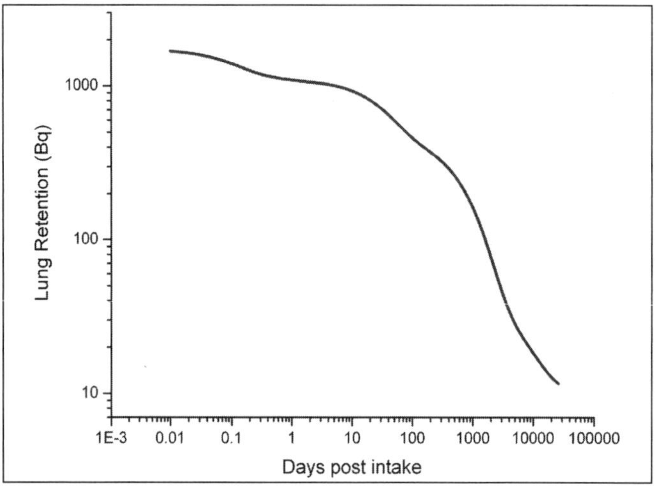

Figure 6.2. Retention of DU in Lung After a 5-Min Exposure

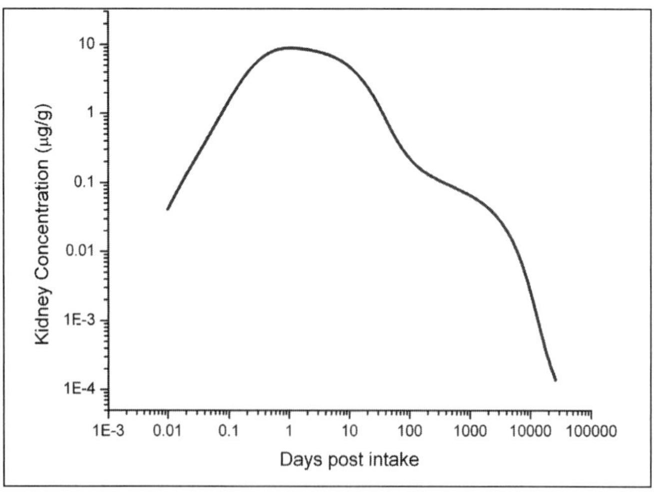

Figure 6.3. Retention of DU in Kidney After a 5-Min Exposure

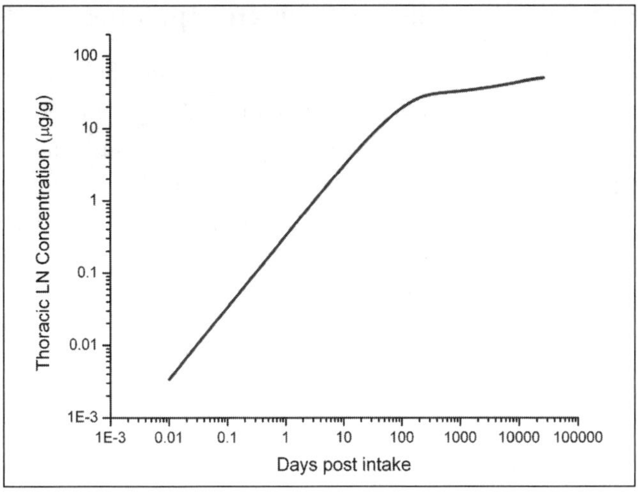

Figure 6.4. Retention of DU in Thoracic Lymph Nodes After a 5-Min Exposure

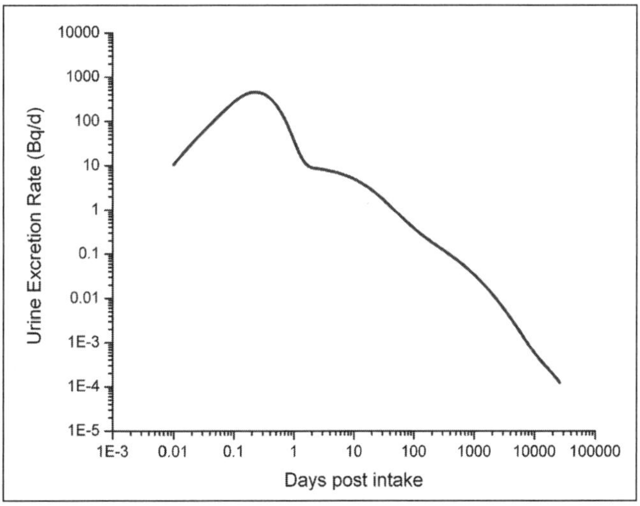

Figure 6.5. Urinary Excretion Rates of DU Resulting from a 5-Min Exposure

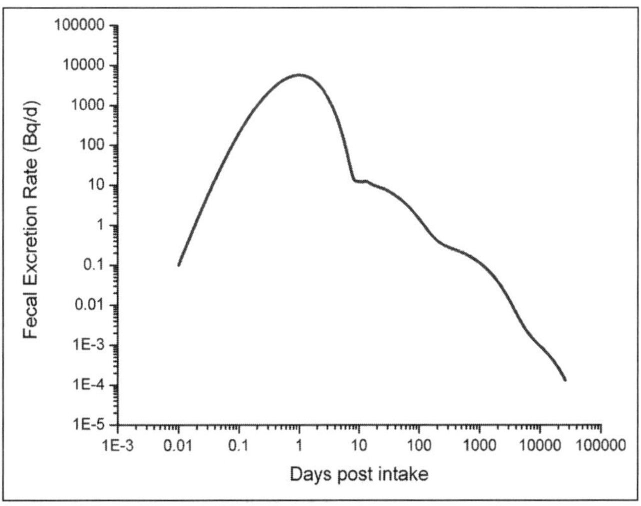

Figure 6.6. Fecal Excretion Rates of DU Resulting from a 5-Min Exposure

6.2.2 Organ Distribution of Uranium in Human Populations

In a number of studies, the content and distribution of uranium have been measured in human organs and tissues from inhalation and ingestion from natural sources. Harley et al. (1999) summarized the published data from 17 studies around the world. The total mass of uranium in the human body ranged from 2 to 62 μg, based on postmortem studies (BEIR IV 1988). The distribution of uranium in key organs of normal individuals exposed to normal environmental levels is shown in Table 6.1.

These individuals were exposed to uranium primarily through food and water. Small differences were observed in uranium concentrations among the soft tissues and bone. The concentrations show equilibrium of uranium concentrations in various organs, most likely the result of a relatively low intake over a long period of time.

For comparison, Table 6.2 shows the uranium content in the organs of workers occupationally exposed to relatively high backgrounds levels of uranium and dying of natural causes. All the workers were healthy and were not involved in accidental high exposures.

Of interest is the high concentration of uranium in the lung and tracheobronchial lymph nodes of workers in the uranium milling and processing industries and, to some extent, in the uranium mining industry. These high concentrations are probably caused by chronic exposure to elevated concentrations of uranium in the air. The kidney uranium concentrations were also higher in all the occupationally exposed individuals (Table 6.2) compared with unexposed individuals (Table 6.1). Because uranium has a relatively short half-time in the kidney, this may reflect continual exposure from the

Table 6.1. Uranium in Normal Humans—Environmental Exposures (based on Harley et al. 1999)

Organ/Tissue	Number of Studies	Total Mass μg U/kg[(a,b)]	Mass per Organ μg U[(c)]
Skeleton	26	2.4 2.3	25.2
Lung	7	2.5 2.5	3.0
Kidney	9	2.3 2.3	0.70
Liver	8	2.3 4.0	4.1
Muscle	4	3.7 4.0	107.3
Fat	2	1.3 0.9	19

(a) Wet weight.
(b) Mean ± standard deviation.
(c) Organ weights from ICRP 89 (2003): skeleton=10.5 kg, lung=1.2 kg, =0.31 kg, liver=1.8 kg, muscle=29 kg, fat=14.6 kg.

Table 6.2. Uranium in Normal Humans—Occupational Exposures

Organ/Tissue	Miners[a] (mean of 3)		Millers[a] (mean of 2)		Processors[b] (1)		Shop Workers[c] (1)	
	µg U/ kg[d]	µg U/ organ[e]	µg U/ kg[d]	µg U/ organ[e]	µg U/ kg[d]	µg U/ organ[e]	µg U/ kg[d]	µg U/ organ [e]
Skeleton	4.9	50	10.8	113	485	5,090	90	945
Lung	31.8	3.8	326	390	1,020	1,220	1,200	1,440
Kidney	6.1	1.9	5.5	1.7	162	50	120	37
Liver	3.4	6.1	4.3	7.7	93	170	30	54
Gonads	2.9	0.10	4.8	0.17	--	--	--	--
Lymph nodes[f]	--[g]	--	3,600	54	--	--	1,800	27

(a) Singh et al. 1987.
(b) Kathren et al. 1989.
(c) Donoghue et al. 1972.
(d) Wet weight.
(e) Organ weights from ICRP 89 (2003) skeleton=10.5 kg, lung=1.2 kg, kidney=0.31 kg, liver=1.8 kg, testes=0.035 kg, lymph nodes=0.015 kg.
(f) Lymph nodes = tracheobronchial.
(g) Hyphens indicate no data.

high lung burdens or from the occupational environment. The concentration of uranium in the skeleton is also higher in the occupationally exposed individuals, and is dramatically higher in uranium process workers. Although the skeletal concentration is not dramatically higher in most cases, the skeleton, because of its mass, holds a greater amount of uranium compared to other organs. Concentrations in the muscle of these individuals were not reported.

6.3 CHEMICAL TOXICITY OF URANIUM

Uranium has been demonstrated to affect several organ systems following exposure to sufficiently high concentrations; the most important of these organs are the kidney and the lung (ATSDR 1999). The dose, or concentration of uranium in the tissues, is the primary factor in the chemical toxicity of uranium. Which organs and tissues are affected is determined by the route of exposure and the solubility of the chemical form of uranium compound. Inhalation is the most common route of occupational exposure to uranium. Compounds can be categorized based on how rapidly they translocate from the lung to the blood and then to other organs. Type F compounds have fast dissolution, Type M have moderate dissolution, and Type S have slow dissolution from the lung (ICRP [71] 1995b). Type F compounds are water-soluble, and include uranyl acetate, uranyl nitrate, and uranium hexafluoride (UF_6). Type M and Type S compounds are less water soluble, and include uranium trioxide (UO_3), uranium dioxide (UO_2), and triuranium octaoxide

(U_3O_8). Uranium oxides are the principal forms of uranium found following impacts of DU munitions with armor.

6.3.1 Effects in the Kidney

The kidney is considered to be the target organ for uranium chemical toxicity. Upon entering the bloodstream, approximately 40% of the uranium in plasma is present as a uranium-transferrin complex. The remaining 60% is found in stable, low-molecular-weight complexes with carbonate or bicarbonate, which are filtered in the kidney at the glomerulus. As the glomerular filtrate passes along the proximal tubules, the complexed uranium dissociates with decreasing pH. This dissociation liberates the reactive uranyl ion, which can interact with other complexing species in the filtrate or with components of the proximal tubule membrane. Some of the uranium that remains complexed in the lumen, and a portion of the freed uranyl ions, may traverse the length of the tubules and enter the bladder, thus resulting in a high rate of urinary excretion of uranium soon after exposure. Heavy metals, including uranium, have a great affinity for ionic sites of the brush-border membrane of the proximal tubule. The suggested mechanism for renal damage is that the binding of uranium to these cells may alter cellular permeability to sodium, which, in turn, interferes with the transport of glucose, amino acids, and phosphates, thus resulting in the increased release of these compounds into the urine.

A *de facto* occupational guideline for the maximum permissible concentration of uranium of 3 μg/g kidney has been accepted for years. As calculated by Spoor and Hursh (1973), this value was based on an extrapolation of the radiological limit and other data contained in ICRP Publication 2 (1959) and the results of experiments described in Voegtlin and Hodge (1953), in which animals were exposed to soluble compounds of uranium for one year. Because it is commonly associated with ICRP Publication 2, references to the 3 μg U/g kidney guideline throughout the rest of this document are cited as ICRP (1959).

6.3.1.1 Renal Effects in Humans after Acute Exposures to Uranium

Soldiers with embedded DU fragments are also likely to have received an acute intake of DU oxides. It cannot be determined if there were any acute renal effects resulting from this exposure, as acute effects would be resolved after a few days or weeks. However, seven years after ODS, no renal effects had been observed in any of the wounded personnel (McDiarmid et al.

2000). The most recent follow up of these veterans showed that there were subtle, but clinically insignificant, changes in measures related to renal tubular function (McDiarmid et al. 2004).

The Royal Society (2002) reviewed renal effects occurring within a few days after brief (i.e., less than 1 d) uranium exposures. This review and 13 additional cases described by Fisher et al. (1990a) form the basis of the following discussion. A summary of acute uranium exposures and their associated renal effects in 27 individuals is listed in Table 6.4, which was modified from The Royal Society's report. The 27 cases were selected because documentation was provided both of the clinical symptoms or the biochemical indicators of renal dysfunction and the peak renal concentrations of uranium in the kidney as reported by Fisher et al. (1990b) or The Royal Society. The biochemical indicators used to identify renal dysfunction were from tests conducted at the time of the exposure and reported in the original literature. These techniques have improved through the years, so subtle effects from exposures long ago may not have been detected.

This review covers nine different exposure situations in which 27 individuals were exposed. Six of the nine exposures resulted from industrial accidents involving a total of 20 workers in the uranium processing industry. The accidents involved exposure by inhalation, ingestion, or dermal burns with incorporation of uranium nitrate, uranium tetrafluoride, or uranium hexafluoride. The other seven individuals either ingested or were injected with uranium nitrate or uranium acetate.

As the results in Table 6.3 show, the intake of uranium does not correlate well with the concentration of uranium in the kidney in these situations, probably because of the different solubilities of the chemical forms of the uranium, the different routes of exposure, and in some cases, complications caused by other injuries. Each exposed individual is noted in the table, although several accidents involved the exposure of more than one worker.

An independent evaluation was conducted of the 27 cases drawn from The Royal Society's review (2002) and by Fisher et al. (1990b). These cases are addressed in the following discussions (Table 6.3).

- The individual with the highest intake deliberately ingested 15 g of uranium acetate (equivalent to 8.5 g uranium) and consequently suffered severe illness (Pavlakis et al. 1996). The individual suffered acute renal failure (requiring dialysis for 2 weeks), refractory anemia, and effects in the muscles, heart, liver, and intestines. The peak uranium concentration was calculated to be 100 μg U/g kidney. He was discharged

Table 6.3. Acute Human Exposures to Uranium Resulting in Effects in the Kidney[a]

Intake Route	Chemical Form	Subjects	Intake mg U	Peak Kidney µg U/g	Effect[b]	Reference
Ingestion	Acetate	1	8500	100	+++	Pavlakis et al. (1996)
Dermal (burn)	Nitrate	1	130	35	+++	Zhao and Zhao (1990)
Inhalation	Tetrafluoride	1	920	10	++	Zhao and Zhao (1990)
Injection	Nitrate	1 1	16 11	6 4	+ +	Luessenhop et al. (19580
Dermal (burn)	Nitrate	1	10	3	++	Butterworth (1955)
Inhalation	Hexafluoride	1	24	2.5	+	Fisher et al. (1990a)
Injection	Nitrate	1 1 1	5.9 5.5 4.3	2 2 1.5	+ - -	Luessenhop et al. (1958)
Inhalation	Hexafluoride	1 1 1	40-50	4 4 1.2	+ + +	Kathren and Moore (1986)
Inhalation	Hexafluoride	1 1 1 1 1 1 1	18 18 17 15 12 11 11	1.9 1.9 1.8 1.5 1.2 1.1 1.1	- - - - - - -	Fisher et al. (1990a)
Ingestion	Nitrate	1	470	1	+	Butterworth (1955)
Inhalation	Hexafluoride	1	20	1	-	Boback (1975)
Inhalation	Hexafluoride	1 1 1 1 1	8.7 8.4 7.4 6.0 6.0	0.90 0.87 0.76 0.62 0.62	- - - - -	Fisher et al. (1990a)

(a) Modified from The Royal Society (2002).
(b) Clinical symptoms of renal dysfunction: Severe = +++.
Biochemical indicators of renal dysfunction: Protracted = ++, Transient = + .
No detectable effects = -.

from the hospital 32 days after initial presentation. At 6 months, the biochemical signs of kidney dysfunction persisted, but signs of muscle, liver, and cardiac toxicity had resolved. This case report indicates that the acute intake of uranium with a peak concentration at 1 day of 100 µg U/g kidney has a serious effect on kidney function, requiring hemodialysis, and results in prolonged illness.

- An industrial accident in which a man was extensively burned with a hot solution of uranyl nitrate and uranium oxide resulted in severe symptoms related to the skin burns and renal dysfunction (Zhao and

Zhao 1990). Renal dysfunction began 1 day after exposure and became worse over the next 6 days. By day 7, the patient was in critical condition. On day 8, renal function began to improve and returned to normal at 1 month after exposure. The uranium in the kidney was primarily from uranyl nitrate absorbed through the burned skin. This absorption resulted in a calculated renal peak concentration of 35 μg U/g kidney, which occurred 5 days after exposure.

- Renal dysfunction, as noted by biochemical tests, was reported after an accidental inhalation exposure to UF_4 powder for 5 min (Zhao and Zhao 1990). The lung deposition, (recalculated in The Royal Society report [www.royalsoc.ac.uk/policy/du/du2_a.pdf]) was 915 mg. One week after exposure, the patient was ill with abdominal pain and diarrhea and was treated with antibiotics. Biochemical indications of renal dysfunction, including proteinuria, elevated amino acid nitrogen/creatinine, elevated non-protein nitrogen, and decreased retention of phenolsulfonphthalein, appeared 78 days after the exposure and continued until about 590 days after exposure. The levels of uranium in the urine increased over the first 2 months after exposure and then gradually decreased to background at about 3 years after exposure. This intake resulted in a calculated peak uranium concentration of 10 μg U/g in the kidney, which occurred 60 days after exposure. The urinary excretion of uranium and the indicators of renal dysfunction peaked about 2 months after the inhalation exposure, much later than predicted from ICRP biokinetic models for uranium tetrafluoride Thus, this case is more of a chronic, rather than acute, exposure of the kidney.
- The toxicity of hexavalent uranium nitrate administered intravenously was examined in a group of five terminally ill patients with brain tumors (Luessenhop et al. 1958). Injected doses resulted in calculated peak concentrations ranging from 1.5 to 6 μg U/g kidney. The peak concentrations occurred 4 to 8 days after injection. The three patients with the largest doses showed striking elevated concentrations of urinary catalase and albumin from 2 to 8 days after administration. In addition, those with the highest two doses showed elevations of non-protein nitrogen and appearance of casts in the urine. Glucose was not elevated in any of the patients.
- Another case of dermal exposure to hot uranium nitrate was reported by Butterworth (1955). The calculated peak concentration of uranium was about 3 μg U/g kidney and occurred about 9 days after the accident.

Albumin was present in the urine shortly after exposure and persisted for about 3 weeks.

- A group of 31 workers was exposed to uranium hexafluoride and hydrolysis products following the rupture of a large shipping container. Urine bioassay measurements for uranium and medical laboratory tests were conducted to document any health effects and to correlate with exposure (Fisher et al. 1990a). The peak kidney uranium concentrations, calculated from a modified biokinetic model, ranged from 0.05 to 2.5 μg U/g kidney (Fisher et al. 1990b). Although uranium hexafluoride can be toxic when inhaled, none of the workers in this accident showed clinical symptoms. The worker with the highest kidney uranium concentration, however, had elevated concentrations of protein (>200 mg/dL) and glucose (1+ to 3+) in the urine in the first two samples submitted 4 days after the accident. This response is considered a transient renal effect (Table 6.4). The original report of these data did not consider this a toxicologic effect. None of the other workers had renal effects related to this accident. In a follow-up 2 years after the accident, none of the 31 workers had observable health effects (Fisher et al. 1990a). Thirteen individuals from this accident, with kidney uranium concentrations ranging from 0.62 to 2.5 μg U/g, are included in Table 6.3. They were selected because their exposure was brief, the dose and response data were well characterized, and the kidney uranium concentrations extended into the 1 μg U/g kidney range.

- Three workers, who survived a serious accident involving a massive uranium hexafluoride release in 1944, were reported to have renal injury in addition to other injuries (Kathren and Moore 1986). All three had albumin and casts in the urine, which was indicative of renal injury. No other medical details were reported. The uranium burdens were based on limited data. Based on urinary excretion data, the peak kidney uranium concentrations were 1.5, 2, and 2 μg U/g for the three workers and occurred 2 days after the accident. Two of these individuals had no clinical findings attributable to the uranium exposure when given medical examinations 38 years after the accident (Kathren and Moore 1986).

- An experiment in which a healthy volunteer ingested 1 g of uranyl nitrate to help establish excretion rates was reported by Butterworth (1955). The person experienced vomiting, diarrhea, and albuminuria. Clinical recovery was rapid and completed by 24 hours. The peak kidney uranium concentration of 1 μg U/g occurred at 1 day after ingestion.

- An accidental release of UF_6 occurred when a worker inadvertently removed a valve from a heated UF_6 cylinder (Boback 1975). He was immediately engulfed in a cloud of hydrolyzed UF_6 (UO_2F_2) and hydrofluoric acid. He was hospitalized for observation for 6 days. Clinical findings, including urinalysis, were negative. He returned to work 9 days after the accident. The peak kidney uranium concentration of 1 μg U/g occurred at 1 day after the accident.

The renal effects were classified based on the appearance of severe symptoms related to renal damage (Severe, +++) or the detection of serum or urinary biochemical indicators of renal dysfunction. The latter were characterized as either Protracted, ++ or Transient, +. Individuals with severe clinical symptoms (+++) required hospital treatment; while those with biochemical indicators of renal dysfunction for protracted times (++) were not treated and apparently were not ill. Those with transient biochemical indicators of renal dysfunction (+) were also not treated and apparently were not ill.

The follow-up on the health of the individuals with the most severe renal effects from uranium has been reported (Table 6.4). The individual that did not have a complete clinical recovery had the highest intake and renal dose of uranium (as uranium acetate). However, only a 6-month follow-up time has been reported. Other individuals followed over a period of years were reported to have normal physical examinations and kidney function tests. These results suggest that acute renal toxicity from uranium does not necessarily lead to chronic toxicity. Studies in animals also support this conclusion. Rats that were injected with uranyl fluoride developed kidney necrosis and marked renal dys-

Table 6.4. Follow-Up to Acute Human Exposures to Uranium Resulting in Effects in the Kidney

Intake mg U	Kidney μg U/g	Acute Effect[a]	Follow-up Time	Residual Renal Effect	Reference
8500	100	+++	6 mon	Clinical symptoms, under treatment	Pavlakis et al. 1996
130	35	+++	7 yr	Normal physical examination	Zhao and Zhao 1990
920	10	++	7yr	Normal physical examination Normal kidney function tests	Zhao and Zhao 1990
24	2.5	+	2 yr	Normal physical examination	Fisher et al. 1990a
40-50	4 1.2	+ +	38 yr 38 yr	Normal physical examination	Kathren and Moore 1986

(a) = Clinical symptoms of renal dysfunction: Severe = +++
Biochemical indicators of renal dysfunction: Protracted = ++, Transient = +
No detectable effects = -

function within a matter of days, but the damage was repaired within 35 days (Diamond et al. 1989). Similarly, rats exposed to UO_2F_2 had severe tubular necrosis, which was essentially resolved within 28 days, but residual effects were noted at 60 days (Morrow et al. 1982). The severity of these effects varied between animals and included compressed glomeruli and tubules, and stroma focally infiltrated with chronic inflammatory cells.

6.3.1.2 Dose Response Analysis of Renal Effects in Humans after Acute Exposures to Uranium: Development of a Risk Model

A risk model was developed from the data in Table 6.3. A prediction was made of the renal effect scores, based on the kidney burdens of uranium. The 27 cases of uranium exposure and their kidney burdens were used in this analysis to determine the Renal Effects Group (REG) and prediction of REGs for future exposures. The number of cases for each REG was as follows:

Renal Effects Group	Number of Cases
0 (-)	15
1 (+)	8
2 (++)	2
3 (+++)	2

The kidney burdens were not normally distributed, and a log-transformation was used to normalize the data. The log-transformed data were used through the remainder of the analyses. Preliminary analyses were conducted to determine significant differences between REGs using an analysis of variance. A discriminant analysis was used to build a model to predict the REG based on the log-transformed amount of uranium in the kidney. This procedure generated a discriminant function based on a linear combination of the predictor variable, peak uranium concentration in the kidney (μg U/g kidney), that provided the best discrimination among the REGs. The function was generated from the sample of 27 human cases in Table 6.3. The function can also be used to predict the REG of new cases in which peak uranium concentrations in the kidney have been measured or predicted.

The discriminant analysis was able to correctly identify 85.2% of the cases to their respective REG using the log-transformed value of the peak kidney uranium burden as the predictor variable. The discriminant function was:

$$D = 4.4^{*} \log (\mu g\ U/g\ kidney) - 1.5$$

The REGs can be defined by the following discriminant score, D, associated with kidney uranium concentrations (Table 6.5).

Table 6.5. Renal Effects Groups Based on Peak Kidney Uranium Concentration

REG	Effects	D (range)	µg U/g kidney
0	No Effect	≤0.0	≤2.2
1	Transient	>0.0 to 2.0	>2.2 to ≤6.4
2	Protracted	>2.0 to 4.0	>6.4 to ≤18
3	Severe Symptoms	>4.0	>18

Because there was a large difference between REGs 2 and 3 in their respective kidney uranium burdens and there were only two cases for each REG, it was somewhat difficult to accurately predict the cutoff point for the D score. However, recognizing these limitations, it was determined to be between 4.0 and 5.0, and the D score of 4.0 is used here as a conservative estimate. The cutoff kidney uranium concentrations for the REGs (or risk categories) were determined using the discriminate function, with D being equal to 0, 2.0, or 4.0.

A graphic representation of the REGs, based on Table 6.5, shows the group number plotted against the log of the kidney uranium concentration (Figure 6.7). The median values are represented by the lines within the boxes. The 25th and 75th percentiles are shown as the lower and upper bounds, respectively, of the boxes. The minimum and maximum values are represented by

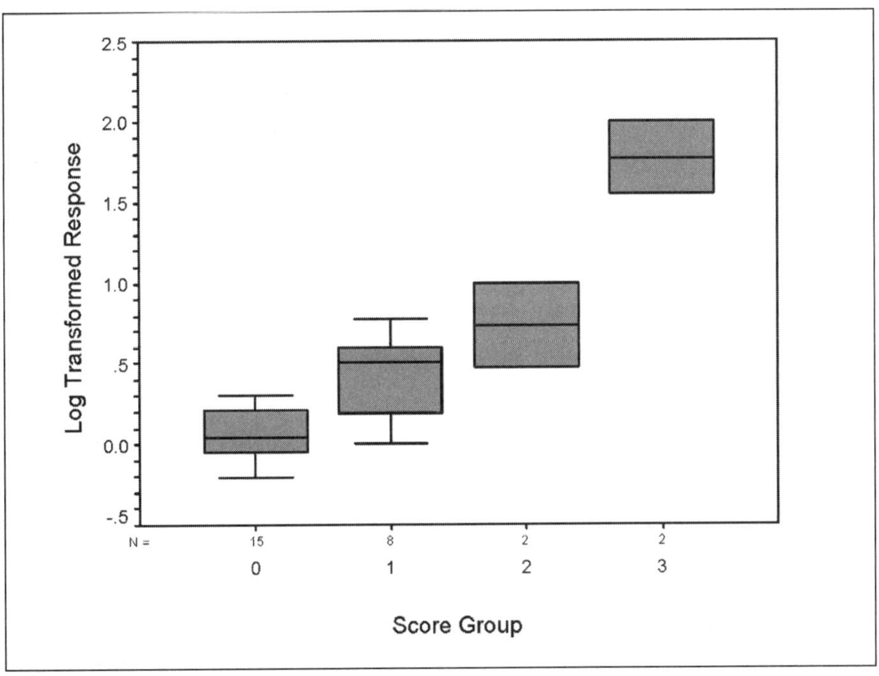

Figure 6.7. Distribution of Acute Kidney Toxicity by REGs

the lower and upper bars. Because of the small sample size for REGs 2 and 3, the minimum and maximum concentrations are the same as the 25th and 75th percentiles.

Each phase and scenario was modeled to estimate kidney uranium burden, and these kidney uranium burdens were compared with the REGs and assigned a score. The concentrations relative to the likely acute renal effects and predicted outcomes are categorized by REG in Table 6.6.

Table 6.6. Risk Characterization—Chemical Risk to the Kidney

REG	Kidney Uranium Concentration (µg U/g kidney tissue)	Acute Renal Effect	Predicted Outcome
0	≤2.2	No Detectable Effects	No Detectable Effects
1	>2.2 to ≤6.4	Possible Transient Indicators of Renal Dysfunction	Not Likely to Become Ill
2	>6.4 to ≤18	Possible Protracted Indicators of Renal Dysfunction	May Become Ill
3	>18	Possible Severe Clinical Symptoms of Renal Dysfunction	Likely to Become Ill

Based on the analysis of acute exposure data in humans, the range of kidney doses for each REG as well as the predicted outcomes are as follows:

- *Renal Effects Group 0*: Kidney uranium burdens ≤2.2 µg U/g kidney will not result in detectable biochemical changes indicative of renal toxicity, or in any detectable clinical effects.
- *Renal Effects Group 1*: Kidney uranium burdens >2.2 µg U/g kidney and ≤6.4 µg U/g kidney may result in transient indicators of renal dysfunction. Biochemical changes may include enzymuria, amino aciduria, proteinuria, albuminuria, and glucosuria. There may also be the presence of casts in the urine. These effects should be transient; in human and animal studies, biochemical changes typically returned to normal levels over a period of days to a few weeks. Overt symptoms of toxicity were not usually seen in humans with kidney burdens in this range.
- *Renal Effects Group 2*: Kidney uranium burdens >6.4 µg U/g kidney and ≤18.0 µg U/g kidney may result in protracted indicators of renal dysfunction. Biochemical changes would be the same as those with lower kidney burdens; however, it may be a period of weeks or months before they are resolved. Additionally, there may be overt symptoms of toxicity, such as diarrhea or nausea. (These symptoms may not be due to renal effects, but may be other responses to the exposure.)

- *Renal Effects Group 3*: Kidney burdens >18 μg U/g kidney may result in severe clinical symptoms of renal dysfunction, and a person so exposed may suffer from acute renal failure and exhibit secondary effects to other organs.

6.3.1.3 Renal Effects in Humans from Chronic Exposure to Uranium

Renal effects occurring after chronic (months to years) uranium exposures have been reviewed by the Royal Society (2002). This review and additional cases from Boback (1975) and Kurttio et al. (2002) form the basis for this discussion. Reports of chronic uranium exposures and their renal effects are listed in Table 6.7. The table is modified from the Royal Society's report and uses the concentrations of uranium in the kidney from that report.

Three intake routes were associated with these exposures—inhalation of yellowcake or ore concentrate by uranium process workers, embedment of depleted uranium metal in wounded Soldiers, and ingestion of drinking water by the general population. With one exception, the exposures were chronic, lasting for years. The highest renal concentration of 3 μg U/g was calculated from a report of one worker being exposed to moderate concentrations of an ore concentrate aerosol over a period of about one month (Boback 1975). No biochemical indicators of renal dysfunction were seen in this worker. Uranium millers, exposed over a period of years to aerosols of yellowcake, had calculated kidney uranium concentrations up to ~1 μg U/g (Thun et al. 1985). Biochemical indicators of renal effects, but not clinical symptoms, were noted in some of these workers.

The lowest kidney concentrations of uranium reported to result in renal effects were related to high concentrations (~80 μg/L) of uranium in the drinking water seen in two studies (Zamora et al. 1998; Kurttio et al. 2002). In both of these studies, biochemical indicators of subtle renal effects of undetermined significance were seen. The trend in these chronic exposures is toward greater renal effects with lower kidney concentrations of uranium, an unexpected result. The length of time of the exposure, however, may be an important factor. The groups with the longer exposure times have the greatest effects (Table 6.7).

The National Academy of Sciences (NAS), Institute of Medicine, recently reviewed data on the toxicity of uranium, including six studies of uranium processing facilities and one study of a phosphate fertilizer plant (which mined and processed ores containing uranium) (Fulco et al. 2000). The NAS committee concluded from these epidemiological studies that there was

Table 6.7. Chronic Human Exposures to Uranium Resulting in Effects on the Kidney[(a)]

Intake Route	Chemical Form	Subjects	Kidney µg U/g	Effect[(b)]	Length of Exposure (years)	Reference
Inhalation	Ore Concentrate	1	3	-	1.1 to 23	Boback (1975)
Inhalation	Yellowcake	27	up to ~1	++	~1 month	Thun et al. (1985)
Intramuscular	Uranium Metal	15	up to ~0.5	-	3	Hooper et al. (1999)
Ingestion	Drinking water	30	up to ~0.1 (~80 µg/L)	++	3-59	Zamora et al. (1998)
Ingestion	Drinking water	101	(<100µg/L)	++	1-34	Kurttio et al. (2002)

(a) Modified from The Royal Society (2002).
(b) Biochemical indicators of renal dysfunction: Protracted = ++, Transient = +, No detectable effects = -.

limited evidence of no association between exposure to uranium and clinically significant long-term renal dysfunction.

Soldiers with DU fragments embedded in their muscles for 3 years after ODS had calculated kidney uranium concentrations of ~0.5 µg U/g (Hooper 1999). Seven years after ODS, no renal effects had been observed in any of the wounded personnel (McDiarmid et al. 2000). The most recent follow up of these veterans showed that there were subtle, but clinically insignificant, changes in measures related to renal tubular function (McDiarmid et al. 2004).

6.3.2 Effects on the Lung

Depleted uranium penetrators create inhalable uranium oxide particles when they perforate armor. These uranium oxide particles are relatively insoluble and, when inhaled and deposited, may be retained in the lung and regional lymph nodes for weeks to years. Alveolar macrophages are resident cells of the lung that are involved in the phagocytosis and removal of these particles. Activation, damage, or death of these cells by the phagocytosed uranium oxide particles may result in the release of compounds that initiate inflammation and stimulate fibroblasts leading to pulmonary fibrosis (Tasat and deRey 1987).

6.3.2.1 Pulmonary Effects after Inhalation Exposure to Uranium

The acute pulmonary toxicity of inhaled uranium is highly dependant on the chemical form of the uranium. Uranium hexafluoride is the only uranium compound that has been associated with acute effects in humans after inhalation. Two accidents involving uranium hexafluoride have resulted in the deaths of three workers in the U.S. uranium processing industry. The initial accident was the sudden rupture of a cylinder containing gaseous uranium hexafluoride. The released uranium hexafluoride, plus steam, produced a dense cloud of uranium hexafluoride and its hydrolysis products. Twenty one workers were exposed resulting in the death of two, shortly after exposure, with extensive chemical burns and pulmonary edema (Howland 1949). Hydrogen fluoride gas was associated with the release and was considered the major factor in the sudden deaths. Three workers were seriously injured and required hospitalization for over a week. The estimated total uranium lung burden of these workers was 40 to 50 mg. Thirteen slightly injured individuals were discharged from the hospital in 48 hrs. A study of two of the seriously injured survivors 38 years after the accident revealed no detectable deposition of uranium in the lung or any physical findings attributable to uranium exposure (Moore and Kathren 1985).

A second accident involved 32 workers exposed to aerosols of uranium hexafluoride and hydrolysis products (hydrogen fluoride gas) following the rupture of a shipping cylinder (Fisher et al. 1990a). One worker died of massive pulmonary edema and another was treated for skin irritation and burn; all of these injuries resulted from exposure to hydrogen fluoride. Twenty-one other workers were examined at the hospital. Eighteen were released in 2 days with minimal treatment for potential renal damage, and three were observed for possible lung damage. Pulmonary function was not affected as determined by clinical tests (forced expiratory volume in 1 sec and total expiratory volume) out to 2 years after exposure. The calculated alveolar deposition of uranium in the survivors ranged from 0.47 to 24 mg with a mean of 6.8 mg. These results suggest that inhalation of up to 24 mg of uranium hexafluoride does not cause acute or chronic pulmonary effects.

The chronic exposure to aerosols of uranium compounds has not been related to chronic pulmonary health effects. The NAS Institute of Medicine, recently reviewed eight epidemiologic studies of the workers in six major uranium processing facilities in the United States (Fulco et al. 2000). These were studies of workers who were occupationally exposed to uranium compounds over a period of years. The evaluation of these studies was complicated because of ambiguities in the classification of disease on the death certificates, as well as confounding factors such as exposure to other respiratory toxins in the work place and, most importantly, cigarette smoking. The committee did conclude that there was not sufficient evidence to establish an association between exposure to uranium and chronic pulmonary disease.

Studies in animals show that high lung burdens are required before significant chronic noncancer lung disease is seen. In one study, dogs and monkeys were exposed to aerosols of 5 mg/m^3 uranium dioxide for 5 years (Leach et al. 1970), and then observed for another 6.5 years after exposure (Leach et al. 1973). Uranium lung concentrations ranged from 1200 to 2400 μg U/g lung tissue and 3130 to 3830 μg U/g lung tissue for dogs and monkeys, respectively, at the end of exposure. The lung concentrations were 1000 times higher than the <1.5 μg U/g lung tissue reported in uranium processing workers exposed for their working lifetime (see Table 6.2). At the end of the 6.5-yr observation period, the dogs had a moderate, patchy fibrosis, noticeable only with microscopic examination. The monkeys, with approximately two times higher uranium burdens than dogs, had a severe, pulmonary fibrosis. This condition was attributed to a radiation effect rather than a chemical effect; it was "dose-dependent and more marked in the monkeys than in dogs" (Leach et al. 1973).

6.3.3 Effects to Other Systems

The study of a group of veterans from ODS exposed to DU aerosols and fragments from combat vehicle perforations provides information concerning potential effects of uranium in other organs as well as the kidney (Hooper et al. 1999). Approximately 70 Soldiers have been evaluated at least once, of which 30 have participated in at least three clinical surveillance visits over the past 10 years (McDiarmid et al. 2000, 2001, 2004). Approximately one-third of this cohort has embedded fragments. All in this group had injuries associated with being wounded; most had major soft tissue trauma and many had burns. Other injuries included fractures, amputations, and concussions. Outcome measures used in the periodic health evaluations were urinary uranium determinations, serum chemistry and hematology values, psychiatric and neurocognitive assessments, and reproductive health measures. The outcomes were compared with those for a comparable control group of military veterans.

Of those with embedded fragments, elevated uranium urine concentrations have persisted for most in the cohort, which were related to the persistence of the DU fragments in the tissues. However, clinical outcomes and renal function related to urine uranium measures have revealed few abnormalities and no health effects. Subtle, but clinically insignificant, changes in measures related to renal tubular function were most recent findings. Subtle changes in neurocognitive function and reproductive health were reported at one clinical examination period, but have not persisted. These clinical evaluations have revealed no solid evidence to date of any deleterious effects related to the chronic low level exposure to depleted uranium embedded in the soft tissues.

6.3.3.1 Central Nervous System

Uranium has been demonstrated to cross the blood-brain barrier following exposure to DU implants as well as injections of uranium salts in rats. Pellmar and colleagues (1999a) studied the organ distribution of uranium dissolved from DU fragments embedded in the muscles. Uranium concentrations in the brain were elevated at 1 day and 1, 6, and 18 months after implantation. The concentrations were similar to those in soft tissues. More importantly, the uranium was not uniformly distributed in the brain. Synaptic potentials were altered in hippocampal slices from rats exposed to DU implants for 12 months in this study (Pellmar et al. 1999b). Barber and Kopplin (2003) reported that following exposure to uranyl acetate, uranium distributed to the hippocampus, striatum, cerebellum, and frontal cortex; concentrations in the brain were reduced by stress prior to exposure. Further ongoing studies are assessing the neurotoxic potential of acute and chronic exposure to DU as well as the contribution of stress to the expression of DU toxicity (Jortner and Barber 2003). These data demonstrate that, once in the bloodstream, uranium is able to cross the blood-brain barrier and concentrate in certain portions of the brain. The implications of these higher concentrations in localized areas of the brain and the effect on neural function are not understood. Changes in neurocognitive function were reported in some veterans who participated in ODS, but these effects have not persisted (McDiarmid et al. 2001).

6.3.3.2 Reproductive System

Uranium has been demonstrated to cross the placental barrier in rats, rabbits, and dogs following exposure either orally, intratracheally, or via implanted pellets (Kurlyandskaya 1970; Benson and McBride 1997). Additionally, uranium was found in the milk of female test animals following oral or intratracheal exposures (Kurlyandskaya 1970). Adverse effects to the pups were not noted in any of these studies. However, additional multigenerational studies in rats are being conducted with DU implants (Arfsten et al. 2001). Endpoints assessed in these studies will include effects on reproduction and the development of the F1 (first generation) and F2 (second generation) offspring.

Exposure of rats fed uranium nitrate has resulted in decreases in numbers of litters and numbers of offspring per litter (Maynard and Hodge 1949). In more recent studies, developmental and teratogenic effects have been reported in rats and mice following parental exposure to uranium acetate (Domingo et al. 1989; Llobot et al. 1991; Domingo 1994). All of these exposures were relatively high, in the mg/kg body-weight range.

6.3.3.3 Immunity

No studies have demonstrated a link between exposure to uranium and immune system functions. However, uranium oxide particles deposited in the lung are taken up by alveolar macrophages, which have a role in the immune response of the lung, and are concentrated in the lymph nodes draining the lung. *In vitro* studies have demonstrated that uranium dioxide is cytotoxic to alveolar macrophages (Tasat and deRey 1987). Some killing of lymphocytes by alpha particles from retained particles might occur as the lymphocytes pass through the lung-associated lymphoid tissues, but these are unlikely to lead to significant reductions in the ability of the body to combat infections (The Royal Society 2002). Reductions in immunity would require continuous effects on the mature leukocytes or on the precursor cells in the lympho-hematopoietic organs, including the red bone marrow.

6.4 CHEMICAL HEALTH RISKS OF INHALED DU AEROSOLS

The chemical health risks to the kidney from inhaled DU aerosols were estimated using the peak uranium kidney burdens calculated from the Capstone data (Chapter 5). Kidney burdens of uranium were modeled using the ICRP 78 Biokinetic Model (ICRP 1997). The risks from the kidney uranium burdens were determined using the risk model developed in Section 6.3.1.2. Kidney uranium burdens are also compared to the *de facto* occupational guideline of 3 μg U/g kidney. The results are presented for each phase (Abrams tank through non-DU armor, Phase I; Bradley vehicle, Phase II; Abrams tank through DU armor, Phase III; and a fully operational Abrams tank through DU armor, Phase IV) and each scenario (Table 6.8). Estimates

Table 6.8. Summary of Exposure Scenario Parameters

Level I Scenario	Time of Exposure	Exposure Duration	Breathing Rate
Crew Inside Vehicle			
Scenario A	From impact to exit 1 min post shot	1 min	3 m^3/h
Scenario B	From impact to exit 5 min post shot	5 min	3 m^3/h
Scenario C	From impact to exit up to 1 h post shot	1 h	3 m^3/h for first 15 min, 1.5 m^3/h thereafter
Scenario D	From impact to exit up to 2 h post shot	2 h	3 m^3/h for first 15 min, 1.5 m^3/h thereafter
First Responder			
Scenario E	Entry 5 min post shot, exit 10 min later	10 min	3 m^3/h

are also provided for the percentage of peak kidney uranium burdens (calculated at 24 h) falling into the different REGs discussed in Section 6.3.1.2 (REG 0, no renal effects; REG 1, transient biochemical effects, no overt symptoms; REG 2, possible protracted indicators of renal dysfunction, possible overt symptoms; REG, possible severe symptoms of renal dysfunction). The Capstone data are based on sampling arrays installed near the crew positions, which for the Abrams BHT included the commander, driver, gunner, and loader. In the Bradley BHT, sampling arrays were installed near the commander, driver, left scout (denoted loader), and right scout (denoted gunner).

Aerosols from all LC-DU shots from each Capstone test were evaluated in this analysis of kidney uranium concentrations. The double shots performed with the Abrams tank through non-DU armor and through the Bradley vehicle were evaluated as single shots. The contribution from the first shot in each of the double shots was used to evaluate DU concentrations for scenarios A, B, and E for which actual measurements were made. Scenario E required a small extrapolation to extend the aerosol concentrations to 15 min from the 10 min of actual sampling. All scenarios were modeled for the second shot in each of the two double shots. The minimum and maximum peak kidney burdens at 24 h based on the DU aerosol measurement data are presented by phase and scenario. Also presented are data for the median, 10th and 90th percentiles.

The summary tables of each phase provide the predicted minimum, median, and maximum peak kidney uranium concentrations and associated REGs as well as the 10th and 90th percentile results to three significant figures. These values are summarized to two significant figures in the discussions of the results.

6.4.1 Abrams Tank Perforated Through Non-DU Armor (Phase I)

In Phase I, seven shots were fired across the turret through conventional (non-DU) armor in three shot-line variations:

- through the side armor
- through the side armor and into the breech
- through the hull.

Table 6.9 summarizes of the minimum, 10th percentile, median (50th percentile), 90th percentile, and maximum peak kidney uranium concentrations for all Phase-I shots. The asterisk (1*) indicates a REG 1 that exceeded the 3 μg U/g kidney level guideline. The range for REG 1 is between 2.2 and 6.4 μg U/g kidney.

Table 6.9. Predicted Peak Kidney Uranium Concentrations and REGs, Abrams Tank Perforated Through Non-DU Armor

	Minimum		10th Percentile		Median		90th Percentile		Maximum	
Scenario	µg U/g Kidney	REG	µg U/g Kidney	REG	µg U/g Kidney	REG	µg U/g Kidney	REG	µg U/g Kidney	REG
A	0.419	0	1.94	0	2.99	1	4.20	1*	5.33	1*
B	1.60	0	4.64	1*	6.44	1/2	9.97	2	14.9	2
C	2.48	1	5.34	1*	8.17	2	14.0	2	16.8	2
D	2.74	1	5.12	1*	8.03	2	16.1	2	17.8	2
E	0.567	0	0.815	0	1.54	0	4.77	1*	5.79	1*

The estimated median kidney uranium burden to two significant figures within Scenario A was 3.0 µg U/g kidney; the range within this scenario was 0.42 to 5.4 µg U/g kidney. The estimated kidney uranium burdens met the risk model criteria for either REG 0 (18%, n=4) or 1 (82%, n=18). The estimated median kidney uranium burden within Scenario B was 6.4 µg U/g kidney (at the upper limit of REG 1 and the lower limit of REG 2); the range in this scenario was 1.6 to 15 µg U/g kidney. The estimated kidney uranium burdens met the risk model criteria for REG 0 (5%, n=1), 1 (45%, n=10) or 2 (50%, n=11). The estimated median kidney uranium burden within Scenario C was 8.2 µg U/g kidney; the range in this scenario was 2.5 to 17 µg U/g kidney. The estimated kidney uranium concentrations met the risk model criteria for either REG 1 (15%, n=3) or 2 (85%, n=17). Predicted kidney uranium burdens for individuals in Scenario D were similar to those in Scenario C, with a median of 8.0 µg U/g kidney and ranging from 2.7 to 18 µg U/g kidney (at the upper end of REG 2). The estimated kidney uranium burdens met the risk model criteria for either REG 1 (17%, n=3) or 2 (83%, n=15). The median estimated kidney uranium burden in Scenario E was 1.5 µg U/g kidney; the range within this scenario was from 0.57 to 5.8 µg U/g kidney. The estimated kidney uranium concentrations met the risk model criteria for REG 0 (64%, n=10) or 1 (36%, n=8).

6.4.2 Bradley Vehicle Perforated Through Conventional Armor (Phase II)

Phase II consisted of three shots including a double shot through the passenger compartment and a single shot from the back of the vehicle and into the 25-mm gun feeder (intended to maximize the DU aerosol in a Bradley vehicle) in the turret. Table 6.10 summarizes the minimum, 10th percentile, median (50th percentile), 90th percentile and maximum peak kidney uranium concentrations for all Phase-II shots. The asterisk (1*) indicates an REG 1 that exceeded the 3 µg U/g kidney level guideline. The range for REG 1 is >2.2 to 6.4 µg U/g kidney.

Table 6.10. Predicted Peak Kidney Uranium Concentrations and REGs, Bradley Vehicle

	Minimum		10th Percentile		Median		90th Percentile		Maximum	
Scenario	μg U/g Kidney	REG	μg U/g Kidney	REG	μg U/g Kidney	REG	μg U/g Kidney	REG	μg U/g Kidney	REG
A	0.249	0	0.271	0	0.988	0	1.63	0	3.81	1*
B	0.536	0	0.611	0	2.88	1	4.96	1*	8.66	2
C	1.43	0	1.43	0	3.53	1*	11.5	2	11.5	2
D	1.83	0	1.83	0	4.02	1*	11.9	2	11.9	2
E	0.326	0	0.364	0	1.41	0	1.98	0	2.75	1

The estimated median kidney burden within Scenario A was 1.0 μg U/g kidney; the range within this scenario was 0.25 to 3.8 μg U/g kidney. The estimated kidney burdens met the risk model criteria for either REG 0 (91%, n=10) or Group 1 (9%, n=1). The estimated median kidney burden within Scenario B was 2.9 μg U/g kidney; the range in this scenario was 0.54 to 8.7 μg U/g kidney. The estimated kidney burdens met the risk model criteria for REG 0 (36%, n=4), Group 1 (55%, n=6), or Group 2 (9%, n=1). The estimated median kidney burden within Scenario C was 3.5 μg U/g kidney; the range in this scenario was 1.4 to 12 μg U/g kidney. The estimated kidney uranium concentrations met the risk model criteria for either REG 0 (50%, n=4), REG 1 (38%, n=3), or REG 2 (13%, n=1). The estimated median kidney burden within Scenario D was 4.0 μg U/g kidney; the range in this Scenario was from 1.8 to 12 μg U/g kidney. The estimated kidney burdens met the risk model criteria for either REG 0 (25%, n=2), REG 1 (50%, n=4), or REG 2 (25%, n=2). The median estimated kidney burden in Scenario E was 1.4 μg U/g kidney; the range within this Scenario was from 0.33 to 2.8 μg U/g kidney. The estimated kidney uranium concentrations met the risk model criteria for REG 0 (91%, n=10) or REG 1 (9%, n=1).

6.4.3 Abrams Tank Perforated Through DU Armor (Phase III)

Phase-III shots included two single shots, both of which were fired into the turret through DU armor. These were essentially duplicate shots, but a hydraulic fluid flash fire occurred immediately after perforation in the first shot and may have affected aerosol particle coagulation. This event did not appear to have significantly affected the doses. Table 6.11 summarizes the minimum, 10th percentile, median (50th percentile), 90th percentile and maximum peak kidney uranium concentrations for all Phase-III shots. Because of the small sample size, the minimum and 10th percentile values are the same and the maximum and 90th percentile numbers are the same. The

Table 6.11. Predicted Peak Kidney Uranium Concentrations and REGs, Abrams Tank Perforated Through DU Armor

	Minimum		10th Percentile		Median		90th Percentile		Maximum	
Scenario	μg U/g Kidney	REG	μg U/g Kidney	REG	μg U/g Kidney	REG	μg U/g Kidney	REG	μg U/g Kidney	REG
A	0.566	0	0.566	0	1.08	0	2.91	1	2.91	1
B	1.72	0	1.72	0	2.58	1	7.98	2	7.98	2
C	2.44	1	2.44	1	3.55	1*	10.8	2	10.8	2
D	2.70	1	2.70	1	3.66	1*	11.2	2	11.2	2
E	0.344	0	0.344	0	0.672	0	2.32	1	2.32	1

asterisk (1*) indicates a REG 1 that exceeded the 3 μg U/g kidney level guideline. The range for REG 1 is >2.2 to 6.4 μg U/g kidney.

The estimated median kidney uranium burden within Scenario A was 1.1 μg U/g kidney; the range within this scenario was 0.57 to 2.9 μg U/g kidney. The estimated kidney uranium burdens met the risk model criteria for either REG 0 (88%, n=7) or Group 1 (13%, n=1). The estimated median kidney uranium burden within Scenario B was 2.6 μg U/g kidney; the range in this scenario was 1.7 to 8.0 μg U/g kidney. The estimated kidney uranium burdens met the risk model criteria for REG 0 (38%, n=3), Group 1 (50%, n=4), or Group 2 (13%, n=1). The estimated median kidney uranium burden within Scenario C was 3.6 μg U/g kidney; the range in this scenario was 2.4 to 11 μg U/g kidney. The estimated kidney uranium concentrations met the risk model criteria for either REG 1 (88%, n=7) or REG 2 (13%, n=1). Predicted kidney burdens for individuals in Scenario D were similar to those in Scenario C, with a median of 3.7 μg U/g kidney and ranging from 2.7 to 11 μg U/g kidney. The estimated kidney burdens met the risk model criteria for either REG 1 (88%, n=7) or REG 2 (13%, n=1). The median estimated kidney burden in Scenario E was 0.67 μg U/g kidney; the range within this scenario was from 0.34 to 2.3 μg U/g kidney. The estimated kidney uranium concentrations met the risk model criteria for REG 0 (88%, n=7) or REG 1 (13%, n=1).

6.4.4 Fully Operational Abrams Tank Perforated Through DU Armor (Phase IV)

The only LC-DU cartridge fired in Phase IV was fired through DU armor. The data from this shot were more limited than in the previous phases because only four driver's position samplers (one for each of four time intervals between 10 sec and 27 min) collected aerosol samples. The vacuum lines of the four loader's position samplers were severed, and these samplers provided no data. (To model Scenarios C and D, which extended beyond the last measured aerosol concentration, concentrations at 1 and 2 h were set to equal that the concentration from the 27 min sampler rather than continuing to diminish with time. As a result, these numbers are believed to be conservative.) The EC/NBC ventilation system was operating during and after this shot and significantly reduced kidney doses compared to Phase-III shots in which there no ventilation systems were present. Applying the available data, predicted kidney burdens are below 2.2 μg U/g kidney (Table 6.12), and these individuals would be assigned to REG 0 (i.e., no changes to renal function).

Table 6.12. Predicted Peak Kidney Uranium Concentrations, Fully Operational Abrams Tank Perforated Through DU Armor

Scenario	μg U/g Kidney
A	0.0542
B	0.225
C	0.458
D	0.556
E	0.140

6.4.5 Comparison of Risks of Kidney Dysfunction by Phase and Scenario

The median peak kidney uranium concentrations for Phases I, II, and III using the five scenarios are illustrated in Figure 6.8. Also illustrated are the data for Phase IV. The horizontal lines at 2.2, 6.4, and 18 μg U/g kidney show the boundaries between the Effect Groups 0, 1, 2, and 3 (Table 6.5). The

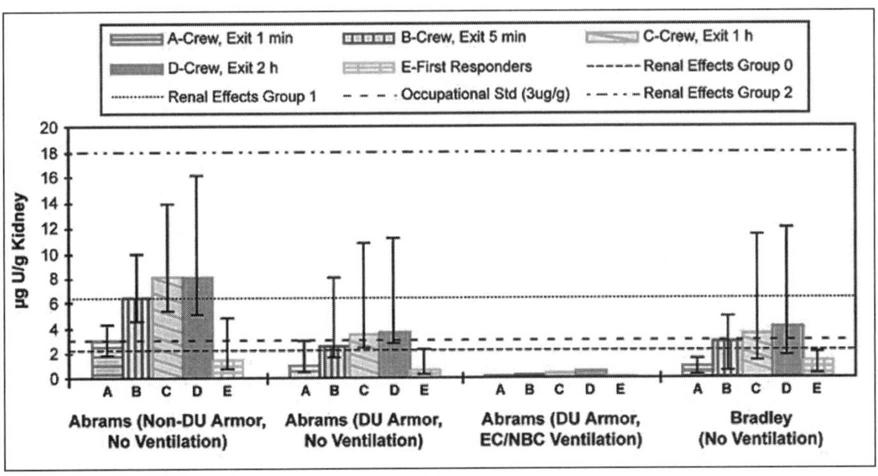

Figure 6.8. Median Peak Kidney Uranium Concentrations and 10th and 90th Percentiles for All Phases and Scenarios

dotted line represents 3 μg U/g kidney, which is the *de facto* occupational guideline for the maximum permissible concentration in the kidney. Bars represent the 10th and 90th percentiles.

SigmaStat[a] was used for all data analysis. For normally distributed data, a one-way analysis of variance (ANOVA) was used to determined differences among groups, followed by a Tukey Test to determine differences between group pairs. For non-normally distributed data, Kruskal-Wallis one-way ANOVA on ranks was used, followed by Dunn's Method to determine differences between group pairs.

In Scenarios A, B, C, and D, peak kidney uranium burdens were statistically higher in the Abrams tank without DU armor than in either the Bradley vehicle or the Abrams tank with DU armor. In Scenario E, the predicted kidney burdens were statistically higher in the Abrams without DU than the Abrams tank with DU armor. In general, however, the Bradley vehicle numbers were similar to the Abrams with DU armor results and a factor of 2 to 3 less than the Abrams tank without DU armor except for Scenario E. These data suggest that the type of vehicle struck may not greatly affect the amount of uranium reaching the kidney. The concentrations for exposures in an Abrams tank with an operating ventilation system are predicted to be at least an order of magnitude lower than in an Abrams tank without ventilation. This reduction in concentration would also occur in a Bradley vehicle, though the difference may not be as large. As shown by the Phase IV results, operating ventilation reduces DU aerosol concentrations, and in this case, reduced the associated kidney burdens by an order of magnitude.

In a worked example, predicted clearance times for the highest peak kidney uranium burden per scenario were calculated and are listed in Table 6.13. (Final calculations of peak kidney uranium concentrations for these positions are shown in parentheses.) Estimates were made for the length of time for the peak kidney burdens to decrease to the various threshold levels for the different REGs, as well as the length of time to reach the ICRP occupational guideline of 3 μg U/g kidney. For those kidney uranium concentrations in REG 1, the time to reach the ICRP guideline was up to 12 d, and the time to reach the lowest threshold of 2.2 μg U/g kidney was 2 to 3 wks. For individuals with higher kidney uranium concentrations, time to the ICRP guideline was approximately 30 days, and the time to the lowest threshold was 30 to 40 d.

---

[a]SPSS SigmaStat for Windows Version 2.03. SPSS Inc. 1992–1997.

Table 6.13. Clearance Time for Peak Uranium Kidney Burdens

Scenario	Phase	Shot	Position	Peak Conc. μg U/g kidney	Days to 6.4 μg U/g kidney	Days to 3 μg U/g kidney[(a)]	Days to 2.2 μg U/g kidney
A	1	3	Gunner	5.4 (5.3)	NA	11.7	17
B	1	3	Gunner	15 (14.9)	16	30.8	39
C	1	4	Loader	16 (16.8	14.9	27	33.1
D	1	4	Loader	17 (17.8)	15.7	28	34.3
E	1	3	Gunner	5.7 (5.8)	NA	12.4	17.5

(a) ICRP guideline for maximum occupational exposure

6.4.6 Nonradiological Risks to Lung from Inhaled DU

The potential risk to the lung from the nonradiological effect of inhaled DU is related to the total mass of uranium in the lung. The calculated total mass of DU in the scenario and position with the highest exposure (PIII-2) was about 46 μg U/g alveolar lung (Table 6.14). This concentration is considerably above those of individuals exposed in the environment or in the uranium processing industry (Table 6.15). However, the exposure is only for minutes to hours for personnel exposed in a tank compared with months to years for workers in the uranium industry. No pulmonary effects (radiological or nonradiological) have been seen in uranium processors who experience longer exposure durations than those experienced in the Capstone scenarios (Section 6.3.2.1).

Another useful value is the total mass of dust deposited and retained in the lung. For this shot, DU was 65% of the mass of the particles. Thus, the total mass particulate burden was approximately 90 μg of dust per gram lung. This dust burden can be compared to that of coal miners who frequently attain

Table 6.14. Lung Burdens of DU or Dust in Exposed Populations

Exposure Groups	DU Concentration (μg DU/g Lung)	Particle Mass μg/g Lung
Highest DU Exposure (PIII-2, gunner)[(a)]	46	90[(f)]
Environmental Exposure[(b)]	0.0025	--
Occupational Exposure[(c))]	0.72	--
Coal Miners (w/macules)[(d)]	--	5,400
Coal Miners (w/nodules)[(d)]	--	11,000
Coal Miners (w/PMF)[(d,e)]	--	15,000

(a) From Section 5.3.
(b) From Table 6.1.
(c) From Table 6.2, mean of four cases.
(d) Douglas et al (1986).
(e) Progressive Massive Fibrosis.
(f) Determined by assuming dust mass is twice the DU mass.

large lung burdens of coal dust. Although there is much dissimilarity related to exposure times and inherent toxicities of the dusts, the comparison gives a perspective on the potential health effects. One study of 430 British coal miners correlated the effects in the lung with the dust burdens (Douglas et al. 1986). Miners with burdens averaging 5400 µg/g lung had no fibrotic lesions (macules) larger than 1 mm. In contrast, those with an mean of 15,000 µg/g lung had fibrotic lesions greater that 10 mm, diagnostic of progressive massive fibrosis, a significant clinical lung disorder. If the dust resulting from the perforation of an armored vehicle with a DU munition is similar to or less toxic than coal dust, no clinical lung disease would be anticipated in the highest exposed group in the Capstone study.

6.5 RADIATION TOXICITY OF URANIUM

Natural uranium has not been shown to cause cancer in occupationally exposed humans, although a very small increased risk or a risk for some relatively uncommon cancers cannot be ruled out (Fulco et al. 2000). The fact that uranium emits radiation in the form of alpha particles and its immediate progeny emit beta and gamma radiation has created some concern for the potential radiation induction of cancers in exposed individuals. This concern exists in spite of the lack of epidemiologic evidence demonstrating a relationship between uranium and cancer.

6.5.1 Epidemiology of Uranium Exposure

Current epidemiologic evidence concerning the chronic toxicity, delayed toxicity, and carcinogenicity of uranium comes from studies of workers in facilities that processed and refined raw uranium ore into U-235-enriched uranium metal and oxides for use in nuclear weapons and nuclear reactors. The principal exposure in these workers is to uranium oxides and derivative uranium compounds produced during the refinement process. Uranium miners are not included because they worked in environments with potentially high radiation dose received from radon gas and its progeny that emanated from the uranium ore, not from the uranium itself. These epidemiologic studies are constrained to worker populations that were chronically or episodically exposed to uranium rather than acutely exposed as in the military scenarios that are the focus of this report.

The Royal Society also reviewed epidemiologic studies of workers in the uranium processing industry (The Royal Society 2001). These workers were exposed to low concentrations of uranium dust aerosols over many years, with potential acute spikes in exposure. They analyzed 14 major health studies of workers, 11 from the United States and three from the United Kingdom. The cause of death was determined in 33,502 workers who died of natural causes (Table 6.15). The observed number of deaths was compared with that expected in the general population. There was no increase in deaths from any cause, or from all cancers combined, or in individual types of cancer, or urogenital disease, including chronic urinary dysfunction. In fact, the central estimate of deaths for each cause-of-death category was less than the general population (the ratio of observed to expected less than 1). The reduced number of observed deaths in the uranium workers may be the result of the "healthy-worker" effect.

However, the lack of excess health effects in these occupationally exposed populations is clear. If exposure to uranium were a substantial hazard, it

Table 6.15. Summary of Mortality for Various Causes in Uranium Workers Compared with the General Population
(The Royal Society 2001, used with permission)

Cause of death[(a)]	Total number of deaths	O/E (95% CI)*	Estimated ratio of observed (O) to expected (E) deaths & 95% CI*
All causes	33502	0.86 (0.79–0.93)	
All cancers	7442	0.91 (0.85–0.97)	
Stomach cancer	365	0.76 (0.62–0.89)	
Colorectal cancer	728	0.91 (0.78–1.04)	
Liver cancer	123	0.84 (0.62–1.07)	
Lung cancer	2846	0.94 (0.83–1.05)	
Bone cancer	30	0.93 (0.53–1.33)	
Prostate cancer	490	0.98 (0.89–1.07)	
Bladder cancer	196	0.83 (0.71–0.96)	
Kidney cancer	151	0.78 (0.59–0.96)	
Brain cancer	223	0.91 (0.65–1.17)	
Thyroid cancer	7	0.38 (0.00–0.81)	
Non-Hodgkin lymphoma	266	0.82 (0.71–0.92)	
Hodgkin's disease	68	0.83 (0.61–1.06)	
Leukemia	295	0.90 (0.67–1.14)	
All genito-urinary diseases	318	0.70 (0.54–0.87)	

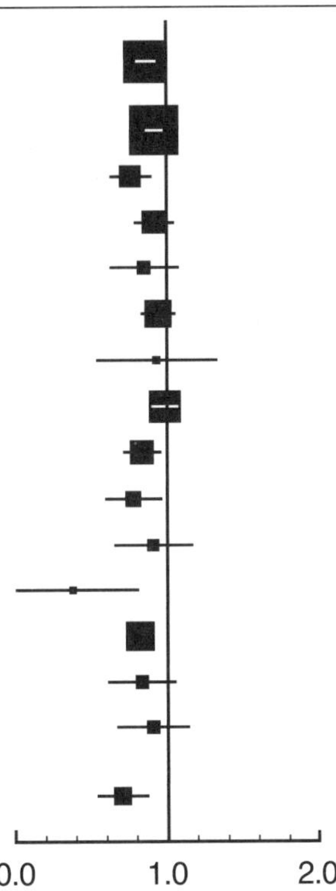

[(a)]Detailed results that contribute to each cause of death shown here are given in annexe I.
*Estimated ratio of observed number of deaths (O) to that expected for the general population (E) with 95% confidence intervals —■—.

should be apparent from these epidemiologic studies that involve more than 33,000 workers.

6.5.2 Effects in the Lung

Irradiation of the lung has been convincingly associated with lung cancer, but not from uranium exposure. The NAS Institute of Medicine, recently reviewed 11 epidemiologic studies of the workers in six major uranium-processing facilities in the United States (Fulco et al. 2000). These 11 studies were all of workers occupationally exposed to uranium compounds over a period of years. Of particular interest were two studies that detailed the risk of lung cancer in these workers and related this to radiation dose to the lung. In one report, the patterns of cancer mortality were studied in a cohort of 4014 uranium processing workers (Ritz 1999). These workers were exposed to insoluble compounds of uranium and small amounts of thorium and radium. Lung doses were estimated from routine urinalysis. Using the expected cancer mortality in 112 deceased workers, an increase in lung-cancer mortality was present in workers with 100 to 200 mSv (10 to 20 rem) cumulative dose from internal alpha radiation. The relative risk ratio was 1.92 (95% confidence interval of 0.53 to 6.96). A second study was of workers employed in four uranium processing or fabrication operations (Dupree et al. 1995). The primary exposures were to dust containing insoluble natural uranium. In addition, three of the facilities processed thorium, and one facility had enriched uranium present. A case-control study was conducted with 787 lung cancer cases and matched controls. Individual lung doses were determined from sums of the annual occupational intakes estimated from contemporary monitoring data. Odds ratios for lung cancer mortality did not show an increasing risk with increasing dose. An odds ratio of 2.0 was found for workers exposed to doses ≥25 rem, although there was great uncertainty in this estimate.

The NAS committee concluded from these 11 epidemiologic studies that there was "limited evidence indicating no association between exposure to uranium and lung cancer at cumulative doses lower that 200 mSv [20 rem]," and that there was "insufficient evidence to determine an association at higher doses" (Fulco et al. 2000).

6.5.3 Effects in the Bones

There is no direct evidence that natural uranium will cause bone cancers in humans, although enriched uranium will induce bone sarcomas in rats and mice (ATSDR 1999). However, uranium is retained in bone and distributes initially on bone surfaces, where radiosensitive cells reside, and then to the bone volume. Because of these characteristics, the potential for an increased risk of bone cancer has been explored (BEIR IV 1988). The risk from uranium in humans was inferred from the risk for a long-lived, bone-volume-seeking Ra-226. The estimate was 1.5 bone sarcomas per million persons ingesting an environmentally relevant dose of 0.18 Bq U per day (Mays et al. 1985). Recent studies have shown that there may be an effective threshold for the induction of bone sarcomas by alpha-emitting radionuclides, indicating that virtually no cancer would be expected from environmental concentrations of uranium (Rowland 1995; Lloyd et al. 2000).

In its review of epidemiology studies of workers in uranium processing facilities in the United States, the NAS Institute of Medicine concluded that there was insufficient evidence to determine an association between uranium exposure and bone cancer (Fulco et al. 2000).

6.5.4 Other Organs

The bone marrow might be considered a target organ because of the known deposition of uranium in the bone. However, the relevant epidemiologic experiences with alpha-emitting radionuclides in human bone have been reviewed with the conclusion that the risk from radiation-induced leukemias is insignificant relative to bone sarcoma (Mays et al. 1985).

The thoracic lymph nodes potentially accumulate large burdens of insoluble uranium compounds if lung burdens are large. There are no data from studies of humans to suggest that large burdens might result in cancers of the lymph nodes. In fact, studies of exposures to dogs suggest that cancers do not develop in the lymph nodes after high burdens and radiation dose from uranium dioxide inhaled for 5 yr (Leach et al. 1973). In addition, the NAS Institute of Medicine, in its review of epidemiology studies of workers in uranium processing facilities in the United States concluded that there was insufficient evidence to determine an association between uranium exposure and lymphatic cancer (Fulco et al. 2000).

6.6 RADIOLOGICAL HEALTH RISKS OF INHALED DU AEROSOL

The concern that DU may increase cancer is based on the knowledge that radiation doses from inhaled DU can be delivered to several organs of the body and that radiation is a known carcinogen. Biokinetic models predict that the dose of internally deposited DU is primarily to the lung and that other organs may accumulate much smaller doses. There is, however, no compelling evidence from human epidemiologic studies that uranium is associated with an increased cancer risk.

The basis approach used to determine the radiological health risks was to multiply the radiation doses for individual organs, as presented in Chapter 5, by the risk coefficients for fatal cancers. Two sets of risk coefficients were used to estimate the cancer risk. The first was the ICRP and the Interagency Steering Committee on Radiation Standards (ISCORS) methodology using committed effective whole body dose and risk coefficient to estimate the whole body cancer risk. The second approach used organ-specific risk coefficients developed by ICRP and U.S. Environmental Protection Agency (EPA) to estimate cancer risks for various organs of the body. Both of these approaches assume a linear, no-threshold dose response for cancer. This dose response is traditionally used for setting radiation protection guidelines. The validity of the approach has not been determined for estimating risks at low doses, such as those calculated for most organs after inhalation of DU, and may overestimate the risks.

6.6.1 Risk Coefficients for Total Cancer Mortality

The Interagency Steering Committee on Radiation Standards addressed the issue of estimating cancer risk from total effective dose equivalent for whole body irradiation (ISCORS 2002). The report provides guidance for estimating cancer risks for environmental exposures to radiation. The report recommends a risk coefficient of 6×10^{-4}/rem for fatal cancer from external whole-body radiation and for estimating risks from internally deposited radionuclides when there is uncertainty concerning the nuclide. The ICRP approach is comparable but uses a separate risk coefficient of 5×10^{-4}/rem for workers (ICRP [60] 1991).

6.6.2 Risk Coefficients for Lung Cancer

Based on the radiation dose to the lung, the primary radiological health risk for inhaled depleted uranium is a potential increase in lung cancer. Fatal lung cancer risk coefficients were examined from a number of sources. These are summarized in Table 6.16. There are no lung cancer risk coefficients for uranium that have been derived directly from studies of people exposed to uranium aerosols. Thus, risk coefficients based on other types of exposures were used instead to estimate lung cancer risks.

The ICRP and NCRP risk coefficients for fatal lung cancer for workers were taken directly from the cited sources and are based on epidemiologic studies of populations exposed to low-LET radiations (atomic bomb survivors and radiation therapy patients). The EPA risk coefficients were obtained using the uranium data from the EPA database (EPA 2002), which provides the risk per Bq intake for specified particle size and solubility (F, M, and S) for the lung for a specified age at intake. The age groups are 0 to 5 yr, 5 to 15 yr, 15 to 25 yr, 25 to 70 yr, and 0 to 110 yr. The conversion factor to risk per rem was made by applying the adult Bq to rem conversion factor to the fatal lung cancer risk per Bq for the 15 to 25 yr and 25 to 75 yr age groups. The values of moderate and slow for solubilization were averaged to obtain the risk coefficients in the table for the age groups listed. The mean risk coefficient in Table 6.16 is the population-weighted mean risk coefficient.

Table 6.16. Lung Cancer Lifetime Mortality Risk Coefficients

Reference	Radiation Type	Population	Lifetime Risk (risk/10^{-4} person-rem)
ICRP 60 (1991)	All Types	Public	0.85
		Workers	0.68
NCRP 115 (1993)	All Types	Public	0.85
		Workers	0.68
EPA (2002)	DU (estimated from EPA (2002) uranium data)	15-25 yr	1.00
		25-75 yr	0.60
		Population (weighted avg.)	0.80
BEIR IV (1988)	Internal Alpha	N/A	0.35
Koshurnikova et al. (1998)	Plutonium Dioxide	Workers (male smokers)	1.20
Grogan et al. (2001)	Plutonium Dioxide	Workers (male smokers)	0.65
		Workers (male non-smokers)	0.20
Mean of Grogan and Koshurnikova	Plutonium Dioxide	Smoking (weighted avg.)	0.40

The BEIR IV (1988) risk coefficient is for internally deposited transuranic elements; BEIR IV did not give a lung cancer risk coefficient specifically for inhaled uranium. The authors noted that there is "suggestive but not convincing evidence of deleterious human effects of chronic exposure to uranium dust." The risk coefficient is based primarily on the risk from inhaled radon and radon progeny. The risk coefficient from Koshurnikova et al. (1998) is based on an epidemiologic study of plutonium production workers in Russia. Essentially all the workers smoked cigarettes and were exposed to relatively high levels of plutonium over their lifetime. Grogan et al. (2001) developed risk coefficients for inhaled plutonium using an approach that combined the results of epidemiologic studies of the Russian plutonium workers, of persons exposed to other alpha emitters or low-LET radiation and of animals exposed to plutonium aerosols. A best estimate of the coefficients for lung cancer risk from plutonium dioxide was derived from Grogan et al. (2001) and Koshurnikova et al. (1998).

The EPA database (EPA 2002) provides an estimate of risk based on age, but none of the age groups match the military age range. The impact of the difference in age categories is illustrated in Figure 6.9. This graph is representative for all uranium isotopes and adsorption Types M and S. The younger age groups (0 to 5yr, 5 to 15 yr) were inappropriate for the population under consideration. The risks for these age groups are significantly higher than the age groups found in the military. The EPA's risk coefficient for the age range of 0 to 110 yr was not used for the best estimate because the high values of risk for the younger age groups would also overestimate the risk for U.S. military personnel. There is no easy method to adjust the 15- to 25-yr age bracket to remove the impact of the 15- and 16-yr-old subjects on the number calculated. However, any adjustment will lower the estimate and will bring the ICRP/NCRP and EPA (2002) risk coefficients into closer agreement.

The Grogan et al. (2001) and Koshurnikova et al. (1998) data were not used in this analysis because the data were not adjusted for smoking. The smoking rates for the Russian plutonium worker population were significantly higher than the rates in the U.S. general population or the U.S. Army population, both of which are about 30%. Averaging the risk coefficients using smoking percentages yielded a number that was lower than the ICRP/NCRP risk coefficients. The validity of this averaging technique is suspect without understanding how smoking influenced the results of the study. The second

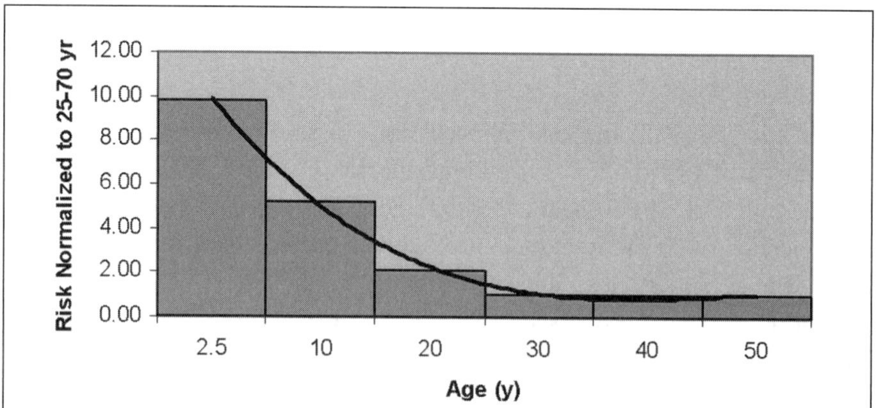

Figure 6.9. Normalized Risk from the Intake of U-238 (EPA 2002), Which Estimates Isotope-Specific Risks

set of confounders deals with attempts to relate a low-dose, low-dose-rate DU exposure to high-dose, high-dose-rate plutonium exposures.

The ICRP/NCRP lung cancer risk coefficient for worker populations ($0.68/10^{-4}$ person-rem) was used to estimate the risk for lung cancer mortality from inhaled DU. This risk coefficient is similar to others in the Table 6.17. In addition, the age range for worker populations, starting at 18 yr and continuing for a working lifetime, is close to the age structure of the U.S. Army. In this analysis, it is assumed that the workers' ages start at 18 yr (assumes that any 17-yr-old recruits would be 18 by the time they were trained and deployed) and continue through the working life of the individual. Recent data from the U.S. Army showed that 16% of U.S. Army personnel were between 17 to 20 yr; 26% were in the 21 to 24 age range; and the remainder (58%) was 25 years and older (Maxfield undated).

Table 6.17. Nominal Probability Coefficients for Fatal Cancer (Worker Population)

Organ/Tissue	Probability of Fatal Cancer[(a,)] $(10^{-4}\ \text{rem})^{-1}$
Bladder	0.24
Bone Marrow	0.40
Bone Surface	0.04
Breast	0.16
Colon	0.68
Extrathoracic Tissues	0.10[(b]
Kidney	0.11[(c)]
Liver	0.12
Lung	0.68
Oesophagus	0.24
Ovary	0.08
Skin	0.02
Stomach	0.88
Thyroid	0.06
Remainder	0.19
Whole Body	4.00

(a) Based on ICRP 60, Table 4 (ICRP 1991)
(b) See Section 6.6.3.
(c) Based on EPA GMC model (Puskin and Nelson 1995)

6.6.3 Risk Coefficients for Cancer in Other Organs

Organs with higher radiation doses and known radiation sensitivity were evaluated for fatal cancer risks. Based on previous analyses, this included the following organs and tissues: lung, liver, bone surface, and red marrow (The Royal Society 2001). The ICRP/NCRP-recommended risk coefficients for workers were used to estimate cancer risks for bone surface and red marrow.

These two tissues had low calculated radiation doses from inhaled DU and a known high susceptibility to radiation-induced cancer. The specific risk coefficients for these two tissues were taken from ICRP 60 and are noted in Table 6.17 (ICRP 1991). The kidney and extrathoracic tissues (which include the anterior nose, posterior nasal passages, pharynx, larynx, mouth, and thoracic lymph nodes) are not given specific risk coefficients by ICRP 60, but are included in the remainder group of tissues. The risk coefficient for the kidney was taken from the EPA Geometric Mean Coefficient (GMC) model (Puskin and Nelson 1995). No specific risk coefficients are published for the ET, for which relatively high doses were calculated (Table 5.2). These tissues are not particularly sensitive to radiation-induced cancer, one reason no risk coefficients have been developed for them. Studies in dogs with high burdens of inhaled, highly radioactive plutonium have shown that lymph nodes that accumulate large burdens of plutonium are not target organs for cancer (Hahn et al. 1999). The other ET tissues are tubular structures that are only irradiated as the DU passes through them on the way to the lung or digestive tract. Their sensitivity to radiation is most likely similar to skin. A coefficient of 0.1×10^{-4} rem^{-1} for fatal cancer in the ET tissues was chosen as a default value. This value is equivalent to the coefficients for liver and kidney and five times the coefficient for skin. Thus the coefficient is probably conservative and over predicts the risk in the ET region.

Although not addressed in earlier publications on radiological protection (ICRP [60] 1991), tissue-weighting factors for the ET tissues, which represent the relative contribution of these tissues to the total cancer risk from uniform irradiation of the whole body, were published in ICRP 69 (1994). In the procedures presented in ICRP 69, the ET tissues were separated into three compartments with different weighting factors. A weighted mean of 0.025 was used in the dose estimation in this health risk assessment.

A number of the organs and tissues listed in Table 6.17 have very low calculated doses from inhaled DU (The Royal Society 2001). No cancer risks were calculated for these tissues (bladder, breast, colon, esophagus, ovary, skin, stomach, thyroid, and remainder tissues).

6.6.4 Comparison of Whole Body Cancer Risks and Summed Organ-Specific Cancer Risks

The doses to the organs and tissues of interest are noted Table 6.18. The organs and tissues listed are those with the highest calculated committed equivalent doses and a high sensitivity to radiation carcinogenesis. Mean committed effective doses (50-yr doses) are listed for whole body, and

Table 6.18. Mean Radiation Doses to Various Organs (50-yr committed dose, rem)

Phase	Whole Body[(a)]	Lung[(b)]	Extra-thoracic[(b)]	Bone Surface[(b)]	Red Marrow[(b)]	Kidney[(b)]	Liver[(b)]
Scenario A							
Phase I	2.00	14.3	8.87	0.987	0.102	0.361	0.135
Phase II	0.676	5.21	1.22	0.382	0.0395	0.140	0.0524
Phase III	2.60	17.5	19.3	0.381	0.0394	0.140	0.0523
Phase IV	0.0900	0.662	0.373	0.0226	0.00233	0.00828	0.00310
Scenario B							
Phase I	4.42	32.1	18.1	2.21	0.229	0.810	0.303
Phase II	1.81	14.0	3.04	1.03	0.107	0.378	0.141
Phase III	6.71	44.1	54.6	0.913	0.0943	0.335	0.125
Phase IV	0.436	3.32	1.28	0.100	0.0103	0.0367	0.0137
Scenario C							
Phase I	5.32	38.3	23.2	2.73	0.282	1.00	0.374
Phase II	2.60	20.0	4.39	1.52	0.157	0.557	0.209
Phase III	9.08	60.0	72.8	1.25	0.129	0.459	0.172
Phase IV	1.02	7.58	4.12	0.181	0.0187	0.0664	0.0248
Scenario D							
Phase I	5.40	39.0	23.0[(c)]	2.79	0.289	1.02	0.383
Phase II	2.88	22.2	4.97	1.70	0.175	0.622	0.233
Phase III	9.36	61.3	77.5	1.31	0.135	0.479	0.179
Phase IV	1.20	8.65	5.88	0.210	0.0217	0.0771	0.0288
Scenario E							
Phase I	1.20	8.81	4.24	0.647	0.0668	0.237	0.0887
Phase II	0.856	6.67	1.20	0.485	0.0501	0.178	0.0665
Phase III	2.01	13.6	14.4	0.263	0.0271	0.0962	0.0360
Phase IV	0.412	3.14	1.30	0.0515	0.00532	0.0189	0.00707

(a) Committed effective dose
(b) Committed equivalent dose
(c) The sampler data used to calculate Scenarios C and D differed slightly and was responsible for the lower dose for the longer exposure time.

committed equivalent doses (H_T) are listed for the various organs after inhalation exposure to aerosols of DU. (Medians were also calculated for lung and kidney for use with risk evaluations.) The lung has the highest committed equivalent doses in all scenarios for all phases.

Risks based on the summed tissue/organ doses were obtained by multiplying the organ dose by an organ-specific fatal cancer risk factor. This approach was used because it effectively deals with the inherent non-uniformity in dose distribution among the organs, with the majority of the dose being delivered to the lung. Table 6.19 compares the *mean* E(50) values calculated using standard ICRP procedures with the value that the E(50) values would have to be to represent a risk equivalent to the risk obtained by summing the risks for the individual organs, $E(50)_{\Sigma Org.}$ $E(50)_{\Sigma Org}$ was calculated by dividing the summed organ risk by the ICRP risk per rem for whole body exposure. The $E(50)_{\Sigma Org}$ doses range from 28 to 46% higher than the E(50) values

Table 6.19. Comparison of Mean E(50) and $E(50)_{\Sigma Org}$ for Use in Risk Analysis

Scenario	Abrams Tank, Non-DU Armor, No Ventilation		Abrams Tank, DU Armor No Ventilation		Abrams Tank, DU Armor, EC/NBC On		Bradley Vehicle, No Ventilation	
	E(50) (rem)	$E(50)_{\Sigma Org}$ (rem)	E(50) (rem)	$E(50)_{\Sigma Org}$ (rem)	E(50) (rem)	$E(50)_{\Sigma Org}$ (rem)	E(50) (rem)	$E(50)_{\Sigma Org}$ (rem)
A - Crew, exit in 1 min	1.97	2.68	2.16	2.88	0.0900	0.124	0.589	0.841
B - Crew, exit in 5 min	3.66	5.00	6.00	7.92	0.436	0.620	1.73	2.47
C - Crew, exit in 1 h	4.85	6.67	8.33	10.9	1.02	1.42	2.08	2.98
D - Crew, exit in 2 h	4.95	6.90	8.68	11.1	1.20	1.62	2.44	3.50
E - First responders	0.916	1.26	1.94	2.59	0.412	0.587	0.890	1.30

for these scenarios, which means the summed organ risks also ranged from 28 to 46% higher than the risks based upon whole body committed effective doses.

Organ-specific cancer risks were summed to estimate the total cancer risk because the risk to the whole body using the ICRP 60 approach was less than the sum of the organ risks. The ICRP states that, in cases in which the radionuclide is not uniformly distributed in the body, the risk estimation procedure noted above may underestimate the risk when compared to the summation of the risks to each of the target organs (ICRP [60] 1991). This observation is true for internally deposited DU.

6.6.5 Cancer Risks by Phase and Scenario

The ICRP 60/NCRP 115 approach and risk coefficients for cancer mortality were applied to all phases and scenarios. Risks associated with the doses to the crewmembers inside the vehicle at the time of perforation and the first responders were evaluated. The risks for cancer mortality are based on the sum of the individual organ risks. The lifetime cancer mortality risk from inhalation of DU aerosols was calculated using generic ICRP risk coefficients and the conservative Linear, No-Threshold theory of effect that may overestimate risks at the levels predicted in this study and are, therefore, thought to be protective of health.

Table 6.20 shows the mean risks for fatal cancers from DU inhalation in various organs and tissues based on the mean doses in Table 6.18. The results are the best estimates of risks to the crewmembers for each armored vehicle type and each selected scenario for initial vehicle configurations that will maximize the aerosol concentrations within each vehicle. For all shots, except for Phase IV, no on-board ventilation or EC/NBC ventilation systems were present, and the hatches were closed at the time of perforation, though most shots forced a hatch open at least partially. For Phase IV, the hatches

Table 6.20. Risks for Lifetime Cancer Mortality After DU Aerosol Inhalation (Probability of cancer mortality [$\times 10^{-4}$])

Phase	Whole Body	Lung	Extra-thoracic	Bone Surface	Red Marrow	Kidney	Liver	Organ Sum
Scenario A								
Phase I	7.98	9.75	0.887	0.0395	0.0408	0.0398	0.0162	10.8
Phase II	2.71	3.54	0.122	0.0153	0.0158	0.0154	0.00629	3.72
Phase III	10.4	11.9	1.93	0.0153	0.0157	0.0154	0.00627	13.9
Phase IV	0.360	0.450	0.0373	0.000902	0.000932	0.000911	0.000372	0.491
Scenario B								
Phase I	17.7	21.8	1.81	0.0886	0.0914	0.0892	0.0364	23.9
Phase II	7.23	9.50	0.304	0.0413	0.0426	0.0415	0.0170	9.94
Phase III	26.8	30.0	5.46	0.0365	0.0377	0.0368	0.0150	35.7
Phase IV	1.74	2.26	0.128	0.00400	0.00413	0.00404	0.00165	2.40
Scenario C								
Phase I	21.3	26.1	2.32	0.109	0.113	0.110	0.0449	28.8
Phase II	10.4	13.6	0.439	0.0609	0.0629	0.0613	0.0250	14.3
Phase III	36.3	40.8	7.28	0.0501	0.0517	0.0505	0.0206	48.3
Phase IV	4.09	5.15	0.412	0.0072	0.00747	0.0073	0.00298	5.60
Scenario D								
Phase I	21.6	26.5	2.30	0.112	0.115	0.113	0.0460	29.2
Phase II	11.5	15.1	0.497	0.0680	0.0702	0.0684	0.0279	15.8
Phase III	37.4	41.7	7.75	0.0523	0.0540	0.0527	0.0215	49.7
Phase IV	4.78	5.88	0.588	0.00841	0.00868	0.00848	0.00346	6.51
Scenario E								
Phase I	4.79	5.99	0.424	0.0259	0.0267	0.0260	0.0106	6.50
Phase II	3.42	4.53	0.120	0.0194	0.0200	0.0195	0.00798	4.72
Phase III	8.04	9.28	1.44	0.0105	0.0109	0.0106	0.00432	10.8
Phase IV	1.65	2.14	0.130	0.00206	0.00213	0.0021	0.000848	2.28

were closed at the time of perforation, and the on-board ventilation systems were off, but the EC/NBC ventilation system was on.

The lung is the organ that by far is associated with the highest cancer mortality risk, accounting for 97% or greater of the total risks from the summed organs. The cancer risks for the other organs are comparatively insignificant.

The various scenarios are based on the time in the vehicle, which has a marked influence on the intake of DU and the subsequent dose and risk. The majority of the intake in the vehicle occurs in less than an hour after perforation. The difference in the risk associated with that intake between 5 min and 1 h is about 36%, and between 1 h and 2 h is about 3% for the Abrams tank (mean of results with and without DU armor and with no ventilation). The increase between 1 and 2 h for the Abrams tank with DU armor and with the EC/NBC ventilation system on is at least 14%, and the increase for the Bradley vehicle is about 21%. The overall risk for the latter two vehicle configurations is well below the risks for the Abrams tank with the EC/NBC ventilation system off. Particle size distribution differences may explain the difference. In general, the particle sizes found in the Abrams tank with the EC/NBC ventilation system on and in the Bradley vehicle were smaller as time progressed.

The risks for the first responders are lower than any of the crewmembers, except those crewmembers that exit within 1 min. With regard to an increased air exchange rate caused by hatch openings, the risks estimated for the first responders may be similar to most of the actual shots during which a hatch was forced open. By definition the hatches would need to be open for the first responders to enter or exit. Opening the hatches would foster air exchange within the vehicle and reduce the concentration from the initial DU strike more rapidly than estimated from these data.

The risks for increased lifetime cancer mortality resulting from the minimum 10th percentile, median (50th percentile), 90th percentile, and maximum doses calculated for the crewmembers in various scenarios are given in Table 6.21. These doses are based on results from single shots. The maximum and minimum doses were used as examples of the range of the data and resulting risks.

Table 6.21. Risks for Increased Lifetime Cancer Mortality from DU Aerosol Inhalation (Probability of cancer mortality [$\times 10^{-4}$])

Phase	Minimum	Percentile 0.1	Median	0.9	Maximum
Scenario A					
Phase I	0.996	6.90	10.7	14.3	23.8
Phase II	0.806	0.866	3.36	5.86	13.9
Phase III	8.00	8.00	11.5	24.1	24.1
Phase IV	NA	NA	0.495	NA	NA
Scenario B					
Phase I	4.92	16.5	20.0	40.1	61.1
Phase II	1.79	1.86	9.87	18.0	31.1
Phase III	24.3	24.3	31.7	51.8	51.8
Phase IV	NA	NA	2.48	NA	NA
Scenario C					
Phase I	8.18	17.3	26.7	44.1	51.7
Phase II	5.71	5.71	11.9	39.8	39.8
Phase III	31.8	31.8	43.7	64.0	64.0
Phase IV	NA	NA	5.67	NA	NA
Scenario D					
Phase I	9.51	16.0	27.6	44.2	53.9
Phase II	6.50	6.50	14.0	41.1	41.1
Phase III	33.3	33.3	44.5	65.3	65.3
Phase IV	NA	NA	6.47	NA	NA
Scenario E					
Phase I	1.81	2.52	5.04	10.5	18.0
Phase II	1.28	1.55	5.20	7.74	9.59
Phase III	5.85	5.85	10.4	15.7	15.7
Phase IV	NA	NA	2.35	NA	NA

na = not applicable

6.7 SUMMARY OF HEALTH RISKS

Review of the human epidemiologic studies of workers occupationally exposed to uranium compounds indicates that kidney toxicity is possible for high acute exposures and that there is no clear association with long-term effects, such as cancer. Organ concentrations and radiological doses associated with the Capstone Level I scenarios were used to estimate the human health risks from inhaled DU. The health risks are low and relate to kidney dysfunction and lung cancer.

6.7.1 Chemical Health Effects

Data obtained from a literature review of acute uranium exposures in humans and the associated renal effects were used to characterize chemical risk to the kidney. These data formed the basis of the assessment of probable adverse health effects to individuals exposed to DU aerosols consistent with Level I scenarios. Biochemical effects were first given a score according to severity of the effect. A discriminant analysis was then used to build a predictive model, in which a REG was determined based on the score and the predicted peak kidney uranium burden for each case.

Each phase and scenario was modeled to estimate peak kidney uranium burden, and these kidney uranium burdens were compared with the REGs and assigned a score. The concentrations, relative to the likely acute renal effects and predicted outcomes, are categorized by REGs in Table 6.22.

For all scenarios, median, 10th-, and 90th-percentile kidney uranium burdens were determined across all shots within that phase and scenario. The median (50th), 10th-, and 90th- percentile kidney uranium concentrations per shot and their assigned REGs by phase and scenario are listed in Table 6.23 and illustrated in Figure 6.8 (p. 147). The asterisk (1*) indicates an REG 1 that exceeded the *de facto* occupational 3 μg U/g kidney guideline.

Table 6.22. Risk Characterization—Chemical Risk to the Kidney

REG	Kidney Uranium Concentration (μg U/g kidney tissue)	Acute Renal Effect	Predicted Outcome
0	≤2.2	No Detectable Effects	No Detectable Effects
1	>2.2 to ≤6.4	Possible Transient Indicators of Renal Dysfunction	Not Likely to Become Ill
2	>6.4 to ≤18	Possible Protracted Indicators of Renal Dysfunction	May Become Ill
3	>18	Possible Severe Clinical Symptoms of Renal Dysfunction	Likely to Become Ill

Table 6.23. Peak Kidney Uranium Concentrations, Percentiles and Their Respective REGs

Scenario	10th Percentile (µg U/g Kidney)	REG	Median (µg U/g Kidney)	REG	90th Percentile (µg U/g kidney)	REG
Abrams Tank through non-DU Armor (Phase I)						
A. Crew, Exit in 1 min	1.94	0	2.99	1	4.20	1*
B. Crew, Exit in 5 min	4.64	1*	6.44	1/2	9.97	2
C. Crew, Exit in 1 h	5.34	1*	8.17	2	14.0	2
D. Crew, Exit in 2 h	5.12	1*	8.03	2	16.1	2
E. First Responders	0.815	0	1.54	0	4.77	1*
Bradley Fighting Vehicle (Phase II)						
A. Crew, Exit in 1 min	0.271	0	0.988	0	1.63	0
B. Crew, Exit in 5 min	0.611	0	2.88	1	4.96	1*
C. Crew, Exit in 1 h	1.43	0	3.53	1*	11.5	2
D. Crew, Exit in 2 h	1.83	0	4.02	1*	11.9	2
E. First Responders	0.364	0	1.41	0	1.98	0
Abrams Tank through DU Armor (Phase III)						
A. Crew, Exit in 1 min	0.566	0	1.08	0	2.91	1
B. Crew, Exit in 5 min	1.72	0	2.58	1	7.98	2
C. Crew, Exit in 1 h	2.44	1	3.55	1*	10.8	2
D. Crew, Exit in 2 h	2.70	1	3.66	1*	11.2	2
E. First Responders	0.344	0	0.672	0	2.32	1
Fully Operating Abrams Tank, through DU Armor (ventilating system functioning) (Phase IV)[(a)]						
A. Crew, Exit in 1 min	NA	NA	0.0542	0	NA	NA
B. Crew, Exit in 5 min	NA	NA	0.225	0	NA	NA
C. Crew, Exit in 1 h	NA	NA	0.458	0	NA	NA
D. Crew, Exit in 2 h	NA	NA	0.556	0	NA	NA
E. First Responders	NA	NA	0.140	0	NA	NA
(a) There was only one data point for each Phase IV scenario.						

Even for the 90th percentile, none of the estimated kidney uranium burdens exceeded REG 2. The figure also illustrates the significant impact of ventilation (Phase IV) on the peak kidney uranium concentration. The Phase IV results are well below both the current *de facto* occupational guideline for the kidney and the lowest value from Phases I, II, and III for each scenario tested. Activating the Bradley vehicle ventilation system would also reduce the kidney uranium concentration for the Phase-II scenarios, especially for the longer stay-times.

6.7.2 Radiological Health Effects

The primary radiological health risk for internally deposited DU is an increased potential for developing cancer. Risks for lifetime cancer mortality were estimated by summing the risks for the organs with relatively high radiation doses and known radiosensitivity. Risk coefficients for lung, liver, bone surface, and marrow cancer mortality were taken from ICRP and coefficients from EPA were used in the calculation of kidney cancer mortality.

The highest *median* risk for lifetime cancer mortality from DU inhalation found for any of the vehicle configurations tested and any of the five scenarios 44.5×10^{-4} (from Table 6.20). This risk was for Scenario D, a crewmember staying for 1 to 2 h in an unventilated Abrams tank perforated through DU armor. A risk of 44.5×10^{-4} (or 0.00445) implies that, on the average, over an individual's lifetime, their chance of dying of cancer caused by the exposure is about 0.45%. This means the risk of dying from cancer would increase from the natural or background risk in the U.S. of 23.6% (Ries et al. 2003) to about 24.0%. The *maximum* lifetime cancer risk in any of the vehicle configurations and scenarios was 0.65%, resulting from the estimated exposures in Scenario D (a 2-h stay in an unventilated Abrams tank perforated through DU; in this case the 90th percentile and maximum were the same). This raises an individual's risk from 23.6% to about 24.3%.

The median risks for each of the tested configurations for each of the five scenarios are presented in Table 6.24. The risks of cancer mortality from DU aerosol inhalation are based on the sum of the individual organ risks. The highest DU aerosol concentration and associated risks on a rate basis occur immediately after perforation, and for occupants remaining in the vehicle, the majority of the intake in the vehicle occurs in less than 1 h after perforation. Exiting the vehicle within the first few minutes minimizes doses and risks, although these accrue more slowly with time. The generally low risks to an occupant of an unventilated Abrams tank in which the DU armor is perforated and who exits within 1 min are estimated at a 0.12% lifetime increased risk of cancer mortality from DU exposure. The estimated risk of remaining in this vehicle for 1 to 2 h increases to 0.45%.

The results are the best estimates of the crew risks in this study resulting from a single shot for each type of armored vehicle. The equipment configurations

Table 6.24. Median Risk Estimates for Lifetime Cancer Mortality from DU Aerosol Inhalation

Scenario	Abrams Tank – Through Non-DU Armor (Phase I)[a]	Abrams Tank – Through DU Armor (Phase III)[a]	Abrams Tank Through DU Armor NBC On (Phase IV)[b]	Bradley Vehicle (Phase II)[a]
Crewmember Risks				
A. Exits 1 min post shot	0.107%	0.115%	0.00495%	0.0336%
B. Exits 5 min post shot	0.200%	0.317%	0.0248%	0.0987%
C. Exits 1 h post shot	0.267%	0.437%	0.0567%	0.119%
D. Exits 2 h post shot	0.276%	0.445%	0.0647%	0.140%
First Responder Risks				
E. Entry 5 min post shot and stays 10 min	0.0504%	0.104%	0.0235%	0.0520%

(a) Hatches closed, no ventilation or EC/NBC ventilation systems operating
(b) Hatches closed, ventilation system off, EC/NBC ventilation system on

for the vehicles (except Phase IV) maximized the aerosol concentrations within each vehicle. No ventilation or EC/NBC ventilation overpressure systems were operating, and the hatches were closed at the time of perforation.

The effectiveness of the EC/NBC ventilation systems in markedly reducing the cancer risk is shown in Table 6.24. For Phase IV, the hatches were closed at the time of perforation, the onboard ventilation systems were off, but the EC/NBC ventilation system was on. The estimated lifetime cancer risk from DU aerosol inhalation from this shot was 4 to 23% that of the unventilated tank perforated similarly through DU armor. The estimated cancer risks in the Bradley vehicle were one-third to one-half that of the DU armored Abrams tank.

Among all organs, the lung had the highest lifetime risk for fatal cancer, accounting for greater that 97% of the total cancer risk. The highest *mean* probability of lifetime lung cancer mortality from DU aerosol inhalation predicted in any of the scenarios was 41.7×10^{-4} or 0.42% (from Table 6.20). This risk was predicted for Scenario D, a crewmember in the vehicle at the time of perforation who remains inside for 2 h an unventilated Abrams tank that has been perforated through DU armor. A risk of fatal cancer of 0.42% implies that the chance of dying of lung cancer is increased from the natural or background risk of 7.35% for U.S. males (Ries et al. 2003) to 7.77%. This background rate includes lung cancer due to smoking. The fraction of Soldiers who smoke–approximately 30% (U.S DoD 1998)—is roughly equivalent to the fraction in the general population of approximately 24% (Freid 2003).

These results of this chemical and radiological health risk assessment indicate that the estimated risks for kidney damage and cancer mortality from inhaled DU are relatively low based on the results of the Capstone study. Even so, these risks can be significantly reduced by using existing ventilation systems and by reducing time in the vehicle immediately after a shot.

6.8 REFERENCES

Agency for Toxic Substances and Disease Registry (ATSDR). 1999. *Toxicological Profile for Uranium.* Report TP 90-29, Atlanta, Georgia.

Arfsten DP, KR Still, and GD Ritchie. 2001. "A Review of the Effects of Uranium and Depleted Uranium Exposure on Reproduction and Fetal Development." *Toxicology and Industrial Health* 17:180191.

Benson KA and SA McBride. 1997. "Uranium Levels in the Fetus and Placenta of Female Rats Implanted with Depleted Uranium Pellets Prior to Breeding." *The Toxicologist* 36:258.

Barber DS and MJ Kopplin. 2003. "Regional Distribution of Uranium in Rat Brain." *The Toxicologist* 72:(S-1).

Biological Effects of Ionizing Radiations (BEIR IV). 1988. *Chapter 6. Uranium* In *Health Risks of Radon and Other Internally Deposited Alpha-Emitters*. Committee on the Biological Effects of Ionizing Radiations, National Research Council, National Academy Press, Washington DC.

Boback MW. 1975. "A Review of Uranium Excretion and Clinical Urinalysis Data in Accidental Exposure Cases." In *Conference Proceedings on Occupational Health Experience with Uranium*, April 28–30, 1975, Arlington Virginia, Document ERDA 93, pp. 226-243, National Technical Information Service, Springfield Virginia.

Butterworth A. 1955. "Significance and Value of Uranium in Urine Analysis." *Transactions of the Association of Industrial Medical Officers* 5:36-43.

Diamond GL, PE Morrow, BJ Panner, RM Gelein, and RB Baggs. 1989. "Reversible Uranyl Fluoride Nephrotoxicity in the Long Evans Rat." *Fundamental and Applied Toxicology* 13:65-78.

Domingo JL, JL Peternain, JM Llobet, and J Corbella. 1989. "The Developmental Toxicity of Uranium in Mice." *Toxicology* 55:143-152.

Domingo JL. 1994. "Metal-Induced Developmental Toxicity in Mammals: A Review." *Journal of Toxicology and Environmental Health* 42:123-141.

Donoghue JK, ED Dyson, JS Hislop, AM Leach, and NL Spoor. 1972. "Human Exposure to Natural Uranium." *British Journal of Industrial Medicine* 29:81-89.

Douglas AN, A Robertson, JS Chapman, and VA Ruckley. 1986. "Dust Exposure, Dust Recovered from the Lung, and Associated Pathology in a Group of British Coalminers." *British Journal of Industrial Medicine* 43:795-801

Dupree EA, JP Watkins, JN Ingle, PW Wallace, CM West, and WG Tankersley. 1995. "Uranium Dust Exposure and Lung Cancer Risk in Four Uranium Processing Operations." *Epidemiology* 6:370-375.

Durbin PW. 1984. "Metabolic Models for Uranium." In *Biokinetics and Analysis of Uranium in Man*, ed. TH Moore, U.S. Transuranium Registry, USUR-05, HEHF-47, F1-F65, National Technical Information Service, Springfield, Virginia.

Fisher DR, MJ Swint, and RL Kathren. 1990a. "Evaluation of Health Effects in Sequoyah Fuels Corporation Workers from Accidental Exposure to Uranium Hexafluoride." NUREG/CR-5566, US Nuclear Regulatory Commission, Washington, DC.

Fisher DR, RL Kathren, and MJ Smint. 1990b. "Modified Biokinetic Model for Uranium from Analysis of Acute Exposure to UF_6." *Health Physics* 60:335-342.

Freid V, Prager K, MacKay A, and H Xia. 2003. *Chartbook on Trends in the Health of Americans. Health, United States, 2003*. National Center for Health Statistics, Hyattsville, Maryland.

Fulco CE, CT Liverman, and HC Sox, eds. 2000. "Depleted Uranium." In *Gulf War and Health*, Vol. 1, Chapter 4, National Academy Press, Washington, DC.

Grogan HA, WK Sinclair, and PG Voilleque. 2001. "Risks of Fatal Cancer from Inhalation of ^{239,240}Plutonium by Humans: A Combined Four-Method Approach with Uncertainty Evaluation." *Health Physics* 80:447-461.

Hahn FF, BA Muggenburg, RA Guilmette, and BB Boecker. 1999. "Comparative Stochastic Effects of Inhaled Alpha- and Beta-Emitting Radionuclides in Beagle Dogs." *Radiation Research 152*:S19-S22.

Harley NH, FC Foulkes, LH Hilborne, A Hudson, and CR Anthony. 1999. *A Review of the Scientific Literature As It Pertains to Gulf War Illnesses: Volume 7, Depleted Uranium.* MR-1018/7-OSD, RAND, Santa Monica, Calfornia. Online report available at URL: www.gulflink.osd.mikl in the Environmental Exposure Reports Section.

Hodge HC, JN Stannard, and JB Hursh, eds. 1973. "Uranium, Plutonium, Transplutonium Elements." *Handbook of Experimental Pharmacology*, Vol 36, Springer-Verlag, New York.

Hooper FJ, KS Squibb, EL Siegel, K McPhaul, and JP Keogh. 1999. "Elevated Urine Uranium Excretion by Soldiers with Retained Shrapnel." *Health Physics* 77:512-519.

Howland JW. 1949. "Chapter 16. Studies on Human Exposures to Uranium." In *The Pharmacology and Toxicology of Uranium Compounds*, eds. Voegtlin C and HC Hodge, Natl. Nuclear Energy Series, McGraw-Hill Book Co., New York.

Interagency Steering Committee on Radiation Standards (ISCORS). 2002. *A Method for Estimating Radiation Risk from TEDE.* ISCORS Technical Report 1, Washington, DC.

International Commission on Radiological Protection (ICRP). 1959. *Report on the Committee II on Permissible Dose for Internal Radiation.* ICRP Publication 2, Pergamon Press, Oxford, United Kingdom.

International Commission on Radiological Protection (ICRP). 1991. *Recommendations of the International Commission on Radiological Protection.* ICRP Publication 60, Pergamon Press, Oxford, United Kingdom.

International Commission on Radiological Protection (ICRP). 1994. *Human Respiratory Tract Model for Radiological Protection.* ICRP Publication 66, Pergamon Press, Oxford, United Kingdom.

International Commission on Radiological Protection (ICRP). 1995a. *Age-Dependent Doses to Members of the Public from Intake of Radionuclides: Part 3 Ingestion Dose Coefficients.* ICRP Publication 69, Pergamon Press, Oxford, United Kingdom.

International Commission on Radiological Protection (ICRP). 1995b. *Age-Dependent Doses to Members of the Public from Intake of Radionuclides: Part 4 Inhalation Dose Coefficients.* ICRP Publication 71, Pergamon Press, Oxford, United Kingdom.

International Commission on Radiological Protection (ICRP). 1997. *Individual Monitoring for Internal Exposure of Workers.* ICRP Publication 78, Pergamon Press, Oxford, United Kingdom.

International Commission on Radiological Protection (ICRP). 2003. *Report of the Task Force on Reference Man.* ICRP Publication 89, Pergamon Press, Oxford, United Kingdom.

Interagency Steering Committee on Radiation Standards (ISCORS). 2002. *A Method for Estimating Radiation Risk from TEDE.* ISCORS Technical Report 1, Washington, DC.

Jortner B and D Barber. 2003. "Multifactorial Assessment of Depleted Uranium Neurotoxicity." *Govt. Reports Announcements & Index*, Issue 24.

Kathren RL, JF McInroy, RH Moore, and SE Dietert. 1989. "Uranium in the Tissues of an Occupationally Exposed Individual." *Health Physics* 57:17-21.

Kathren RL and RH Moore. 1986. "Acute Accidental Inhalation of U: A 38-Year Follow-Up." *Health Physics* 51:609-619.

Koshurnikova NA, MG Bolotnikova, LA Ilyin, IB Keirim-Markus, ZS Menshikh, PV Okatenko, SA Romanov, VI Tsvetkov, and NS Shilnikova. 1998. "Lung Cancer Risk Due to Exposure to Incorporated Plutonium." *Radiation Research* 149:366-371.

Kurlyandskaya EB. 1970. "The Effect of U_3O_8 on the Body of the Pregnant Female and the Foetus." In *The Toxicology of Radioactive Substances*, eds. AA Letavet and EB Kurlyandskaya, translated by A Crozy, translation edited by GW Dolphin, pp. 109-119, Pergamon Press, Oxford, United Kingdom.

Kurttio P, A Auvinen, L Salonen, H Saha, J Pekkannen, I Mäkeläinen, SB Väisänen, IM Penttilä, and H Komulainen. 2002. "Renal Effects of Uranium in Drinking Water." *Environmental Health Perspectives* 110:337342.

Leach LJ, EA Maynard, HC Hodge, JK Scott, CL Yuile, GE Sylvester, and HB Wilson. 1970. "A Five-Year Inhalation Study with Natural Uranium Dioxide (UO_2) Dust-I. Retention and Biologic Effect in the Monkey, Dog, and Rat." *Health Physics* 18:599-612.

Leach LJ, CL Yuile, HC Hodge, GE Sylvester, and HB Wilson. 1973. "A Five-Year Inhalation Study with Natural Uranium Dioxide (UO_2) Dust-II. Postexposure Retention and Biologic Effect in the Monkey, Dog, and Rat." *Health Physics* 25:239-258.

Luessenhop AJ, JC Gallimore, WH Sweet, EG Struxness, and J Robinson. 1958. "The Toxicity in Man of Hexavalent Uranium following Intravenous Administration." *American Journal of Roentgenology* 79:83-100.

Llobet JM, JJ Sirvent, A Ortega, and JL Domingo. 1991. "Influence of Chronic Exposure to Uranium on Male Reproduction in Mice." *Fundamental and Applied Toxicology* 16:821-829.

Lloyd RD, GN Taylor, DR Fisher, RA Schlenker, and SC Miller. 2000. "Effective Thresholds for Induction of Skeletal Malignancies by Radionuclides." *Health Physics* 79:722-727.

Maxfield BD. Undated. *Army Demographics: FY02 Army Profile," Headquarters Department of Army, Deputy Chief of Staff of Personnel, G-1.* Accessed online at www.armyg1.army.mil during January 2004.

Maynard EA and HC Hodge. 1949. "Studies of the Toxicity of Various Uranium Compounds When Fed to Experimental Animals." In *Pharmacology and Toxicology of Uranium Compounds*, eds. Voegtlin C and HC Hodge, pp. 309-376, McGraw-Hill, New York.

Mays CW, RE Rowland, and AF Stehney. 1985. "Cancer Risk from the Lifetime Intake of Ra and U Isotopes." *Health Physics* 48:635-647.

McDiarmid MA, JP Keogh, FJ Hooper, K McPhaul, K Squibb, R Kane, R DiPino, M Kabat, B Kaup, L Anderson, D Hoover, L Brown, M Hamilton, D Jacobson-Kram, B Burrows, and M Walsh. 2000. "Health Effects of Depleted Uranium on Exposed Gulf War Veterans." *Environmental Research* 82(2):168-180.

McDiarmid MA, K Squibb, S Engelhardt, M Oliver, P Gucer, PD Wilson, R Kane, M Kabat, B Kaup, L Anderson, D Hoover, L Brown, and D Jacobson-Kram. 2001. "Surveillance of Depleted Uranium Exposed Gulf War Veterans: Health Effects Observed in an Enlarged 'Friendly Fire' Cohort." *Journal of Occupational and Environmental Medicine* 43:991-1000.

McDiarmid MA, S Engelhardt, M Oliver, P Gucer, PD Wilson, R Kane, M Kabat, B Kaup, L Anderson, D Hoover, L Brown, B Handwerger, R Albertini, D Jacobson-Kram, C Thorne, and K Squibb. 2004. "Health Effects of Depleted Uranium on Exposed Gulf War Veterans: A 10-Year Follow-Up." *Journal of Toxicology and Environmental Health* 67:277-296.

Moore RH, and RL Kathren. 1985. "A World War II Uranium Hexafluoride Inhalation Event with Pulmonary Implications for Today." *Journal of Occupational Medicine* 10:753-756.

Morrow PE, LJ Leach, FA Smith, RM Gelein, JB Scott, HD Beiter, FJ Amato JJ Picano, CL Yuile, and TG Consler. 1982. *Metabolic fate and evaluation of injury in rats and dogs following exposure to the hydrolysis products of uranium hexafluoride. Implications for a bioassay program related to potential releases of uranium hexafluoride.* NUREG/CR-2268, US Nuclear Regulatory Commission, Washington, DC.

National Council on Radiation Protection and Measurements (NCRP). 1993. *Risk Estimates for Radiation Protection.* NCRP No. 115, Bethesda, Maryland.

Pavlakis N, CA Pollock, G McLean, and R Bartrop. 1996. "Deliberate Overdose of Uranium." *Nephron* 72:313-317.

Pellmar TC, AF Fuciarelli, JW Ejnik, M Hamilton, J Hogan, S Strocko, C Emond, HM Mottaz, and MR Landauer. 1999a. "Distribution of Uranium in Rats Implanted with Depleted Uranium Pellets." *Toxicological Sciences* 49:29 39.

Pellmar TC, DO Keyser, C Emery, and JB Hogan. 1999b. "Electrophysiological Changes in Hippocampal Slices Isolated from Rats Embedded with Depleted Uranium Fragments." *Neurotoxicology* 20(5):785-792.

Puskin JS and CB Nelson. 1995. "Estimates of Radiogenic Cancer Risks." *Health Physics* 69:93-101.

Ries LAG, MP Eisner, CL Kosary, BF Hankey, BA Miller, L Clegg, A Mariotto, MP Fay, EJ Feuer, and BK Edwards (eds). 2003. SEER Cancer Statistics Review, 1975-2000, National Cancer Institute. Bethesda, Maryland. Accessed online in January 2004 at http://seer.cancer.gov/csr/1975_2000,2003.

Ritz B. 1999. "Radiation Exposure and Cancer Mortality in Uranium Processing Workers." *Epidemiology* 10:531-538.

Rowland RE. 1995. "Dose-Response Relationships for Female Radium Dial Workers: A New Look." In *Health Effects of Internally Deposited Radionuclides: Emphasis on Radium and Thorium*. Van Kaick G, A Karaoglou, and AM Keller, World Scientific, pp. 135-144, Singapore.

Singh NP, DD Bennett, and ME Wrenn. 1987. "Concentrations of a-emitting Isotopes of U and Th in Uranium Miners' and Millers' Tissues." *Health Physics* 53:261-265.

Spoor NL and JB Hursch. 1973. "Protection Criteria." In "Uranium, Plutonium, Transplutonium Elements." Hodge HC, JN Stannard, and JB Hursh, eds., pp. 241-269, *Handbook of Experimental Pharmacology*, Vol 36, Springer-Verlag, New York.

Tannenbaum A, ed. 1951. *Toxicology of Uranium*. National Nuclear Energy Series, McGraw-Hill Book Company, New York.

Tasat DR and BM deRey. 1987. "Cytotoxic Effect of Uranium Dioxide on Rat Alveolar Macrophages." *Environmental Research* 44:71-81.

The Royal Society. 2001. The Health Hazards of Depleted Uranium Munitions Part I. Policy Document 6/01, London, United Kingdom. Online report available at www.royalsoc.ac.uk in the Science Policy Section.

The Royal Society. 2002. *The Health Hazards of Depleted Uranium Munitions Part II*, Policy Document 5/02, London, United Kingdom. Online report available at www.royalsoc.ac.uk in the Science Policy Section.

Thun MJ, DB Baker, K Steenland, AB Smith, W Halperin, and T Berl. 1985. "Renal Toxicity in Uranium Mill Workers." *Scandinavian Journal of Work, Environment and Health* 11:83-90.

US Environmental Protection Agency (EPA). 2002. *U.S. Environmental Protection Agency, Federal Guidance Report 13 Cancer Risk Coefficients for Environmental Exposure to Radionuclides: CD Supplement*. EPA 402-C-99-001, Rev. 1, Prepared by Oak Ridge National Laboratory, Oak Ridge, Tennessee, for the US Environmental Protection Agency, Washington, DC.

Voegtlin C, and HC Hodge, eds. 1949. *The Pharmacology and Toxicology of Uranium Compounds*. National Nuclear Energy Series, Volumes 1 and 2, McGraw-Hill Book Company, New York.

Voegtlin C, and HC Hodge, eds. 1953. *The Pharmacology and Toxicology of Uranium Compounds*. National Nuclear Energy Series, Volumes 3 and 4, McGraw-Hill Book Company, New York.

US Department of Defense (DoD). 1998. *1998 DoD Survey of Health Related Behaviors*. Accessed on January 27, 2004 at http://www.tricare.osd.mil/analysis/surveys/98survey/survey6.html, 1998.

World Health Organization (WHO). 2001. *Depleted Uranium: Sources, Exposure and Health Effects*. Department of the Protection of the Human Environment, Geneva, Switzerland. Online report available at URL: www.who.int/environmental_information/radiation/depleted_uranium.htm.

Zamora ML, BL Tracy, JM Zielinski, DP Meyerhof, and MA Moss. 1998. "Chronic Ingestion of Uranium in Drinking Water: a Study of the Kidney Bioeffects in Humans." *Toxicological Sciences* 43:68-77.

Zhao S, and F Zhao. 1990. "Nephrotoxic Limit and Annual Limit on Intake for Natural U." *Health Physics* 58:619-623.

7.0 Management of DU Aerosol Risks

The overall purpose of the Capstone test series (field studies and Human Health Risk Assessment [HHRA]) was to characterize DU aerosols from perforated armored vehicles to support the HHRA for Level I personnel. The DoD community could then use this risk assessment to determine if DoD's medical procedures and policies and personnel protective measures are adequate for Level I personnel. In addition to evaluating the chemical and radiological doses and the associated risks for Level I personnel, this report also provides estimates of risk in a context that will allow commanders to determine how the risks from exposure to DU aerosols relate to other operational risks. Risk estimates provided in this context can be used by commanders to weigh decisions regarding continued exposure of personnel exposed to DU aerosols against the risks of implementing protective practices. For example, the risk to crewmembers and other personnel from continued exposure to DU in a perforated armored vehicle could possibly be less than the risk from donning protective masks (reduced visibility that could lead to an accident) or protective clothing (overheating in hot climates that could lead to other physiological stresses), or from abandoning the vehicle in an active combat area (increased exposure to hostile fire). This chapter focuses on placing the risks from DU exposure into operational perspective so they can be used for operational risk management and used by others to assess the adequacy of current policy and doctrine.

The military radiation exposure control philosophy for balancing operational risk and the risks from radiation and chemical exposures for all operational environments, including high-intensity combat operations, is discussed in Section 7.1. Section 7.2 describes the rationale, approach, and assumptions used to evaluate and categorize the risks. Section 7.3 compares the exposure levels calculated in Chapter 5 with the risks identified in Chapter 6 to assign/categorize the relative risk with risk-effect guidelines. Section 7.4 summarizes these chemical and radiological risks by phase and scenario. Section 7.5 identifies recommended actions based on the assigned risks.

7.1 MILITARY RISK MANAGEMENT

The total risk to personnel in combat includes operational risks, risks of both battle and non-battle injuries, risks associated with protective procedures if implemented, and risks from DU aerosols and other toxicants. The issue of how to incorporate the chemical and radiological effects that serve as the basis for toxicity standards and long-term, post-conflict health effects into the commander's risk management process is complex because of the difficulty in balancing immediate risk (morbidity and mortality from combat, operational accidents, and diseases) with risks that may or may not manifest for months or even years after the operation.

The Institute of Medicine (IOM) also recognized that military personnel, including commanders, Soldiers, Marines, and medical personnel, had different information needs with respect to the risks from DU internalization (Mettler1999). The approach used in this assessment responds to that need. Field commanders require the best estimate of risk so they can appropriately include the predominantly long-term potential risks associated with DU exposure in their mission risk analysis and in their mission risk-management process. The ultimate goal of these processes is to ensure that the benefits of the mission outweigh the risks and, most importantly, ensure that appropriate actions are taken to balance the total risk, including combat mortality/morbidity as well as disease and non-battle injuries. These data will also aid the commander in fulfilling due diligence in ensuring personnel are aware of the risks (Mettler 1999).

Combat medical personnel need to know the best estimate risk for clinically significant health effects so appropriate immediate treatment decisions can be made and timely and accurate medical information can be given to the injured person. Best estimate information is also required for medical planners so that appropriate measures are in place in the theater of operations and in the United States. After conflict, medical personnel require dose and risk information so appropriate occupational medicine follow-up can be initiated in accordance with the standards of care for non-combat/non-hostile environments.

The U.S. Army has developed a system for managing risks associated with all operations, including combat, outlined in U.S. Army Field Manuals (FMs) (FM [100-14] 1998; FM [3-101.12] 2001). This system is used by U.S. military services to identify, assess, and mitigate the tactical, accident, and other risks during all military operations from peacetime to combat. This process provides a framework that commanders can use to perform a standard cost-benefit analysis with an important difference—the primary "cost" in combat operations is casualties or lives lost (Joint Chiefs of Staff Publication (JP) [3-11] 2000; FM [3-101.12] 2001; FM [100-14] 1998).

Applicability of Occupational Risk Standards

> "Military operations are inherently complex, dynamic, dangerous and, by nature, involve the acceptance of risk." (FM 3-100.12 [2001])

Military risk management "…is a process that assists decision makers in reducing or offsetting risk (by systematically identifying, assessing, and controlling risk arising from operational factors) and making decisions that weigh risks against mission benefits" (FM [3-100.12] 2001). The total risk to military personnel includes operational risks, risk of non-battle injury, risks associated with protective procedures, and risk from DU and other battlefield materials that are toxic at some level.

Often, sub-clinical chemical effects serve as the basis of occupational toxicity standards, and the long-term probability of developing cancer using the statistically conservative linear, no threshold approach serves as the basis for some radiological limits. The issue of how to incorporate these standards into the commander's risk management process is straightforward for garrison, peacetime environments. In this case, the risks are comparable to civilian industry, and peacetime standards and practices are generally applied. However, applying appropriate limits becomes complex during military combat during which personnel may be exposed to acute, life-threatening circumstances. In these cases, the operational risks are greater and the occupational standards for peacetime have less direct applicability.

The IOM reviewed the peacetime occupational standards for applicability to combat situations and concluded that applying peacetime radiological risk standards to combat situations is neither applicable nor practical because the peacetime safety standards and procedures were designed for relatively safe occupations, not for combat operations that generate Level I exposures (Mettler 1999). The IOM also stressed the need to include long-term risks in the commander's risk assessment procedures (Mettler 1999). The DoD recognized this gap and developed policy and doctrine from technical, legal, and ethical perspectives to fill this void by establishing guidance levels for exposure that are based on the operational risks posed by the mission (JP [3-11] 2000, FM [3-11.4] 2003; FM [4-02.7] 2002; North Atlantic Treaty Organization [NATO] STANAG [Standardized Agreement]-2083 1996).

For high-intensity combat operations that generate Level I exposures, key differences from peacetime operations include the following:

- The risks associated with the military task can be much higher than those found in the industrial setting.
- The risks associated with the application of a protective practice may be considerably higher than the risks posed in the industrial setting.
- The adversary (rather than the employer) controls most of the level of risk associated with the occupational task by deciding when, where, how, how much, and what types of weapons will be employed, including weapons that present a radiological, chemical, or biological hazard.
- Most importantly, the cost associated with an improper risk balance or task failure is measured in terms of human loss (immediate, measurable morbidity, and mortality).

The application of current peacetime practices and standards to stability and support operations and high-intensity combat presents unacceptable risks to battlefield personnel (Mettler 1999). NATO addressed this issue and established radiation exposure guidance for commanders for all operational environments. The NATO document, STANAG 2083, and the implementing U.S. doctrine (JP [3-11] 2000; FM [3-11] 2003; FM [3-11.4] 2003; FM [4-02.7] 2002) established the criteria for when these levels could be exceeded and provided recommended protection practices for use when these levels were established. This approach was reviewed by IOM from technical, legal, and ethical perspectives (Mettler 1999).

No similar U.S. military policy has been developed to address chemical toxicity. This report proposes a framework for the chemical effects of DU that follows the same approach used for the radiation system described below.

7.2 MILITARY RADIATION EXPOSURE CONTROL PHILOSOPHY

The U.S. military radiation control philosophy provides a mechanism for balancing military operational risk and the risks from radiation exposure for all operational environments. The IOM analysis (Mettler 1999) and military doctrine (JP [3-11] 2000) define three operational environments for military operations: peacetime operations, high-intensity combat operations, and stability and support operations. For peacetime operations, the occupational risks are comparable to industrial settings for which the radiation safety systems were designed. In high-intensity combat, operations range from situations similar to ODS and Operation Iraqi Freedom to all-out global war with the employment of weapons of mass destruction. Stability and support operations (formerly known as Military Operations Other than War [MOOTW]) fill the gap between these two extremes and are defined rather loosely as those operations for which operational risks are greater than peacetime risks but less than high-intensity combat risks. Examples of stability and support operations range from humanitarian operations, to peacekeeping, to disaster response, to limited, short-term combat operations (JP [3-11] 2000).

7.2.1 "Peacetime" Occupational Radiation Protection Standards

The basic framework established in the guidance provided by ICRP (ICRP [37] 1982, [55] 1989, and [60] 1991), NCRP (NCRP [115] 1993, [116] 1993, [126] 1997, and [127] 1998), the International Atomic Energy Agency (IAEA) [RSG-1.1] 1999, [SRS 21] 2002, and [SS 115] 1996), and U.S. regulatory agencies (NRC [10 CFR Part 20]; EPA [53 FR 172822]) for peacetime ionizing radiation protection from nuclear materials is intended to provide an appropriate standard of protection against ionizing radiation without unduly limiting the beneficial practices giving risk to the exposure. The basic intent of controlling the exposure is to prevent deterministic effects[a] by keeping the radiation doses below the relevant threshold and making all reasonable efforts to minimize or reduce the stochastic effects[b] to an acceptable level. These systems also require actions to maintain the actual exposure as low as reasonably achievable (ALARA). Current U.S. protection standards and those recommended by the ICRP/NCRP are summarized in Table 7.1 (using limits as published). The intent of the approach, which has proven to be effective, is to ensure that exposures approaching the annual limit are rare events (53 FR 172822). In 2001, the average radiation dose for people with measurable doses was 0.24 rem with only 1 person out of 154,000 exceeding the 5 rem/yr limit (NRC 2002). This system applies to U.S. armed forces during peacetime, garrison-like operations.

---

[a]The NRC (Title 10 CFR Part 20) defines a deterministic effect or *nonstochastic effect* as health effects, the severity of which varies with the dose and for which a threshold is believed to exist. Radiation-induced cataract formation and radiation burns are examples of deterministic effects.

[b]The NRC (Title 10 CFR Part 20) defines *stochastic effects* as health effects where radiation exposure increases the probability of the effect occurring, rather than its severity, is assumed to be a linear function of dose without threshold. Hereditary and cancer incidence are examples of stochastic effects.

Table 7.1. Radiation Exposure Limits for Peacetime Operations

Radiation Limits	10 CFR Part 20		NCRP 116		ICRP 60	
	Worker	Public	Worker	Public	Worker	Public
Total effective dose equivalent (rem/yr) or equivalent dose (mSv/yr)	5 rem/yr (0.05 Sv/y)	0.1 rem/yr (0.001 Sv/y) and, with justification to the NRC, 0.5 rem/yr.	0.050 Sv/yr and 0.01 Sv times age (yr) cumulative effective dose.	0.001 Sv/yr annual effective dose and 0.005 Sv annual effective dose for infrequent exposures.	0.020 Sv/yr and 0.1 Sv in 5 yr cumulative effective dose.	0.001 Sv/yr effective dose. If needed, higher values as long as 5 yr average is not >0.001 mSv.
Single organ	50 rem/yr (0.5 Sv/y)	N/A	0.15 Sv/yr to the lens of the eye and 0.5 Sv/yr to skin, hands, and feet.	0.05 Sv annual equivalent dose limit to the skin and extremities.	0.15 Sv/yr to the lens of the eye and 0.5 Sv/yr to skin, hands, and feet.	0.015 Sv/yr to the lens and 0.05 Sv/yr to skin, hands, and feet.

7.2.2 The U.S./NATO Operational Radiation Control System: Operational Exposure Guidance

The U.S. Army uses the Operational Exposure Guidance (OEG) system to implement this system and control the radiation exposure to its personnel during combat and MOOTW (FM [3-3.1] 1994; FM [3-11] 2003; FM [3-11.4] 2003; FM [4-02.7] 2002). The system has two components. The first is the OEG. The OEG it is the maximum radiation exposure a unit is allowed to receive during an operation, and it is set by the commander. The OEG is a risk-based dose level that is not to be exceeded by subordinate commanders. The second is the Radiation Exposure Status (RES), which is the radiation dose received by a Field Unit (JP [3-11] 2000).

This system was initially developed during the Cold War for use in the event of a global war. The radiation levels for OEG were based on the onset of deterministic effects (primarily the onset of emesis) that impair the performance of an individual or a unit. Long-term health effects, effects that would manifest after the operation, were not considered during the Cold War period because the doctrine was developed for all-out global nuclear war with the Soviet Union and assumed U.S. national survival was at stake.

The system was designed specifically for exposures from external gamma radiation. It did not include the effect of internal emitters primarily because, in all cases considered during the development of this system, external dose was the primary hazard and the internal dose was negligible. For operational reasons, dose was expressed in terms of centigray (rad) rather than in terms of dose equivalent. Since the system was designed for external gamma radiation, the quality factors converting rad to rem were unity and therefore the values of the dose and dose equivalent quantities were numerically equal.

Cold-War doctrine had only three OEG categories: *Negligible Risk* (doses less than 75 rad); *Moderate Risk* (doses greater than 75 rad but less than 100 rad);

and *Emergency Risk* (doses greater than 125 rad) (FM [3-11.4] 2003). (U.S. Army doctrine does not address the gap between 100 rad and 125 rad.) In this case, risk refers to the risk that, because of exposure to radiation, the unit's performance would be degraded to the point that it could not complete its mission. The doctrine also provides guidance for circumstances in which each of these risk levels is appropriate and the precautions required for each level. The principles of justification and optimization are a part of this doctrine.

The RES is used to track and report the average dose to the members of a combat unit. There are four RES categories that roughly parallel the OEG: RES-0 is a unit with no exposure; RES-1 is a unit that has received some radiation exposure (greater than RES-0) that is less than or equal to 75 rad; RES-2 covers exposures greater than 75 rad and equal to or less than 125 rad; and RES-3 includes exposures over 125 rad. The RES of a unit and the doses associated with the risk levels above are used to establish the OEG for a mission. For example, a unit at RES-1 can receive up to 75 rad and not exceed negligible risk for a combat operation. A unit with a higher RES can receive correspondingly less exposure for the operation.

The Cold-War OEG doctrine was still appropriate for post-Cold-War, high-intensity combat operations but not for stability and support operations, where the associated risk ran the gamut from peacetime to just below high-intensity combat. Current doctrine does not allow exposure to a unit involved in stability and support operations to exceed RES-1. Additionally, six sub-categories have been established within RES-1 to better allow the matching of operational risk with radiation risk (Mettler 1999). The RES sub-categories were established based upon radiation levels that are currently in use for peacetime practice in the United States.

Table 7.2 lists the RES categories and sub-categories for RES-1 and the upper dose limit for these categories. The peacetime equivalent is a listing of standards or guidelines that are currently in place for either occupational exposures or for exposure guidelines during radiological emergencies. Each of the RES-1 subcategories is based on peacetime standards or regulatory guidelines (shown in the references in the table).

7.3 RISK ASSESSMENT APPROACH AND ASSUMPTIONS

The DoD radiation guidance (JP [3-11] 2000) relates to whole-body exposures of low-LET radiation and for surface contamination for alpha emitters. It does not specifically address inhaled or ingested internal emitters such as DU. The approach used here to bridge this gap was to convert the NATO

Table 7.2. Comparison of RES Categories with "Peacetime" Standards or Guidelines

RES Level	Maximum Dose[(a)] (rad or rem)	"Peacetime" Equivalent	Reference
RES-0	0.05	One-half of the annual limit for the public (0.1 rem/yr, 10 CFR 20)	JP 3-11 (2000)
RES-1A	0.5	Annual dose limit for a member of the public when prior authorization is obtained	10 CFR 20
RES-1B	5	Annual dose limit for occupational workers	10 CFR 20
		EPA Protective Action Guide (PAG) level for all workers in a radiological emergency	EPA (1992)
RES-1C	10	Annual dose limit for occupational workers with a planned special exposure	10 CFR 20
		EPA PAG level for protecting valuable property during an emergency	EPA (1992)
RES-1D	25	EPA PAG level for protecting large populations or saving lives	EPA (1992)
RES-1E	75	No equivalent; EPA PAG for exposures greater than 25 rem for protecting large populations or saving lives on a voluntary basis – NATO	EPA (1992) JP 3-11 (2000)
RES-2	125	No equivalent – NATO	JP 3-11 (2000)
RES-3	>125	No equivalent – NATO	JP 3-11 (2000)

(a) RES categories are defined in terms of rad (cGy). For gamma exposures in which the quality factor is 1, the dose in rad (cGy) is equal to the effective dose in rem (0.01Sv) for whole body exposure.

guidance levels to levels of increased fatal cancer risk. This risk was then compared with the increased risk of fatal cancer estimated for DU.

For this evaluation of Level I DU aerosol exposures, the approach used to assess the radiological risks was to adapt the military's system (OEG) for managing radiation risk during non-peacetime operations to the DU exposure scenarios. A comparable military system for chemical toxicity was not available, so the Renal Effects Group (REG) system developed in Chapter 6 was used to estimate these risks and help manage these risks.

Vehicle perforation is assumed for this dose assessment. The risks to crewmembers in Abrams tanks or Bradley vehicles that are struck but not perforated are not addressed. Other key elements of the approach include:

- *Using Best Estimates of Risk.* The risk estimates in this report represent an analysis of most probable risks to crewmembers rather than worst-case estimates. This analysis benefited from the robust data developed in the Capstone study in which the shot lines were selected to generate reasonable upper-bound DU aerosols. In each case, the median value for each phase was used as the best estimate of risk. Variances in physiological parameters (nose breather versus mouth breather) were discussed in Chapter 5, and the most probable conditions were used in this risk analysis. The risk estimates can be used for operational and medical planning and suggest levels at which the immediate care of exposed individuals is most likely warranted. These estimates may be used to provide tentative estimates of individual exposure levels and risk, but the actual evaluation of DU intake, dose, and the associated risks must be made using follow-up bioassays (e.g., analysis of uranium in urine) to measure DU clearance from the body and relate it to exposure and risk.

- *Basing Scenarios on Reasonable Upper Bounds.* The risk assessments are based on the test conditions of the Capstone study and the scenarios described earlier. The scenarios were designed to produce reasonable upper-bound aerosol concentrations, and these concentrations were used to estimate radiation doses and kidney concentrations. The scenarios did not consider several key conditions that would reduce the aerosol concentration such as the use of ventilation systems in either vehicle type, the activation of the fire-suppression systems (Halon), or the activation of the EC/NBC ventilation systems in the Abrams tank BHTs in the Phase-I and Phase-III results. The Phase-IV test results showed the significant reduction in aerosol concentration that occurs when the Abrams tank EC/NBC ventilation system is operating. The effect of hatch openings on reducing the DU aerosol is qualitatively discussed but not quantitatively measured or calculated.
- *Basing Radiation Risks on the Summed Organ Risks.* The summed organ risks were used rather than the whole-body risks because 1) the dose distribution in the body was extremely non-uniform with the lung receiving substantially more dose than any other organ, and 2) radiation-specific risk estimates for the lung are available. The summed organ risks were slightly greater (between 10 to 30%) than those determined using the ICRP/NCRP methodology for whole-body risk.

7.3.1 Application of Radiation Exposure Status to the Risk Estimates

The RES risk-based system was specifically developed to allow the establishment of radiation risk levels that were appropriate for the operational risks faced by the commander. The system was peer reviewed by IOM (Mettler 1999) and adopted by NATO. The six subcategories for RES-1 were established to provide guidance to commanders on how to incorporate the long-term health effects of radiation exposure (primarily increased risk of cancer and potential for genetic effects) into their risk management assessment procedures for military operations.

Each of these dose levels represents a level of risk. The RES doses were intended for external whole body radiation and cannot be directly compared to the non-uniform internal doses estimated for DU. In the case of DU aerosols, the external whole body radiation exposure from these aerosols is negligible. A method was developed to apply the RES categories to relate risk from internal DU deposition and its related dose. This was accomplished first by converting the RES category doses into levels of increased risk and then by comparing the calculated risks of cancer from Chapter 6 to risks represented by these RES categories. Two risk coefficients could be used to convert the RES categories to estimates of increased fatal cancers. The first is the risk coefficient recommended by ISCORS (2002), and the second is the risk coefficient recommended by ICRP Publication 60 (1991) and NCRP 115 (1993). The ICRP/NCRP risk coefficients were selected to convert the radiation doses to risk because the RES risk system was developed using the ICRP/NCRP guidelines.

Table 7.3 shows the conversion of the RES categories for combat operations and operations other than combat into estimates of increased risk of fatal cancer using the EPA risk estimates in its Federal Guidance Report 13 (1999) and the ICRP/NCRP increased risk of fatal cancer for workers. The dose and dose-rate effectiveness factor (DDREF) of 2 was applied to the doses less than 20 rad (RES-0, -1A, -1B, and -1C) but not to the doses greater than 20 rad (RES-1D, -1E, -2, and -3). The ICRP/NCRP's system applies the DDREF to all absorbed doses less than 20 rad and any absorbed dose delivered at dose rates less than 10 rad/h. The RES system places no limitation on

Table 7.3. Conversion of DoD and NATO RES Categories into Risk of Fatal Cancers Due to Radiation Exposure

DoD and NATO		Fatal Cancer Risk for Workers	
Total Cumulative Dose (rad - Whole Body, Low-LET)	Radiation Exposure Status (RES) Category	ISCORS	ICRP/NCRP
0	NA	0.00	0.00
0.05[(a)]	0	2.50×10^{-5}	2.00×10^{-5}
0.5[(a)]	1A	2.50×10^{-4}	2.00×10^{-4}
5[(a)]	1B	2.50×10^{-3}	2.00×10^{-3}
10[(a)]	1C	5.00×10^{-3}	4.00×10^{-3}
25	1D	2.50×10^{-2}	2.00×10^{-2}
75	1E	7.50×10^{-2}	6.00×10^{-2}
125	2	1.25×10^{-1}	1.00×10^{-1}
>125	3	NA	NA

(a) The DDREF of 2 was applied to the risk per rem for these RES categories.

the dose rate or time to achieve the dose, so the doses can be received at any rate. For doses less than 20 rad, the ICRP/NCRP risk factor for workers is 4×10^{-4} rem^{-1}. For RES categories greater than 20 rad, the risk factor is 8×10^{-4} rem^{-1}. The same is true for the ISCORS risk factors. For doses that are less than 20 rad, a risk factor of 5×10^{-4} rem^{-1} was used, and for higher doses, a risk factor of 1×10^{-3} rem^{-1} was applied to convert the RES doses to risk of fatal cancer for workers.

7.3.2 Comparison of DU Best Estimate Risk with U.S./NATO RES Categories

Figure 7.1 compares the risks for exposure to DU for each of the vehicle types and scenarios with the risks associated with each of the RES categories. The graph plots the 50th percentile risk for each of the vehicles for each of the scenarios in this study. The upper bar represents the 90th percentile risk and the lower bar represents the 10th percentile. The Abrams tank test in which the EC/NBC ventilation system was operating before and after perforation (Phase IV) has no bars because it was based on only one shot. Note that scale on the risk axis is logarithmic, representing increasing factors of 10.

The vehicle configurations—BHT with hatches closed at perforation, ventilation systems off—and shot lines used in Phases I through III provide a reasonable upper estimate of the risks from DU aerosol exposures inside each of the vehicles.

All the median risks and 90th percentile risks for each vehicle (phase) and scenario were below the guidelines for workers responding to peacetime

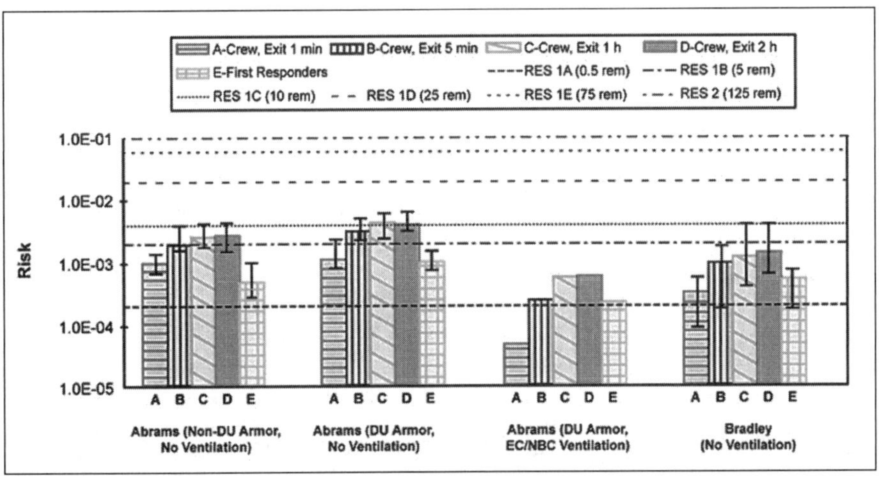

Figure 7.1. Best Estimate Risks for Perforation of the Vehicles (BHTs) Shown Compared to the RES Categories (bars represent the 90th and 10th percentiles)

radiological emergencies including the EPA PAGs of 25-rem, life-saving level (EPA 1992), a level that corresponds to RES-1D. All median risks were at or below the annual occupational exposure allowed for a planned, special exposure and the EPA PAG for protecting valuable property (RES-1C, 10 rem). None of the best estimate (median) risks exceed the DoD/NATO RES-1 category. The highest median risk for any phase and scenario is within the RES-1C range (between 5 and 10 rem). Several scenarios for Phase III (the Abrams tank fired through DU armor) exceeded the RES-1C range at the 90th percentile, but none of the risks exceeded RES-1D (10 to 25 rem).

Table 7.4 provides a direct comparison of the risks and the RES levels for each of the phases. The median/best estimate risk of fatal cancers to the crew is highest for the Abrams tank when the DU armor is perforated. This risk ranges from values below the peacetime occupational annual limit (5 rem/yr,

Table 7.4. Fatal Cancer Risk from DU Inhalation Exposures and the Associated RES Categories

Scenario	10th Percentile		Median		90th Percentile	
	Cancer Risk	RES	Cancer Risk	RES	Cancer Risk	RES
Abrams Tank without DU Armor (Phase I)						
A. Crew, Exit in 1 min	0.0690%	1B	0.107%	1B**	0.143%	1B**
B. Crew, Exit in 5 min	0.165%	1B**	0.200%	1B**	0.401%	1C/1D
C. Crew, Exit in 1 h	0.173%	1B**	0.267%	1C	0.441%	1D
D. Crew, Exit 2 h[(a)]	0.160%	1B**	0.276%	1C	0.442%	1D
E. First Responders	0.0252%	1B	0.0504%	1B	0.105%	1B**
Bradley Fighting Vehicle (Phase II)						
A. Crew, Exit in 1 min	0.00866%	1A*	0.0336%	1B	0.0586%	1B
B. Crew, Exit in 5 min	0.0186%	1A*	0.0987%	1B**	0.180%	1B**
C. Crew, Exit in 1 h	0.0571%	1B	0.119%	1B**	0.398%	1C
D. Crew, Exit 2 h	0.0650%	1B	0.140%	1B**	0.410%	1D
E. First Responders	0.0155%	1A*	0.0520%	1B	0.0774%	1B
Abrams Tank with DU Armor (Phase III)						
A. Crew, Exit in 1 min	0.0800%	1B	0.115%	1B**	0.241%	1C
B. Crew, Exit in 5 min	0.243%	1C	0.317%	1C	0.518%	1D
C. Crew, Exit in 1 h	0.318%	1C	0.437%	1C/1D	0.640%	1D
D. Crew, Exit 2 h	0.333%	1C	0.445%	1C/1D	0.653%	1D
E. First Responders	0.0585%	1B	0.104%	1B**	0.157%	1B**
Fully Operating Abrams Tank (ventilating system functioning) (Phase IV)[(b)]						
A. Crew, Exit in 1 min	NA	NA	0.00495%	1A*	NA	NA
B. Crew, Exit in 5 min	NA	NA	0.0248	1B	NA	NA
C. Crew, Exit in 1 h	NA	NA	0.0567	1B	NA	NA
D. Crew, Exit 2 h	NA	NA	0.0647	1B	NA	NA
E. First Responders	NA	NA	0.0235	1B	NA	NA

(a) Risks comparable to Scenario C but based on a slightly different dataset.
(b) No percentile values are given for Phase IV because there was only one data point.
1A* Indicates that the risk associated with U.S. non-occupational exposure limit if 0.1 rem/yr may be exceeded
1B** Indicates that the risk associated with the ICRP occupational limit of 2 rem may be exceeded.

RES-1B) for exits within 1 min to slightly greater (approximately 10%) than the values for planned special exposures or the EPA PAG for protecting valuable property (RES-1C, 10 rem), and the risk is about 0.45% for stay-times of 1 to 2 h. The dose and associated risk can be significantly reduced if the EC/NBC ventilation system is on at the time of the perforation and remains on, or can be turned on within the first minute after perforation. If the EC/NBC ventilation system is on at the time of the perforation and remains running after perforation, the best estimate crew risk is less than the peacetime limit for occupational workers (0.5 to 5 rem/yr, RES-1B). (The Abrams tank circulation system is an internal system that is used to move internal air as a part of the heating system in the vehicle. Although not as powerful as the EC/NBC ventilation system, it does increase the air exchange rates and would decrease dose.)

The best estimate radiological risk to the crew of an Abrams tank perforated through non-DU armor ranges from less than the peacetime occupational limit (5 rem/yr, RES-1B) for exiting within 1 min to below the annual limit for a planned special exposure or the EPA PAG to protect valuable property (10 rem, RES-1C) for longer stay-times if the EC/NBC ventilation system is not running at the time of perforation and is not turned on shortly after perforation. The 90th percentile doses for this circumstance are just over RES-1C. If the EC/NBC ventilation system is running at the time of perforation, the crew risk is less than the annual limit to a member of the population with NRC approval (0.5 rem/yr, RES-1A), but will probably be above RES 0.

In all cases for the Bradley vehicle, the median/best estimate crew risk is RES-1B (0.5 to 5 rem), or less than the annual peacetime occupational exposure limit. The Bradley vehicle is not equipped with an EC/NBC ventilation system. It does, however, have ventilation fans in the driver's and crew compartments that draw air from outside the vehicle through an air filter and into the crew compartment. As discussed previously, activation of this system would rapidly increase the air exchange rate, and consequently reduce the concentration of DU in the air, resulting in significantly reduced doses to the crew.

The impact of hatch opening depends on wind and weather conditions. It is likely to impact aerosol concentration in the Abrams tank more than the Bradley vehicle because the former is relatively air tight. The hatches need to be opened for the first responder to enter and exit the vehicle. Aerosol leakage and dilution in the Bradley vehicles, which are much less air tight, are not as likely to be as significant.

In all cases, the additional risk incurred as a function of stay-time is relatively consistent from vehicle to vehicle because the aerosol intake rates in the vehicles decrease similarly with time. Figure 7.2 shows the risk normalized to the accumulated maximum risk incurred through 2 h (Scenario D; 2-h stay-time) for each of the Phase I, II, and III vehicles. Each time point represents the stay-times for each of the scenarios for Phases I through III beginning with a 1-min stay-time (Scenario A). For each vehicle in which the ventilation system did not operate, about 70% of the maximum risk is incurred within the first 5 min after perforation. The additional risk incurred for longer stay-times is, therefore, comparatively small.

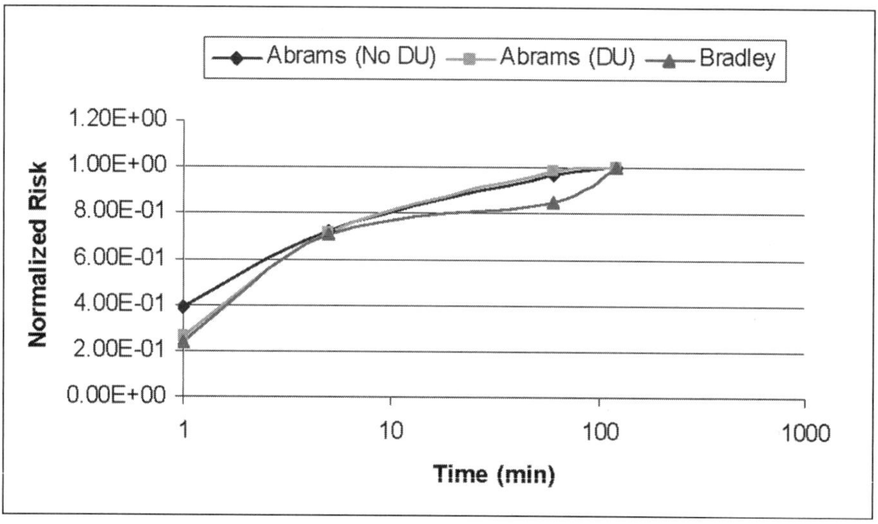

Figure 7.2. Normalized Accumulated Risk as a Function of Time

7.3.3 Risks of Adverse Effects to the Kidney

The United States regulates chemical risks on the basis of deterministic and stochastic effects. However, the differences in the approaches for risk regulation are significant (NCRP [96] 1989). There is general agreement that the target organ for uranium toxicity is the kidney. However, there has been significant discussion concerning the *de facto* occupational guideline of 3 μg U/g kidney (ICRP [2] 1959). These discussions center on estimates for the kidney uranium concentration at which a change in kidney function is observed and are based on animal studies in which biochemical indicators of kidney function are affected at kidney uranium concentrations below the ICRP guideline. Some of these studies were conducted with species more sensitive to the effects of uranium than humans, and are sub-chronic or chronic studies. Similar to radiation protection, the safety community is entering into these discussions to ensure that the occupational limits are set well below the level at which any deterministic effects may occur in any individual in the affected population. For reasons already discussed, the safety construct that is appropriate for the normal occupational environment is not appropriate for military operations where the operational risks are greater than the occupational risks for which the safety system was developed.

A doctrinal construct similar to the OEG/RES system noted above is not available for uranium chemical toxicity. The REG risk model developed for this assessment (presented in Section 6.3.1.2) performs a similar function (see Table 6.13). This model categorizes the chemical effects of uranium exposure in a manner that is useful for the operational commander doing operational risk assessments and for medical personnel assessing the clinical situation and providing the clinician evidence-based recommendations for potential clinical outcomes. This grouping was possible because of the availability of human exposure data that correlated estimated uranium kidney burdens with measured biochemical and clinical effects, thus allowing the development of a dose-response model. Use of the Capstone aerosol data (Parkhurst et al. 2004) and the uranium biokinetic models allowed for the estimation of kidney uranium dose for the personnel in the scenarios assessed. This provides a framework for obtaining best estimate risks of kidney toxicity that are population-specific and have meaningful endpoints for the users of these data.

The REG analysis showed that the predicted outcomes for chemical effects range from no effects to protracted indicators of renal dysfunction where personnel may feel ill, but may not necessarily need treatment. It also showed the significant reduction in kidney burdens (and therefore potential adverse effects) that can be gained if the EC/NBC ventilation system is turned on in an Abrams tank or the ventilation systems are turned on in a Bradley vehicle following perforation.

7.3.4 Multiple Perforations

This report assesses risk from single perforation events. The Capstone tests provided data for two tests in which two perforations were made 13 to 14 min apart. As discussed in Chapter 5, in both cases, the DU aerosol concentrations occurring after the second perforation were lower or could not be distinguished from the first. With this time separation, the aerosol concentrations for each perforation could be said to be independent, i.e, the residual aerosol or deposited aerosol from the first shot that was resuspended by the second shot did not appear to contribute significantly to the DU aerosol concentration due to the second shot. Although the data are limited, it appears that the radiation doses and kidney concentrations for two shots separated by 10 to 15 min or more can be approximated by doubling the doses and kidney concentrations. No data were collected to characterize aerosols from multiple perforations occurring within a shorter time frame.

If the vehicles are perforated less than a minute apart, the aerosol concentration in the vehicle would be expected to double if the shot line and structures perforated were essentially the same. It is possible that additional aerosol may increase particle agglomeration and deposition and more rapidly reduce aerosol concentrations. In addition, the overpressure from the second shot may clear some of the suspended aerosol from the first shot from the vehicle while resuspending deposited aerosol, mechanisms that may counteract each other. Therefore, doubling the concentrations and doses may be conservative.

7.4 SUMMARY OF RISKS

The summary in Table 7.5 shows the differences in risks from DU aerosol inhalation between an Abrams tank and a Bradley vehicle, the influence of the type of armor (DU versus non-DU) perforated in an Abrams tank, and the importance of functioning EC/NBC ventilation systems in an Abrams tank to reduce dose.

The results indicate that the chemical effects would be more likely to have a detrimental effect on the mission than the radiological effects. The maximum chemical effects predicted, represented by the REG 2, were not exceeded even at the 90th percentile level in any of the scenarios. REG 2 represents the level at which protracted, but not irreversible, indicators of renal dysfunction and possible overt kidney symptoms may be experienced. None of the calculations of peak kidney concentrations from the four phases or various scenarios resulted in levels reaching REG 3, which represents the level at which more serious effects may occur.

Table 7.5 also shows that none of the radiological risks exceed those allowed for peacetime occupational exposures or exceed the PAG for peacetime radiological emergencies. The highest RES category is RES-1D for the 90th percentile doses for long exposure times. The risks were less than those for the EPA PAG for saving lives during peacetime emergencies (risks associated with 25-rem effective dose). The majority of the estimates of risk in the table are less than the peacetime occupational limit—RES-1B or 5 rem.

RES Category 1D, the risk level for emergency peacetime lifesaving (25 rem effective dose), was never exceeded. RES Category 1C was exceeded in the Abrams tank when the DU armor was penetrated for the 90th percentile groups

Table 7.5. Summary of the 10th, 50th, and 90th Percentile Radiological and Chemical Health Risk Categories

Scenario	Percentile	Abrams Tank (Through Non-DU Armor, No Ventilation)		Abrams Tank (Through DU Armor, No Ventilation)		Abrams Tank (Through DU Armor, EC/NBC On)		Bradley Fighting Vehicle (No Ventilation)	
		RES	REG	RES	REG	RES	REG	RES	REG
Most Likely--Crewmembers									
A-Crew, Exit 1 min	10th	1B	0	1B	0	NA	NA	1A*	0
	50th	1B**	1	1B**	0	1A*	0	1B	0
	90th	1B**	1*	1C	1	NA	NA	1B	0
B-Crew, Exit 5 min	10th	1B**	1*	1C	0	NA	NA	1A*	0
	50th	1B**	1/2	1C	1	1B	0	1B**	1
	90th	1C/1D	2	1D	2	NA	NA	1B**	1*
Upper Bound-Crewmembers									
C-Crew, Exit 1 h	10th	1B**	1*	1C	1	NA	NA	1B	0
	50th	1C	2	1C/1D	1*	1B	0	1B**	1*
	90th	1D	2	1D	2	NA	NA	1C	2
D-Crew, Exit 2 h	10th	1B**	1*	1C	1	NA	NA	1B	0
	50th	1C	2	1C/1D	1*	1B	0	1B**	1*
	90th	1D	2	1D	2	NA	NA	1D	2
First Responders									
E-10-min Exposure	10th	1B	0	1B	0	NA	NA	1A*	0
	50th	1B	0	1B**	0	1B	0	1B	0
	90th	1B**	1*	1B**	1	NA	NA	1B	0

1* REG 1 that exceeded the *de facto* occupational guideline of 3 μg U/g kidney.
1A* RES-1A that exceeded the risk associated with the 0.1 rem/yr limit for the general public.
1B** RES 1B that exceeded the risk associated with ICRP 2 rem/yr occupational level, averaged over 5 yr.

for stay-times greater than 5 min and in the 50th percentile for stay-times greater than 1 h. RES Category 1C is the risk level (10 rem effective dose) that NRC (1992) allows occupational workers to receive in 1 yr for a planned special exposure and is the peacetime PAG for saving valuable property.

These risks are further summarized below for each type of vehicle.

- **Crew Risks—Abrams Tank Perforated Through DU Armor with the EC/NBC Ventilation System Operating.** The most important finding of this study was the substantial impact of an operating EC/NBC ventilation system in the Abrams tank. Activation of the EC/NBC ventilation system significantly reduced the risks to well below the maximum risks allowed during normal peacetime operations. The RES category was reduced to RES-1B from RES-1C for the 5-min stay-time and from RES-1C to RES-1B for the 1-h stay-time. The REG was reduced from REG 1 or 2 to REG 0.
- **Crew Risks—Bradley Fighting Vehicle with No Ventilation System Operating.** The predicted radiation and chemical risks for the Bradley vehicle were lower than for the more heavily armored Abrams tank. In

each instance, the median radiological risk for the Bradley vehicle was less than the peacetime occupational limit of 5 rem (RES-1B) for all percentiles and all scenarios. The chemical risk was within REG 2 for only the 90th percentile at the stay-times longer than 1 min. The median REG in each case did not exceed REG 1. The risks can be substantially reduced if the ventilation fans in the vehicle are turned on immediately after perforation.

- **Crew Risks—Abrams Tank Perforated Through Non-DU Armor with Neither EC/NBC Ventilation Nor Air Circulation Systems Operating.** The radiological risk for perforations through non-DU armor were greater than those in the Bradley vehicle but less than those that would occur if an Abrams tank with DU armor were perforated. The same was not true for the kidney effects. The kidney effects levels for the non-DU perforations were higher than when the DU armor was perforated. This finding was attributed to particle size and solubility differences.

  Both the median and 90th percentile chemical effects were REG 2 or less for the Abrams tank with non-DU armor for all stay-times greater than 5 min. The primary impact for this effect level is the potential for decreased crew performance if they become ill due to the levels of uranium internalized. Studies of human cases with similar predicted uranium kidney burdens suggest that the illness is temporary, but it is possible that this level of intake could decrease performance. None of the radiation risks, including the 90th percentile risks, exceeded RES Category 1D. The median risks were Category 1C or lower for all stay-times examined.

  The risks can be substantially reduced if the ventilation fans in the vehicle are turned on immediately after perforation.

- **Crew Risk—Abrams Tank Perforated Through DU Armor with Neither the EC/NBC Ventilation Nor the Air Circulation System Operating.** The radiological risk was the highest for this configuration but did not exceed RES Category 1D (EPA PAG) for protecting large populations or saving lives, 25 rem effective dose) for either the single or double perforations. The median risk exceeded Category 1C—PAG for saving valuable property or the maximum annual dose for planned special exposures (10 rem effective dose) by only 10% for the longer stay-times (1 and 2 h).

  For most of the modeled exposures, the chemical risks fell within REGs 0 or 1. REGs greater that 1 were obtained only at the 90th percentile for stay-times of up to 1 h or longer. The primary impact of the highest level of risk posed to these crew members is the potential for decreased crew performance if they become ill due to the levels of uranium internalized.

- **First Responders.** In all cases, first responder RES categories were 1B (annual limit for occupational workers, 5 rem effective dose) or less, and the kidney REGs were 1 or less. First responder risks will also be reduced if the ventilation and EC/NBC ventilation systems are activated.

  Based on DoD current policy (JP [3-11] 2000) and U.S. Army Medical Doctrine (FM [4-02.7] 2002), the levels of DU aerosols to which crewmembers in each of the vehicles for each scenario could conceivably be exposed would require pre-risk, post-risk, and risk-mitigation training. To satisfy these policies, post-exposure or post-conflict measures would need to be implemented to obtain individual radiation dose assessments. Kidney function may be assessed if deemed appropriate by a physician.

7.5 RECOMMENDED ACTIONS BASED ON RISKS

The DoD and U.S. Army guidance (JP [3-11] 2000, FM [4-02.7] 2002) provides the actions required for each of these RES levels for whole-body, low-LET radiation. Table 7.6 (taken from FM [4-02.7] 2002; AFRRI 2003) shows the recommended actions taken from U.S. Army medical doctrine for each of the RES categories considered applicable to DU. This guidance requires modification because of the differences between DU exposure and whole-body, low-LET exposures. Military personnel can be placed into an initial category based upon the description of the scenario. However, final placement should be made based on the follow-up bioassay results obtained to determine the amount of DU actually internalized.

The recommendations for monitoring and recording the doses are applicable for DU except that the final dose assessment cannot be conducted until after the bioassay results have been received.

The medical impacts and recommendations need to be modified because of the differences between DU radiation type (low-LET from beta and a small gamma component from uranium progeny; high-LET from uranium emissions of alpha particles) and whole-body, low-LET exposure. Many of the medical recommendations are based on possible compromises of the immune system compromise that occur from the whole-body, low-LET exposure (AFRRI 2003). This cannot reasonably be expected for DU exposure because the doses are so low, particularly to immune cell populations. Specifically, the recommendation for no live virus for three months, the increased morbidity from other illnesses, and the need to evacuate from the theater are not applicable to the DU exposure because the immune system is spared relative to an equivalent whole-body, low-LET, exposure.

The risks from the initial perforation are less than the risks allowable during peacetime emergencies. The NRC allows workers to receive up to a total of 10 rem (RES-1C) in a year for a planned special exposure (PSE) to avert non-life-threatening situations (10CFR 20, NRC [REG Guide 8.35] 1992). The criteria established by NRC (REG Guide 8.35, 1992) for a PSE include, ". . . a severe problem in the licensee's operations. Problems might include unscheduled facility shutdowns, high radiation levels that impeded operations important to safety." The best estimate risks are also below those associated with the emergency limit of 25 rem for saving a life (RES-1D, whole-body, low-LET exposure) (Mettler 1999).

Table 7.6. Stochastic Risks, Medical Implication, and Recommended Medical Actions for Each of the RES Categories (for Whole-Body, Low-LET Radiation; medical actions appropriate for exposures to penetrating, gamma radiation but not DU)

Radiation Exposure Status	Total Cumulative Dose	Stochastic Risk Long-Term Health Effects	Medical Note	Medical Actions
0	<0.05 rad	Normal risk.	U.S. baseline 20% lifetime risk of fatal cancer.	Record in exposure record if normally monitored personnel.
1A	0.05–0.5 rad	Up to 0.04% increased risk lifetime fatal cancer.	None (0.001 Sv U.S. annual general population exposure limit).	Record as history in medical record-tactical operation exposure.
1B	0.5–5 rad	U.S. radiation occupational risk. 0.04% to 0.4% increased risk lifetime cancer.	Reassurance (0.05 Sv U.S. annual occupational limit).	Record in medical record-tactical operation exposure.
1C	5–10 rad	0.4% to 0.8% increased risk lifetime fatal cancer.	Counsel regarding increased long-term risk. No live virus vaccines for 3 months.	Record in medical record-tactical operation exposure.
1D	10–25 rad	0.8% to 2% cancer increased risk lifetime fatal cancer.	Potential for increased morbidity of other injuries or incidental disease. <2% increased lifetime risk of fatal cancer.	Record in medical record-tactical operation exposure. Consider routine evacuation from theater in accordance with commander's operational guidance.
1E	25–75 rad	2% to 5.6% increased risk lifetime fatal cancer.	Increased morbidity of other injuries or incidental disease <5% increased lifetime risk of fatal cancer.	Record in medical record-tactical operational exposure. Consider expedited evacuation from theater in accordance with commander's operational guidance.

The counseling of affected personnel and health care providers is especially important for DU exposures because of the perceived risks associated with DU. There is also a need to counsel family members to aid in the patient's recovery and to reduce the morbidity associated with the concern and fear that can occur after exposure to DU. Prior training should stress the actual risks of DU exposure to prevent unnecessary actions that may lead to greater risks, such as dismounting an operational vehicle while still under enemy fire or in open terrain. The DU training tape, U.S. Army Training Video Tape (U.S. Army 2000), is a general awareness training aid that outlines the procedures required when personnel encounter DU on the battlefield. The tape emphasizes the need to consider the tactical situation before taking protective actions. The underlying theme is the need to make sure that the operational risks are minimal before mitigation actions are taken. Improper training or exaggerating the risks will skew decision making and counter the intent of the training film and the policy.

The current DU training and current DU operational guidance recommends that mitigation procedures, (e.g., masking, evacuation, cleaning the vehicle) occur only when the operational condition allows. The results of this analysis support this recommendation.

It is worth emphasizing the importance of performing follow-up bioassays required to estimate the actual dose to personnel. The variability in the exposures is significant and dependent upon stay-time, the actual operational configuration of the vehicle, and many other variables too numerous to model. The magnitude of the exposure from the analysis done in this study indicates that the bioassay can be delayed for a period of time and still have adequate sensitivity to detect exposures that exceed the occupational exposure standards.

NATO, DoD, and U.S. Army doctrine require recording individual exposures for whole-body, low-LET exposure; this requirement should be extended to include the measurements needed to quantify the individual exposures for each Soldier. U.S. Army policy (U.S. Army Medical Command 1999, 2004) is in place to implement this requirement. The policy needs to be placed into doctrine.

Exposures can be minimized without increased risk to the crewmembers or other personnel in these vehicles by turning on the ventilation systems in Bradley vehicles or the EC/NBC ventilation systems in Abrams tanks even for brief periods. Modifying crew reaction drills to incorporate the activation of these systems will reduce exposures significantly without requiring any other changes in crew procedures or protective posture. Activation of the Abrams tank EC/NBC ventilation system for 5 min will reduce the airborne concentration of DU by several orders of magnitude. This is a good practice whether the vehicle is struck by either DU or non-DU munitions, because it will increase the quality of the air for breathing and reduce crew risks from the other non-DU material in the air.

The kidney concentration data show that for each of the instances in which a DU munition perforated a vehicle, anticipated uranium kidney concentrations in individuals present in perforated unventilated vehicles may be high enough to retain the requirements in U.S. Army policy to examine kidney function (MEDCOM 1999). The data also indicate that the doctrine needs to be modified to include the impact of a potential for decreased performance and protracted kidney dysfunction that may occur for Category 2 (REG 2) personnel.

7.6 REFERENCES

10 CFR 20. Code of Federal Regulations, Title 10, *Energy*, Part 20, "Standards for Protection Against Radiation."

53 FR 172822. January 27, 1987. "Radiation Protection Guidance to Federal Agencies for Occupational Exposure; Approval of Environmental Protection Agency Recommendations." *Federal Register*.

Armed Forces Radiobiology Research Institute (AFRRI). 2003. *Medical Management of Radiological Casualties,* Second Edition. Bethesda, Maryland.

International Atomic Energy Agency (IAEA). 1996. *IAEA Safety Series: International Basic Safety Standard for Protection Against Ionizing Radiation and for the Safety of Radiation Sources.* Safety Series No. 115, Vienna, Austria.

International Atomic Energy Agency (IAEA). 1999. *IAEA Safety Standards Series: Occupational Radiation Protection*. Safety Guide No. RSG-1.1, Vienna, Austria.

International Atomic Energy Agency (IAEA). 2002. *IAEA Safety Reports Series: Optimization of Radiation Protection in the Control of Occupational Exposure.* SRS No. 21, Vienna, Austria.

International Commission on Radiological Protection (ICRP). 1959. *Report of Committee II on Permissible Dose for Internal Radiation.* ICRP Publication 2, Pergamon Press, Elmsford, New York.

International Commission on Radiological Protection (ICRP). 1982. *Cost-Benefit Analysis in the Optimization of Radiation Protection*. ICRP Publication 37, Pergamon Press, Oxford, United Kingdom.

International Commission on Radiological Protection (ICRP). 1989. *Statement from the 1989 Paris Meeting of the ICRP: Optimization and Decision-Making in Radiological Protection*. ICRP Publication 55, Pergamon Press, Oxford, United Kingdom.

International Commission on Radiological Protection (ICRP). 1991. *1990 Recommendations of the International Commission on Radiological Protection*. ICRP Publication 60, Pergamon Press, Oxford, United Kingdom.

Interagency Steering Committee on Radiation Standards (ISCORS). 2002. *A Method for Estimating Radiation Risk from TEDE*. ISCORS Technical Report 1, Washington, DC.

Joint Chiefs of Staff Publication, Joint Publication 3-11, "Joint Doctrine for Operations in Nuclear, Biological, and Chemical (NBC) Environments," JP 3-11, 11 July 2000.

Mettler FA, Jr., Committee Chairman. 1999. *Potential Radiation Exposure in Military Operations: Protecting the Soldier Before, During, and After*. Committee on Battlefield Radiation Exposure Criteria, Institute of Medicine, National Academy Press, Washington, DC.

National Council on Radiation Protection and Measurements (NCRP). 1989. *Comparative Carcinoenicity of Ionizing Radiation and Chemicals*. NCRP No. 96, Bethesda, Maryland.

National Council on Radiation Protection and Measurements (NCRP). 1993. *Risk Estimates for Radiation Protection*. NCRP No. 115, Bethesda, Maryland.

National Council on Radiation Protection and Measurements (NCRP). 1993. *Limitation of Exposure to Ionizing Radiation*. NCRP No. 116, Bethesda, Maryland.

National Council on Radiation Protection and Measurements (NCRP). 1997. *Uncertainties in Fatal Cancer Risk Estimates Used in Radiation Protection*. NCRP No. 126, Bethesda, Maryland.

National Council on Radiation Protection and Measurements (NCRP). 1998. *Operational Radiation Safety Program*. NCRP No. 127, Bethesda, Maryland.

North Atlantic Treaty Organization (NATO). 1996. *Commander's Guide on Nuclear Radiation Exposure of Groups. North Atlantic Treaty Organization (NATO) Standardization Agreements*. STANAG-2083, Edition 5, (originally issued September 19, 1986 with latest amendment issued June 26, 1994), Brussels, Belgium.

Parkhurst MA, F Szrom, RA Guilmette, TD Holmes, YS Cheng, JL Kenoyer, JW Collins, TE Sanderson, RW Fliszar, K Gold, JC Beckman, and JA Long. 2004. *Capstone Depleted Uranium Aerosols: Generation and Characterization, Volumes 1 and 2*. PNNL-14168, Prepared for the US Army by Pacific Northwest National Laboratory, Richland, Washington.

US Army. 2000. "Tier I, Depleted Uranium (DU) General Awareness Training." Training Video Tape (TVT) 3-120, AV Product Number 711314, Washington, DC.

US Army Field Manual (FM) 3-3.1, "Nuclear Contamination Avoidance," September 1994.

US Army Field Manual (FM) FM 3-11, "Multiservice Tactics, Techniques, and Procedures for Nuclear, Biological and Chemical Defense Operations," March 2003.

US Army Field Manual (FM) FM 3-11.4, "Multiservice Tactics, Techniques, and Procedures for Nuclear, Biological and Chemical (NBC) Protection," June 2003.

US Army Field Manual (FM) 3-100.12, "Risk Management: Multiservice, Tactics, Techniques, and Procedures," February 2001.

US Army Field Manual (FM) 3-101.12, "Multi-Service Tactics, Techniques and Procedures Manual, Risk Management," FM3-101-12, MCRP 5-12.1C, NTTP 5-03.5, AFTTP(I) 3-2.34, February 2001.

US Army Field Manual (FM) 100-14, "Risk Management,"23 April 1998.

US Army Field Manual (FM) 4-02.7. 2002. "Health Service Support in a Nuclear, Biological and Chemical Environment: Tactics, Techniques, and Procedures."

US Army Medical Command (USAMEDCOM). 1999. "Policy for the Treatment of Personnel Wounded by Depleted Uranium Munitions." Memorandum, 9 April 1999, Washington, DC.

US Army Medical Command. 2004. OTSG/MEDCOM Policy Memo 03-007, 13 January 2004, *Medical Management of Army Personnel Exposed to Depleted Uranium (DU),* Office of the US Army Surgeon General, US Army Medical Command Policy Memorandum, Ft. Sam Houston, Texas.

US Environmental Protection Agency (EPA). 1992. *Manual of Protective Action Guides and Protective Actions for Nuclear Incidents*, EPA 400-R-92-001, Office of Radiation Programs, Washington, DC.

US Environmental Protection Agency (EPA). 1999. *Cancer Risk Coefficients for Environmental Exposure to Radionuclides.* Federal Guidance Report 13, EPA 402-R-99-001, Prepared by Oak Ridge National Laboratory, Oak Ridge, Tennessee, for the US Environmental Protection Agency, Washington, DC.

US Nuclear Regulatory Commission (NRC). 2002. *Occupational Radiation Exposure at Commercial Nuclear Power Reactors and Other Facilities: Thirty-Fourth Annual Report.* NUREG 0713, Vol. 23, Washington, DC.

US Nuclear Regulatory Commission (NRC). 1992. *Planned Special Exposures.* NRC Regulatory Guide 8.35, Washington, DC.

8.0 Summary, Conclusions, and Recommendations

Depleted uranium (DU) aerosols formed when Abrams tanks and Bradley Fighting Vehicles (referred to here as Bradley vehicles) were perforated with large-caliber (LC)-DU munitions were characterized in the Capstone DU Aerosol Study (Parkhurst et al. 2004). The measured quantities of DU aerosol and some of the aerosol characteristics were used to model the DU concentrations, intakes, and associated organ and tissue doses to crewmembers who were in one of these vehicles at the time the crew compartments (the turrets in Abrams tanks or Bradley vehicles or the passenger compartment in Bradley vehicles) were perforated by LC-DU rounds. Additionally, the DU concentrations and doses were modeled for first responders who entered either type of vehicle shortly after perforation. The overall intent of this risk assessment was to estimate the risks to human health from exposure to these DU aerosols. Knowledge of these risks will improve the evaluation of past exposures, decisions on future policy on medical management and post-conflict follow up, and tactical and operational combat decisions by commanders.

Four vehicle configurations were tested in the Capstone study, and a total of 13 LC-DU shots were fired into and perforated the test vehicles. The shot lines were selected to 1) simulate shot lines experienced in Operation Desert Storm (ODS) and 2) maximize the amount of penetrator erosion. With the exception of the Phase-IV shot (discussed below), the target vehicles were ballistic turrets and hulls (BHTs) of Abrams tanks and Bradley vehicles. These BHTs were actual vehicle structures in which the instrumentation, unnecessary structures (e.g., seats), and flammable materials were removed or had not been installed. Unlike operational vehicles, there were no ventilation systems of any kind in these BHTs. In the four test phases, the vehicle configurations included:

- Phase I—Abrams tank BHT (seven shots were fired through non-DU armor)
- Phase II—Bradley vehicle BHT (three shots were fired)
- Phase III—Abrams tank BHT (two shots were fired through DU armor)
- Phase IV—A fully functional Abrams tank equipped with an operating ventilation system (one DU shot was fired into the DU armor).

Based on the DU concentrations collected in the aerosols, it is estimated that a maximum of 7% of the penetrator was aerosolized inside the heavily armored Abrams tank and a maximum of 1% in the less heavily armored Bradley vehicle.

Scenarios were developed for Level I inhalation exposures. A Level I exposure refers to 1) those crew-members inside an Abrams tank or a Bradley vehicle at the time a crew compartment is perforated and 2) those personnel entering immediately afterward. The scenario elements for those in the vehicle at the time of perforation included 1) the time that elapsed (duration) between the time of perforation and the time crewmembers exited the vehicle and 2) the degree of exertion (light or heavy) in terms of breathing rate for the duration of the exposure. Four scenarios were developed for those inside the vehicle at the time of the incident and who remained inside for up to 1 min (Scenario A), up to 5 min (Scenario B), up to 1 h (Scenario C), and up to 2 h (Scenario D) (Table 8.1). A fifth scenario (Scenario E) was developed for first responders. In this scenario, the first responders entered the vehicle 5 min after perforation, exited 10 min later, and heavily exerted themselves the entire time they were in the vehicle.

Table 8.1. Summary of Exposure Scenario Parameters

Level I Scenario	Time of Exposure	Exposure Duration	Breathing Rate
Crewmembers Inside Vehicle			
Scenario A	From impact to exit 1 min post shot	1 min	3 m^3/h
Scenario B	From impact to exit 5 min post shot	5 min	3 m^3/h
Scenario C	From impact to exit 1 h post shot	1 h	3 m^3/h for first 15 min, 1.5 m^3/h thereafter
Scenario D	From impact to exit 2 h post shot	2 h	3 m^3/h for first 15 min, 1.5 m^3/h thereafter
First Responder			
Scenario E	Entry 5 min post shot, exit 10 min later	10 min	3 m^3/h

8.1 ADEQUACY OF MATHEMATICAL MODELING

Mathematical models developed by the International Commission on Radiological Protection (ICRP) were used to model the organ doses and concentration. These models included the Human Respiratory Tract Model (ICRP [66] 1994a, [68] 1994b, [71] 1995b, [78] 1997), the Gastrointestinal Tract Model (ICRP [30] 1979), and the Uranium Systemic Biokinetic Model (ICRP [69] 1995a, [78] 1997). The respiratory tract model implemented for the dosimetric analysis of the Capstone data is that described in ICRP 66 (1994a) with three important distinctions. The deposition model is identical to the ICRP 66 model, but the clearance model has been modified to better describe the experimental data. The "bound" subcompartment of the ICRP 66 model has been renamed the "very slowly dissolving" or "vsd" compartment. In the standard ICRP 66 respiratory tract model, the activity in the "bound" subcompartment does not transfer between compartments. In the modified model, the activity in the "vsd" subcompartment does transfer between compartments with the same rate constants as for the "initial" and "transformed" subcompartments. Also, for purposes of dose calculations, the "vsd" compartment is the same source tissue as the "initial" and "transformed" subcompartments. The transfer coefficients between subcompartments have been renamed and are numerically different than the transfer coefficients between the subcompartments of the ICRP 66 respiratory tract model. The transfer coefficients used to calculate the doses in Table A.2 are shown in Table 4.9.

Measured aerosol parameters used from the Capstone study included the time-dependent DU aerosol concentration, DU particle size, and *in vitro* solubility. DU aerosol concentrations were taken from Marple eight-stage cascade impactor (CIs) with data obtained at eight separate time intervals. The CIs also provided the particle size distributions as a function of time. To improve accuracy and reduce uncertainty in the dose calculations, stage-specific concentration data were used directly in the analyses rather than using the fitted unimodal and bimodal lognormal distributions, which for many samples were poorly fit. *In vitro* solubility characteristics were determined from the powder samples collected in a five-stage cyclone that collected aerosols during the first 2 h after the vehicle was perforated. The residues were relatively insoluble; however, most had a small soluble component, and their dissolution rates were classified as moderate, slow, or very slow.

The following conclusions were reached regarding the adequacy of modeling.

- The data from the Capstone study provided a sound basis for dose and risk assessment and reduced many of the uncertainties in estimating DU exposures that existed prior to that study.

- The development of exposure-dose models that use the size-specific stage data from CIs was an important improvement in dose assessment, allowing the quality and variability of the original measurement data to be maintained in the calculations. Although this technique was driven by the high degree of heterogeneity of the DU aerosols, making lognormal fitting unreliable, the technique should be useful for any dose assessment in which size-selective aerosol data are available.
- The transformation of multi-component exponential functions obtained from *in vitro* dissolution study into the ICRP 66 solubility model structure enabled direct application of the *in vitro* results. Although the mechanisms likely responsible for DU dissolution are not compatible with the description of the compartments in the ICRP 66 model, the transformation is mathematically correct and makes the best use of the data.
- Implementation of the Markov chain Monte Carlo method for uncertainty estimation made good use of the extensive aerosol measurement data and allowed propagation of uncertainties in a robust manner. As further analysis of uncertainties in model structures and parameter values become available, they can be incorporated into more extensive uncertainty analysis approaches that will be able to take advantage of the Bayesian statistical approaches.
- Although the uncertainty in the data underlying the uncertainty analysis was about 40%, the uncertainty on E(50) was about 10 to 20%. This is due to the relatively large number of measurements (30 to 40) on which the E(50) was calculated.

8.2 PREDICTED HUMAN HEALTH EFFECTS

The Capstone field test results and the exposure scenarios evaluated were selected to support estimates of radiological and chemical exposures for Level I crewmembers and passengers in, on, or near either an Abrams tank or a Bradley vehicle at the time the vehicle was perforated. Biokinetic models were used to calculate relevant chemical and radiological doses; these doses were converted into chemical and radiological risks using available human data to assess the chemical risks of uranium exposure and the radiological risks of internally deposited alpha-emitting radioactive materials.

Modeling intakes, uranium concentrations, and the associated doses to the organs of personnel that meet the Level I exposure criterion led to predictions about health effects from chemical effects and radiological doses for these personnel. These predictions assume one LC-DU round perforates the crew compartment and that no ventilation systems are operating during or after perforation, except where indicated.

8.2.1 DU Mass Intakes

The predicted median intakes of DU mass (0.38 nCi/mg) from the aerosols for the five scenarios are listed for each vehicle configuration in Table 8.2. The intakes range from a low of 10 mg for a 1-min exposure in the ventilated Abrams tank with DU armor to a high of 1000 mg for a 2-h exposure in an unventilated Abrams tank with DU armor. These data were used as input to the models for calculations of DU concentrations to tissues and organs and radiological doses.

Table 8.2. Summary of Median DU Intakes by Scenario

	DU Intake (mg)			
	Abrams Tank			Bradley Vehicle, Through Conventional Armor, No Ventilation
Scenario	Through Conventional Armor, No Ventilation	Through DU Armor, No Ventilation	Through DU Armor, EC/NBC Ventilation	
A - Crew, exit in 1 min	280	250	10	83
B - Crew, exit in 5 min	590	710	43	220
C - Crew, exit in 1 h	760	970	91	330
D - Crew, exit in 2 h	780	1000	110	380
E - First responders, entry at 5 min, exit in 10 min	160	200	27	99

8.2.2 Chemical Toxicity Effects

The chemical toxicity of uranium has been evaluated in numerous animal studies and through human studies of workers in the uranium processing field over many years. These studies have identified the kidney as the target organ for uranium toxicity. For this reason, occupational safety levels are based on the kidney as the target organ. A review of renal effects in humans occurring after acute exposures to uranium was used to develop a set of effects groups correlating uranium (U) concentrations in the kidneys with renal effects. These Renal Effects Groups (REGs) were used to predict the potential effects of the current exposure concentrations.

Table 8.3 summarizes the predicted median peak kidney uranium concentrations for exposure Scenarios A through E to two significant figures. For comparison, the *de facto* occupational guideline is 3 μg U/g kidney (ICRP [2] 1959). The concentrations predicted for exposures in an Abrams tank with an

Table 8.3. Summary of Median Peak Kidney Uranium Concentrations by Scenario

Scenario	Peak Kidney Uranium Concentrations (μg U/g Kidney)			
	Abrams Tank			Bradley Vehicle, Through Conventional Armor, No Ventilation
	Through Conventional Armor, No Ventilation	Through DU Armor, No Ventilation	Through DU Armor, EC/NBC Ventilation	
A - Crew, exit in 1 min	3.0	1.1	0.05	1.0
B - Crew, exit in 5 min	6.4	2.6	0.23	2.9
C - Crew, exit in 1 h	8.2	3.5	0.46	3.5
D - Crew, exit in 2 h	8.0	3.7	0.56	4.0
E - First responders, entry at 5 min, exit in 10 min	1.5	0.67	0.14	1.4

operating ventilation system are about an order of magnitude below the ICRP occupational guideline and are well below levels at which adverse health effects might be expected. The concentrations predicted for exposures in unventilated vehicles range from those that are not predicted to cause kidney dysfunction to concentrations where there may be transient or protracted indicators of kidney dysfunction. For Scenarios A and B, the predicted kidney uranium concentrations were either within REG 0 (≤2.2 μg U/g kidney; no kidney dysfunction) or REG 1 (>2.2 to 6.4 μg U/g kidney; transient renal dysfunction) with the exception of Scenario B for the Abrams tank through conventional armor, which is at the interface between REG 1 and REG 2 (>6.4 to 18 mg U/g kidney; protracted kidney dysfunction). In Scenarios C and D, the predicted kidney uranium concentrations were in REG 1 for the Bradley vehicle and the Abrams tank through DU armor and in REG 2 for the Abrams tank through conventional armor. Based on the modeling results, the highest peak concentrations of DU in the kidney would be reduced to 3 mg U/g kidney or less within a month. Even at the 90th percentile level, no scenarios reached REG 3 levels (>18 g U/g kidney), for which more severe symptoms are predicted.

8.2.3 Radiological Effects

The radiation doses from internally deposited uranium and other alpha-emitting radionuclides have been studied in both humans and animals. These studies provide robust biokinetic models for estimating the distribution of uranium in the body and associated radiation doses to various organs of the body. The human data are robust enough to allow the use of risk factors derived from internal alpha emitters and organ-specific factors rather than using the E(50) and the cancer risk factors used to convert E(50) to risk. The results of this study indicated that the risk from using organ-specific risks was greater than the risk obtained using the E(50) methodology by about 35%. The potential for this was discussed by the ICRP in Publication 66 (1994a). Consequently, this assessment was based on organ-specific risks, which are driven primarily by the dose to the lung.

The median committed effective doses, E(50)s, are presented in Table 8.4 by scenario. Doses for all scenarios in the Abrams/conventional armor tank, the ventilated Abrams/DU armor tank, and the Bradley vehicle were within U.S. Nuclear Regulatory Commission's (NRC's) annual dose limit of 5 rem/yr for occupational workers. The E(50) for Scenario B in the unventilated Abrams tank perforated through DU armor was predicted to be 6 rem with a 5 min exposure, increased to nearly 9 rem for a 2-h exposure. These doses are within NRC's annual dose limit of 10 rem/yr for occupational workers with a planned, special exposure (e.g., protecting valuable property during an emergency). Radiation doses at the levels predicted are not expected to cause adverse health effects.

Table 8.4. Summary of the Median Committed Effective Doses for Most Likely Scenarios

	E(50), rem			
	Abrams Tank			Bradley Vehicle, Through Conventional Armor, No Ventilation
Scenario	Through Conventional Armor, No Ventilation	Through DU Armor, No Ventilation	Through DU Armor, EC/NBC On	
A - Crew, exit in 1 min	2.0	2.2	0.090	0.59
B - Crew, exit in 5 min	3.7	6.0	0.44	1.7
C - Crew, exit in 1 h	4.8	8.3	1.02	2.1
D - Crew, exit in 2 h	5.0	8.7	1.20	2.4
E - First responders, entry at 5 min, exit in 10 min	0.92	1.9	0.41	0.89

Committed equivalent doses, $H_T(50)$, to organs were also calculated for the lung, bone surface, kidney, red marrow, and liver. The doses to the lung were higher by at least a factor of 10 than the doses to these tissues/organs. The predicted *median* $H_{Lung}(50)$s calculated for the lung are summarized in Table 8.5 by scenario. For reference, the U.S. annual occupational limit to any individual organ or tissue other than the lens of the eye is 50 rem (10 CFR 20). Except for the case in which an Abrams tank was perforated through DU armor and the stay-time was 1 to 2 h, the predicted doses to the organs were within this annual occupational limit, although the first-year doses, which relate better to immediate health effects, were less than the 50 rem organ dose limit, even for the unventilated Abrams tank. The $H_{Lung}(50)$s at the levels predicted are not expected to cause adverse health effects.

Table 8.5. Summary of the Median Committed Equivalent Lung Doses for Most Likely Scenarios

Most Likely Scenarios	$H_{Lung}(50)$, rem			
	Abrams Tank			Bradley Vehicle, Through Conventional Armor, No Ventilation
	Through Conventional Armor, No Ventilation	Through DU Armor, No Ventilation	Through DU Armor, EC/NBC On	
A - Crew, exit in 1 min	14	18	0.66	5.2
B - Crew, exit in 5 min	32	44	3.3	14
C - Crew, exit in 1 h	38	60	7.6	20
D - Crew, exit in 2 h	39	61	8.7	22
E - First responders, entry at 5 min, exit in 10 min	8.8	14	3.1	6.7

The long-term radiation health effects of inhaled uranium compounds have been difficult to identify with any certainty in human populations. Epidemiologic studies of workers in the uranium processing industry suggest that there may be a very weak association between exposure to inhaled uranium and cancer. Because of the relative absence of data on increased cancer incidence in exposed populations, no risk coefficients for uranium-induced cancer are available from these populations. The increase in lifetime cancer mortality risk was calculated using generic ICRP risk coefficients and the conservative linear, no-threshold theory of effect thought to be protective of health. The risks for total cancer mortality are based on the sum of the individual organs risks.

Table 8.6 summarizes the estimated median lifetime cancer mortality from inhaled, deposited DU for the five exposure scenarios evaluated. Using conservative factors that may overestimate risks, the highest median probability of lifetime cancer mortality calculated in any of the scenarios was 0.0045 for a crewmember staying for 2 h (Scenario D) in an **unventilated** Abrams tank BHT that had been perforated through DU armor. A risk of 0.0045 implies

Table 8.6. Summary by Scenario of Median Lifetime Risk of Fatal Cancer from DU Aerosol Inhalation

Scenario	Lifetime Risk of Fatal Cancer (%)			
	Abrams Tank			Bradley Vehicle, Through Conventional Armor, No Ventilation
	Through Conventional Armor, No Ventilation	Through DU Armor, No Ventilation	Through DU Armor with EC/NBC On	
A - Crew, exit in 1 min	0.11	0.12	0.005	0.034
B - Crew, exit in 5 min	0.20	0.32	0.025	0.099
C - Crew, exit in 1 h	0.27	0.44	0.057	0.12
D - Crew, exit in 2 h	0.28	0.45	0.065	0.14
E - First responders, entry at 5 min, exit in 10 min	0.050	0.10	0.023	0.052

that, on average, the chance over an individual's lifetime of dying of cancer from this exposure is 0.45% greater than the chance of that same individual dying from the natural or background rate, which is 23.6% for males and 20.0% for females (Ries et al. 2003). If a crewmember exited this same tank within 1 min after perforation, the lifetime risk of cancer mortality is predicted to be 0.0012 or 0.12%. The risks were slightly lower for crewmembers exposed to DU aerosols in an unventilated Abrams BHT penetrated through non-DU armor, and less still for crewmembers exposed in an unventilated Bradley vehicle, and least to a crewmember in a ventilated Abrams tank penetrated through DU armor.

8.3 EFFECT OF VENTILATION AND NBC OVERPRESSURE SYSTEMS

The Capstone data were robust enough to provide upper bound estimates of Level I exposures for vehicles in which ventilation systems and, for the Abrams BHT, the EC/NBC system were off and hatches were closed at the time of perforation. The redundancies in the experiment were sufficient to ensure reasonable certainties in the data and confidence in the conclusions drawn from the data. Operation of the EC/NBC system in the Abrams tank or the ventilation systems in the Bradley vehicle can significantly reduce the DU aerosol concentrations and its associated radiological and toxicological risks. The intake of DU and the doses can be significantly reduced without increasing the operational risks to the crew if:

- these ventilation systems are operating at the time of perforation and remain operating after perforation, or
- these ventilation systems are activated shortly after impact (within the first couple of minutes post-perforation).

The effect of a functional EC/NBC overpressure system on DU aerosols and dose reduction in an Abrams tank was significant as demonstrated in the Phase-IV test in which the EC/NBC system in the Abrams tank was on and in the modeled results with the Abrams BHT using differences in air exchange rates. The chemical and radiological risks were reduced by more than an order of magnitude for a stay-time of 5 min (Scenario B). The impact of the ventilation systems on dose reduction to a Bradley vehicle crew was not experimentally measured. Modeling suggests that ventilation will significantly reduce the toxicological risks to the Bradley crew and passengers, but by a factor less than for the Abrams tank. Activating ventilation systems would also reduce the risks to first responders and to other personnel subsequently entering the vehicle. The opening of hatches would be expected to add to the rate of aerosol dispersion for either vehicle.

8.4 APPLYING THE RISK ASSESSMENT RESULTS

The ultimate application of the risk assessment to prospective situations is to assess the risk of Level I personnel and provide a best estimate of risk that could be used by operational commanders and medical personnel before, during, and after the conflict for risk assessment, risk management, treatment, and training. The use of the best estimate risks provide data that are robust enough for use in the commander's risk assessment during combat and can provide initial estimates of individual risks in terms of Radiation Exposure Status (RES) categories and initial estimates of the toxicological effects. The estimated variability in dose and in individual characteristics and actions is large enough that biomonitoring would be required to determine individual doses.

Translating the estimated risks into operationally relevant terms for the military was accomplished using NATO and DoD doctrine for radiological risks and developing an REG system for the chemical risks. The NATO/DoD/Army RES categories can be used to categorize DU exposures in a manner that is useful to field commanders and medical personnel. The RES system provides a means for placing the potential risk of increased incidence of fatal cancer into operational perspective. It was developed specifically for use in those military environments where the operational risks exceed those in the peacetime environments, for which the peacetime safety system was designed and is applicable to Level I exposures because by definition, Level I exposures can only occur during combat operations.

8.4.1 Categories of Human Health Effects

Radiological risks were compared with NATO (NATO STANAG-2083 [1996]) and DoD (JP 3-11 [2000]) RES levels to establish levels of risk of increased fatal cancer for use in all military operations including combat. The fatal cancer risks calculated for the DU exposures were used to place the exposures into one of the RES categories. Because a comparable chemical effects system was not already in place, the chemical REG strategy was employed based upon clinically relevant endpoints from human uranium toxicology data.

Table 8.7 summarizes the best estimate radiological and chemical risks for perforations of these vehicles based on median values to reflect the most probable risks for each of the vehicles and each of the configurations.

The following health effects conclusions can be drawn from this table:

- Crewmembers exiting an unventilated Abrams tank perforated through DU armor or a perforated Bradley vehicle within 1 min are unlikely to exhibit observable chemical or radiological effects from inhaling DU aerosols. For crewmembers in an unventilated Abrams tank in which the perforation is through non-DU armor, a 1-min exposure may cause transient renal effects.
- Crewmembers remaining in an unventilated vehicle through DU or non-DU armor for longer periods of time (greater than 1 min and up to 2 h) may experience transient or protracted but reversible renal dysfunction.
- For all scenarios, even with the ventilation systems off or inoperative, the radiological risks do not exceed those risks that are allowed for occupational limits for planned special exposures–10 rem or RES 1C.
- In all vehicles, exposures to first responders are not projected to exceed RES 1B (the risk equivalent to NRC's occupational dose limit of 5 rem).
- The radiation doses and dose rates from any of the five scenarios and in either vehicle are too low for deterministic effects to occur, including damage to the immune system.
- All of the risks were sufficiently high that post-exposure biomonitoring is recommended to determine the actual exposure and the associated radiation dose and toxicological impact. The radiological requirement

Table 8.7. Median Radiological and Chemical Risk Groups by Vehicle and Scenario

Scenario	Abrams Tank (Through Non-DU Armor, No Ventilation)		Abrams Tank (Through DU Armor, No Ventilation)		Abrams Tank (Through DU Armor, with EC/NBC Ventilation)		Bradley Fighting Vehicle (No Ventilation)	
	RES	REG	RES	REG	RES	REG	RES	REG
A - Crew, exit in 1 min	1B**	1	1B**	0	1A*	0	1B	0
B - Crew, exit in 5 min	1B**	1/2	1C	1	1B	0	1B**	1
C - Crew, exit in 1 h	1C	2	1C/1D	1*	1B	0	1B**	1*
D - Crew, exit in 2 h	1C	2	1C/1D	1*	1B	0	1B**	1*
E - First responders, entry at 5 min, exit in 10 min	1B	0	1B**	0	1B	0	1B	0

* For RES, this indicates that the dose exceeded the U.S. average annual limit for the general public (0.1 rem) but did not exceed the risk associated with RES 1A–equivalent to 0.5 rem.

* For REG, this indicates that the kidney concentration within REG 1 exceeded the ICRP occupational guideline of 3 g U/g kidney.

** For RES, this indicates the risk exceeded the risk associated with the ICRP occupational limit (5-yr average) of 2 rem.

for individual dose assessment is a part of the RES doctrine. The need for kidney-effects monitoring should be considered a part of post-deployment occupational health screening based on the results of the biomonitoring.

8.4.2 Multiple Perforations

The Capstone data provided DU aerosol concentrations as a function of time for two perforations where the second perforation occurred 13 to 14 min after the first perforation. These data showed that the events were essentially independent of one another, and both the peak kidney uranium concentrations and radiation doses from the second shot were approximately the same as from the first shot. Although the data are limited, it appears that the radiation doses and kidney uranium concentrations for two shots can be roughly approximated by doubling the doses and kidney concentrations. The DU aerosol concentration from three or more perforations has not been evaluated.

As the time delay between perforations is reduced, more residual aerosol from the first shot would be mixed in with aerosols produced in the subsequent shot. In this circumstance and with multiple perforations, the doses from higher concentrations may be offset if particle agglomeration increases in proportion to the amount of aerosol created. Increased particle agglomeration would increase deposition amounts and decrease the doses associated with aerosol inhalation.

8.4.3 Relationship of Risk Assessment Results to DoD/Army Doctrine and Policy

Based on this analysis using Capstone data, current DoD and U.S. Army doctrine and policies appear to be appropriate for the risks incurred by perforation of armored vehicles with DU munitions. The following comments and conclusions are provided:

- The DoD/Army RES system provides a means for placing the potential risk of increased incidence of fatal cancer into operational perspective. The system is derived from peacetime occupational and contingency doses that were adapted to the military operational environments including combat. The system is applicable to Level I exposures because, by definition, Level I exposures can only occur during combat operations.

- Activation of available ventilation or EC/NBC systems is a quick, effective method for removing DU aerosols from the interior of armored vehicles, thereby reducing health risks to crewmembers.
- The doctrine may need to be evaluated to determine if temporary illness and diminished performance may occur for personnel whose intake of uranium aerosol concentrations is consistent with REG 2.
- The medical and operational recommendations in U.S. Army (FM 4.02-7 2002) and DoD (AFRRI 2003; JP 3-11 2000) doctrine for actions recommended for each of the RES levels need to be reviewed and modified for DU exposure.

Consideration should be given to reviewing and modifying the RES doctrine in the following areas:

- Using risk levels rather than radiation dose as the standard for each category. Translating the RES to risk levels allows the use of this table for all carcinogens, both radiological and chemical. This will provide more consistent risk assessments by using a well-established, peer-reviewed framework for determining risk.
- Modifying the medical outcomes and recommended medical actions in current DoD policies for each RES level to take into account the differences between whole-body, low-LET radiation and the internal exposures from DU. The primary difference to consider is the difference in injury to the immune system. Injury to the immune system from DU exposure is negligible when compared to exposure to whole-body, low-LET radiation.
- Adopting a REG-type toxicological framework for deterministic effects from radiation or chemical exposure and using the RES structure for stochastic chemical effects.

8.5 HUMAN HEALTH RISK CONCLUSIONS

The Capstone data were robust enough to provide upper estimates of Level I exposures within vehicles with non-operating or inoperative ventilation systems (including the EC/NBC ventilation system in the Abrams tank). Although variations would be expected from shot to shot, the redundancies in the experiment were sufficient to ensure reasonable certainties in the data and in the conclusions drawn from the data.

The activation of ventilation systems in either vehicle or the EC/NBC ventilation system in the Abrams tank can significantly reduce the DU aerosol concentration and associated health risks without affecting the operational status of the crew or increasing operational risk. The Capstone data from a single-shot perforation of an Abrams tank through DU armor and with an operating EC/NBC ventilation system showed that the risks were reduced to below the risk limits for peacetime occupational exposures. The same is likely to also be true for two perforations through DU armor of a tank with an operating EC/NBC system. Theoretical calculations indicate that activation of the Bradley vehicle ventilation system will also rapidly reduce the aerosol concentrations and associated risks when compared with an unventilated Bradley vehicle.

General results of the Level I risk assessment include the following:

- The level of chemical and radiological risks to crewmembers from DU intake following perforation of an Abrams tank or a Bradley vehicle will depend on whether or not one or more vehicle ventilation systems were operating at the time of and after armor perforation.
- For exposure in an unventilated vehicle, the most probable health effects predicted from analyses of the highest doses are for acute transient kidney injury and an increase in the long-term risk of lung cancer.
- The kidney is the organ that has the highest chemical risk from uranium.
- None of the chemical risks exceeded REG 2. The Bradley vehicle and Abrams tank with DU armor (with or without ventilation) were within REG 1.
- The lung is the organ of consequence predicted to receive the highest radiation dose (extrathoracic airways and lymph nodes receive more in some cases but are not particularly sensitive to chemical effects nor are they targets for radiation).
- The highest median lifetime risk of cancer mortality from any single-perforation scenario and shot was 0.45%, which should be compared to the current 23.6% risk of cancer mortality in U.S. males.
- For the Abrams tank with no operating ventilation system, the majority of the aerosol intake occurs in the first 5 to 10 min after perforation. This finding reinforces the importance of early activation of ventilation systems.
- In all vehicles, exposures to first responders are not projected to exceed REG 0 (which is below the *de facto* occupational guideline of 3 μg U/g kidney).
- Although the chemical and radiation risks are estimated to be low, counseling of exposed personnel and their family members is very important because of the perceived risks associated with DU exposure.

- Although the radiation risks are estimated to be low, all of the radiation risks exceeded the risks for which individual dose assessments are required. Because of differences in individual exposure for a given crewmember in a perforated vehicle, a DU bioassay is essential for estimating individual doses.
- Activation of the Abrams tank EC/NBC System will reduce the estimated intake of DU and the chemical dose significantly if it is activated shortly (i.e., 2 to 3 min) after perforation. Likewise, operating the Bradley vehicle ventilation system during and after perforation by a DU munition will result in lower chemical doses.

8.6 RECOMMENDATIONS FOR FOLLOW-ON ACTIONS

The results of this work highlighted several areas that may require follow-on work. The following recommendations are provided for consideration:

- Review and revise, as necessary, DU bioassay procedures. The need for and the mechanisms for conducting and timing DU bioassays should be reviewed to ensure that the data needed to determine individual doses are obtained and archived appropriately.
- Conduct epidemiology studies on crewmembers with positive DU bioassay results. The primary data gap is the human health effects from acute exposures to DU aerosols. The recently implemented DU bioassay program (U.S. Army Medical Command 2004) could provide the exposure measurement data lacking in previous studies.
- Conduct experimental studies in laboratory animals to clarify the relationship of kidney toxicity to the dose of uranium following acute inhalation of DU. Because acute kidney toxicity is the predominant predicted risk for staying 1 to 2 h in an unventilated vehicle immediately after it is perforated by DU munitions, a better understanding of the clinical course of the effects from inhaled uranium is needed to assist management of combat situations. The studies should correlate kidney uranium burden over time with biochemical and pathological changes to the kidney. Uranium oxides aerosols with solubility characteristics similar to those observed in the Capstone study are recommended.
- It is unlikely that Soldiers will be exposed to DU alone. Therefore, there should be an evaluation of other metals and chemicals likely to be present in conjunction with DU when a combat vehicle is perforated, and the potential effects of DU in combination with these compounds. (An analysis of other metals in the Capstone DU aerosols is discussed in Parkhurst et al. 2004.)

- Re-examine crew immediate-action drills to ensure that appropriate actions are included. Immediate activation of the Abrams tank EC/NBC ventilation system or the Bradley vehicle ventilation systems should be considered for all perforations of armored vehicles in which these systems remain operational. These systems are effective at removing dust and particulates from the crew compartment and significantly reducing DU aerosol intake.
- Review and modify, as appropriate, the DU awareness training using the results of this work as the bases for the modifications.

8.7 REFERENCES

Armed Forces Radiation Research Institute (AFRRI). 2003. *Medical Management of Radiological Casualties*, Second Edition. Bethesda, Maryland.

FM 4-02.7. 2002. "Health Service Support in a Nuclear, Biological and Chemical Environment: Tactics, Techniques, and Procedures," October 2002.

International Commission on Radiological Protection (ICRP). 1959. *Report on the Committee II on Permissible Dose for Internal Radiation.* ICRP Publication 2, Pergamon Press, Oxford, United Kingdom.

International Commission on Radiological Protection (ICRP). 1979. *Limits for Intakes of Radionuclides by Workers*. ICRP Publication 30, Part 1, Pergamon Press, Oxford, United Kingdom.

International Commission on Radiological Protection (ICRP). 1994a. *Human Respiratory Tract Model for Radiological Protection*. ICRP Publication 66, Pergamon Press, Oxford, United Kingdom.

International Commission on Radiological Protection (ICRP). 1994b. *Dose Coefficients for Intakes of Radionuclides by Workers*. ICRP Publication 68, Pergamon Press, Oxford, United Kingdom.

International Commission on Radiological Protection (ICRP). 1995a. *Age-dependent Doses to Members of the Public from Intake of Radionuclides: Part 3 Ingestion Dose Coefficients*. ICRP Publication 69, Pergamon Press, Oxford, United Kingdom.

International Commission on Radiological Protection (ICRP). 1995b. *Age-dependent Doses to Members of the Public from Intake of Radionuclides: Part 4 Inhalation Dose Coefficients*. ICRP Publication 71, Pergamon Press, Oxford, United Kingdom.

International Commission on Radiological Protection (ICRP). 1997. *Individual Monitoring for Internal Exposure of Workers*. ICRP Publication 78, Pergamon Press, Oxford, United Kingdom.

Joint Publication 3-11. 2000. "Joint Doctrine for Operations in Nuclear, Biological, and Chemical (NBC) Environments," JP 3-11, 11 July 2000.

NATO STANAG-2083. 1986. Commander's Guide on Nuclear Radiation Exposure of Groups. North Atlantic Treaty Organization (NATO) Standardization Agreements (STANAG)-2083, Edition 5. 19 September 1986. (Latest Amendment, 26 June 1994.)

Parkhurst MA, F Szrom, RA Guilmette, TD Holmes, YS Cheng, JL Kenoyer, JW Collins, TE Sanderson, RW Fliszar, K Gold, JC Beckman, and JA Long. 2004. *Capstone Depleted Uranium Aerosols: Generation and Characterization, Volumes 1 and 2*. PNNL-14168, Prepared for the US Army by Pacific Northwest National Laboratory, Richland, Washington.

Ries LAG, MP Eisner, CL Kosary, BF Hankey, BA Miller, L Clegg, A Mariotto, MP Fay, EJ Feuer, and BK Edwards (eds). 2003. SEER Cancer Statistics Review, 1975-2000, National Cancer Institute. Bethesda, Maryland. Accessed online in January 2004 at http://seer.cancer.gov/csr/1975_2000,2003.

US Army Medical Command. 2004. OTSG/MEDCOM Policy Memo 03-007, 13 January 2004, *Medical Management of Army Personnel Exposed to Depleted Uranium (DU),* Office of the US Army Surgeon General, US Army Medical Command Policy Memorandum, Ft. Sam Houston, Texas.

Abbreviations and Acronyms

AD—aerodynamic diameter (also known as aerodynamic equivalent diameter, AED)
AED—aerodynamic equivalent diameter
AEPI—Army Environmental Policy Institute
AI—alveolar-interstitial (region)
ALARA—as low as reasonably achievable
AMAD—activity median aerodynamic diameter
AMTD—activity median thermodynamic diameter
APFSDS—armor piercing fin-stabilized discarding sabot
APG—Aberdeen Proving Ground, Maryland
ATC—Aberdeen Test Center
ATSDR—Agency for Toxic Substances and Disease Registry

BDAR—Battlefield Damage Assessment and Repair
BDAT—Battle Damage Assessment Team
BHT—ballistic hull and turret
BK—Biokinetic Model (computer code)
BMI—body mass index
Bq/µg—becquerel(s) per microgram
BU—backup filter

CD—committed equivalent dose
CDE—committed equivalent organ dose
CED—committed effective dose
CEDE—committed effective dose equivalent
cfm—cubic feet per minute
cGy—centiGray (equivalent to rad)
cGy/h—centiGray per hour
CI—cascade impactor
cm—centimeter(s)
CV—coefficient of variation
CY—cyclone

d—day
DDREF—dose and dose-rate effectiveness factor
DoD—U.S. Department of Defense
DOE—U.S. Department of Energy
DQO—data quality objective
DU—depleted uranium
DUR-IPT—Depleted Uranium Research – Integrated Process Team

E(50)—50-yr committed effective dose (see glossary for relationship to CEDE)

ECD—effective cutoff diameter
EC/NBC—environmental control/nuclear, biological, chemical ventilation system
EPA—U.S. Environmental Protection Agency
ET—extrathoracic

F—uranium mass fraction in first peak of bimodal distribution
FM—U.S. Army Field Manual
FS—filter sampler (or sample)

g—gram
GI—gastrointestinal
GSD—geometric standard deviation

h—hour(s)
H_T—dose equivalent to tissue T
$H_T(50)$—50-yr committed equivalent dose
HRTM—Human Respiratory Tract Model
HVAC—heating, ventilation, and air conditioning

ICRP—International Commission on Radiological Protection
IDLH—concentrations immediately dangerous to life or health
IOM—Institute of Occupational Medicine (used as an abbreviation for filter cassettes)
ISCORS—U.S. Interagency Steering Committee on Radiation Standards

JP—U.S. Army Joint Publication

L—liter(s)
LARs—Logistics Assistance Representatives
LANL—Los Alamos National Laboratory
LC—large caliber (referring to depleted uranium munition)
LET—linear energy transfer (of ionizing particles)
Lpm or L m^{-1}—liter(s) per minute
LRRI—Lovelace Respiratory Research Institute
LUDUC—Lung Dose Uncertainty Calculation (computer code)

m—meter
MCMC—Markov chain Monte Carlo (statistical technique)
mCi—microcurie
µm—micrometer
µg—milligram(s)
min—minute(s)
MMAD—mass median aerodynamic diameter
MOOTW—military operations other than war
MVF—moving filter

NAS—National Academy of Sciences
NATO—North Atlantic Treaty Organization
NBC—nuclear-biological-chemical
nCi—nanocurie
NCRP—National Council on Radiation Protection and Measurements
NIOSH—National Institute of Occupational Safety and Health
NIST—National Institute of Standards and Technology
NPN—non-protein nitrogen
NRC—U.S. Nuclear Regulatory Commission
NRPB—National Radiological Protection Board

ODS—Operation Desert Storm
OEG—Operational Exposure Guidance
OSAGWI—Office of the Special Assistant for Gulf War Illnesses, Medical Readiness, and Military Deployment

PAG—protective action guides
PNNL—Pacific Northwest National Laboratory
PPE—personal protective equipment
ppm—parts per million

QA—quality assurance
QC—quality control

rad—unit of absorbed dose equal to 0.01 J/kg (U.S. conventional unit)
RADCON—radiation control
RBC—red blood corpuscles
REG—Renal Effects Group
rem—special unit of dose equivalent (U.S. conventional unit)
RES—radiation exposure status
RSO—Radiation Safety Office(r)
RT—respiratory tract

sec—second(s)
SEE—specific effective energy
SEM—scanning electron microscopy
SI—international system of units
SID—System for Internal Dosimetry (computer code)
ST—soft tissue
STANAG—standardized agreement
Sv—sievert (special unit of dose equivalent, SI unit)

TACOM-ARDEC—Tank-automotive and Armaments Command, Armament Research, Development & Engineering Center
TRU—transuranic

U—uranium
ULI—upper large intestine
UNEP—United Nations Environmental Program
U.S.—United States
USACHPPM—U.S. Army Center for Health Promotion and Preventive Medicine
UXO—unexploded ordnance

V&V—validation and verification
vsd—very slow dissolution

WHO—World Health Organization

yr—year

Conversion Factors

In the United States, absorbed dose is traditionally stated in terms of rad and other radiation protection quantities, such as equivalent dose and effective dose, are given in rem. However, the National Council on Radiation Protection and Measurements (NCRP 1985) supports the use of the International System (*Système International* [SI]) of Units, in which absorbed dose is given in gray (Gy) and other radiation protection quantities are given in sievert (Sv). In this document, a combination of the two unit systems is used depending on the discussion. The following tables are provided to assist conversion between the conventional and SI units.

Table C.1. Conversions for Absorbed Dose

Conventional Units		SI Units	
0.001 rad	= 1.0 mrad	= 0.01 mGy	--
0.01 rad	= 10 mrad	= 0.1 mGy	--
0.1 rad	= 100 mrad	= 1.0 mGy	= 0.001 Gy
1.0 rad	= 1,000 mrad	= 10 mGy (or 1 cGy)	= 0.01 Gy
10 rad		= 100 mGy	= 0.1 Gy
100 rad		= 1000 mGy	= 1.0 Gy
1000 rad		--	= 10 Gy

Table C.2 Conversions for Effective Dose, Equivalent Dose, Dose Equivalent, and Ambient Dose Equivalent

Conventional Units		SI Units	
0.001 rem	= 1.0 mrem	= 0.01 mSv	--
0.01 rem	= 10 mrem	= 0.1 mSv	--
0.1 rem	= 100 mrem	= 1.0 mSv	= 0.001 Sv
1.0 rem	= 1,000 mrem	= 10 mSv (or 1 cSv)	= 0.01 Sv
10 rem		= 100 mSv	= 0.1 Sv
100 rem		= 1000 mSv	= 1.0 Sv
1000 rem		--	= 10 Sv

The conversion between radioactivity and uranium mass for U.S. DoD depleted uranium is 0.38 nCi/mg (14 Bq/mg U).

REFERENCE

National Council on Radiation Protection and Measurements (NCRP). 1985. *SI Units in Radiation Protection and Measurements, Appendix C.* NCRP No. 82, Bethesda, Maryland.

Glossary

Abrams Main Battle Tank—A full-tracked, armored, land combat vehicle with a 105-mm (M1) or 120-mm (M1A1/M1A2) gun operated by a four-man crew consisting of a commander, gunner, loader, and driver. The Abrams tank is the principal weapon of tank battalions of the U.S. Army during all types of combat operations.

absorbed dose—The energy absorbed per unit mass; the physical dose quantity is given by:

$$D = \frac{d\bar{\varepsilon}}{dm}$$

where $d\bar{\varepsilon}$ is the mean energy imparted by ionizing radiation to matter in a volume element and dm is the mass of the matter in the volume element. The SI unit for absorbed dose is J/kg and the special name is the gray (Gy). 1 Gy equals 100 rad.

absorption—The movement of material to blood regardless of the mechanism. Applies to the dissociation of particles and the uptake into blood of soluble substances and material dissociated from particles.

activity median aerodynamic diameter (AMAD)—The diameter where 50% of the radioactivity in the aerosol is associated with particles of aerodynamic diameter (AED) greater than the AMAD. This term is used when deposition depends upon inertial impaction and sedimentation, typically when the AMAD is greater than 0.5 µm.

aerodynamic diameter (AD) or aerodynamic equivalent diameter (AED)—AD and AED are both used to describe the diameter of a sphere, in µm, of unit density (1 g/cm^3) that has the same terminal settling velocity in air as the particle of interest. A 1-µm-AED particle has 1000 times the volume of a 0.1-µm-AED particle). Discussion of AED measurements refers to individual particles.

aerosol—An assemblage of liquid or solid particles suspended in a gaseous medium long enough to be observed and measured; generally about 0.001-µm to 100-µm AED.

ballistic hull and turret (BHT)—A production Abrams or Bradley structure without any operational components. The turret is mounted on the hull via a race ring, but no other internal or external components are present (i.e., no power train, fire control system, ventilation system, etc). A BHT may contain a gun, road wheels, and track if the specific test requires these. A BHT is typically used to reduce test cost when an operational vehicle is not required to meet test objectives.

becquerel (Bq)—The SI unit of radioactivity. 1 Bq is equal to 1 nuclear transformation per second.

Biokinetic (metabolic, retention, or pharmacokinetic) Model—A mathematical description of the physiologically time-dependent compartmental transport, distribution, and excretion of a substance from the human body after its intake and absorption into body fluids. The use of biokinetic models is central to the estimation of organ and tissue doses resulting from the internal deposition of radionuclides in the human body.

bioassay—An analysis of body fluids, tissue, or excreta to determine the absence, the degree, or presence of specific materials. Used as an index of radioactivity in the body.

Bradley Fighting Vehicle—A full-tracked, medium-armored fighting vehicle that provides protected, cross-country mobility and vehicular-mounted firepower to infantry/cavalry units. The Bradley Fighting Vehicle System family consists of infantry and cavalry versions.

clearance—The removal of material from the respiratory tract via particle transport. The main routes are:

- into blood by absorption
- into the gastrointestinal tract via the pharynx
- into the regional lymph nodes via the lymphatic channels.

committed effective dose, E(τ)—The sum of the products of the committed organ or tissue equivalent doses and the appropriate organ or tissue weighting factor (W_T) where τ is the integration time in years following an intake. The integration time for an adult worker is 50 yr and is represented as E(50) This term replaces committed effective dose equivalent (CEDE) as defined in ICRP Publication 26.

committed equivalent dose, $H_T(\tau)$—The time integral of the equivalent dose rate in a particular organ or tissue that will be received by an individual following an intake of radioactive material into the human body, where τ is the integration time in years following an intake. The integration time for an adult worker is 50 yr and is represented as $H_T(50)$. This term replaces committed dose equivalent (CDE) as defined in ICRP Publication 26.

Data Quality Objective—A systematic strategic planning tool based on the scientific method that identifies and defines the type, quality, and quantity of data needed to satisfy a specified use.

depleted uranium—Depleted uranium (DU) is the primary material used in the large-caliber (LC) penetrators fired at vehicle targets in this study. In 10 CFR 40.4, the Nuclear Regulatory Commission defines depleted uranium as any uranium with less that 0.711 weight percent U-235. Natural uranium has approximately 0.72% by weight U-235. The depleted uranium used by the U.S. military contains approximately 0.2% by weight U-235 and is 40% less radioactive than natural uranium.

deposition—The initial process that determines how much of the material in the inspired air remains in the respiratory tract after expiration. Deposition of material may occur during both inspiration and expiration.

deterministic effects (or non-stochastic effects)—Those effects that increase in severity with an increase in dose and have a practical threshold dose below which they are not observed. The type and severity of the deterministic effects depends upon the type of ionizing radiation involved, the magnitude of the dose, and the rate at which the dose is accumulated (dose rate). Examples of deterministic effects are cataracts, temporary sterility, permanent sterility, temporary hair loss, and depression of blood cell formation process.

detriment—The mathematical expectation of the amount of harm in the exposed individual, taking into account both the probability and the severity of the different possible harmful health effects. Harmful health effects include fatal cancer, non-fatal cancer, and hereditary effects.

dissolution rate—The rate of change of a solid into a liquid form by immersion in a fluid of suitable chemical composition or character.

environmental control/nuclear-biological-chemical ventilation system (EC/NBC)—A system found on the Abrams Main Battle tank that conditions air for breathing (filtering out nuclear, biological, and chemical agents) as well as personal heating and cooling as required, while crew members are wearing protective suits and masks. The EC/NBC system on the Abrams tank also provides positive air pressure within the turret and driver's locations to prevent diffusion of NBC contaminants.

equivalent dose (H_T)—A weighted absorbed dose in a tissue or organ. The equivalent dose in tissue, T. It is given by the expression:

$$H_T = \sum_R W_R * D_{T,R}$$

Where:

H_T = Equivalent dose in J/kg (or Sv)
$D_{T,R}$ = The sum of the doses over all radiation type R absorbed dose over tissues T
T = Organ or tissue
W_R = Radiation weighting factor.

effective dose (E)—The sum of the weighted equivalent doses in all the tissues and organs of the body. It is given by the expression:

$$E = \sum_T W_T * H_T$$

Where:

E = Effective dose, in J/kg (or Sv)
H_T = Equivalent dose in organ or tissue T
W_T = The tissue or organ weighting factor.

fractional absorption in the gastrointestinal tract (f_1)—The fraction amount of the ingested element that is directly absorbed into body fluids from the gastrointestinal tract. Absorption is assumed to occur in thc small intestine.

first responder—Personnel who enter damaged vehicles shortly after DU perforation to evacuate personnel or recover equipment.

gray (Gy)—The SI unit of absorbed dose, 1 Gy is equal to 1 J/kg (or 100 rad).

inhalability—The ratio of the number concentration of particles with a particular AED inspired through the nose or mouth to the number concentration of particles with the same AED in the inspired volume of ambient air.

inhalable fraction—The fraction of aerosolized material of a particular size (usually particles 100 μm or less AED) available for intake via the respiratory tract.

large-caliber DU munitions—Rounds with large-caliber depleted uranium (LC-DU) penetrators that are fired from the Abrams platform (M1A1 and M1A2 series tanks). These heavy metal, long-rod penetrators use kinetic energy to penetrate a target.

Marple cascade impactor (CI)—A personal cascade impactor consisting of several stages used to separate particles of different aerodynamic sizes. Marple CIs used in this test had a nominal flow rate of 2 Lpm through an eight-stage (25-mm-diameter) unit capable of collecting particles ranging from about 0.5-to 20-μm AED.

notional—Theoretical or speculative rather than actual, used particularly in modeling vernacular.

particle size distribution—The number concentration of particles as a function of particle diameter, usually aerodynamic diameter, and expressed in micrometers, μm.

particle transport—The movement of material from one part of the respiratory tract to another and the processes that clear materials from the respiratory tract to the gastrointestinal tract and to the lymph nodes.

penetrated—Used here to convey the piercing of the armor by the DU penetrator that may or may not enter a turret, driver, or passenger (Bradley) compartment.

perforated—Used here to convey the breach by the DU penetrator through vehicle armor into the turret, driver, or passenger (Bradley) compartment.

posterior distribution—In the context of Bayesian statistical methods, the posterior distribution gives the probability distribution of the inferred quantities after the measurement. This is the answer resulting from a Bayesian statistical analysis. Bayesian methods result in probabilistic expressions of answers from which central values like a mean or median can be extracted, as well as upper and lower confidence limits.

prior distribution—In the context of Bayesian statistical methods, a prior distribution is a way to express the initial knowledge of the inferred quantities (e.g., intake amounts or doses). It gives the probability distribution of the inferred quantities before the measurement. The prior may be based on previous measurements, general principles (e.g. intakes are never negative), or may be assumed based on prior knowledge and experience.

quality assurance—An integrated system of management activities involving planning, implementation, assessment, reporting, and quality improvement to ensure that a process, item, or service is of the type and quality needed for the application.

quality control—The overall system of technical activities that measure the attributes and performance of a process, item, or service against defined standards to verify that they meet the stated requirements for the application, operational techniques, and activities that are used to fulfill requirements for quality.

rad—The special unit of absorbed dose. One rad is equal equal to an absorbed dose of 100 ergs/g or 0.01 J/kg (0.01 gray).

Renal Effects Groups (REGs)—A model to address chemical risks to the kidney from uranium exposure, based on mg U/g kidney. This model is described in Section 6.3.1.2.

rem—The special unit of any quantities expressed as dose equivalent. The dose equivalent in rem (the plural is referred to as rem or rems) is equal to the absorbed dose in rads multiplied by the quality factor (1 rem=0.01 sievert).

respirable fraction—The fraction of aerosolized material of a particle size (usually particles 4 µm or less AED) available for intake and deposition into the deep lung.

radiation weighting factor (W_R)—A dimensionless factor that is used to derive the equivalent dose from the absorbed dose averaged over an organ or tissue and is based on the quality of the radiation.

Reference Man—An adult male with the anatomical, physiological, and metabolic characteristics defined in ICRP Publication 23 and updated by ICRP Publications 66, 70, and 89.

risk—The probability that a given individual will incur a particular adverse health effect as a result of the chemical or radiation dose received.

scenario—An outline of a projected chain of events, which as it relates to this risk assessment, includes the time and duration of exposure and the breathing rate.

shot line—The trajectory (attack azimuth and elevation) of a penetrator and the impact point relative to a target.

shape factor— A dimensionless constant use to relate the drag force experienced by an irregular particle moving in air to the particle's equivalent volume diameter. The default value for an irregular shaped particle is 1.5.

sievert (Sv)—The SI unit of dose. 1 Sv equivalent dose is equal to 1 J/kg (or 100 rad) times the appropriate radiation weighing factor, W_R. 1 Sv effective dose is equal to 1 J/kg (or 100 rem) and includes the radiation weighting factor, W_R, and the tissue or organ weighting factor, W_T.

solubility—The ability of a substance to form a solution with another substance. Normally lung or tissue fluid is considered the fluid of choice for evaluation of residence time of a chemical in the body with regard to inhalation intake.

source term—The amount of radionuclide or chemical released from a source or site to the environment over a specific period for use in dose assessment or exposure assessment.

stochastic effects—Those effects for which the probability of occurrence is deemed to be proportional to the committed effective dose. These effects appear to be related to random events. Cancer and hereditary effects are examples of stochastic effects.

target organ—An organ or tissue in which radiation is absorbed or where a chemical compound accumulates, resulting in a clinical or sub-clinical change.

tissue weighting factor (W_T)—A dimensionless factor by which the equivalent dose in an organ or a tissue is weighted to represent the relative contribution of that organ or tissue to the total detriment resulting from uniform irradiation of the human body.

transfer compartment—The compartment that is introduced (for mathematical convenience) into most of the biokinetic models used in this report to account for the translocation of the radioactive material through the body fluids from where they are deposited in organs or tissues.

turret—Revolving armored structure (located on top of the tank hull) that primarily houses the main gun and fire control system.

APPENDIX A Doses and Concentrations

The Human Health Risk Assessment (HHRA) was conducted to estimate the doses and risks to crewmembers and first responders from inhaling depleted uranium (DU) aerosols generated inside armored vehicles perforated by DU munitions. The risk assessment relied on the results of the Capstone DU Aerosol Study (Parkhurst et al. 2004) for input data. In the Capstone study, DU aerosols were collected inside ballistic hulls and turrets (BHTs—vehicles stripped of instrumentation and combustible materials) of an Abrams tank and a Bradley Fighting Vehicle, both of which were perforated by large-caliber DU rounds under controlled test conditions. Aerosols generated inside these BHTs were collected and characterized. In this HHRA final report, the scenarios selected for modeling are described in Chapter 2. The Capstone test phases (Phases I through IV) are described in Chapter 3. A description of the modeling approach and the computer codes used is presented in Chapter 4.

Capstone data specifically used in the risk assessment included the aerosol concentration from 13 shots as a function of time, the aerosol particle size distribution as unimodal and bimodal activity median aerodynamic diameters (AMADs) as a function of time, and the dissolution *in vitro* of the aerosols as a function of particle size. For the first 12 of the shots, at least three and usually four sampling arrays collected aerosols in accordance with a predetermined time sequence. The thirteenth shot used two sampling sites. The sampling arrays consisted of filter cassettes and eight-stage cascade impactors (CIs). For the thirteenth shot, a smaller sampling array consisting only of CIs was used. The CI sample data was selected for dose analysis because these data provided concentration and size information and because it collected aerosols that generally fell within the respirable and inhalable size range. Rather than use AMAD values as input to the computer codes, actual DU concentrations by CI stage were used in dose and organ concentration modeling as discussed in Chapter 3.

The seven tables in this appendix provide modeling input data listings of DU concentrations by particle size and modeling output. The information included in each table is briefly described below.

- Table A.1 provides information about the sampler arrays used for each scenario modeled.
- Table A.2 (in three parts) summarizes the calculated intakes, doses, and concentrations for all phases and scenarios in which Bayesian statistics were used. These summary results are the means of the individual results (presented in Table A.3) except where the data are identified as medians or as 10th or 90th percentile values.
- Table A.3 lists the individual intakes, doses, and concentrations by phase, shot, crew position (sampling array position), and scenario. Bayesian statistics were used in calculating these values.
- Table A.4 (in three parts) summarizes the calculated intakes, doses, and concentrations for all phases and scenarios using conventional statistics. These summary results are the means of the individual results (presented in Table A.5) except where identified as medians or as 10th or 90th percentile values.
- Table A.5 lists the individual intakes, doses, and concentrations by phase, shot, crew position (sampling array position), and scenario. Conventional statistics were used in calculating these values.
- Table A.6 provides the output of the software assurance analysis conducted with six sampler scenarios. The results using the System for Internal Dosimetry (SID) code correspond to the Biokinetic (BK) model results presented in Table A.5. See Section 4.5 and 4.6 for more information about these models.
- Table A.7 provides input data including the time midpoint of each sampling interval, the calculated median diameter of particles collected on each stage, and the DU concentration collected on each stage. The concentrations presented are adjusted for CI wall loss (discussed in Section 3.2.3.3). This table also lists sample identifications for cross-referencing with the data reported by Parkhurst et al. (2004) in Appendix A of the Capstone study final report.

These tables include combinations of the following column headings:

- Phase—Relates to the type of BHT/vehicle or armor fired at (phase number in Arabic numerals rather than the Roman numerals used in the report text)

  —Phase I was an Abrams BHT perforated through non-DU armor.

  —Phase II was a Bradley vehicle.

  —Phase III was an Abrams tank perforated through DU armor.

  —Phase IV was a fully functioning Abrams tank perforated through DU armor.

- Shot—The number of the test shot within a phase. Seven shots were fired in Phase I, three in Phase II, two in Phase III, and one in Phase IV.

- Position—The sampling arrays were set up near crew positions and are named for those positions. In the Abrams tanks (Phases I, III, and IV), these positions were commander, driver, loader, and gunner. In the Bradley vehicle (Phase II), the positions have the same designation in the table to reflect the identification stamped on the samplers, but the gunner sampler array was placed near the right scout position and the loader array was located near the left scout position.

- Scenario—The set of parameters related to time and duration of exposure and breathing rate as described in Chapter 2 and as summarized in the following table:

Scenario	Population Affected	Exposure Description
A	Crewmember	0-1 min heavy exercise
B	Crewmember	0-5 min heavy exercise
C	Crewmember	0-15 min heavy, 15-60 min light exercise
D	Crewmember	0-15 min heavy, 15-120 min light exercise
E	First responder	5-15 min heavy exercise

  Scenarios A through D represent cases in which crewmembers present in a vehicle at the time of perforation exit some time from 1 min to 2 h after the perforation . Scenario E represents first responders who enter the vehicle 5 min after perforation and exit 10 min later.

- Samplers Operated—The number of CIs used in the scenario analysis.

- Number of Components—The number of individual calculated components of the dose quantities. The dose quantities are sums of this number of component parts. For example, for scenarios with only one set of breathing parameters (i.e., Scenarios A, B, and E), there are 18 components coming from the nine aerosol components (from the eight-stage substrates and backup filter) and two nuclides, U-238 and U-234 (other components are ignored in the calculation). Scenarios C and D consist of 54 components, being the sum of 18 components each from Scenarios B and E plus 18 components from the light exercise breathing parameters for the identified time period.

- Number Avg'd—The number of CIs operating during or immediately before or after the scenario interval that were included in the calculation of mean or median intakes, doses, and concentrations.
- AMAD—The activity median aerodynamic diameter calculated as the median of the discretized, nine-component distribution of aerosol activity over particle diameters. (Provided only for comparison. AMADs were not used in the dose/concentration assessment.) The AMAD is calculated as the AMAD of the first component for which the cumulative probability exceeds 0.5 to be reasonably consistent with assuming a mixture of nine monodisperse aerosol components. This column is followed by a column showing the AMAD standard deviation.
- sd—The standard deviation (one sigma) of the parameter.
- Intake—The intake amount provided variously in Bq, nCi, and mg DU.
- E(50)—The committed effective dose (equivalent), in rem. The first column is the mean. This is followed by columns that include the 10th percentile, the median, the 90th percentile E(50), and the ratio of median to mean. (Phase IV had only one result per scenario.)
- Kidneys—The first column of kidney results is the calculated mean concentration of uranium in the kidneys at 1 day (peak concentration) in μg U per gram kidney tissue. Columns that provide data for the 10th percentile, the median, the 90th percentile, and the ratio of median to mean follow the first column. (Phase IV had only one result per scenario.)
- Mean Organ Doses and Standard Deviations—Listed by organ in rem. The committed dose equivalent $H_T(50)$ for lung, bone surfaces, extrathoracic airways, kidney, red marrow, liver, and lymph nodes. The 10th percentile, median, 90th percentile, and median/mean ratio are listed for the lung dose. In Tables A.3 and A.5 intakes and organ doses are expressed as sums, which refers to summing of the CI stages.
- SumOfLN-Th200d (at the end of Table A.5)—The concentration of uranium in the thoracic lymph nodes 200 days post intake in μg per gram tissue. The lymph node concentration usually peaks at about 200 days post intake.

- MidTimeMin—The middle point of the time interval in minutes.
- ConcMicroG/m3—The concentration of DU in μg/m^3 collected on the CI substrate.
- SDConcMicroG/m3—The standard deviation in the DU concentration in μg/m^3.
- MedDiaMic—The median diameter in micrometers for that CI stage.
- SampleID—Refers to the CI substrate sample, cross-references to Appendix A tables in Volume II of the Capstone study (Parkhurst et al. 2004).

REFERENCE

Parkhurst MA, F Szrom, RA Guilmette, TD Holmes, YS Cheng, JL Kenoyer, JW Collins, TE Sanderson, RW Fliszar, K Gold, JC Beckman, and JA Long. 2004. *Capstone Depleted Uranium Aerosols: Generation and Characterization, Volumes 1 and 2.* PNNL-14168, Prepared for the U.S. Army by Pacific Northwest National Laboratory, Richland, Washington.

Table A.1. Sample Arrays Included in Each Scenario—Listed by Phase, Shot, and Position

Phase	Shot	Position	Scenario	Samplers Operated	# Components	Comments
1	1	Commander	a	1	18	
1	1	Driver	a	1	18	
1	1	Gunner	a	1	18	
1	1	Loader	a	1	18	
1	1	Commander	b	1	18	
1	1	Driver	b	1	18	
1	1	Gunner	b	1	18	
1	1	Loader	b	1	18	
1	1	Commander	c	3	54	
1	1	Driver	c	3	54	
1	1	Gunner	c	3	54	
1	1	Loader	c	3	54	
1	1	Commander	d	3	54	
1	1	Driver	d	3	54	
1	1	Gunner	d	3	54	
1	1	Loader	d	3	54	
1	1	Commander	e	1	18	
1	1	Driver	e	1	18	
1	1	Gunner	e	1	18	
1	1	Loader	e	1	18	
1	2	Commander	a	1	18	
1	2	Driver	a	1	18	
1	2	Gunner	a	1	18	
1	2	Loader	a	1	18	
1	2	Commander	b	1	18	
1	2	Driver	b	1	18	
1	2	Gunner	b	1	18	
1	2	Loader	b	1	18	
1	2	Commander	c	3	54	
1	2	Driver	c	3	54	
1	2	Gunner	c	3	54	
1	2	Loader	c	3	54	
1	2	Commander	d	3	54	
1	2	Driver	d	3	54	
1	2	Gunner	d	3	54	
1	2	Loader	d	3	54	
1	2	Commander	e	1	18	
1	2	Driver	e	1	18	
1	2	Gunner	e	1	18	
1	2	Loader	e	1	18	

Table A.1. *(continued)*

Phase	Shot	Position	Scenario	Samplers Operated	# Components	Comments
1	3	Commander	all	Sampler operated only in second of double shots		
1	3	Driver	a	1	18	
1	3	Gunner	a	1	18	
1	3	Loader	all	Sampler operated only in second of double shots		
1	3	Driver	b	1	18	
1	3	Gunner	b	1	18	
1	3	Driver	c and d	Sampler operated during shots 3 and 4;		
1	3	Gunner	c and d	no unused samplers remained during 1 to 2 h interval		
1	3	Driver	e	1	18	
1	3	Gunner	e	1	18	
1	4	Commander	a	1	18	
1	4	Driver	a	1	18	
1	4	Gunner	a	1	18	
1	4	Loader	a	1	18	
1	4	Commander	b	1	18	
1	4	Driver	b	1	18	
1	4	Gunner	b	1	18	
1	4	Loader	b	1	18	
1	4	Commander	c	3	54	
1	4	Driver	c	3	54	
1	4	Gunner	c	3	54	
1	4	Loader	c	3	54	
1	4	Commander	d	3	54	
1	4	Driver	d	3	54	
1	4	Gunner	d	3	54	
1	4	Loader	d	3	54	
1	4	Commander	e	1	18	
1	4	Driver	e	1	18	
1	4	Gunner	e	1	18	
1	4	Loader	e	1	18	
1	5	Commander	a	1	18	
1	5	Driver	a	1	18	
1	5	Gunner	all	No sampler at gunner's position to avoid line of fire		
1	5	Loader	a	1	18	
1	5	Commander	b	1	18	
1	5	Driver	b	1	18	
1	5	Loader	b	1	18	
1	5	Commander	c	3	54	
1	5	Driver	c	3	54	
1	5	Gunner	all	No sampler at gunner's position to avoid line of fire		
1	5	Loader	c	3	54	
1	5	Commander	d	3	54	

Table A.1. *(continued)*

Phase	Shot	Position	Scenario	Samplers Operated	# Components	Comments
1	5	Driver	d	3	54	
1	5	Loader	d	3	54	
1	5	Commander	e	1	18	
1	5	Driver	e	1	18	
1	5	Loader	e	1	18	
1	6	Commander	a	1	18	
1	6	Driver	a	1	18	
1	6	Loader	a	1	18	Used for resuspension analysis
1	6	Commander	b	1	18	
1	6	Driver	b	1	18	
1	6	Loader	b	1	18	Used for resuspension analysis
1	6	Commander	c	3	54	
1	6	Driver	c	3	54	
1	6	Loader	c	3	54	Used for resuspension analysis
1	6	Commander	d	3	54	
1	6	Driver	d	3	54	
1	6	Loader	d	3	54	Used for resuspension analysis
1	6	Commander	e	1	18	
1	6	Driver	e	1	18	
1	6	Loader	e	1	18	Used for resuspension analysis
1	7	Commander	a	1	18	
1	7	Driver	a	1	18	
1	7	Gunner	a	1	18	
1	7	Loader	a	1	18	Used for resuspension analysis
1	7	Commander	b	1	18	
1	7	Driver	b	1	18	
1	7	Gunner	b	1	18	
1	7	Loader	b	1	18	Used for resuspension analysis
1	7	Commander	c	3	54	
1	7	Driver	c	3	54	
1	7	Gunner	c	3	54	
1	7	Loader	c	3	54	Used for resuspension analysis
1	7	Commander	d	3	54	
1	7	Driver	d	3	54	
1	7	Gunner	d	3	54	
1	7	Loader	d	3	54	Used for resuspension analysis
1	7	Commander	e	1	18	
1	7	Driver	e	1	18	
1	7	Gunner	e	1	18	
1	7	Loader	e	1	18	Used for resuspension analysis
2	1	Commander	a	1	18	
2	1	Driver	a	1	18	

Table A.1. *(continued)*

Phase	Shot	Position	Scenario	Samplers Operated	# Components	Comments
2	1	Right Scout	all	Sampler operated only in second of double shots		
2	1	Left Scout	a	1	18	
2	1	Commander	b	1	18	
2	1	Driver	b	1	18	
2	1	Left Scout	b	1	18	
2	1	Commander	c and d	Sampler operated during shots 3 and 4		
2	1	Driver	c and d	no unused samplers remained during 1 to 2 h interval		
2	1	Left Scout	c and d	no unused samplers remained during 1 to 2 h interval		
2	1	Commander	e	1	18	
2	1	Driver	e	1	18	
2	1	Left Scout	e	1	18	
2	2	Commander	a	1	18	
2	2	Driver	a	1	18	
2	2	Right Scout	a	1	18	
2	2	Left Scout	a	1	18	
2	2	Commander	b	1	18	
2	2	Driver	b	1	18	
2	2	Right Scout	b	1	18	
2	2	Left Scout	b	1	18	
2	2	Commander	c	3	54	
2	2	Driver	c	3	54	
2	2	Right Scout	c	3	54	
2	2	Left Scout	c	3	54	
2	2	Commander	d	3	54	
2	2	Driver	d	3	54	
2	2	Right Scout	d	3	54	
2	2	Left Scout	d	3	54	
2	2	Commander	e	1	18	
2	2	Driver	e	1	18	
2	2	Right Scout	e	1	18	
2	2	Left Scout	e	1	18	
2	3	Commander	a	1	18	
2	3	Driver	a	1	18	
2	3	Right Scout	a	1	18	
2	3	Left Scout	a	1	18	
2	3	Commander	b	1	18	
2	3	Driver	b	1	18	
2	3	Right Scout	b	1	18	
2	3	Left Scout	b	1	18	
2	3	Commander	c	3	54	
2	3	Driver	c	3	54	
2	3	Right Scout	c	3	54	

Table A.1. *(continued)*

Phase	Shot	Position	Scenario	Samplers Operated	# Components	Comments
2	3	Left Scout	c	3	54	
2	3	Commander	d	3	54	
2	3	Driver	d	3	54	
2	3	Right Scout	d	3	54	
2	3	Left Scout	d	3	54	
2	3	Commander	e	1	18	
2	3	Driver	e	1	18	
2	3	Right Scout	e	1	18	
2	3	Left Scout	e	1	18	
3	1	Commander	a	1	18	
3	1	Driver	a	1	18	
3	1	Gunner	a	1	18	
3	1	Loader	a	1	18	
3	1	Commander	b	1	18	
3	1	Driver	b	1	18	
3	1	Gunner	b	1	18	
3	1	Loader	b	1	18	
3	1	Commander	c	3	54	
3	1	Driver	c	3	54	
3	1	Gunner	c	3	54	
3	1	Loader	c	3	54	
3	1	Commander	d	3	54	
3	1	Driver	d	3	54	
3	1	Gunner	d	3	54	
3	1	Loader	d	3	54	
3	1	Commander	e	1	18	
3	1	Driver	e	1	18	
3	1	Gunner	e	1	18	
3	1	Loader	e	1	18	
3	2	Commander	a	1	18	
3	2	Driver	a	1	18	
3	2	Gunner	a	1	18	
3	2	Loader	a	1	18	
3	2	Commander	b	1	18	
3	2	Driver	b	1	18	
3	2	Gunner	b	1	18	
3	2	Loader	b	1	18	
3	2	Commander	c	3	54	
3	2	Driver	c	3	54	
3	2	Gunner	c	3	54	
3	2	Loader	c	3	54	
3	2	Commander	d	3	54	

le A.1. *(continued)*

hase	Shot	Position	Scenario	Samplers Operated	# Components	Comments
3	2	Driver	d	3	54	
3	2	Gunner	d	3	54	
3	2	Loader	d	3	54	
3	2	Commander	e	1	18	
3	2	Driver	e	1	18	
3	2	Gunner	e	1	18	
3	2	Loader	e	1	18	
4	4	Driver	a	1	18	
4	4	Driver	b	1	18	
4	4	Driver	c	3	54	Extrapolated from 25-27 min sample
4	4	Driver	d	3	54	Extrapolated from 25-27 min sample
4	4	Driver	e	1	18	
4	4	Loader	all	Power or vacuum lines damaged; no samples collected		
se 4, Shot 4--No samplers used at commander and gunner positions						

Table A.2. Summary of Intakes, Doses, and Concentrations Using Bayesian Statistics (Part I)

Scenario	Phase	Number Avg'd	AMAD	sd AMAD	Intake (Bq)	sd Intake (Bq)	Intake (nCi)	Intake (mg)	E(50) (rem)	sd E(50) (rem)	10% E(50) (rem)	Median E(50) (rem)	90% E(50) (rem)	Median/ Mean
A	1	22	8.18	8.35	3881	1718	104.9	275.3	1.995	0.820	1.328	1.965	2.777	0.985
A	2	11	4.18	3.03	1172	926	31.7	83.2	0.676	0.645	0.155	0.589	1.008	0.872
A	3	8	7.50	4.17	3556	2524	96.1	252.3	2.603	1.228	1.494	2.159	5.256	0.829
A	4	1	1	NA	137	NA	3.7	9.7	0.090	NA	0.090	0.090	0.090	1
B	1	22	5.73	6.29	8337	4132	225.3	591.4	4.422	2.100	3.031	3.664	7.141	0.829
B	2	11	3.55	3.08	3063	2281	82.8	217.3	1.807	1.525	0.335	1.728	3.089	0.956
B	3	8	11.50	12.11	9946	8239	268.8	705.5	6.711	2.619	4.522	5.997	12.690	0.894
B	4	1	0.6	NA	610	NA	16.5	43.3	0.436	NA	0.436	0.436	0.436	1
C	1	20	5.65	6.59	10757	5873	290.7	763.0	5.323	1.994	3.153	4.846	8.556	0.910
C	2	8	1.45	1.10	4689	3371	126.7	332.6	2.595	2.020	1.002	2.078	6.837	0.801
C	3	8	11.13	12.31	13690	11442	370.0	971.2	9.084	3.182	5.923	8.332	16.367	0.917
C	4	1	2	NA	1284	NA	34.7	91.0	1.022	NA	1.022	1.022	1.022	1
D	1	18	5.39	6.86	11003	6454	297.4	780.5	5.400	1.992	2.888	4.955	9.036	0.918
D	2	8	1.08	0.40	5299	3299	143.2	375.9	2.880	1.988	1.126	2.440	7.064	0.847
D	3	8	10.88	12.49	14716	11978	397.7	1043.9	9.360	3.246	6.327	8.680	16.848	0.927
D	4	1	2	NA	1594	NA	43.1	113.1	1.196	NA	1.196	1.196	1.196	1
E	1	22	4.48	6.77	2203	1577	59.6	156.3	1.198	0.803	0.451	0.916	2.236	0.765
E	2	11	1.35	0.65	1401	787	37.9	99.4	0.856	0.493	0.263	0.890	1.328	1.040
E	3	8	7.75	9.81	2857	2781	77.2	202.7	2.011	0.695	1.052	1.944	3.080	0.967
E	4	1	2	NA	386	NA	10.4	27.4	0.412	NA	0.412	0.412	0.412	1

Table A.2. Summary of Intakes, Doses, and Concentrations Using Bayesian Statistics (Part 2)

Scenario	Phase	Avg. KID Conc 1d (µg U)	sd KID1d (µg U)	10% KID1d (µg U)	Median KID1d (µg U)	90% KID1d (µg U)	Median/ Mean	Lung (rem)	sd Lung (rem)	10% Lung (rem)	Median Lung (rem)	90% Lung (rem)	Median/ Mean
A	1	3.035	1.200	1.936	2.991	4.203	0.986	14.34	6.13	9.24	14.34	19.14	1.000
A	2	1.114	1.017	0.271	0.988	1.632	0.887	5.21	5.06	1.16	4.50	7.83	0.864
A	3	1.312	0.753	0.566	1.083	2.912	0.826	17.50	6.98	10.70	15.42	32.20	0.882
A	4	0.054	NA	0.054	0.054	0.054	1	0.66	NA	0.66	0.66	0.66	1
B	1	6.905	2.881	4.644	6.441	9.974	0.933	32.09	15.75	22.13	26.75	53.64	0.833
B	2	2.987	2.451	0.611	2.878	4.955	0.964	13.97	11.94	2.49	13.20	24.06	0.945
B	3	3.267	2.082	1.725	2.578	7.977	0.789	44.14	11.26	32.49	42.38	69.36	0.960
B	4	0.225	NA	0.225	0.225	0.225	1	3.32	NA	3.32	3.32	3.32	1
C	1	8.877	3.482	5.336	8.172	13.956	0.921	38.31	13.77	23.18	35.71	59.05	0.932
C	2	4.329	3.435	1.433	3.526	11.481	0.815	20.04	15.74	7.64	15.96	53.19	0.796
C	3	4.379	2.728	2.439	3.550	10.785	0.811	59.97	12.61	42.50	58.50	85.57	0.975
C	4	0.458	NA	0.458	0.458	0.458	1	7.58	NA	7.58	7.58	7.58	1
D	1	9.070	3.719	5.123	8.028	16.106	0.885	38.96	13.66	21.43	36.90	59.12	0.947
D	2	4.790	3.433	1.826	4.020	11.926	0.839	22.21	15.51	8.70	18.74	54.93	0.844
D	3	4.604	2.803	2.697	3.656	11.227	0.794	61.26	12.61	44.59	59.54	87.36	0.972
D	4	0.555	NA	0.555	0.555	0.555	1	8.65	NA	8.65	8.65	8.65	1
E	1	2.152	1.511	0.815	1.543	4.765	0.717	8.81	5.90	3.37	6.74	14.04	0.765
E	2	1.372	0.814	0.364	1.411	1.980	1.029	6.67	3.84	2.07	6.96	10.35	1.045
E	3	0.872	0.616	0.344	0.672	2.322	0.770	13.64	4.12	7.83	13.88	21.03	1.017
E	4	0.140	NA	0.140	0.140	0.140	1	3.14	NA	3.14	3.14	3.14	1

Table A.2. Summary of Intakes, Doses, and Concentrations Using Bayesian Statistics (Part 3)

Scenario	Phase	BS (rem)	sd BS (rem)	ET (rem)	sd ET (rem)	Kidney (rem)	sd KID (rem)	RM (rem)	sd RM (rem)	Liver (rem)	sd Liver (rem)	LN (rem)	sd LN (rem)
A	1	0.987	0.419	8.87	4.81	0.361	0.153	0.1019	0.0432	0.1353	0.0574	13.37	4.89
A	2	0.382	0.369	1.22	0.74	0.140	0.135	0.0395	0.0381	0.0524	0.0506	2.58	2.20
A	3	0.381	0.186	19.33	15.49	0.140	0.068	0.0394	0.0192	0.0523	0.0255	307.78	148.56
A	4	0.023	NA	0.37	NA	0.008	NA	0.0023	NA	0.0031	NA	7.33	NA
B	1	2.214	1.087	18.12	10.74	0.810	0.398	0.2286	0.1122	0.3034	0.1490	29.14	12.14
B	2	1.032	0.880	3.04	1.87	0.378	0.322	0.1065	0.0909	0.1414	0.1206	6.88	5.12
B	3	0.913	0.412	54.58	50.88	0.335	0.151	0.0943	0.0426	0.1252	0.0565	817.55	236.81
B	4	0.100	NA	1.28	NA	0.037	NA	0.0103	NA	0.0137	NA	43.32	NA
C	1	2.732	1.082	23.22	17.08	1.000	0.396	0.2821	0.1118	0.3744	0.1483	35.15	12.42
C	2	1.523	1.219	4.39	2.66	0.557	0.446	0.1572	0.1259	0.2087	0.1670	11.20	6.45
C	3	1.252	0.503	72.79	70.32	0.459	0.184	0.1293	0.0519	0.1716	0.0689	1104.30	273.77
C	4	0.181	NA	4.12	NA	0.066	NA	0.0187	NA	0.0248	NA	122.42	NA
D	1	2.794	1.118	23.02	18.65	1.023	0.409	0.2885	0.1155	0.3830	0.1532	35.40	11.65
D	2	1.699	1.218	4.97	2.53	0.622	0.446	0.1754	0.1258	0.2329	0.1669	12.87	6.14
D	3	1.308	0.515	77.53	72.39	0.479	0.189	0.1350	0.0532	0.1793	0.0706	1121.78	276.67
D	4	0.210	NA	5.88	NA	0.077	NA	0.0217	NA	0.0288	NA	141.30	NA
E	1	0.647	0.451	4.24	4.16	0.237	0.165	0.0668	0.0466	0.0887	0.0617	7.37	4.61
E	2	0.485	0.281	1.20	0.72	0.178	0.103	0.0501	0.0290	0.0665	0.0384	3.37	1.48
E	3	0.263	0.107	14.40	17.62	0.096	0.039	0.0271	0.0111	0.0360	0.0147	256.62	76.56
E	4	0.052	NA	1.30	NA	0.019	NA	0.0053	NA	0.0071	NA	59.74	NA

Phase	Shot	Position	Scenario	AMAD	Intake (Bq)	E(50) (rem)	Lung (rem)	Bone Surface (rem)	ET (rem)	Kidney (rem)	LN-TH (rem)	Red Marrow (rem)	Liver (rem)	Kidney Conc. 1d (μg U)
1	1	Commander	a	4	3038	1.743	12.75	0.856	6.7	0.3134	12.07	0.0884	0.1173	2.745
1	1	Driver	a	31	679	0.215	1.33	0.108	2.0	0.0394	1.44	0.0111	0.0147	0.419
1	1	Gunner	a	1	3058	1.930	14.55	0.971	5.3	0.3555	13.54	0.1002	0.1331	2.661
1	1	Loader	a	2	3162	1.993	14.89	1.013	6.1	0.3709	13.39	0.1046	0.1389	3.251
1	1	Commander	b	2	5620	3.375	25.31	1.712	9.9	0.6267	23.08	0.1767	0.2346	5.156
1	1	Driver	b	7	2044	0.934	6.58	0.465	4.8	0.1702	6.41	0.0480	0.0637	1.599
1	1	Gunner	b	1	5680	3.449	25.95	1.757	9.7	0.6435	24.29	0.1814	0.2409	5.004
1	1	Loader	b	2	5948	3.688	27.55	1.827	11.4	0.6688	26.03	0.1887	0.2505	5.869
1	1	Commander	c	1	7754	4.658	35.17	2.403	12.4	0.8799	31.35	0.2481	0.3294	7.154
1	1	Driver	c	4	2985	1.508	10.95	0.760	6.2	0.2782	10.22	0.0785	0.1042	2.478
1	1	Gunner	c	1	7431	4.470	33.82	2.301	11.6	0.8428	31.16	0.2375	0.3155	6.475
1	1	Loader	c	2	7839	4.827	36.27	2.446	13.8	0.8953	33.60	0.2526	0.3353	7.748
1	1	Commander	d	1	8070	4.875	36.86	2.517	12.7	0.9216	32.80	0.2598	0.3450	7.417
1	1	Driver	d	4	3348	1.737	12.72	0.876	6.6	0.3206	11.76	0.0904	0.1200	2.736
1	1	Gunner	d	1	7753	4.665	35.33	2.400	11.9	0.8792	32.49	0.2478	0.3291	6.697
1	1	Loader	d	2	8109	5.035	37.91	2.549	14.1	0.9333	35.04	0.2633	0.3495	7.962
1	1	Commander	e	1	1618	1.023	7.88	0.542	1.9	0.1986	6.55	0.0560	0.0743	1.589
1	1	Driver	e	2	588	0.351	2.64	0.178	1.0	0.0653	2.29	0.0184	0.0245	0.605
1	1	Gunner	e	0.6	1262	0.744	5.75	0.399	1.3	0.1461	4.98	0.0412	0.0547	1.125
1	1	Loader	e	1	1417	0.868	6.64	0.470	1.8	0.1721	5.72	0.0486	0.0645	1.496
1	2	Commander	a	4	2835	1.683	12.39	0.827	6.1	0.3026	11.38	0.0853	0.1133	2.539
1	2	Driver	a	7	3170	1.813	13.15	0.875	7.5	0.3204	12.36	0.0904	0.1200	2.914
1	2	Gunner	a	4	2927	1.712	12.54	0.824	6.5	0.3017	11.66	0.0851	0.1130	2.683
1	2	Loader	a	2	2837	1.627	11.98	0.825	5.8	0.3022	11.99	0.0852	0.1131	2.464
1	2	Commander	b	2	5151	3.325	24.98	1.668	9.6	0.6108	21.94	0.1723	0.2287	4.997
1	2	Driver	b	7	5260	3.038	22.13	1.474	12.1	0.5394	20.73	0.1522	0.2020	4.943
1	2	Gunner	b	4	5651	3.216	23.66	1.615	11.6	0.5911	21.60	0.1667	0.2213	5.094
1	2	Loader	b	2	5015	3.031	22.57	1.520	9.7	0.5567	21.90	0.1570	0.2084	4.644

Table A.3. *(continued)*

Phase	Shot	Position	Scenario	AMAD	Intake (Bq)	E(50) (rem)	Lung (rem)	Bone Surface (rem)	ET (rem)	Kidney (rem)	LN-TH (rem)	Red Marrow (rem)	Liver (rem)	Kidney Conc. 1d (µg U)
1	2	Commander	c	2	7001	4.541	34.34	2.315	11.9	0.8478	29.40	0.2391	0.3174	6.808
1	2	Driver	c	4	6455	3.603	26.30	1.775	14.1	0.6497	24.45	0.1833	0.2433	5.853
1	2	Gunner	c	4	7359	4.270	31.70	2.167	14.0	0.7934	28.51	0.2237	0.2970	6.625
1	2	Loader	c	2	6886	4.261	32.02	2.156	12.2	0.7898	29.90	0.2227	0.2956	6.527
1	2	Commander	d	2	7463	4.873	36.94	2.486	12.4	0.9104	31.66	0.2567	0.3408	7.188
1	2	Driver	d	4	6725	3.776	27.64	1.867	14.4	0.6835	25.61	0.1928	0.2560	6.072
1	2	Gunner	d	4	7809	4.503	33.46	2.291	14.6	0.8388	30.04	0.2365	0.3140	6.938
1	2	Loader	d	1	7539	4.704	35.48	2.381	12.8	0.8724	32.97	0.2459	0.3265	7.021
1	2	Commander	e	1	1200	0.847	6.53	0.448	1.6	0.1642	5.02	0.0463	0.0615	1.318
1	2	Driver	e	2	679	0.329	2.42	0.173	1.2	0.0632	2.18	0.0178	0.0237	0.567
1	2	Gunner	e	1	1112	0.725	5.54	0.375	1.6	0.1373	4.74	0.0387	0.0514	1.080
1	2	Loader	e	2	1068	0.751	5.74	0.385	1.7	0.1408	4.70	0.0397	0.0527	1.284
1	3	Driver	a	2	3934	2.473	18.67	1.099	6.9	0.4031	16.00	0.1134	0.1507	1.960
1	3	Gunner	a	2	6788	4.217	31.90	2.142	11.0	0.7850	23.36	0.2212	0.2938	5.329
1	3	Driver	b	2	10880	7.141	54.08	3.401	18.8	1.2470	45.25	0.3511	0.4664	7.853
1	3	Gunner	b	2	17310	10.820	81.75	5.581	28.5	2.0440	61.00	0.5762	0.7652	14.875
1	3	Driver	e	2	4481	3.092	23.46	1.703	7.4	0.6229	17.85	0.1759	0.2333	5.589
1	3	Gunner	e	4	5467	3.218	24.09	1.807	9.2	0.6611	17.56	0.1866	0.2476	5.792
1	4	Commander	a	16	4276	2.380	17.42	1.165	9.1	0.4270	14.57	0.1203	0.1598	2.830
1	4	Driver	a	11	2792	1.328	9.24	0.619	7.5	0.2268	9.10	0.0639	0.0849	1.936
1	4	Gunner	a	11	6031	2.777	19.14	1.296	16.4	0.4747	20.34	0.1338	0.1777	4.101
1	4	Loader	a	31	7596	2.209	13.05	1.082	23.4	0.3961	14.72	0.1117	0.1483	3.969
1	4	Commander	b	4	10230	5.757	42.46	2.933	20.2	1.0750	34.55	0.3028	0.4021	7.366
1	4	Driver	b	7	7031	3.539	25.05	1.734	17.6	0.6349	23.38	0.1791	0.2377	5.503
1	4	Gunner	b	7	13410	6.331	44.11	3.001	35.1	1.0990	45.86	0.3099	0.4114	9.235
1	4	Loader	b	31	17690	5.538	33.87	2.755	53.0	1.0090	36.21	0.2845	0.3777	9.974
1	4	Commander	c	4	14598	7.833	57.56	4.122	28.3	1.5100	46.72	0.4256	0.5650	11.114

Phase	Shot	Position	Scenario	AMAD	Intake (Bq)	E(50) (rem)	Lung (rem)	Bone Surface (rem)	ET (rem)	Kidney (rem)	LN-TH (rem)	Red Marrow (rem)	Liver (rem)	Kidney Conc. 1d (µg U)
1	4	Gunner	c	7	17644	8.532	60.54	4.197	41.7	1.5369	59.97	0.4334	0.5753	12.369
1	4	Loader	c	31	29428	8.581	51.88	4.414	84.7	1.6157	54.19	0.4558	0.6049	16.847
1	4	Commander	d	4	15280	8.052	59.12	4.266	29.2	1.5629	47.90	0.4404	0.5848	11.521
1	4	Driver	d	4	12348	5.747	40.73	3.007	28.0	1.1008	35.75	0.3106	0.4121	9.635
1	4	Gunner	d	4	18946	9.137	65.10	4.503	43.4	1.6490	63.67	0.4649	0.6173	13.015
1	4	Loader	d	31	32084	9.036	54.35	4.669	90.5	1.7092	56.26	0.4822	0.6400	17.816
1	4	Commander	e	4	2971	1.582	11.56	0.879	5.9	0.3215	9.40	0.0907	0.1204	2.894
1	4	Driver	e	7	3047	1.419	10.00	0.786	7.1	0.2876	7.98	0.0812	0.1077	2.798
1	4	Gunner	e	1	2828	1.575	11.74	0.851	4.8	0.3117	10.35	0.0879	0.1167	2.347
1	4	Loader	e	16	6465	2.236	14.04	1.166	19.6	0.4264	14.62	0.1204	0.1597	4.765
1	5	Commander	a	4	3593	2.000	14.44	0.948	8.7	0.3473	16.34	0.0979	0.1300	3.150
1	5	Driver	a	7	4369	2.222	15.56	1.105	11.8	0.4040	14.85	0.1141	0.1513	4.138
1	5	Loader	a	7	3842	1.717	11.90	0.844	9.8	0.3087	12.89	0.0871	0.1156	3.086
1	5	Commander	b	4	7410	4.215	30.59	1.974	17.5	0.7223	33.40	0.2038	0.2705	7.015
1	5	Driver	b	11	8578	4.332	30.27	2.069	23.6	0.7565	28.20	0.2137	0.2835	8.161
1	5	Loader	b	4	6754	3.342	23.77	1.608	16.2	0.5885	25.42	0.1661	0.2204	5.940
1	5	Commander	c	4	9189	5.007	36.25	2.390	21.2	0.8745	38.81	0.2468	0.3275	8.596
1	5	Driver	c	11	11587	5.322	36.84	2.615	30.5	0.9561	34.07	0.2700	0.3583	10.354
1	5	Loader	c	4	8650	4.234	30.24	2.095	19.8	0.7666	31.28	0.2163	0.2871	7.641
1	5	Commander	d	4	9613	5.188	37.59	2.498	21.8	0.9140	39.99	0.2579	0.3423	8.923
1	5	Driver	d	11	12344	5.502	38.00	2.728	31.9	0.9974	34.99	0.2817	0.3737	10.763
1	5	Loader	d	4	9303	4.475	31.98	2.241	20.7	0.8200	32.77	0.2314	0.3071	8.093
1	5	Commander	e	4	1105	0.596	4.34	0.291	2.4	0.1064	4.28	0.0301	0.0399	1.144
1	5	Driver	e	31	1461	0.661	4.58	0.334	3.8	0.1221	4.00	0.0345	0.0458	1.355
1	5	Loader	e	4	1185	0.607	4.41	0.315	2.4	0.1154	4.04	0.0326	0.0432	1.170
1	6	Commander	a	11	5224	2.315	15.83	1.101	14.3	0.4028	16.09	0.1137	0.1509	4.203
1	6	Driver	a	4	558	0.304	2.20	0.143	1.3	0.0525	2.45	0.0148	0.0196	0.520
1	6	Loader	a	1	5	0.002	0.02	0.001	0.0	0.0005	0.02	0.0001	0.0002	0.004

Table A.3. *(continued)*

Phase	Shot	Position	Scenario	AMAD	Intake (Bq)	E(50) (rem)	Lung (rem)	Bone Surface (rem)	ET (rem)	Kidney (rem)	LN-TH (rem)	Red Marrow (rem)	Liver (rem)	Kidney Conc. 1d (μg U)
1	6	Commander	b	4	9173	4.531	32.11	2.219	22.4	0.8117	30.53	0.2292	0.3041	8.302
1	6	Driver	b	4	3372	2.042	15.14	0.995	6.9	0.3640	13.13	0.1028	0.1364	3.661
1	6	Loader	b	2	25	0.012	0.09	0.006	0.0	0.0023	0.08	0.0006	0.0009	0.021
1	6	Commander	c	4	11954	5.817	41.54	2.923	27.1	1.0694	38.15	0.3019	0.4006	10.716
1	6	Driver	c	4	4611	2.703	20.06	1.344	9.0	0.4917	17.13	0.1388	0.1842	4.818
1	6	Loader	c	2	238	0.095	0.68	0.054	0.4	0.0198	0.62	0.0056	0.0074	0.174
1	6	Commander	d	7	13281	6.085	43.13	3.081	29.8	1.1270	39.54	0.3182	0.4222	11.291
1	6	Driver	d	4	5052	2.888	21.43	1.451	9.6	0.5307	18.34	0.1498	0.1988	5.123
1	6	Loader	d	2	546	0.207	1.46	0.119	1.0	0.0436	1.32	0.0123	0.0163	0.375
1	6	Commander	e	2	1607	0.964	7.29	0.520	2.4	0.1903	5.70	0.0537	0.0713	1.808
1	6	Driver	e	4	750	0.451	3.37	0.230	1.4	0.0840	2.61	0.0237	0.0315	0.815
1	6	Loader	e	7	53	0.023	0.16	0.012	0.1	0.0043	0.16	0.0012	0.0016	0.042
1	7	Commander	a	11	5647	2.370	16.63	1.307	12.2	0.4783	14.43	0.1350	0.1791	4.180
1	7	Driver	a	4	5203	2.932	21.70	1.551	9.8	0.5677	18.89	0.1602	0.2126	4.617
1	7	Gunner	a	4	3822	1.937	14.24	1.015	7.0	0.3713	12.24	0.1048	0.1390	3.068
1	7	Loader	a	4	6	0.003	0.02	0.002	0.0	0.0006	0.01	0.0002	0.0002	0.006
1	7	Commander	b	11	9366	3.641	25.13	2.037	20.6	0.7452	22.13	0.2104	0.2791	6.942
1	7	Driver	b	4	13390	7.242	53.64	3.960	23.7	1.4490	45.57	0.4089	0.5426	11.926
1	7	Gunner	b	4	8455	4.767	35.37	2.397	15.8	0.8770	30.54	0.2476	0.3285	7.855
1	7	Loader	b	4	39	0.019	0.14	0.012	0.1	0.0042	0.09	0.0012	0.0016	0.036
1	7	Commander	c	11	12694	4.866	33.68	2.764	27.0	1.0113	29.58	0.2855	0.3788	9.439
1	7	Driver	c	2	17317	9.300	69.21	5.255	28.5	1.9228	56.60	0.5426	0.7200	15.543
1	7	Gunner	c	4	12334	6.773	49.99	3.422	23.6	1.2517	44.39	0.3534	0.4689	11.391
1	7	Loader	c	4	255	0.113	0.84	0.069	0.4	0.0252	0.52	0.0071	0.0094	0.206
1	7	Commander	d	7	13272	5.111	35.50	2.923	27.8	1.0695	30.81	0.3019	0.4006	9.837
1	7	Driver	d	2	18092	9.672	72.05	5.497	29.3	2.0114	58.51	0.5676	0.7532	16.106
1	7	Gunner	d	4	12919	7.015	51.80	3.577	24.3	1.3085	45.64	0.3694	0.4901	11.763

Phase	Shot	Position	Scenario	AMAD	Intake (Bq)	E(50) (rem)	Lung (rem)	Bone Surface (rem)	ET (rem)	Kidney (rem)	LN-TH (rem)	Red Marrow (rem)	Liver (rem)	Kidney Conc. 1d (µg U)
1	7	Commander	e	4	2388	0.976	6.84	0.553	5.0	0.2021	6.36	0.0571	0.0757	1.972
1	7	Driver	e	1	2953	1.617	12.27	0.996	3.7	0.3645	8.85	0.1029	0.1365	2.878
1	7	Gunner	e	4	2823	1.715	12.67	0.835	6.0	0.3054	12.42	0.0863	0.1144	2.960
1	7	Loader	e	4	135	0.071	0.54	0.041	0.2	0.0150	0.33	0.0042	0.0056	0.122
2	1	Commander	a	2	1498	0.904	6.99	0.511	1.5	0.1869	2.89	0.0528	0.0700	1.480
2	1	Driver	a	2	1592	1.008	7.83	0.579	1.5	0.2119	3.44	0.0598	0.0794	1.632
2	1	Loader	a	2	1078	0.494	3.75	0.306	1.1	0.1119	2.23	0.0316	0.0419	0.854
2	1	Commander	b	2	4589	2.837	22.00	1.621	4.4	0.5929	9.30	0.1674	0.2221	4.720
2	1	Driver	b	2	4691	3.089	24.06	1.753	4.3	0.6411	10.58	0.1810	0.2401	4.955
2	1	Loader	b	1	3222	1.758	13.55	1.058	3.0	0.3872	7.78	0.1092	0.1449	2.878
2	1	Commander	e	2	1968	1.295	10.06	0.674	2.1	0.2465	4.11	0.0696	0.0923	1.961
2	1	Driver	e	2	2507	1.644	12.83	0.943	2.2	0.3448	5.67	0.0973	0.1292	2.754
2	1	Loader	e	2	1923	1.328	10.35	0.697	2.0	0.2548	5.03	0.0719	0.0954	1.980
2	2	Commander	a	1	3562	2.384	18.65	1.360	3.0	0.4976	8.55	0.1404	0.1864	3.807
2	2	Driver	a	4	1261	0.737	5.65	0.381	1.6	0.1394	2.54	0.0394	0.0522	1.173
2	2	Gunner	a	2	1113	0.589	4.50	0.347	1.2	0.1269	2.63	0.0358	0.0475	0.988
2	2	Loader	a	4	1264	0.653	4.97	0.362	1.5	0.1326	2.66	0.0374	0.0497	1.085
2	2	Commander	b	1	8224	5.325	41.59	3.070	6.9	1.1230	18.97	0.3170	0.4206	8.659
2	2	Driver	b	2	3482	2.022	15.57	1.099	3.8	0.4021	7.17	0.1135	0.1506	3.276
2	2	Gunner	b	1	2849	1.563	11.99	0.927	3.0	0.3394	6.91	0.0957	0.1271	2.607
2	2	Loader	b	4	3294	1.728	13.20	0.982	3.7	0.3594	6.95	0.1014	0.1346	2.902
2	2	Commander	c	1	11495	6.837	53.20	4.050	9.6	1.4818	24.64	0.4182	0.5549	11.481
2	2	Driver	c	1	6638	3.601	27.82	2.116	6.0	0.7742	13.89	0.2185	0.2900	6.042
2	2	Gunner	c	1	5312	2.931	22.63	1.785	4.8	0.6534	12.84	0.1843	0.2447	4.889
2	2	Loader	c	1	5805	3.071	23.60	1.861	5.6	0.6813	13.12	0.1922	0.2551	5.288
2	2	Commander	d	1	11968	7.064	54.93	4.216	9.9	1.5423	25.75	0.4353	0.5776	11.926
2	2	Driver	d	1	7100	3.810	29.42	2.268	6.4	0.8301	14.89	0.2342	0.3109	6.450
2	2	Gunner	d	1	5904	3.277	25.33	2.018	5.2	0.7385	14.65	0.2083	0.2765	5.469

Table A.3. *(continued)*

Phase	Shot	Position	Scenario	AMAD	Intake (Bq)	E(50) (rem)	Lung (rem)	Bone Surface (rem)	ET (rem)	Kidney (rem)	LN-TH (rem)	Red Marrow (rem)	Liver (rem)	Kidney Conc. 1d (μg U)
2	2	Loader	d	1	6331	3.306	25.39	2.029	6.0	0.7425	14.30	0.2095	0.2780	5.736
2	2	Commander	e	1	2109	1.092	8.45	0.655	1.7	0.2397	3.74	0.0677	0.0898	1.904
2	2	Driver	e	0.6	2075	1.157	9.05	0.700	1.3	0.2563	4.71	0.0723	0.0960	1.896
2	2	Gunner	e	1	1459	0.890	6.96	0.515	1.1	0.1886	3.56	0.0532	0.0706	1.379
2	2	Loader	e	1	1455	0.802	6.22	0.505	1.1	0.1847	3.36	0.0521	0.0692	1.411
2	3	Commander	a	7	439	0.207	1.56	0.111	0.6	0.0407	1.12	0.0115	0.0152	0.350
2	3	Driver	a	7	337	0.155	1.16	0.081	0.5	0.0296	0.86	0.0084	0.0111	0.271
2	3	Gunner	a	4	234	0.141	1.08	0.073	0.3	0.0267	0.66	0.0075	0.0100	0.249
2	3	Loader	a	11	518	0.167	1.19	0.094	0.8	0.0343	0.79	0.0097	0.0129	0.369
2	3	Commander	b	11	776	0.335	2.49	0.181	1.1	0.0663	1.69	0.0187	0.0248	0.611
2	3	Driver	b	4	664	0.317	2.39	0.166	0.8	0.0608	1.77	0.0172	0.0228	0.536
2	3	Gunner	b	4	659	0.403	3.10	0.214	0.8	0.0785	2.00	0.0222	0.0294	0.710
2	3	Loader	b	7	1239	0.497	3.68	0.278	1.6	0.1018	2.55	0.0287	0.0381	0.998
2	3	Commander	c	2	1997	1.002	7.64	0.546	2.3	0.2001	5.92	0.0564	0.0749	1.563
2	3	Driver	c	0.6	1939	1.008	7.74	0.534	2.0	0.1958	6.32	0.0552	0.0733	1.433
2	3	Gunner	c	1	1825	1.088	8.41	0.605	1.9	0.2214	6.30	0.0624	0.0829	1.771
2	3	Loader	c	4	2499	1.224	9.29	0.683	2.9	0.2499	6.58	0.0705	0.0936	2.164
2	3	Commander	d	1	2566	1.358	10.40	0.743	2.8	0.2722	8.30	0.0767	0.1019	2.045
2	3	Driver	d	0.6	3582	1.602	12.15	0.856	3.9	0.3137	10.16	0.0884	0.1174	2.299
2	3	Gunner	d	1	1917	1.126	8.70	0.625	2.0	0.2289	6.55	0.0645	0.0857	1.826
2	3	Loader	d	2	3026	1.494	11.36	0.838	3.5	0.3067	8.33	0.0865	0.1148	2.570
2	3	Commander	e	2	496	0.226	1.72	0.124	0.5	0.0454	1.28	0.0128	0.0170	0.364
2	3	Driver	e	0.6	364	0.263	2.07	0.139	0.3	0.0508	1.67	0.0143	0.0190	0.326
2	3	Gunner	e	0.6	493	0.364	2.87	0.199	0.4	0.0728	2.18	0.0205	0.0273	0.526
2	3	Loader	e	2	566	0.352	2.73	0.188	0.6	0.0686	1.70	0.0194	0.0257	0.586
3	1	Commander	a	7	2150	1.893	13.11	0.213	12.3	0.0782	264.7	0.0221	0.0293	0.771
3	1	Driver	a	4	2760	2.208	15.78	0.501	11.5	0.1843	164.8	0.0518	0.0688	1.416

Phase	Shot	Position	Scenario	AMAD	Intake (Bq)	E(50) (rem)	Lung (rem)	Bone Surface (rem)	ET (rem)	Kidney (rem)	LN-TH (rem)	Red Marrow (rem)	Liver (rem)	Kidney Conc. 1d (µg U)
3	1	Loader	a	4	1496	1.494	10.70	0.169	8.0	0.0619	212.3	0.0174	0.0232	0.566
3	1	Commander	b	7	7347	6.259	43.37	0.703	40.7	0.2576	862.2	0.0726	0.0964	2.514
3	1	Driver	b	7	6788	5.848	41.39	0.912	33.3	0.3349	656.7	0.0942	0.1252	2.835
3	1	Gunner	b	4	6076	6.147	44.39	0.679	31.4	0.2486	893.1	0.0701	0.0931	2.147
3	1	Loader	b	4	4426	4.929	36.35	0.566	21.5	0.2072	710.5	0.0584	0.0775	1.725
3	1	Commander	c	7	11085	9.502	66.72	1.104	57.5	0.4045	1296.9	0.1140	0.1514	3.748
3	1	Driver	c	4	9603	8.155	58.19	1.269	44.2	0.4658	945.8	0.1311	0.1742	3.788
3	1	Gunner	c	4	8622	8.509	61.58	0.995	42.6	0.3643	1209.4	0.1028	0.1364	3.037
3	1	Loader	c	4	6912	7.462	55.25	0.910	31.4	0.3333	1055.2	0.0939	0.1247	2.653
3	1	Commander	d	7	11288	9.610	67.43	1.124	58.4	0.4120	1307.5	0.1161	0.1542	3.811
3	1	Driver	d	4	10831	8.374	58.91	1.318	49.3	0.4835	953.4	0.1361	0.1808	4.007
3	1	Gunner	d	4	10278	8.986	63.24	1.072	53.6	0.3926	1245.0	0.1108	0.1470	3.448
3	1	Loader	d	2	7026	7.545	55.84	0.925	31.8	0.3388	1064.0	0.0955	0.1268	2.697
3	1	Commander	e	4	2982	2.879	21.03	0.334	13.5	0.1225	396.6	0.0345	0.0458	1.027
3	1	Driver	e	4	1858	1.907	14.28	0.261	7.2	0.0957	255.4	0.0270	0.0358	0.694
3	1	Gunner	e	2	1762	1.982	14.88	0.235	7.3	0.0859	281.9	0.0242	0.0322	0.649
3	1	Loader	e	2	1959	2.189	16.52	0.273	7.7	0.1002	308.5	0.0282	0.0375	0.744
3	2	Commander	a	11	5185	3.541	23.19	0.490	29.3	0.1797	423.4	0.0506	0.0672	1.788
3	2	Driver	a	4	2478	2.110	15.07	0.417	11.2	0.1532	203.6	0.0431	0.0573	1.199
3	2	Gunner	a	16	9201	5.256	32.20	0.721	54.1	0.2643	613.7	0.0745	0.0989	2.912
3	2	Loader	a	7	2291	1.919	13.40	0.264	11.9	0.0966	241.8	0.0272	0.0361	0.876
3	2	Commander	b	31	14620	7.879	47.59	1.132	84.3	0.4148	884.7	0.1169	0.1552	4.445
3	2	Driver	b	4	6621	5.413	38.22	0.890	31.2	0.3267	614.5	0.0919	0.1221	2.642
3	2	Gunner	b	31	28780	12.690	69.36	1.821	170.6	0.6670	1321.0	0.1881	0.2496	7.977
3	2	Loader	b	4	4906	4.522	32.49	0.604	23.6	0.2212	597.7	0.0623	0.0827	1.851
3	2	Commander	c	31	17248	9.568	58.82	1.365	97.4	0.5003	1093.9	0.1410	0.1872	5.229
3	2	Driver	c	4	8604	7.186	51.15	1.153	39.4	0.4230	837.8	0.1191	0.1581	3.351
3	2	Gunner	c	31	40774	16.367	85.58	2.410	238.6	0.8826	1622.2	0.2489	0.3303	10.785

Table A.3. *(continued)*

Phase	Shot	Position	Scenario	AMAD	Intake (Bq)	E(50) (rem)	Lung (rem)	Bone Surface (rem)	ET (rem)	Kidney (rem)	LN-TH (rem)	Red Marrow (rem)	Liver (rem)	Kidney Conc. 1d (µg U)
3	2	Loader	c	4	6676	5.923	42.50	0.806	31.2	0.2955	773.2	0.0833	0.1106	2.439
3	2	Commander	d	31	17810	9.789	60.16	1.410	99.7	0.5167	1112.2	0.1456	0.1933	5.375
3	2	Driver	d	4	9111	7.397	52.52	1.207	41.3	0.4427	851.3	0.1246	0.1655	3.502
3	2	Gunner	d	31	43251	16.848	87.36	2.505	249.0	0.9175	1645.7	0.2588	0.3434	11.227
3	2	Loader	d	4	8135	6.327	44.59	0.899	37.1	0.3295	795.1	0.0928	0.1233	2.769
3	2	Commander	e	11	2093	1.453	9.78	0.191	10.8	0.0698	187.6	0.0197	0.0261	0.637
3	2	Driver	e	4	1635	1.548	11.33	0.206	7.1	0.0754	208.3	0.0213	0.0282	0.561
3	2	Gunner	e	31	9601	3.080	13.48	0.479	57.5	0.1751	262.8	0.0494	0.0656	2.322
3	2	Loader	e	4	968	1.052	7.83	0.123	4.2	0.0451	151.9	0.0127	0.0169	0.344
4	4	Driver	a	1	137	0.090	0.66	0.023	0.4	0.0083	7.3	0.0023	0.0031	0.054
4	4	Driver	b	0.6	610	0.436	3.32	0.100	1.3	0.0367	43.3	0.0103	0.0137	0.225
4	4	Driver	c	2	1284	1.022	7.58	0.181	4.1	0.0664	122.4	0.0187	0.0248	0.458
4	4	Driver	d	2	1594	1.196	8.65	0.210	5.9	0.0771	141.3	0.0217	0.0288	0.555
4	4	Driver	e	2	386	0.412	3.14	0.052	1.3	0.0189	59.7	0.0053	0.0071	0.140

Scenario	Phase	Number Avg'd	AMAD	sd AMAD	Intake (Bq)	sd Intake (Bq)	Intake (nCi)	Intake (mg)	E(50) (rem)	sd E(50) (rem)	10% E(50) (rem)	Median E(50) (rem)	90% E(50) (rem)	Median/ Mean
A	1	22	8.18	8.35	3739	1641	101.1	265.3	2.007	0.824	1.323	1.965	2.785	0.979
A	2	11	4.18	3.03	1110	884	30.0	78.7	0.678	0.653	0.143	0.590	1.023	0.871
A	3	8	7.50	4.17	3429	2416	92.7	243.2	2.628	1.219	1.494	2.228	5.249	0.848
A	4	1	1	NA	116	NA	3.1	8.2	0.074	NA	0.074	0.074	0.074	1
B	1	22	5.73	6.29	7972	3965	215.5	565.5	4.446	2.130	3.030	3.644	7.253	0.820
B	2	11	3.55	3.08	2893	2202	78.2	205.2	1.811	1.559	0.293	1.735	3.167	0.958
B	3	8	11.50	12.11	10330	9719	279.2	732.8	6.873	2.873	4.544	6.149	13.564	0.895
B	4	1	0.6	NA	539	NA	14.6	38.2	0.384	NA	0.384	0.384	0.384	1
C	1	20	5.65	6.59	10269	5563	277.6	728.5	5.327	1.987	3.162	4.878	8.448	0.916
C	2	8	1.45	1.10	4404	3334	119.0	312.4	2.577	2.083	0.703	2.100	6.885	0.815
C	3	8	11.13	12.31	14289	14035	386.2	1013.6	9.331	3.620	5.935	8.497	17.812	0.911
C	4	1	2	NA	1081	NA	29.2	76.7	0.895	NA	0.895	0.895	0.895	1
D	1	18	5.39	6.86	10558	6150	285.4	749.0	5.406	1.990	2.909	5.020	8.887	0.928
D	2	8	1.08	0.40	4889	3355	132.1	346.8	2.846	2.066	1.113	2.429	7.118	0.854
D	3	8	10.88	12.49	14949	15012	404.0	1060.4	9.535	3.744	6.230	8.568	18.373	0.899
D	4	1	2	NA	1351	NA	36.5	95.8	1.050	NA	1.050	1.050	1.050	1
E	1	22	4.48	6.77	2083	1461	56.3	147.7	1.204	0.814	0.443	0.913	2.183	0.758
E	2	11	1.35	0.65	1352	788	36.5	95.9	0.877	0.525	0.216	0.949	1.387	1.083
E	3	8	7.75	9.81	3207	3800	86.7	227.5	2.119	0.830	1.078	2.012	3.633	0.949
E	4	1	2	NA	305	NA	8.2	21.6	0.362	NA	0.362	0.362	0.362	1

Table A.4. Summary of Intakes, Doses, and Concentrations Using Conventional Statistics (Part 2)

Scenario	Phase	Avg. KID Conc 1d (µg U)	sd KID1d (µg U)	10% KID1d (µg U)	Median KID1d (µg U)	90% KID1d (µg U)	Median/ Mean	Lung (rem)	sd Lung (rem)	10% Lung (rem)	Median Lung (rem)	90% Lung (rem)	Median/ Mean
A	1	2.624	1.027	1.646	2.625	3.562	1.000	14.43	6.17	9.24	14.37	19.26	0.996
A	2	0.946	0.869	0.213	0.846	1.384	0.895	5.23	5.14	1.07	4.52	7.97	0.863
A	3	1.148	0.657	0.486	0.917	2.498	0.799	17.74	6.96	10.72	15.64	32.23	0.882
A	4	0.041	0.000	0.041	0.041	0.041	1.000	0.54	NA	0.54	0.54	0.54	1
B	1	5.956	2.487	3.972	5.487	8.440	0.921	32.31	16.00	22.28	26.87	53.72	0.832
B	2	2.532	2.114	0.457	2.482	4.267	0.980	14.03	12.21	2.18	13.28	24.72	0.947
B	3	2.969	2.126	1.510	2.219	7.901	0.748	44.47	10.95	32.63	43.56	69.01	0.980
B	4	0.170	0.000	0.170	0.170	0.170	1.000	2.95	NA	2.95	2.95	2.95	1
C	1	7.641	2.952	4.626	7.064	11.870	0.924	38.37	13.78	23.19	35.74	59.23	0.931
C	2	3.659	2.987	0.839	3.036	9.781	0.830	19.93	16.24	5.43	16.17	53.61	0.811
C	3	3.997	2.915	2.067	3.068	10.967	0.768	60.46	12.29	42.82	59.59	85.14	0.986
C	4	0.325	0.000	0.325	0.325	0.325	1.000	6.67	NA	6.67	6.67	6.67	1
D	1	7.835	3.168	4.459	6.972	13.929	0.890	39.01	13.68	21.61	36.77	59.45	0.943
D	2	4.037	3.014	1.281	3.517	10.180	0.871	22.01	16.09	8.65	18.64	55.39	0.847
D	3	4.131	3.051	2.230	3.160	11.441	0.765	61.55	12.23	44.90	60.41	86.89	0.981
D	4	0.396	0.000	0.396	0.396	0.396	1.000	7.59	NA	7.59	7.59	7.59	1
E	1	1.851	1.272	0.694	1.414	3.984	0.764	8.89	6.06	3.32	6.84	13.79	0.770
E	2	1.190	0.727	0.298	1.253	1.759	1.053	6.84	4.09	1.65	7.44	10.82	1.089
E	3	0.841	0.744	0.300	0.611	2.637	0.726	13.88	4.26	8.00	14.03	21.31	1.011
E	4	0.092	0.000	0.092	0.092	0.092	1.000	2.79	NA	2.79	2.79	2.79	1

Table A.4. Summary of Intakes, Doses, and Concentrations Using Conventional Statistics (Part 3)

Scenario	Phase	BS (rem)	sd BS (rem)	ET (rem)	sd ET (rem)	Kidney (rem)	sd KID (rem)	RM (rem)	sd RM (rem)	Liver (rem)	sd Liver (rem)	LN (rem)	sd LN (rem)
A	1	0.978	0.417	8.93	4.73	0.358	0.153	0.1009	0.0431	0.1340	0.0572	13.30	4.80
A	2	0.374	0.363	1.19	0.73	0.137	0.133	0.0387	0.0375	0.0513	0.0498	2.53	2.19
A	3	0.387	0.196	19.16	15.34	0.142	0.072	0.0400	0.0202	0.0531	0.0269	302.75	145.09
A	4	0.020	NA	0.30	NA	0.007	NA	0.0021	NA	0.0028	NA	5.06	NA
B	1	2.188	1.084	18.04	10.58	0.801	0.397	0.2260	0.1120	0.3000	0.1487	28.93	12.23
B	2	1.009	0.872	2.95	1.90	0.369	0.319	0.1042	0.0900	0.1383	0.1194	6.74	5.14
B	3	0.930	0.460	59.49	62.66	0.341	0.168	0.0961	0.0475	0.1275	0.0630	811.58	232.17
B	4	0.089	NA	1.02	NA	0.033	NA	0.0092	NA	0.0123	NA	37.56	NA
C	1	2.692	1.066	23.11	16.70	0.985	0.390	0.2780	0.1101	0.3690	0.1461	34.62	12.29
C	2	1.482	1.218	4.24	2.81	0.542	0.445	0.1530	0.1258	0.2031	0.1669	10.86	6.61
C	3	1.273	0.584	80.33	90.36	0.466	0.213	0.1314	0.0603	0.1745	0.0800	1101.97	271.58
C	4	0.146	NA	3.49	NA	0.054	NA	0.0151	NA	0.0201	NA	112.42	NA
D	1	2.759	1.106	23.11	18.36	1.010	0.405	0.2848	0.1142	0.3781	0.1516	34.86	11.54
D	2	1.651	1.225	4.67	2.81	0.604	0.448	0.1704	0.1265	0.2262	0.1679	12.40	6.35
D	3	1.315	0.608	83.16	94.82	0.482	0.222	0.1359	0.0628	0.1803	0.0833	1115.67	273.01
D	4	0.169	NA	5.20	NA	0.062	NA	0.0175	NA	0.0232	NA	130.51	NA
E	1	0.637	0.440	4.14	4.02	0.233	0.161	0.0658	0.0455	0.0873	0.0603	7.30	4.74
E	2	0.481	0.285	1.22	0.77	0.176	0.104	0.0496	0.0295	0.0659	0.0391	3.36	1.59
E	3	0.273	0.135	17.56	25.33	0.100	0.049	0.0282	0.0139	0.0374	0.0185	261.05	78.04
E	4	0.036	NA	1.03	NA	0.013	NA	0.0037	NA	0.0049	NA	56.29	NA

Table A.5. Individual Doses and Concentrations by Scenario Using Conventional Statistics

Phase	Shot	Position	Scenario	Intake (Bq)	E(50) (rem)	Lung (rem)	Bone Surface (rem)	ET (rem)	Kidney (rem)	LN-TH (rem)	Red Marrow (rem)	Liver (rem)	Kidney Conc. 1d (µg U)	LN-TH 200d
1	1	Commander	a	2904	1.757	12.88	0.842	6.67	0.3084	12.14	0.0870	0.1155	2.353	3.79
1	1	Driver	a	668	0.219	1.35	0.107	2.04	0.0392	1.45	0.0111	0.0147	0.366	0.44
1	1	Gunner	a	2903	1.927	14.56	0.956	5.17	0.3502	13.40	0.0987	0.1311	2.272	3.82
1	1	Loader	a	3012	1.995	14.94	0.999	5.97	0.3655	13.21	0.1031	0.1369	2.783	4.35
1	1	Commander	b	5355	3.449	25.97	1.710	9.67	0.6261	23.18	0.1766	0.2344	4.492	7.05
1	1	Driver	b	1976	0.936	6.61	0.461	4.77	0.1687	6.29	0.0476	0.0632	1.379	1.97
1	1	Gunner	b	5362	3.433	25.88	1.725	9.40	0.6319	23.92	0.1781	0.2365	4.264	6.91
1	1	Loader	b	5736	3.717	27.78	1.813	11.50	0.6636	25.90	0.1872	0.2485	5.074	8.39
1	1	Commander	c	7489	4.823	36.53	2.449	12.34	0.8967	31.43	0.2529	0.3358	6.392	9.58
1	1	Driver	c	2949	1.520	11.00	0.760	6.36	0.2781	10.07	0.0784	0.1041	2.166	3.10
1	1	Gunner	c	7048	4.461	33.83	2.268	11.25	0.8308	30.77	0.2341	0.3110	5.536	8.74
1	1	Loader	c	7683	4.933	37.05	2.470	14.21	0.9043	33.72	0.2550	0.3386	6.838	10.75
1	1	Commander	d	7777	5.041	38.25	2.563	12.60	0.9385	32.89	0.2646	0.3514	6.616	9.94
1	1	Driver	d	3310	1.749	12.77	0.875	6.81	0.3205	11.60	0.0904	0.1200	2.392	3.46
1	1	Gunner	d	7398	4.662	35.36	2.370	11.70	0.8681	32.11	0.2446	0.3249	5.738	9.06
1	1	Loader	d	7958	5.149	38.75	2.576	14.49	0.9433	35.22	0.2660	0.3532	7.029	11.10
1	1	Commander	e	1636	1.109	8.56	0.588	2.06	0.2154	6.51	0.0607	0.0807	1.536	2.12
1	1	Driver	e	583	0.356	2.66	0.180	1.06	0.0658	2.27	0.0186	0.0247	0.535	0.77
1	1	Gunner	e	1174	0.748	5.81	0.396	1.17	0.1451	4.96	0.0409	0.0543	0.961	1.39
1	1	Loader	e	1459	0.928	7.08	0.501	2.10	0.1832	5.90	0.0517	0.0686	1.408	1.90
1	2	Commander	a	2714	1.694	12.49	0.819	6.07	0.2997	11.32	0.0845	0.1122	2.190	3.52
1	2	Driver	a	3260	1.879	13.46	0.875	8.69	0.3203	13.16	0.0904	0.1199	2.610	4.30
1	2	Gunner	a	2852	1.758	12.88	0.827	6.71	0.3026	11.78	0.0854	0.1133	2.361	3.91
1	2	Loader	a	2676	1.616	11.94	0.812	5.60	0.2973	11.73	0.0838	0.1113	2.099	3.41

Phase	Shot	Position	Scenario	Intake (Bq)	E(50) (rem)	Lung (rem)	Bone Surface (rem)	ET (rem)	Kidney (rem)	LN-TH (rem)	Red Marrow (rem)	Liver (rem)	Kidney Conc. 1d (μg U)	LN-TH 200d
1	2	Commander	b	4867	3.345	25.20	1.658	9.27	0.6070	21.73	0.1712	0.2272	4.308	6.80
1	2	Driver	b	5174	3.076	22.28	1.445	12.98	0.5288	21.14	0.1492	0.1980	4.301	7.03
1	2	Gunner	b	5451	3.268	24.08	1.613	11.69	0.5904	21.47	0.1665	0.2211	4.446	6.91
1	2	Loader	b	4747	3.030	22.63	1.501	9.39	0.5494	21.58	0.1549	0.2057	3.972	6.48
1	2	Commander	c	6598	4.569	34.68	2.306	11.37	0.8446	29.02	0.2381	0.3162	5.879	8.95
1	2	Driver	c	6295	3.601	26.13	1.723	14.91	0.6306	24.57	0.1779	0.2361	5.057	8.05
1	2	Gunner	c	7176	4.361	32.39	2.183	14.32	0.7991	28.49	0.2253	0.2992	5.839	8.94
1	2	Loader	c	6467	4.251	31.99	2.128	11.94	0.7794	29.33	0.2198	0.2918	5.611	8.86
1	2	Commander	d	7028	4.900	37.28	2.476	11.77	0.9068	31.26	0.2556	0.3394	6.206	9.49
1	2	Driver	d	6589	3.786	27.57	1.822	15.24	0.6671	25.80	0.1882	0.2498	5.266	8.35
1	2	Gunner	d	7622	4.595	34.15	2.306	14.98	0.8441	30.03	0.2380	0.3160	6.112	9.32
1	2	Loader	d	7035	4.676	35.32	2.343	12.53	0.8582	32.25	0.2419	0.3213	6.020	9.54
1	2	Commander	e	1107	0.855	6.64	0.450	1.36	0.1647	4.86	0.0464	0.0617	1.141	1.56
1	2	Driver	e	627	0.295	2.14	0.152	1.19	0.0556	1.94	0.0157	0.0208	0.461	0.66
1	2	Gunner	e	1116	0.753	5.73	0.387	1.78	0.1416	4.81	0.0399	0.0530	0.983	1.49
1	2	Loader	e	1062	0.773	5.89	0.394	1.80	0.1440	4.70	0.0406	0.0539	1.153	1.65
1	3	Driver	a	3887	2.506	18.88	1.105	7.24	0.4055	16.08	0.1141	0.1517	1.734	4.22
1	3	Gunner	a	6498	4.250	32.21	2.141	10.81	0.7843	22.95	0.2210	0.2935	4.665	7.69
1	3	Driver	b	10462	7.336	55.71	3.390	18.81	1.2425	47.22	0.3499	0.4649	6.854	14.35
1	3	Gunner	b	16537	10.930	82.81	5.561	27.86	2.0368	60.44	0.5742	0.7624	12.918	21.10
1	3	Driver	e	4039	3.192	24.48	1.667	6.59	0.6096	19.08	0.1721	0.2284	4.741	7.69
1	3	Gunner	e	5175	3.228	24.24	1.769	9.00	0.6473	17.49	0.1827	0.2424	4.932	7.07
1	4	Commander	a	4757	2.514	17.86	1.215	12.20	0.4451	15.95	0.1254	0.1665	2.780	4.86
1	4	Driver	a	2635	1.323	9.24	0.609	7.24	0.2231	8.90	0.0629	0.0835	1.646	2.93
1	4	Gunner	a	5751	2.785	19.26	1.271	16.20	0.4655	20.07	0.1312	0.1742	3.490	6.43
1	4	Loader	a	7199	2.180	12.94	1.063	22.81	0.3892	14.26	0.1098	0.1457	3.373	4.44
1	4	Commander	b	11197	6.005	43.08	3.012	26.97	1.1036	37.25	0.3110	0.4131	7.054	11.74
1	4	Driver	b	6474	3.514	25.10	1.701	16.48	0.6227	22.87	0.1756	0.2331	4.639	7.95

Table A.5. *(continued)*

Phase	Shot	Position	Scenario	Intake (Bq)	E(50) (rem)	Lung (rem)	Bone Surface (rem)	ET (rem)	Kidney (rem)	LN-TH (rem)	Red Marrow (rem)	Liver (rem)	Kidney Conc. 1d (µg U)	LN-TH 200d
1	4	Gunner	b	12633	6.309	44.16	2.925	34.12	1.0712	45.13	0.3020	0.4010	7.778	14.40
1	4	Loader	b	16716	5.465	33.60	2.700	51.48	0.9886	35.02	0.2788	0.3701	8.440	11.27
1	4	Commander	c	15467	8.069	58.00	4.172	35.43	1.5282	49.26	0.4308	0.5721	10.284	16.11
1	4	Driver	c	10645	5.361	38.14	2.731	25.50	0.9997	33.13	0.2820	0.3743	7.688	11.84
1	4	Gunner	c	16279	8.460	60.46	4.073	39.47	1.4917	59.00	0.4205	0.5584	10.317	18.73
1	4	Loader	c	27719	8.436	51.27	4.306	82.06	1.5758	52.16	0.4447	0.5901	14.203	17.52
1	4	Commander	d	16158	8.277	59.45	4.311	36.51	1.5790	50.34	0.4452	0.5911	10.642	16.42
1	4	Driver	d	11386	5.735	41.01	2.956	26.22	1.0819	35.33	0.3052	0.4051	8.160	12.44
1	4	Gunner	d	17440	9.062	65.04	4.371	40.94	1.6010	62.73	0.4513	0.5993	10.848	19.69
1	4	Loader	d	30251	8.887	53.71	4.557	87.81	1.6678	54.20	0.4706	0.6246	15.031	18.08
1	4	Commander	e	2872	1.584	11.54	0.858	6.13	0.3139	9.36	0.0886	0.1176	2.485	3.63
1	4	Driver	e	2882	1.443	10.24	0.777	6.89	0.2843	8.12	0.0803	0.1065	2.401	3.29
1	4	Gunner	e	2462	1.539	11.65	0.817	3.91	0.2991	10.14	0.0843	0.1120	1.905	3.35
1	4	Loader	e	6034	2.183	13.79	1.127	18.71	0.4122	13.89	0.1164	0.1544	3.984	5.27
1	5	Commander	a	3389	1.996	14.47	0.931	8.40	0.3410	16.09	0.0962	0.1277	2.679	4.79
1	5	Driver	a	4120	2.207	15.51	1.077	11.53	0.3939	14.42	0.1112	0.1476	3.510	5.19
1	5	Loader	a	3608	1.706	11.89	0.832	9.41	0.3044	12.60	0.0859	0.1140	2.633	4.11
1	5	Commander	b	6975	4.208	30.67	1.935	16.96	0.7079	32.85	0.1997	0.2651	5.970	10.47
1	5	Driver	b	8355	4.310	30.01	2.019	24.01	0.7383	27.06	0.2085	0.2766	7.025	10.34
1	5	Loader	b	6325	3.344	23.93	1.586	15.51	0.5803	25.15	0.1638	0.2174	5.073	8.29
1	5	Commander	c	8600	4.980	36.24	2.333	20.28	0.8535	38.06	0.2408	0.3197	7.291	12.20
1	5	Driver	c	11778	5.203	35.24	2.501	33.56	0.9144	31.56	0.2583	0.3426	8.936	12.04
1	5	Loader	c	8077	4.248	30.54	2.062	18.89	0.7545	31.17	0.2129	0.2826	6.516	10.17
1	5	Commander	d	8984	5.162	37.59	2.439	20.84	0.8925	39.24	0.2518	0.3343	7.569	12.49
1	5	Driver	d	12604	5.373	36.26	2.609	35.29	0.9540	32.36	0.2695	0.3574	9.307	12.26
1	5	Loader	d	8709	4.492	32.31	2.209	19.87	0.8083	32.66	0.2281	0.3027	6.915	10.54
1	5	Commander	e	989	0.582	4.28	0.277	2.15	0.1012	4.13	0.0286	0.0379	0.950	1.45

Phase	Shot	Position	Scenario	Intake (Bq)	E(50) (rem)	Lung (rem)	Bone Surface (rem)	ET (rem)	Kidney (rem)	LN-TH (rem)	Red Marrow (rem)	Liver (rem)	Kidney Conc. 1d (µg U)	LN-TH 200d
1	5	Loader	e	1103	0.621	4.55	0.307	2.32	0.1125	4.20	0.0317	0.0421	0.993	1.43
1	6	Commander	a	4925	2.304	15.83	1.073	13.92	0.3924	15.79	0.1108	0.1470	3.562	5.35
1	6	Driver	a	544	0.322	2.34	0.145	1.34	0.0531	2.69	0.0150	0.0199	0.464	0.87
1	6	Loader	a	4	0.002	0.02	0.001	0.01	0.0004	0.01	0.0001	0.0002	0.003	0.00
1	6	Commander	b	8328	4.426	31.64	2.114	20.65	0.7735	29.44	0.2184	0.2897	6.879	10.35
1	6	Driver	b	3265	2.075	15.39	0.996	7.05	0.3644	13.16	0.1029	0.1365	3.214	4.94
1	6	Loader	b	22	0.011	0.08	0.006	0.04	0.0021	0.07	0.0006	0.0008	0.017	0.02
1	6	Commander	c	10995	5.635	40.49	2.763	25.13	1.0107	36.29	0.2853	0.3786	8.854	12.63
1	6	Driver	c	4415	2.724	20.24	1.338	8.99	0.4894	16.92	0.1382	0.1833	4.196	6.18
1	6	Loader	c	219	0.090	0.64	0.051	0.41	0.0185	0.58	0.0052	0.0069	0.142	0.16
1	6	Commander	d	12292	5.900	42.04	2.918	27.92	1.0677	37.62	0.3014	0.3999	9.359	12.97
1	6	Driver	d	4840	2.909	21.61	1.444	9.57	0.5282	18.12	0.1491	0.1979	4.459	6.46
1	6	Loader	d	493	0.194	1.38	0.111	0.92	0.0406	1.24	0.0114	0.0152	0.303	0.32
1	6	Commander	e	1266	0.898	6.96	0.472	1.49	0.1727	5.13	0.0488	0.0647	1.421	1.87
1	6	Driver	e	700	0.443	3.32	0.225	1.31	0.0822	2.40	0.0232	0.0308	0.694	0.92
1	6	Loader	e	49	0.022	0.15	0.011	0.12	0.0040	0.15	0.0011	0.0015	0.034	0.05
1	7	Commander	a	5320	2.339	16.47	1.274	11.82	0.4662	14.00	0.1316	0.1746	3.539	5.39
1	7	Driver	a	5037	2.937	21.72	1.535	9.91	0.5618	18.62	0.1585	0.2104	3.994	7.09
1	7	Gunner	a	3604	1.934	14.27	0.999	6.73	0.3656	11.95	0.1031	0.1369	2.616	4.63
1	7	Loader	a	6	0.003	0.02	0.002	0.01	0.0006	0.01	0.0002	0.0002	0.005	0.01
1	7	Commander	b	8627	3.571	24.84	1.988	19.35	0.7273	20.85	0.2053	0.2724	5.900	8.72
1	7	Driver	b	12908	7.253	53.72	3.916	23.85	1.4330	44.95	0.4044	0.5367	10.292	17.59
1	7	Gunner	b	7915	4.802	35.83	2.377	15.02	0.8695	29.94	0.2455	0.3257	6.759	12.59
1	7	Loader	b	42	0.022	0.16	0.012	0.08	0.0044	0.13	0.0012	0.0016	0.034	0.06
1	7	Commander	c	11709	4.761	33.17	2.688	25.47	0.9833	27.95	0.2776	0.3683	8.021	11.73
1	7	Driver	c	16639	9.325	69.47	5.205	28.38	1.9049	55.77	0.5375	0.7134	13.423	22.08
1	7	Gunner	c	11355	6.811	50.60	3.390	22.28	1.2399	43.71	0.3501	0.4645	9.784	18.15
1	7	Loader	c	285	0.138	1.00	0.071	0.56	0.0258	0.88	0.0073	0.0097	0.199	0.34

Table A.5. *(continued)*

Phase	Shot	Position	Scenario	Intake (Bq)	E(50) (rem)	Lung (rem)	Bone Surface (rem)	ET (rem)	Kidney (rem)	LN-TH (rem)	Red Marrow (rem)	Liver (rem)	Kidney Conc. 1d (μg U)	LN-TH 200d
1	7	Commander	d	12160	4.998	34.97	2.840	25.97	1.0390	29.16	0.2933	0.3892	8.337	12.13
1	7	Driver	d	17438	9.705	72.34	5.451	29.27	1.9950	57.69	0.5629	0.7471	13.929	22.69
1	7	Gunner	d	11897	7.053	52.42	3.543	22.89	1.2960	44.96	0.3659	0.4855	10.102	18.55
1	7	Loader	d	319	0.147	1.06	0.077	0.63	0.0281	0.93	0.0079	0.0105	0.216	0.35
1	7	Commander	e	2213	0.955	6.73	0.535	4.78	0.1956	6.08	0.0553	0.0733	1.686	2.64
1	7	Driver	e	2725	1.622	12.39	0.985	3.28	0.3604	8.63	0.1017	0.1350	2.462	3.76
1	7	Gunner	e	2782	1.743	12.84	0.841	6.27	0.3074	12.44	0.0868	0.1151	2.621	5.11
1	7	Loader	e	157	0.093	0.69	0.044	0.32	0.0161	0.65	0.0045	0.0060	0.124	0.25
2	1	Commander	a	1422	0.905	7.01	0.500	1.49	0.1828	2.83	0.0516	0.0685	1.259	1.83
2	1	Driver	a	1512	1.023	7.97	0.569	1.45	0.2082	3.39	0.0588	0.0780	1.384	2.04
2	1	Loader	a	1025	0.493	3.75	0.302	1.06	0.1104	2.20	0.0311	0.0413	0.731	0.91
2	1	Commander	b	4389	2.880	22.36	1.596	4.43	0.5836	9.27	0.1648	0.2186	4.059	5.95
2	1	Driver	b	4497	3.167	24.72	1.736	4.33	0.6349	10.58	0.1792	0.2378	4.267	6.44
2	1	Loader	b	3082	1.778	13.73	1.049	2.99	0.3842	7.76	0.1084	0.1438	2.488	3.35
2	1	Commander	e	1880	1.326	10.32	0.665	2.08	0.2432	4.12	0.0686	0.0911	1.688	2.86
2	1	Driver	e	2457	1.711	13.36	0.944	2.29	0.3454	5.79	0.0975	0.1294	2.424	3.68
2	1	Loader	e	1871	1.387	10.82	0.699	2.05	0.2556	5.18	0.0721	0.0957	1.759	3.01
2	2	Commander	a	3388	2.406	18.85	1.337	2.91	0.4889	8.47	0.1380	0.1831	3.244	4.94
2	2	Driver	a	1198	0.736	5.64	0.370	1.55	0.1353	2.48	0.0382	0.0507	0.994	1.65
2	2	Gunner	a	1057	0.590	4.52	0.342	1.20	0.1252	2.59	0.0353	0.0469	0.846	1.18
2	2	Loader	a	1198	0.654	4.98	0.357	1.47	0.1307	2.63	0.0369	0.0489	0.927	1.39
2	2	Commander	b	7834	5.362	41.94	3.009	6.76	1.1008	18.69	0.3107	0.4123	7.358	10.99
2	2	Driver	b	3322	2.038	15.72	1.077	3.76	0.3940	7.06	0.1112	0.1476	2.788	4.42
2	2	Gunner	b	2646	1.563	12.03	0.913	2.81	0.3343	6.84	0.0943	0.1252	2.217	3.11
2	2	Loader	b	3122	1.735	13.28	0.970	3.58	0.3549	6.87	0.1001	0.1329	2.482	3.65
2	2	Commander	c	11040	6.885	53.61	3.974	9.58	1.4539	24.27	0.4104	0.5445	9.781	13.93
2	2	Driver	c	6422	3.655	28.26	2.087	6.07	0.7638	13.80	0.2155	0.2861	5.186	7.46

Phase	Shot	Position	Scenario	Intake (Bq)	E(50) (rem)	Lung (rem)	Bone Surface (rem)	ET (rem)	Kidney (rem)	LN-TH (rem)	Red Marrow (rem)	Liver (rem)	Kidney Conc. 1d (µg U)	LN-TH 200d
2	2	Loader	c	5567	3.115	23.97	1.848	5.58	0.6764	13.13	0.1908	0.2533	4.582	6.22
2	2	Commander	d	11518	7.118	55.39	4.142	9.99	1.5155	25.40	0.4277	0.5676	10.180	14.30
2	2	Driver	d	6903	3.879	29.98	2.249	6.48	0.8228	14.85	0.2322	0.3082	5.564	7.82
2	2	Gunner	d	5623	3.345	25.91	2.014	5.13	0.7372	14.70	0.2080	0.2761	4.738	6.37
2	2	Loader	d	6075	3.350	25.77	2.014	6.01	0.7373	14.30	0.2080	0.2761	4.969	6.59
2	2	Commander	e	2063	1.104	8.54	0.644	1.79	0.2357	3.68	0.0665	0.0883	1.630	2.29
2	2	Driver	e	2039	1.196	9.36	0.698	1.41	0.2555	4.76	0.0721	0.0957	1.649	2.38
2	2	Gunner	e	1427	0.949	7.44	0.525	1.12	0.1923	3.70	0.0543	0.0720	1.231	1.93
2	2	Loader	e	1419	0.839	6.52	0.506	1.20	0.1852	3.46	0.0523	0.0694	1.253	1.72
2	3	Commander	a	406	0.204	1.54	0.108	0.53	0.0395	1.08	0.0111	0.0148	0.293	0.39
2	3	Driver	a	305	0.143	1.07	0.074	0.42	0.0272	0.81	0.0077	0.0102	0.213	0.31
2	3	Gunner	a	213	0.138	1.06	0.070	0.28	0.0257	0.62	0.0073	0.0096	0.208	0.29
2	3	Loader	a	486	0.162	1.15	0.089	0.75	0.0327	0.76	0.0092	0.0123	0.307	0.34
2	3	Commander	b	654	0.293	2.18	0.157	0.90	0.0574	1.43	0.0162	0.0215	0.457	0.55
2	3	Driver	b	525	0.236	1.77	0.125	0.65	0.0459	1.49	0.0129	0.0172	0.322	0.46
2	3	Gunner	b	592	0.392	3.03	0.202	0.72	0.0738	1.84	0.0208	0.0277	0.586	0.80
2	3	Loader	b	1156	0.476	3.53	0.266	1.55	0.0973	2.37	0.0274	0.0364	0.827	0.93
2	3	Commander	c	1808	0.950	7.26	0.516	2.07	0.1889	5.55	0.0532	0.0707	1.268	1.63
2	3	Driver	c	1247	0.703	5.43	0.378	1.25	0.1386	4.67	0.0390	0.0519	0.839	1.23
2	3	Gunner	c	1734	1.107	8.57	0.604	1.83	0.2212	6.21	0.0624	0.0828	1.542	2.01
2	3	Loader	c	2376	1.208	9.17	0.670	2.85	0.2452	6.34	0.0691	0.0918	1.841	2.20
2	3	Commander	d	2354	1.307	10.03	0.712	2.62	0.2610	7.92	0.0735	0.0977	1.687	2.21
2	3	Driver	d	1826	1.113	8.65	0.598	1.74	0.2190	7.39	0.0617	0.0820	1.281	1.89
2	3	Gunner	d	1816	1.140	8.82	0.622	1.91	0.2277	6.44	0.0642	0.0853	1.583	2.06
2	3	Loader	d	2997	1.514	11.51	0.855	3.50	0.3130	8.19	0.0883	0.1172	2.295	2.68
2	3	Commander	e	451	0.216	1.65	0.118	0.47	0.0433	1.19	0.0122	0.0162	0.298	0.37
2	3	Driver	e	262	0.186	1.47	0.099	0.16	0.0365	1.26	0.0103	0.0136	0.192	0.31
2	3	Gunner	e	479	0.384	3.03	0.206	0.35	0.0755	2.22	0.0213	0.0283	0.477	0.68

Table A.5. *(continued)*

Phase	Shot	Position	Scenario	Intake (Bq)	E(50) (rem)	Lung (rem)	Bone Surface (rem)	ET (rem)	Kidney (rem)	LN-TH (rem)	Red Marrow (rem)	Liver (rem)	Kidney Conc. 1d (μg U)	LN-TH 200d
2	3	Loader	e	524	0.348	2.71	0.182	0.53	0.0666	1.63	0.0188	0.0249	0.492	0.65
3	1	Commander	a	2021	1.878	13.06	0.208	11.97	0.0763	258.81	0.0215	0.0286	0.653	4.86
3	1	Driver	a	2996	2.501	17.99	0.594	12.49	0.2185	169.35	0.0614	0.0816	1.438	4.52
3	1	Gunner	a	2725	2.375	16.36	0.265	15.94	0.0969	329.84	0.0274	0.0363	0.814	5.94
3	1	Loader	a	1429	1.494	10.72	0.166	7.98	0.0610	209.02	0.0172	0.0228	0.486	3.67
3	1	Commander	b	6960	6.207	43.12	0.685	39.83	0.2508	842.87	0.0707	0.0939	2.130	15.10
3	1	Driver	b	7241	6.257	44.00	1.014	36.90	0.3721	665.20	0.1047	0.1391	2.741	13.79
3	1	Gunner	b	5671	6.090	44.09	0.637	30.64	0.2331	885.00	0.0658	0.0873	1.769	15.78
3	1	Loader	b	4308	5.065	37.37	0.566	22.00	0.2072	722.18	0.0584	0.0775	1.510	12.55
3	1	Commander	c	10589	9.493	66.78	1.075	56.87	0.3937	1280.71	0.1110	0.1473	3.188	22.64
3	1	Driver	c	9785	8.618	61.24	1.369	47.83	0.5025	965.52	0.1414	0.1879	3.573	19.31
3	1	Gunner	c	7871	8.376	61.14	0.937	39.61	0.3432	1190.96	0.0968	0.1285	2.470	21.38
3	1	Loader	c	6867	7.699	56.86	0.908	33.08	0.3328	1077.42	0.0938	0.1245	2.340	18.66
3	1	Commander	d	10752	9.582	67.35	1.087	57.68	0.3983	1291.30	0.1123	0.1491	3.227	22.75
3	1	Driver	d	9935	8.717	61.96	1.396	48.31	0.5122	972.51	0.1442	0.1915	3.629	19.45
3	1	Gunner	d	7944	8.418	61.44	0.947	39.82	0.3468	1194.71	0.0978	0.1298	2.492	21.43
3	1	Loader	d	6959	7.778	57.43	0.921	33.51	0.3375	1086.54	0.0951	0.1263	2.374	18.76
3	1	Commander	e	2883	2.915	21.31	0.326	13.63	0.1194	397.86	0.0337	0.0447	0.881	7.09
3	1	Driver	e	1975	1.999	14.70	0.268	8.81	0.0983	266.05	0.0277	0.0368	0.654	5.03
3	1	Gunner	e	1739	2.024	15.27	0.237	7.20	0.0867	282.25	0.0244	0.0324	0.568	5.23
3	1	Loader	e	1964	2.281	17.16	0.273	8.32	0.1002	318.44	0.0282	0.0375	0.662	5.65
3	2	Commander	a	4928	3.528	23.18	0.482	28.83	0.1767	415.11	0.0498	0.0661	1.527	8.46
3	2	Driver	a	2330	2.081	14.92	0.410	10.75	0.1506	197.49	0.0423	0.0563	1.020	4.67
3	2	Gunner	a	8829	5.249	32.23	0.709	53.72	0.2599	603.34	0.0733	0.0973	2.498	12.32
3	2	Loader	a	2172	1.920	13.46	0.260	11.64	0.0953	239.07	0.0268	0.0356	0.749	4.92
3	2	Commander	b	13961	7.805	47.19	1.109	83.30	0.4062	861.03	0.1145	0.1520	3.793	17.33
3	2	Driver	b	6542	5.454	38.34	0.886	32.29	0.3251	608.61	0.0915	0.1215	2.309	13.00

Table A.5. *(continued)*

Phase	Shot	Position	Scenario	Intake (Bq)	E(50) (rem)	Lung (rem)	Bone Surface (rem)	ET (rem)	Kidney (rem)	LN-TH (rem)	Red Marrow (rem)	Liver (rem)	Kidney Conc. 1d (μg U)	LN-TH 200d
3	2	Loader	b	4746	4.544	32.63	0.594	23.93	0.2175	595.12	0.0613	0.0814	1.596	11.78
3	2	Commander	c	16342	9.427	58.05	1.324	95.59	0.4849	1063.18	0.1367	0.1815	4.422	20.85
3	2	Driver	c	8566	7.291	51.63	1.152	41.39	0.4223	836.62	0.1189	0.1579	2.948	17.41
3	2	Gunner	c	48135	17.812	85.14	2.628	298.04	0.9619	1628.35	0.2715	0.3601	10.967	32.82
3	2	Loader	c	6156	5.935	42.82	0.787	30.24	0.2885	773.00	0.0813	0.1079	2.067	14.89
3	2	Commander	d	17014	9.662	59.38	1.369	98.53	0.5016	1081.93	0.1414	0.1877	4.563	21.11
3	2	Driver	d	9166	7.520	53.03	1.207	43.73	0.4428	850.81	0.1247	0.1656	3.094	17.68
3	2	Gunner	d	51208	18.373	86.89	2.736	311.86	1.0012	1653.11	0.2825	0.3749	11.441	33.21
3	2	Loader	d	6611	6.230	44.90	0.860	31.87	0.3150	794.44	0.0888	0.1178	2.230	15.29
3	2	Commander	e	1921	1.408	9.54	0.178	10.13	0.0653	181.49	0.0184	0.0245	0.517	3.24
3	2	Driver	e	1692	1.615	11.70	0.209	7.96	0.0765	213.34	0.0216	0.0286	0.512	4.10
3	2	Gunner	e	12520	3.633	13.36	0.568	79.93	0.2079	274.51	0.0587	0.0778	2.637	5.28
3	2	Loader	e	958	1.078	8.00	0.121	4.47	0.0443	154.50	0.0125	0.0166	0.300	2.68
4	4	Driver	a	116	0.074	0.54	0.020	0.30	0.0074	5.06	0.0021	0.0028	0.041	0.11
4	4	Driver	b	539	0.384	2.95	0.089	1.02	0.0328	37.56	0.0092	0.0123	0.170	0.65
4	4	Driver	c	1081	0.895	6.67	0.146	3.49	0.0537	112.42	0.0151	0.0201	0.325	1.51
4	4	Driver	d	1351	1.050	7.59	0.169	5.20	0.0621	130.51	0.0175	0.0232	0.396	1.72
4	4	Driver	e	305	0.362	2.79	0.036	1.03	0.0132	56.29	0.0037	0.0049	0.092	0.66

Table A.6. Summary of E(50) and $H_T(50)$ as Calculated Using the SID Code for Selected Cases (2.6 Lpm)

Phase	Shot	Position	Scenario	Intake Bq	E(50) rem	$H_T(50)$, rem						
						Lung	Bone Surf	ET	Kidneys	LN-TH	Red Marrow	Liver
I	1	Commander	a	2906.364	1.756	12.853	0.849	6.733	0.311	12.036	0.088	0.116
I	1	Commander	b	5359.287	3.447	25.923	1.724	9.766	0.631	22.949	0.178	0.236
I	1	Commander	c	7415.080	4.823	36.543	2.470	12.271	0.904	31.119	0.255	0.339
I	1	Commander	d	7703.313	5.046	38.281	2.588	12.589	0.948	32.567	0.267	0.355
I	1	Commander	e	1637.149	1.110	8.551	0.592	2.082	0.217	6.456	0.061	0.081
I	1	Driver	a	669.041	0.219	1.345	0.108	2.064	0.039	1.446	0.011	0.015
I	1	Driver	b	1977.679	0.937	6.596	0.464	4.820	0.170	6.268	0.048	0.064
I	1	Driver	c	2948.450	1.526	11.020	0.768	6.487	0.281	10.056	0.079	0.105
I	1	Driver	d	3309.481	1.758	12.804	0.888	6.987	0.325	11.557	0.092	0.122
I	1	Driver	e	583.267	0.356	2.663	0.181	1.071	0.066	2.286	0.019	0.025
II	3	Driver	a	305.691	0.143	1.070	0.075	0.425	0.027	0.794	0.008	0.010
II	3	Driver	b	525.287	0.236	1.770	0.127	0.652	0.046	1.459	0.013	0.017
II	3	Driver	c	1246.902	0.710	5.462	0.388	1.366	0.142	4.587	0.040	0.053
II	3	Driver	d	1825.882	1.127	8.712	0.615	1.948	0.225	7.254	0.063	0.084
II	3	Driver	e	261.892	0.185	1.465	0.101	0.163	0.037	1.208	0.010	0.014
III	2	Commander	a	8837.544	5.245	32.196	0.713	53.691	0.261	601.226	0.074	0.098
III	2	Commander	b	33237.804	13.553	68.946	1.959	206.860	0.717	1308.071	0.202	0.268
III	2	Commander	c	48172.052	17.800	85.086	2.636	297.754	0.965	1622.872	0.272	0.361
III	2	Commander	d	51247.298	18.369	86.882	2.745	311.623	1.005	1647.741	0.284	0.376
III	2	Commander	e	12531.025	3.630	13.351	0.569	79.842	0.208	273.660	0.059	0.078
III	2	Gunner	a	4931.842	3.525	23.155	0.485	28.817	0.178	413.674	0.050	0.066
III	2	Gunner	b	13974.759	7.800	47.148	1.114	83.254	0.408	858.013	0.115	0.153
III	2	Gunner	c	16357.466	9.426	58.028	1.330	95.582	0.487	1059.655	0.137	0.182
III	2	Gunner	d	17029.897	9.666	59.386	1.377	98.565	0.504	1078.491	0.142	0.189
III	2	Gunner	e	1922.989	1.407	9.535	0.179	10.129	0.065	180.938	0.018	0.024

A.7. Uranium Concentrations by Particle Size

se	Shot	Position	CI stage	Median Dia (µm)	MidTime (min)	U Conc, (µg/m3)	sd U Conc. (µg/m3)	SampleID
	1	Commander	1	30.70	0.25	1.71E+06	2.91E+05	PI-1I-C-1-CI-1
	1	Commander	1	30.86	1	1.91E+05	1.01E+05	PI-1I-C-2-CI-1
	1	Commander	1	31.02	2.5	1.25E+04	3.94E+04	PI-1I-C-3-CI-1
	1	Commander	1	31.59	5.5	-7.05E+04	3.52E+04	PI-1I-C-4-CI-1
	1	Commander	1	31.37	11.5	-3.34E+04	1.67E+04	PI-1I-C-5-CI-1
	1	Commander	1	31.35	23.5	-1.24E+04	6.19E+03	PI-1I-C-6-CI-1
	1	Commander	1	30.79	47.5	-2.87E+03	1.94E+03	PI-1I-C-7-CI-1
	1	Commander	1	31.19	95.5	2.60E+01	1.21E+03	PI-1I-C-8-CI-1
	1	Commander	2	15.73	0.25	3.97E+05	1.61E+05	PI-1I-C-1-CI-2
	1	Commander	2	15.90	1	1.40E+05	6.04E+04	PI-1I-C-2-CI-2
	1	Commander	2	16.06	2.5	2.63E+04	1.27E+04	PI-1I-C-3-CI-2
	1	Commander	2	16.66	5.5	1.02E+04	5.25E+03	PI-1I-C-4-CI-2
	1	Commander	2	16.43	11.5	8.37E+02	9.72E+02	PI-1I-C-5-CI-2
	1	Commander	2	16.41	23.5	-2.94E+02	2.79E+02	PI-1I-C-6-CI-2
	1	Commander	2	15.82	47.5	2.33E+02	2.38E+02	PI-1I-C-7-CI-2
	1	Commander	2	16.24	95.5	1.23E+02	1.26E+02	PI-1I-C-8-CI-2
	1	Commander	3	10.53	0.25	3.13E+05	6.65E+04	PI-1I-C-1-CI-3
	1	Commander	3	10.64	1	6.44E+04	1.41E+04	PI-1I-C-2-CI-3
	1	Commander	3	10.75	2.5	1.88E+04	4.25E+03	PI-1I-C-3-CI-3
	1	Commander	3	11.15	5.5	3.89E+04	8.37E+03	PI-1I-C-4-CI-3
	1	Commander	3	11.00	11.5	1.10E+04	2.40E+03	PI-1I-C-5-CI-3
	1	Commander	3	10.98	23.5	4.87E+02	1.42E+02	PI-1I-C-6-CI-3
	1	Commander	3	10.59	47.5	2.27E+02	6.73E+01	PI-1I-C-7-CI-3
	1	Commander	3	10.87	95.5	1.03E+02	3.34E+01	PI-1I-C-8-CI-3
	1	Commander	4	6.71	0.25	4.91E+05	1.39E+05	PI-1I-C-1-CI-4
	1	Commander	4	6.79	1	2.09E+05	5.96E+04	PI-1I-C-2-CI-4
	1	Commander	4	6.86	2.5	1.22E+04	3.73E+03	PI-1I-C-3-CI-4
	1	Commander	4	7.11	5.5	1.18E+04	3.48E+03	PI-1I-C-4-CI-4
	1	Commander	4	7.01	11.5	7.96E+03	2.33E+03	PI-1I-C-5-CI-4
	1	Commander	4	7.00	23.5	2.65E+03	7.89E+02	PI-1I-C-6-CI-4
	1	Commander	4	6.75	47.5	7.79E+02	2.35E+02	PI-1I-C-7-CI-4
	1	Commander	4	6.93	95.5	6.71E+02	1.99E+02	PI-1I-C-8-CI-4
	1	Commander	5	4.03	0.25	5.46E+05	2.14E+05	PI-1I-C-1-CI-5
	1	Commander	5	4.08	1	2.09E+05	8.24E+04	PI-1I-C-2-CI-5
	1	Commander	5	4.12	2.5	8.50E+04	3.37E+04	PI-1I-C-3-CI-5
	1	Commander	5	4.27	5.5	7.76E+04	3.06E+04	PI-1I-C-4-CI-5
	1	Commander	5	4.21	11.5	2.16E+04	8.58E+03	PI-1I-C-5-CI-5
	1	Commander	5	4.21	23.5	5.90E+03	2.35E+03	PI-1I-C-6-CI-5
	1	Commander	5	4.06	47.5	3.20E+03	1.27E+03	PI-1I-C-7-CI-5
	1	Commander	5	4.17	95.5	1.70E+03	6.73E+02	PI-1I-C-8-CI-5
	1	Commander	6	2.02	0.25	4.09E+05	1.62E+05	PI-1I-C-1-CI-6
	1	Commander	6	2.04	1	2.02E+05	8.07E+04	PI-1I-C-2-CI-6

Table A.7. *(continued)*

Phase	Shot	Position	CI stage	Median Dia (um)	MidTime (min)	U Conc, (ug/m3)	sd U Conc. (ug/m3)	SampleII
1	1	Commander	6	2.06	2.5	6.57E+04	NA[(a)]	PI-1I-C-3-C
1	1	Commander	6	2.14	5.5	3.70E+04	1.49E+04	PI-1I-C-4-C
1	1	Commander	6	2.11	11.5	1.09E+04	4.48E+03	PI-1I-C-5-C
1	1	Commander	6	2.11	23.5	4.32E+03	1.73E+03	PI-1I-C-6-C
1	1	Commander	6	2.03	47.5	1.83E+03	7.37E+02	PI-1I-C-7-C
1	1	Commander	6	2.09	95.5	9.34E+02	3.76E+02	PI-1I-C-8-C
1	1	Commander	7	1.04	0.25	2.84E+05	6.93E+04	PI-1I-C-1-C
1	1	Commander	7	1.05	1	1.75E+05	4.28E+04	PI-1I-C-2-C
1	1	Commander	7	1.06	2.5	1.25E+05	3.03E+04	PI-1I-C-3-C
1	1	Commander	7	1.10	5.5	4.62E+04	1.13E+04	PI-1I-C-4-C
1	1	Commander	7	1.09	11.5	2.28E+04	5.57E+03	PI-1I-C-5-C
1	1	Commander	7	1.08	23.5	5.27E+03	1.31E+03	PI-1I-C-6-C
1	1	Commander	7	1.05	47.5	1.58E+03	4.07E+02	PI-1I-C-7-C
1	1	Commander	7	1.07	95.5	1.21E+03	3.03E+02	PI-1I-C-8-C
1	1	Commander	8	0.61	0.25	2.54E+05	5.83E+04	PI-1I-C-1-C
1	1	Commander	8	0.62	1	3.47E+05	7.73E+04	PI-1I-C-2-C
1	1	Commander	8	0.62	2.5	1.71E+05	3.79E+04	PI-1I-C-3-C
1	1	Commander	8	0.65	5.5	8.68E+04	1.94E+04	PI-1I-C-4-C
1	1	Commander	8	0.64	11.5	3.11E+04	6.93E+03	PI-1I-C-5-C
1	1	Commander	8	0.64	23.5	9.29E+03	2.03E+03	PI-1I-C-6-C
1	1	Commander	8	0.61	47.5	2.27E+03	5.23E+02	PI-1I-C-7-C
1	1	Commander	8	0.63	95.5	1.61E+03	3.65E+02	PI-1I-C-8-C
1	1	Commander	9	0.22	0.25	1.06E+06	3.27E+05	PI-1I-C-1-C
1	1	Commander	9	0.22	1	5.89E+05	1.83E+05	PI-1I-C-2-C
1	1	Commander	9	0.22	2.5	1.69E+05	5.31E+04	PI-1I-C-3-C
1	1	Commander	9	0.22	5.5	1.22E+05	3.79E+04	PI-1I-C-4-C
1	1	Commander	9	0.22	11.5	1.44E+04	4.77E+03	PI-1I-C-5-C
1	1	Commander	9	0.22	23.5	5.85E+03	1.92E+03	PI-1I-C-6-C
1	1	Commander	9	0.22	47.5	8.61E+03	2.65E+03	PI-1I-C-7-C
1	1	Commander	9	0.22	95.5	6.12E+03	1.86E+03	PI-1I-C-8-C
1	1	Driver	1	31.02	0.25	8.66E+05	1.84E+05	PI-1I-D-1-C
1	1	Driver	1	31.58	1	2.19E+05	8.92E+04	PI-1I-D-2-C
1	1	Driver	1	30.75	2.5	1.31E+05	4.66E+04	PI-1I-D-3-C
1	1	Driver	1	30.68	5.5	1.84E+04	1.41E+04	PI-1I-D-4-C
1	1	Driver	1	30.99	11.5	5.28E+03	6.55E+03	PI-1I-D-5-C
1	1	Driver	1	30.79	23.5	6.50E+03	3.91E+03	PI-1I-D-6-C
1	1	Driver	1	30.98	47.5	3.31E+03	2.00E+03	PI-1I-D-7-C
1	1	Driver	1	30.76	95.5	4.90E+03	1.62E+03	PI-1I-D-8-C
1	1	Driver	2	16.07	0.25	3.34E+04	1.64E+04	PI-1I-D-1-C
1	1	Driver	2	16.65	1	4.59E+04	2.50E+04	PI-1I-D-2-C
1	1	Driver	2	15.79	2.5	1.81E+04	1.01E+04	PI-1I-D-3-C
1	1	Driver	2	15.71	5.5	2.38E+03	1.96E+03	PI-1I-D-4-C

e **A.7**. *(continued)*

ase	Shot	Position	CI stage	Median Dia (um)	MidTime (min)	U Conc, (ug/m3)	sd U Conc. (ug/m3)	SampleID
	1	Driver	2	16.03	11.5	5.96E+02	7.30E+02	PI-1I-D-5-CI-2
	1	Driver	2	15.82	23.5	-9.89E+01	2.16E+02	PI-1I-D-6-CI-2
	1	Driver	2	16.03	47.5	9.19E+00	1.25E+02	PI-1I-D-7-CI-2
	1	Driver	2	15.79	95.5	4.31E+02	2.53E+02	PI-1I-D-8-CI-2
	1	Driver	3	10.76	0.25	6.03E+03	2.30E+03	PI-1I-D-1-CI-3
	1	Driver	3	11.14	1	6.08E+04	1.71E+04	PI-1I-D-2-CI-3
	1	Driver	3	10.57	2.5	3.56E+04	9.94E+03	PI-1I-D-3-CI-3
	1	Driver	3	10.52	5.5	1.10E+04	3.13E+03	PI-1I-D-4-CI-3
	1	Driver	3	10.73	11.5	3.79E+03	1.11E+03	PI-1I-D-5-CI-3
	1	Driver	3	10.59	23.5	1.18E+03	3.52E+02	PI-1I-D-6-CI-3
	1	Driver	3	10.73	47.5	3.54E+02	1.13E+02	PI-1I-D-7-CI-3
	1	Driver	3	10.57	95.5	1.87E+02	5.97E+01	PI-1I-D-8-CI-3
	1	Driver	4	6.86	0.25	1.85E+04	7.14E+03	PI-1I-D-1-CI-4
	1	Driver	4	7.11	1	8.97E+04	3.64E+04	PI-1I-D-2-CI-4
	1	Driver	4	6.74	2.5	2.59E+04	1.09E+04	PI-1I-D-3-CI-4
	1	Driver	4	6.71	5.5	1.20E+04	5.11E+03	PI-1I-D-4-CI-4
	1	Driver	4	6.84	11.5	3.82E+03	1.75E+03	PI-1I-D-5-CI-4
	1	Driver	4	6.75	23.5	9.25E+02	5.06E+02	PI-1I-D-6-CI-4
	1	Driver	4	6.84	47.5	3.10E+02	2.01E+02	PI-1I-D-7-CI-4
	1	Driver	4	6.74	95.5	3.57E+02	1.72E+02	PI-1I-D-8-CI-4
	1	Driver	5	4.12	0.25	1.98E+04	8.03E+03	PI-1I-D-1-CI-5
	1	Driver	5	4.27	1	4.87E+04	2.29E+04	PI-1I-D-2-CI-5
	1	Driver	5	4.05	2.5	6.01E+04	2.53E+04	PI-1I-D-3-CI-5
	1	Driver	5	4.03	5.5	2.70E+04	1.14E+04	PI-1I-D-4-CI-5
	1	Driver	5	4.11	11.5	5.13E+03	2.45E+03	PI-1I-D-5-CI-5
	1	Driver	5	4.06	23.5	2.50E+03	1.19E+03	PI-1I-D-6-CI-5
	1	Driver	5	4.11	47.5	1.51E+03	7.02E+02	PI-1I-D-7-CI-5
	1	Driver	5	4.05	95.5	1.19E+03	5.20E+02	PI-1I-D-8-CI-5
	1	Driver	6	2.06	0.25	1.72E+04	4.69E+03	PI-1I-D-1-CI-6
	1	Driver	6	2.14	1	8.09E+04	2.28E+04	PI-1I-D-2-CI-6
	1	Driver	6	2.03	2.5	2.61E+04	7.64E+03	PI-1I-D-3-CI-6
	1	Driver	6	2.02	5.5	1.67E+04	4.73E+03	PI-1I-D-4-CI-6
	1	Driver	6	2.06	11.5	4.40E+03	1.37E+03	PI-1I-D-5-CI-6
	1	Driver	6	2.03	23.5	2.56E+03	7.75E+02	PI-1I-D-6-CI-6
	1	Driver	6	2.06	47.5	7.70E+02	2.58E+02	PI-1I-D-7-CI-6
	1	Driver	6	2.03	95.5	8.05E+02	2.36E+02	PI-1I-D-8-CI-6
	1	Driver	7	1.06	0.25	1.54E+04	3.54E+03	PI-1I-D-1-CI-7
	1	Driver	7	1.10	1	5.22E+04	1.10E+04	PI-1I-D-2-CI-7
	1	Driver	7	1.04	2.5	4.28E+04	8.63E+03	PI-1I-D-3-CI-7
	1	Driver	7	1.04	5.5	2.05E+04	4.14E+03	PI-1I-D-4-CI-7
	1	Driver	7	1.06	11.5	5.89E+03	1.25E+03	PI-1I-D-5-CI-7
	1	Driver	7	1.05	23.5	2.33E+03	4.97E+02	PI-1I-D-6-CI-7

Table A.7. *(continued)*

Phase	Shot	Position	CI stage	Median Dia (um)	MidTime (min)	U Conc, (ug/m3)	sd U Conc. (ug/m3)	SampleID
1	1	Driver	7	1.06	47.5	1.52E+03	3.20E+02	PI-1I-D-7-C
1	1	Driver	7	1.04	95.5	1.05E+03	2.15E+02	PI-1I-D-8-C
1	1	Driver	8	0.62	0.25	1.19E+04	2.14E+03	PI-1I-D-1-C
1	1	Driver	8	0.65	1	7.78E+04	1.16E+04	PI-1I-D-2-C
1	1	Driver	8	0.61	2.5	9.59E+04	1.37E+04	PI-1I-D-3-C
1	1	Driver	8	0.61	5.5	2.53E+04	3.67E+03	PI-1I-D-4-C
1	1	Driver	8	0.62	11.5	7.74E+03	1.16E+03	PI-1I-D-5-C
1	1	Driver	8	0.61	23.5	2.95E+03	4.53E+02	PI-1I-D-6-C
1	1	Driver	8	0.62	47.5	1.88E+03	2.83E+02	PI-1I-D-7-C
1	1	Driver	8	0.61	95.5	1.38E+03	2.03E+02	PI-1I-D-8-C
1	1	Driver	9	0.22	0.25	1.41E+04	2.91E+03	PI-1I-D-1-C
1	1	Driver	9	0.22	1	2.31E+05	4.62E+04	PI-1I-D-2-C
1	1	Driver	9	0.22	2.5	4.99E+04	1.01E+04	PI-1I-D-3-C
1	1	Driver	9	0.22	5.5	9.84E+03	2.02E+03	PI-1I-D-4-C
1	1	Driver	9	0.22	11.5	1.00E+04	2.02E+03	PI-1I-D-5-C
1	1	Driver	9	0.22	23.5	9.98E+03	1.99E+03	PI-1I-D-6-C
1	1	Driver	9	0.22	47.5	8.03E+03	1.60E+03	PI-1I-D-7-C
1	1	Driver	9	0.22	95.5	7.08E+03	1.40E+03	PI-1I-D-8-C
1	1	Gunner	1	30.84	0.25	1.25E+06	2.14E+05	PI-1I-G-1-C
1	1	Gunner	1	31.31	1	1.20E+05	7.83E+04	PI-1I-G-2-C
1	1	Gunner	1	31.00	2.5	1.18E+05	4.67E+04	PI-1I-G-3-C
1	1	Gunner	1	30.98	5.5	1.37E+04	1.66E+04	PI-1I-G-4-C
1	1	Gunner	1	30.75	11.5	-2.36E+04	1.18E+04	PI-1I-G-5-C
1	1	Gunner	1	30.72	23.5	-1.13E+04	5.66E+03	PI-1I-G-6-C
1	1	Gunner	1	31.61	47.5	1.05E+04	3.56E+03	PI-1I-G-7-C
1	1	Gunner	1	31.11	95.5	2.86E+03	1.35E+03	PI-1I-G-8-C
1	1	Gunner	2	15.88	0.25	1.90E+05	7.74E+04	PI-1I-G-1-C
1	1	Gunner	2	16.36	1	2.08E+05	8.89E+04	PI-1I-G-2-C
1	1	Gunner	2	16.05	2.5	5.76E+04	2.55E+04	PI-1I-G-3-C
1	1	Gunner	2	16.02	5.5	1.56E+04	7.43E+03	PI-1I-G-4-C
1	1	Gunner	2	15.79	11.5	1.36E+02	7.47E+02	PI-1I-G-5-C
1	1	Gunner	2	15.75	23.5	-1.25E+02	3.15E+02	PI-1I-G-6-C
1	1	Gunner	2	16.68	47.5	5.95E+01	2.10E+02	PI-1I-G-7-C
1	1	Gunner	2	16.15	95.5	1.09E+02	1.26E+02	PI-1I-G-8-C
1	1	Gunner	3	10.63	0.25	1.96E+05	4.18E+04	PI-1I-G-1-C
1	1	Gunner	3	10.95	1	2.05E+05	4.42E+04	PI-1I-G-2-C
1	1	Gunner	3	10.74	2.5	3.31E+04	7.31E+03	PI-1I-G-3-C
1	1	Gunner	3	10.73	5.5	6.76E+03	1.58E+03	PI-1I-G-4-C
1	1	Gunner	3	10.57	11.5	5.46E+03	1.24E+03	PI-1I-G-5-C
1	1	Gunner	3	10.54	23.5	9.39E+02	2.38E+02	PI-1I-G-6-C
1	1	Gunner	3	11.17	47.5	2.05E+02	6.65E+01	PI-1I-G-7-C
1	1	Gunner	3	10.81	95.5	1.20E+02	3.57E+01	PI-1I-G-8-C

A.7. *(continued)*

se	Shot	Position	CI stage	Median Dia (um)	MidTime (min)	U Conc, (ug/m3)	sd U Conc. (ug/m3)	SampleID
	1	Gunner	4	6.78	0.25	3.23E+05	9.21E+04	PI-1I-G-1-CI-4
	1	Gunner	4	6.98	1	1.98E+05	5.68E+04	PI-1I-G-2-CI-4
	1	Gunner	4	6.85	2.5	1.90E+04	5.64E+03	PI-1I-G-3-CI-4
	1	Gunner	4	6.84	5.5	8.58E+03	2.56E+03	PI-1I-G-4-CI-4
	1	Gunner	4	6.74	11.5	4.00E+03	1.20E+03	PI-1I-G-5-CI-4
	1	Gunner	4	6.72	23.5	1.41E+03	4.30E+02	PI-1I-G-6-CI-4
	1	Gunner	4	7.12	47.5	4.80E+02	1.53E+02	PI-1I-G-7-CI-4
	1	Gunner	4	6.89	95.5	3.22E+02	1.00E+02	PI-1I-G-8-CI-4
	1	Gunner	5	4.07	0.25	3.70E+05	1.45E+05	PI-1I-G-1-CI-5
	1	Gunner	5	4.20	1	2.47E+05	9.69E+04	PI-1I-G-2-CI-5
	1	Gunner	5	4.12	2.5	4.20E+04	1.66E+04	PI-1I-G-3-CI-5
	1	Gunner	5	4.11	5.5	2.98E+04	1.17E+04	PI-1I-G-4-CI-5
	1	Gunner	5	4.05	11.5	1.01E+04	4.02E+03	PI-1I-G-5-CI-5
	1	Gunner	5	4.04	23.5	2.06E+03	8.31E+02	PI-1I-G-6-CI-5
	1	Gunner	5	4.28	47.5	1.04E+03	4.27E+02	PI-1I-G-7-CI-5
	1	Gunner	5	4.14	95.5	7.76E+02	3.10E+02	PI-1I-G-8-CI-5
	1	Gunner	6	2.04	0.25	4.21E+05	1.62E+05	PI-1I-G-1-CI-6
	1	Gunner	6	2.10	1	2.08E+05	8.06E+04	PI-1I-G-2-CI-6
	1	Gunner	6	2.06	2.5	7.88E+04	3.08E+04	PI-1I-G-3-CI-6
	1	Gunner	6	2.06	5.5	2.82E+04	1.10E+04	PI-1I-G-4-CI-6
	1	Gunner	6	2.03	11.5	1.05E+04	4.12E+03	PI-1I-G-5-CI-6
	1	Gunner	6	2.02	23.5	2.58E+03	1.05E+03	PI-1I-G-6-CI-6
	1	Gunner	6	2.14	47.5	1.40E+03	5.64E+02	PI-1I-G-7-CI-6
	1	Gunner	6	2.08	95.5	8.45E+02	3.37E+02	PI-1I-G-8-CI-6
	1	Gunner	7	1.05	0.25	2.58E+05	6.21E+04	PI-1I-G-1-CI-7
	1	Gunner	7	1.08	1	1.31E+05	3.18E+04	PI-1I-G-2-CI-7
	1	Gunner	7	1.06	2.5	9.03E+04	2.17E+04	PI-1I-G-3-CI-7
	1	Gunner	7	1.06	5.5	5.37E+04	1.29E+04	PI-1I-G-4-CI-7
	1	Gunner	7	1.04	11.5	1.53E+04	3.70E+03	PI-1I-G-5-CI-7
	1	Gunner	7	1.04	23.5	3.42E+03	8.59E+02	PI-1I-G-6-CI-7
	1	Gunner	7	1.10	47.5	1.28E+03	3.35E+02	PI-1I-G-7-CI-7
	1	Gunner	7	1.07	95.5	9.00E+02	2.25E+02	PI-1I-G-8-CI-7
	1	Gunner	8	0.62	0.25	6.84E+05	1.46E+05	PI-1I-G-1-CI-8
	1	Gunner	8	0.64	1	1.71E+05	3.76E+04	PI-1I-G-2-CI-8
	1	Gunner	8	0.62	2.5	1.95E+05	4.14E+04	PI-1I-G-3-CI-8
	1	Gunner	8	0.62	5.5	7.70E+04	1.64E+04	PI-1I-G-4-CI-8
	1	Gunner	8	0.61	11.5	2.98E+04	6.38E+03	PI-1I-G-5-CI-8
	1	Gunner	8	0.61	23.5	6.80E+03	1.48E+03	PI-1I-G-6-CI-8
	1	Gunner	8	0.65	47.5	3.78E+03	8.36E+02	PI-1I-G-7-CI-8
	1	Gunner	8	0.63	95.5	2.18E+03	4.80E+02	PI-1I-G-8-CI-8
	1	Gunner	9	0.22	0.25	1.88E+06	NA[(a)]	PI-1I-G-1-CI-9
	1	Gunner	9	0.22	1	4.59E+05	1.38E+05	PI-1I-G-2-CI-9

Table A.7. *(continued)*

Phase	Shot	Position	CI stage	Median Dia (um)	MidTime (min)	U Conc, (ug/m3)	sd U Conc. (ug/m3)	SampleID
1	1	Gunner	9	0.22	2.5	1.84E+05	5.56E+04	PI-1I-G-3-C
1	1	Gunner	9	0.22	5.5	5.50E+04	1.68E+04	PI-1I-G-4-C
1	1	Gunner	9	0.22	11.5	3.45E+04	1.05E+04	PI-1I-G-5-C
1	1	Gunner	9	0.22	23.5	1.15E+04	3.56E+03	PI-1I-G-6-C
1	1	Gunner	9	0.22	47.5	6.19E+03	1.91E+03	PI-1I-G-7-C
1	1	Gunner	9	0.22	95.5	5.99E+03	1.81E+03	PI-1I-G-8-C
1	1	Loader	1	34.40	0.25	7.22E+05	1.25E+05	PI-1I-L-1-C
1	1	Loader	1	31.31	1	1.57E+05	3.79E+04	PI-1I-L-2-C
1	1	Loader	1	32.09	2.5	3.37E+04	1.23E+04	PI-1I-L-3-C
1	1	Loader	1	31.24	5.5	4.11E+02	3.32E+03	PI-1I-L-4-C
1	1	Loader	1	31.09	11.5	-2.46E+03	1.31E+03	PI-1I-L-5-C
1	1	Loader	1	31.10	23.5	-1.56E+03	7.82E+02	PI-1I-L-6-C
1	1	Loader	1	31.38	47.5	2.30E+03	7.53E+02	PI-1I-L-7-C
1	1	Loader	1	31.19	95.5	1.37E+03	4.10E+02	PI-1I-L-8-C
1	1	Loader	2	19.75	0.25	3.69E+05	1.50E+05	PI-1I-L-1-C
1	1	Loader	2	16.37	1	2.47E+05	1.01E+05	PI-1I-L-2-C
1	1	Loader	2	17.19	2.5	7.65E+04	3.16E+04	PI-1I-L-3-C
1	1	Loader	2	16.30	5.5	1.27E+04	5.42E+03	PI-1I-L-4-C
1	1	Loader	2	16.14	11.5	1.94E+03	9.36E+02	PI-1I-L-5-C
1	1	Loader	2	16.14	23.5	3.85E+01	1.13E+02	PI-1I-L-6-C
1	1	Loader	2	16.44	47.5	1.25E+02	8.82E+01	PI-1I-L-7-C
1	1	Loader	2	16.24	95.5	5.08E+01	4.28E+01	PI-1I-L-8-C
1	1	Loader	3	13.22	0.25	4.91E+05	1.05E+05	PI-1I-L-1-C
1	1	Loader	3	10.96	1	2.39E+05	5.11E+04	PI-1I-L-2-C
1	1	Loader	3	11.51	2.5	1.07E+05	2.30E+04	PI-1I-L-3-C
1	1	Loader	3	10.91	5.5	4.90E+04	1.05E+04	PI-1I-L-4-C
1	1	Loader	3	10.80	11.5	5.41E+03	1.19E+03	PI-1I-L-5-C
1	1	Loader	3	10.81	23.5	8.07E+02	1.97E+02	PI-1I-L-6-C
1	1	Loader	3	11.01	47.5	3.53E+02	8.86E+01	PI-1I-L-7-C
1	1	Loader	3	10.87	95.5	7.57E+01	2.39E+01	PI-1I-L-8-C
1	1	Loader	4	8.43	0.25	3.90E+05	1.12E+05	PI-1I-L-1-C
1	1	Loader	4	6.99	1	3.17E+05	9.05E+04	PI-1I-L-2-C
1	1	Loader	4	7.34	2.5	7.06E+04	2.02E+04	PI-1I-L-3-C
1	1	Loader	4	6.96	5.5	1.53E+04	4.40E+03	PI-1I-L-4-C
1	1	Loader	4	6.89	11.5	5.89E+03	1.71E+03	PI-1I-L-5-C
1	1	Loader	4	6.89	23.5	4.11E+03	1.19E+03	PI-1I-L-6-C
1	1	Loader	4	7.02	47.5	6.34E+02	1.87E+02	PI-1I-L-7-C
1	1	Loader	4	6.93	95.5	3.51E+02	1.04E+02	PI-1I-L-8-C
1	1	Loader	5	5.07	0.25	6.96E+05	2.70E+05	PI-1I-L-1-C
1	1	Loader	5	4.20	1	4.01E+05	1.56E+05	PI-1I-L-2-C
1	1	Loader	5	4.41	2.5	5.17E+04	2.03E+04	PI-1I-L-3-C
1	1	Loader	5	4.18	5.5	5.61E+04	2.19E+04	PI-1I-L-4-C

e **A.7**. *(continued)*

se	Shot	Position	CI stage	Median Dia (um)	MidTime (min)	U Conc, (ug/m3)	sd U Conc. (ug/m3)	SampleID
	1	Loader	5	4.14	11.5	1.82E+04	7.12E+03	PI-1I-L-5-CI-5
	1	Loader	5	4.14	23.5	7.68E+03	3.00E+03	PI-1I-L-6-CI-5
	1	Loader	5	4.22	47.5	9.59E+02	3.85E+02	PI-1I-L-7-CI-5
	1	Loader	5	4.16	95.5	2.70E+02	1.11E+02	PI-1I-L-8-CI-5
	1	Loader	6	2.54	0.25	7.20E+05	2.71E+05	PI-1I-L-1-CI-6
	1	Loader	6	2.10	1	2.21E+05	NA[a]	PI-1I-L-2-CI-6
	1	Loader	6	2.21	2.5	1.60E+04	6.85E+03	PI-1I-L-3-CI-6
	1	Loader	6	2.09	5.5	5.11E+04	1.95E+04	PI-1I-L-4-CI-6
	1	Loader	6	2.07	11.5	1.58E+04	6.12E+03	PI-1I-L-5-CI-6
	1	Loader	6	2.07	23.5	2.89E+03	1.19E+03	PI-1I-L-6-CI-6
	1	Loader	6	2.11	47.5	7.93E+02	3.49E+02	PI-1I-L-7-CI-6
	1	Loader	6	2.09	95.5	9.86E+02	3.88E+02	PI-1I-L-8-CI-6
	1	Loader	7	1.31	0.25	3.37E+05	7.99E+04	PI-1I-L-1-CI-7
	1	Loader	7	1.08	1	1.40E+05	3.35E+04	PI-1I-L-2-CI-7
	1	Loader	7	1.14	2.5	8.05E+04	1.92E+04	PI-1I-L-3-CI-7
	1	Loader	7	1.08	5.5	5.88E+04	1.39E+04	PI-1I-L-4-CI-7
	1	Loader	7	1.07	11.5	2.56E+04	6.06E+03	PI-1I-L-5-CI-7
	1	Loader	7	1.07	23.5	4.46E+03	1.10E+03	PI-1I-L-6-CI-7
	1	Loader	7	1.09	47.5	1.35E+03	3.48E+02	PI-1I-L-7-CI-7
	1	Loader	7	1.07	95.5	9.10E+02	2.27E+02	PI-1I-L-8-CI-7
	1	Loader	8	0.77	0.25	7.18E+05	1.49E+05	PI-1I-L-1-CI-8
	1	Loader	8	0.64	1	2.42E+05	5.05E+04	PI-1I-L-2-CI-8
	1	Loader	8	0.67	2.5	2.20E+05	4.54E+04	PI-1I-L-3-CI-8
	1	Loader	8	0.63	5.5	6.59E+04	1.38E+04	PI-1I-L-4-CI-8
	1	Loader	8	0.63	11.5	2.53E+04	5.30E+03	PI-1I-L-5-CI-8
	1	Loader	8	0.63	23.5	6.47E+03	1.39E+03	PI-1I-L-6-CI-8
	1	Loader	8	0.64	47.5	1.63E+03	3.86E+02	PI-1I-L-7-CI-8
	1	Loader	8	0.63	95.5	1.30E+03	2.87E+02	PI-1I-L-8-CI-8
	1	Loader	9	0.24	0.25	1.02E+06	2.97E+05	PI-1I-L-1-CI-9
	1	Loader	9	0.22	1	5.84E+05	1.70E+05	PI-1I-L-2-CI-9
	1	Loader	9	0.22	2.5	9.28E+04	2.78E+04	PI-1I-L-3-CI-9
	1	Loader	9	0.22	5.5	6.13E+04	1.81E+04	PI-1I-L-4-CI-9
	1	Loader	9	0.22	11.5	1.82E+04	5.47E+03	PI-1I-L-5-CI-9
	1	Loader	9	0.22	23.5	1.12E+04	3.35E+03	PI-1I-L-6-CI-9
	1	Loader	9	0.22	47.5	7.88E+03	2.34E+03	PI-1I-L-7-CI-9
	1	Loader	9	0.22	95.5	6.92E+03	1.97E+03	PI-1I-L-8-CI-9
	2	Commander	1	30.69	0.125	2.03E+06	4.33E+05	PI-2I-C-1-CI-1
	2	Commander	1	30.74	0.75	9.26E+04	9.85E+04	PI-2I-C-2-CI-1
	2	Commander	1	30.79	2	-2.43E+04	3.80E+04	PI-2I-C-3-CI-1
	2	Commander	1	31.29	4.5	2.49E+04	2.64E+04	PI-2I-C-4-CI-1
	2	Commander	1	31.28	9.5	8.51E+03	1.25E+04	PI-2I-C-5-CI-1
	2	Commander	1	31.38	23.5	5.86E+03	3.83E+03	PI-2I-C-6-CI-1

Table A.7. *(continued)*

Phase	Shot	Position	CI stage	Median Dia (um)	MidTime (min)	U Conc, (ug/m3)	sd U Conc. (ug/m3)	SampleID
1	2	Commander	1	30.84	55.5	1.13E+03	1.07E+03	PI-2I-C-7-C
1	2	Commander	1	31.12	95.5	1.01E+03	1.08E+03	PI-2I-C-8-C
1	2	Commander	2	15.73	0.125	4.68E+05	NA[a]	PI-2I-C-1-C
1	2	Commander	2	15.77	0.75	3.17E+05	1.66E+05	PI-2I-C-2-C
1	2	Commander	2	15.83	2	3.03E+04	2.22E+04	PI-2I-C-3-C
1	2	Commander	2	16.35	4.5	5.21E+03	6.96E+03	PI-2I-C-4-C
1	2	Commander	2	16.33	9.5	2.23E+03	3.28E+03	PI-2I-C-5-C
1	2	Commander	2	16.44	23.5	2.38E+02	7.01E+02	PI-2I-C-6-C
1	2	Commander	2	15.88	55.5	1.92E+02	2.67E+02	PI-2I-C-7-C
1	2	Commander	2	16.16	95.5	3.07E+02	3.19E+02	PI-2I-C-8-C
1	2	Commander	3	10.53	0.125	3.50E+05	9.60E+04	PI-2I-C-1-C
1	2	Commander	3	10.56	0.75	1.25E+05	3.83E+04	PI-2I-C-2-C
1	2	Commander	3	10.59	2	1.08E+05	3.17E+04	PI-2I-C-3-C
1	2	Commander	3	10.94	4.5	1.20E+04	4.42E+03	PI-2I-C-4-C
1	2	Commander	3	10.93	9.5	7.52E+03	2.62E+03	PI-2I-C-5-C
1	2	Commander	3	11.01	23.5	8.54E+02	3.88E+02	PI-2I-C-6-C
1	2	Commander	3	10.63	55.5	6.54E+02	2.24E+02	PI-2I-C-7-C
1	2	Commander	3	10.82	95.5	2.22E+02	1.12E+02	PI-2I-C-8-C
1	2	Commander	4	6.71	0.125	5.25E+05	1.99E+05	PI-2I-C-1-C
1	2	Commander	4	6.73	0.75	1.65E+05	6.50E+04	PI-2I-C-2-C
1	2	Commander	4	6.76	2	5.85E+04	2.34E+04	PI-2I-C-3-C
1	2	Commander	4	6.98	4.5	2.16E+04	8.80E+03	PI-2I-C-4-C
1	2	Commander	4	6.97	9.5	4.42E+03	2.04E+03	PI-2I-C-5-C
1	2	Commander	4	7.02	23.5	1.41E+03	6.25E+02	PI-2I-C-6-C
1	2	Commander	4	6.78	55.5	1.65E+03	6.48E+02	PI-2I-C-7-C
1	2	Commander	4	6.90	95.5	4.91E+02	2.15E+02	PI-2I-C-8-C
1	2	Commander	5	4.03	0.125	4.13E+05	1.63E+05	PI-2I-C-1-C
1	2	Commander	5	4.05	0.75	4.57E+05	1.82E+05	PI-2I-C-2-C
1	2	Commander	5	4.06	2	9.70E+04	3.91E+04	PI-2I-C-3-C
1	2	Commander	5	4.19	4.5	8.61E+04	3.44E+04	PI-2I-C-4-C
1	2	Commander	5	4.19	9.5	1.72E+04	6.99E+03	PI-2I-C-5-C
1	2	Commander	5	4.22	23.5	7.14E+03	2.87E+03	PI-2I-C-6-C
1	2	Commander	5	4.07	55.5	2.99E+03	1.20E+03	PI-2I-C-7-C
1	2	Commander	5	4.15	95.5	2.21E+03	8.91E+02	PI-2I-C-8-C
1	2	Commander	6	2.02	0.125	2.47E+05	6.42E+04	PI-2I-C-1-C
1	2	Commander	6	2.03	0.75	2.22E+05	5.80E+04	PI-2I-C-2-C
1	2	Commander	6	2.03	2	5.71E+04	1.51E+04	PI-2I-C-3-C
1	2	Commander	6	2.10	4.5	4.10E+04	1.08E+04	PI-2I-C-4-C
1	2	Commander	6	2.10	9.5	1.09E+04	2.90E+03	PI-2I-C-5-C
1	2	Commander	6	2.11	23.5	3.08E+03	8.23E+02	PI-2I-C-6-C
1	2	Commander	6	2.04	55.5	2.10E+03	5.50E+02	PI-2I-C-7-C
1	2	Commander	6	2.08	95.5	1.14E+03	3.02E+02	PI-2I-C-8-C

e A.7. *(continued)*

se	Shot	Position	CI stage	Median Dia (um)	MidTime (min)	U Conc, (ug/m3)	sd U Conc. (ug/m3)	SampleID
	2	Commander	7	1.04	0.125	1.89E+05	3.74E+04	PI-2I-C-1-CI-7
	2	Commander	7	1.04	0.75	1.74E+05	3.47E+04	PI-2I-C-2-CI-7
	2	Commander	7	1.05	2	6.71E+04	1.35E+04	PI-2I-C-3-CI-7
	2	Commander	7	1.08	4.5	6.22E+04	1.22E+04	PI-2I-C-4-CI-7
	2	Commander	7	1.08	9.5	9.62E+03	2.00E+03	PI-2I-C-5-CI-7
	2	Commander	7	1.09	23.5	3.68E+03	7.52E+02	PI-2I-C-6-CI-7
	2	Commander	7	1.05	55.5	2.39E+03	4.73E+02	PI-2I-C-7-CI-7
	2	Commander	7	1.07	95.5	1.96E+03	3.91E+02	PI-2I-C-8-CI-7
	2	Commander	8	0.61	0.125	3.50E+05	4.95E+04	PI-2I-C-1-CI-8
	2	Commander	8	0.61	0.75	2.23E+05	3.22E+04	PI-2I-C-2-CI-8
	2	Commander	8	0.61	2	9.84E+04	1.43E+04	PI-2I-C-3-CI-8
	2	Commander	8	0.63	4.5	3.96E+04	5.82E+03	PI-2I-C-4-CI-8
	2	Commander	8	0.63	9.5	9.72E+03	1.50E+03	PI-2I-C-5-CI-8
	2	Commander	8	0.64	23.5	9.10E+03	1.31E+03	PI-2I-C-6-CI-8
	2	Commander	8	0.62	55.5	3.26E+03	4.68E+02	PI-2I-C-7-CI-8
	2	Commander	8	0.63	95.5	1.74E+03	2.57E+02	PI-2I-C-8-CI-8
	2	Commander	9	0.22	0.125	1.27E+06	2.52E+05	PI-2I-C-1-CI-9
	2	Commander	9	0.22	0.75	7.59E+05	1.53E+05	PI-2I-C-2-CI-9
	2	Commander	9	0.22	2	2.30E+05	4.70E+04	PI-2I-C-3-CI-9
	2	Commander	9	0.22	4.5	1.27E+05	2.61E+04	PI-2I-C-4-CI-9
	2	Commander	9	0.22	9.5	2.88E+04	6.11E+03	PI-2I-C-5-CI-9
	2	Commander	9	0.22	23.5	1.41E+04	2.90E+03	PI-2I-C-6-CI-9
	2	Commander	9	0.22	55.5	1.10E+04	2.22E+03	PI-2I-C-7-CI-9
	2	Commander	9	0.22	95.5	1.02E+04	2.05E+03	PI-2I-C-8-CI-9
	2	Driver	1	30.95	0.125	1.47E+06	2.49E+05	PI-2I-D-1-CI-1
	2	Driver	1	30.62	2	3.29E+04	2.67E+04	PI-2I-D-3-CI-1
	2	Driver	1	30.68	4.5	-3.98E+03	1.07E+04	PI-2I-D-4-CI-1
	2	Driver	1	30.81	9.5	2.29E+04	9.00E+03	PI-2I-D-5-CI-1
	2	Driver	1	30.79	23.5	5.53E+03	2.21E+03	PI-2I-D-6-CI-1
	2	Driver	1	31.11	55.5	1.89E+03	7.66E+02	PI-2I-D-7-CI-1
	2	Driver	1	30.76	95.5	5.00E+02	5.40E+02	PI-2I-D-8-CI-1
	2	Driver	2	15.99	0.125	8.32E+05	3.40E+05	PI-2I-D-1-CI-2
	2	Driver	2	15.66	2	3.23E+04	1.42E+04	PI-2I-D-3-CI-2
	2	Driver	2	15.71	4.5	3.36E+03	2.02E+03	PI-2I-D-4-CI-2
	2	Driver	2	15.85	9.5	5.93E+03	2.68E+03	PI-2I-D-5-CI-2
	2	Driver	2	15.82	23.5	-1.19E+02	NA[a]	PI-2I-D-6-CI-2
	2	Driver	2	16.16	55.5	1.21E+02	7.57E+01	PI-2I-D-7-CI-2
	2	Driver	2	15.79	95.5	6.40E+01	5.46E+01	PI-2I-D-8-CI-2
	2	Driver	3	10.70	0.125	2.81E+05	5.93E+04	PI-2I-D-1-CI-3
	2	Driver	3	10.48	2	8.01E+04	1.72E+04	PI-2I-D-3-CI-3
	2	Driver	3	10.51	4.5	2.71E+04	5.91E+03	PI-2I-D-4-CI-3
	2	Driver	3	10.61	9.5	1.94E+03	5.68E+02	PI-2I-D-5-CI-3

Table A.7. *(continued)*

Phase	Shot	Position	CI stage	Median Dia (um)	MidTime (min)	U Conc, (ug/m3)	sd U Conc. (ug/m3)	SampleID
1	2	Driver	3	10.59	23.5	1.67E+03	3.82E+02	PI-2I-D-6-C
1	2	Driver	3	10.81	55.5	1.01E+03	2.25E+02	PI-2I-D-7-C
1	2	Driver	3	10.57	95.5	2.01E+02	5.76E+01	PI-2I-D-8-C
1	2	Driver	4	6.82	0.125	5.81E+05	1.62E+05	PI-2I-D-1-C
1	2	Driver	4	6.68	2	6.19E+04	1.76E+04	PI-2I-D-3-C
1	2	Driver	4	6.70	4.5	1.41E+04	4.07E+03	PI-2I-D-4-C
1	2	Driver	4	6.77	9.5	2.48E+03	7.85E+02	PI-2I-D-5-C
1	2	Driver	4	6.75	23.5	2.78E+03	7.89E+02	PI-2I-D-6-C
1	2	Driver	4	6.90	55.5	7.15E+02	2.06E+02	PI-2I-D-7-C
1	2	Driver	4	6.74	95.5	3.02E+02	9.21E+01	PI-2I-D-8-C
1	2	Driver	5	4.10	0.125	5.92E+05	2.41E+05	PI-2I-D-1-C
1	2	Driver	5	4.02	2	1.00E+05	4.22E+04	PI-2I-D-3-C
1	2	Driver	5	4.03	4.5	2.08E+04	9.25E+03	PI-2I-D-4-C
1	2	Driver	5	4.07	9.5	6.06E+03	2.84E+03	PI-2I-D-5-C
1	2	Driver	5	4.06	23.5	2.36E+03	1.06E+03	PI-2I-D-6-C
1	2	Driver	5	4.14	55.5	4.10E+03	1.68E+03	PI-2I-D-7-C
1	2	Driver	5	4.05	95.5	9.33E+02	4.06E+02	PI-2I-D-8-C
1	2	Driver	6	2.05	0.125	4.12E+05	1.99E+05	PI-2I-D-1-C
1	2	Driver	6	2.01	2	5.60E+04	2.92E+04	PI-2I-D-3-C
1	2	Driver	6	2.02	4.5	1.68E+04	9.39E+03	PI-2I-D-4-C
1	2	Driver	6	2.04	9.5	1.02E+04	5.48E+03	PI-2I-D-5-C
1	2	Driver	6	2.03	23.5	3.03E+03	1.59E+03	PI-2I-D-6-C
1	2	Driver	6	2.08	55.5	1.95E+03	9.37E+02	PI-2I-D-7-C
1	2	Driver	6	2.03	95.5	1.33E+03	6.57E+02	PI-2I-D-8-C
1	2	Driver	7	1.06	0.125	1.58E+05	4.60E+04	PI-2I-D-1-C
1	2	Driver	7	1.03	2	3.87E+04	1.41E+04	PI-2I-D-3-C
1	2	Driver	7	1.04	4.5	1.60E+04	6.28E+03	PI-2I-D-4-C
1	2	Driver	7	1.05	9.5	3.52E+03	2.15E+03	PI-2I-D-5-C
1	2	Driver	7	1.05	23.5	2.10E+03	8.18E+02	PI-2I-D-6-C
1	2	Driver	7	1.07	55.5	1.72E+03	5.25E+02	PI-2I-D-7-C
1	2	Driver	7	1.04	95.5	9.10E+02	3.23E+02	PI-2I-D-8-C
1	2	Driver	8	0.62	0.125	3.25E+05	1.04E+05	PI-2I-D-1-C
1	2	Driver	8	0.61	2	1.07E+05	3.33E+04	PI-2I-D-3-C
1	2	Driver	8	0.61	4.5	3.14E+04	1.03E+04	PI-2I-D-4-C
1	2	Driver	8	0.62	9.5	2.99E+04	9.09E+03	PI-2I-D-5-C
1	2	Driver	8	0.61	23.5	7.75E+03	2.39E+03	PI-2I-D-6-C
1	2	Driver	8	0.63	55.5	3.75E+03	1.06E+03	PI-2I-D-7-C
1	2	Driver	8	0.61	95.5	2.03E+03	5.95E+02	PI-2I-D-8-C
1	2	Driver	9	0.22	0.125	1.01E+06	3.86E+05	PI-2I-D-1-C
1	2	Driver	9	0.21	2	5.25E+04	3.58E+04	PI-2I-D-3-C
1	2	Driver	9	0.22	4.5	2.02E+04	1.57E+04	PI-2I-D-4-C
1	2	Driver	9	0.22	9.5	-1.24E+03	4.46E+03	PI-2I-D-5-C

e A.7. *(continued)*

se	Shot	Position	CI stage	Median Dia (um)	MidTime (min)	U Conc, (ug/m3)	sd U Conc. (ug/m3)	SampleID
	2	Driver	9	0.22	23.5	6.56E+03	3.46E+03	PI-2I-D-6-CI-9
	2	Driver	9	0.22	55.5	8.60E+03	3.36E+03	PI-2I-D-7-CI-9
	2	Driver	9	0.22	95.5	3.99E+03	1.73E+03	PI-2I-D-8-CI-9
	2	Gunner	1	30.64	0.125	1.00E+06	1.71E+05	PI-1I-G-1-CI-1
	2	Gunner	1	31.16	0.75	3.60E+05	1.00E+05	PI-1I-G-2-CI-1
	2	Gunner	1	30.66	2	1.66E+05	4.65E+04	PI-1I-G-3-CI-1
	2	Gunner	1	30.81	4.5	7.47E+04	2.21E+04	PI-1I-G-4-CI-1
	2	Gunner	1	30.64	9.5	-2.37E+03	4.97E+03	PI-1I-G-5-CI-1
	2	Gunner	1	30.71	23.5	2.89E+03	1.74E+03	PI-1I-G-6-CI-1
	2	Gunner	1	32.77	55.5	4.37E+03	1.24E+03	PI-1I-G-7-CI-1
	2	Gunner	1	31.04	95.5	4.42E+03	1.14E+03	PI-1I-G-8-CI-1
	2	Gunner	2	15.67	0.125	8.12E+05	3.30E+05	PI-1I-G-1-CI-2
	2	Gunner	2	16.21	0.75	4.05E+05	1.69E+05	PI-1I-G-2-CI-2
	2	Gunner	2	15.70	2	9.72E+04	4.17E+04	PI-1I-G-3-CI-2
	2	Gunner	2	15.85	4.5	1.30E+04	6.60E+03	PI-1I-G-4-CI-2
	2	Gunner	2	15.67	9.5	7.03E+01	8.45E+02	PI-1I-G-5-CI-2
	2	Gunner	2	15.75	23.5	1.72E+02	2.59E+02	PI-1I-G-6-CI-2
	2	Gunner	2	17.92	55.5	2.74E+02	1.81E+02	PI-1I-G-7-CI-2
	2	Gunner	2	16.08	95.5	3.01E+02	1.80E+02	PI-1I-G-8-CI-2
	2	Gunner	3	10.49	0.125	4.13E+05	8.75E+04	PI-1I-G-1-CI-3
	2	Gunner	3	10.85	0.75	3.19E+05	6.75E+04	PI-1I-G-2-CI-3
	2	Gunner	3	10.51	2	6.34E+04	1.36E+04	PI-1I-G-3-CI-3
	2	Gunner	3	10.61	4.5	3.48E+04	7.46E+03	PI-1I-G-4-CI-3
	2	Gunner	3	10.49	9.5	8.51E+03	1.87E+03	PI-1I-G-5-CI-3
	2	Gunner	3	10.54	23.5	1.67E+03	3.68E+02	PI-1I-G-6-CI-3
	2	Gunner	3	12.00	55.5	5.11E+02	1.18E+02	PI-1I-G-7-CI-3
	2	Gunner	3	10.77	95.5	2.75E+02	6.72E+01	PI-1I-G-8-CI-3
	2	Gunner	4	6.69	0.125	8.20E+05	2.31E+05	PI-1I-G-1-CI-4
	2	Gunner	4	6.92	0.75	7.71E+04	2.21E+04	PI-1I-G-2-CI-4
	2	Gunner	4	6.70	2	4.79E+04	1.37E+04	PI-1I-G-3-CI-4
	2	Gunner	4	6.76	4.5	1.84E+04	5.30E+03	PI-1I-G-4-CI-4
	2	Gunner	4	6.69	9.5	6.60E+03	1.93E+03	PI-1I-G-5-CI-4
	2	Gunner	4	6.72	23.5	3.81E+03	1.08E+03	PI-1I-G-6-CI-4
	2	Gunner	4	7.65	55.5	1.80E+03	5.12E+02	PI-1I-G-7-CI-4
	2	Gunner	4	6.86	95.5	1.73E+03	4.91E+02	PI-1I-G-8-CI-4
	2	Gunner	5	4.02	0.125	8.65E+05	3.44E+05	PI-1I-G-1-CI-5
	2	Gunner	5	4.16	0.75	3.33E+05	1.33E+05	PI-1I-G-2-CI-5
	2	Gunner	5	4.03	2	1.12E+05	4.47E+04	PI-1I-G-3-CI-5
	2	Gunner	5	4.06	4.5	3.43E+04	1.38E+04	PI-1I-G-4-CI-5
	2	Gunner	5	4.02	9.5	1.96E+04	7.90E+03	PI-1I-G-5-CI-5
	2	Gunner	5	4.04	23.5	7.46E+03	2.97E+03	PI-1I-G-6-CI-5
	2	Gunner	5	4.60	55.5	2.13E+03	8.51E+02	PI-1I-G-7-CI-5

Table A.7. *(continued)*

Phase	Shot	Position	CI stage	Median Dia (um)	MidTime (min)	U Conc, (ug/m3)	sd U Conc. (ug/m3)	SampleID
1	2	Gunner	5	4.12	95.5	2.18E+03	8.69E+02	PI-1I-G-8-C
1	2	Gunner	6	2.01	0.125	2.52E+05	1.08E+05	PI-1I-G-1-C
1	2	Gunner	6	2.08	0.75	1.31E+05	5.89E+04	PI-1I-G-2-C
1	2	Gunner	6	2.02	2	1.05E+05	4.59E+04	PI-1I-G-3-C
1	2	Gunner	6	2.04	4.5	3.36E+04	1.49E+04	PI-1I-G-4-C
1	2	Gunner	6	2.01	9.5	9.45E+03	4.42E+03	PI-1I-G-5-C
1	2	Gunner	6	2.02	23.5	2.55E+03	1.18E+03	PI-1I-G-6-C
1	2	Gunner	6	2.30	55.5	2.43E+03	1.06E+03	PI-1I-G-7-C
1	2	Gunner	6	2.07	95.5	1.20E+03	5.42E+02	PI-1I-G-8-C
1	2	Gunner	7	1.04	0.125	2.34E+05	6.12E+04	PI-1I-G-1-C
1	2	Gunner	7	1.07	0.75	1.12E+05	3.40E+04	PI-1I-G-2-C
1	2	Gunner	7	1.04	2	9.20E+04	2.53E+04	PI-1I-G-3-C
1	2	Gunner	7	1.05	4.5	3.21E+04	9.38E+03	PI-1I-G-4-C
1	2	Gunner	7	1.04	9.5	9.57E+03	3.12E+03	PI-1I-G-5-C
1	2	Gunner	7	1.04	23.5	3.39E+03	1.02E+03	PI-1I-G-6-C
1	2	Gunner	7	1.18	55.5	1.67E+03	4.91E+02	PI-1I-G-7-C
1	2	Gunner	7	1.06	95.5	1.30E+03	3.82E+02	PI-1I-G-8-C
1	2	Gunner	8	0.61	0.125	7.35E+05	1.82E+05	PI-1I-G-1-C
1	2	Gunner	8	0.63	0.75	2.78E+05	7.00E+04	PI-1I-G-2-C
1	2	Gunner	8	0.61	2	1.90E+05	4.68E+04	PI-1I-G-3-C
1	2	Gunner	8	0.62	4.5	7.07E+04	1.80E+04	PI-1I-G-4-C
1	2	Gunner	8	0.61	9.5	1.37E+04	3.65E+03	PI-1I-G-5-C
1	2	Gunner	8	0.61	23.5	6.85E+03	1.74E+03	PI-1I-G-6-C
1	2	Gunner	8	0.70	55.5	4.92E+03	1.21E+03	PI-1I-G-7-C
1	2	Gunner	8	0.62	95.5	2.69E+03	6.73E+02	PI-1I-G-8-C
1	2	Gunner	9	0.21	0.125	1.15E+06	3.82E+05	PI-1I-G-1-C
1	2	Gunner	9	0.22	0.75	5.12E+05	1.72E+05	PI-1I-G-2-C
1	2	Gunner	9	0.21	2	2.38E+05	7.96E+04	PI-1I-G-3-C
1	2	Gunner	9	0.22	4.5	9.34E+04	3.15E+04	PI-1I-G-4-C
1	2	Gunner	9	0.21	9.5	3.18E+04	1.09E+04	PI-1I-G-5-C
1	2	Gunner	9	0.22	23.5	1.55E+04	5.21E+03	PI-1I-G-6-C
1	2	Gunner	9	0.23	55.5	5.61E+03	1.90E+03	PI-1I-G-7-C
1	2	Gunner	9	0.22	95.5	6.46E+03	2.16E+03	PI-1I-G-8-C
1	2	Loader	1	31.51	0.125	1.25E+06	2.67E+05	PI-2I-L-1-C
1	2	Loader	1	31.19	0.75	3.01E+05	8.38E+04	PI-2I-L-2-C
1	2	Loader	1	31.48	2	7.79E+04	2.75E+04	PI-2I-L-3-C
1	2	Loader	1	31.05	4.5	1.56E+04	8.80E+03	PI-2I-L-4-C
1	2	Loader	1	31.00	9.5	1.95E+03	3.35E+03	PI-2I-L-5-C
1	2	Loader	1	31.10	23.5	1.25E+03	9.82E+02	PI-2I-L-6-C
1	2	Loader	1	31.56	55.5	7.97E+02	4.11E+02	PI-2I-L-7-C
1	2	Loader	1	31.41	95.5	3.19E+03	8.94E+02	PI-2I-L-8-C
1	2	Loader	2	16.58	0.125	4.60E+05	2.22E+05	PI-2I-L-1-C

e A.7. *(continued)*

se	Shot	Position	CI stage	Median Dia (um)	MidTime (min)	U Conc, (ug/m3)	sd U Conc. (ug/m3)	SampleID
	2	Loader	2	16.24	0.75	1.70E+05	8.70E+04	PI-2I-L-2-CI-2
	2	Loader	2	16.55	2	6.18E+04	3.22E+04	PI-2I-L-3-CI-2
	2	Loader	2	16.10	4.5	7.63E+03	5.04E+03	PI-2I-L-4-CI-2
	2	Loader	2	16.04	9.5	2.19E+03	1.74E+03	PI-2I-L-5-CI-2
	2	Loader	2	16.15	23.5	2.30E+02	2.90E+02	PI-2I-L-6-CI-2
	2	Loader	2	16.63	55.5	2.25E+02	1.68E+02	PI-2I-L-7-CI-2
	2	Loader	2	16.47	95.5	2.35E+02	1.71E+02	PI-2I-L-8-CI-2
	2	Loader	3	11.10	0.125	6.04E+05	1.65E+05	PI-2I-L-1-CI-3
	2	Loader	3	10.87	0.75	1.70E+05	4.80E+04	PI-2I-L-2-CI-3
	2	Loader	3	11.08	2	4.55E+04	1.33E+04	PI-2I-L-3-CI-3
	2	Loader	3	10.77	4.5	3.54E+04	1.00E+04	PI-2I-L-4-CI-3
	2	Loader	3	10.74	9.5	1.17E+04	3.38E+03	PI-2I-L-5-CI-3
	2	Loader	3	10.81	23.5	1.09E+03	3.51E+02	PI-2I-L-6-CI-3
	2	Loader	3	11.13	55.5	7.22E+02	2.14E+02	PI-2I-L-7-CI-3
	2	Loader	3	11.02	95.5	3.00E+02	1.01E+02	PI-2I-L-8-CI-3
	2	Loader	4	7.08	0.125	2.64E+05	1.00E+05	PI-2I-L-1-CI-4
	2	Loader	4	6.93	0.75	2.31E+05	8.82E+04	PI-2I-L-2-CI-4
	2	Loader	4	7.06	2	6.88E+04	2.66E+04	PI-2I-L-3-CI-4
	2	Loader	4	6.87	4.5	3.96E+04	1.53E+04	PI-2I-L-4-CI-4
	2	Loader	4	6.85	9.5	7.39E+03	2.92E+03	PI-2I-L-5-CI-4
	2	Loader	4	6.89	23.5	3.07E+03	1.19E+03	PI-2I-L-6-CI-4
	2	Loader	4	7.10	55.5	1.39E+03	5.35E+02	PI-2I-L-7-CI-4
	2	Loader	4	7.03	95.5	8.90E+02	3.49E+02	PI-2I-L-8-CI-4
	2	Loader	5	4.25	0.125	4.09E+05	1.61E+05	PI-2I-L-1-CI-5
	2	Loader	5	4.17	0.75	1.82E+05	7.31E+04	PI-2I-L-2-CI-5
	2	Loader	5	4.24	2	1.02E+05	4.11E+04	PI-2I-L-3-CI-5
	2	Loader	5	4.13	4.5	6.71E+04	2.69E+04	PI-2I-L-4-CI-5
	2	Loader	5	4.11	9.5	2.61E+04	1.05E+04	PI-2I-L-5-CI-5
	2	Loader	5	4.14	23.5	3.97E+03	1.62E+03	PI-2I-L-6-CI-5
	2	Loader	5	4.27	55.5	2.97E+03	1.19E+03	PI-2I-L-7-CI-5
	2	Loader	5	4.22	95.5	1.69E+03	6.86E+02	PI-2I-L-8-CI-5
	2	Loader	6	2.13	0.125	5.29E+05	1.37E+05	PI-2I-L-1-CI-6
	2	Loader	6	2.09	0.75	1.83E+05	4.84E+04	PI-2I-L-2-CI-6
	2	Loader	6	2.13	2	7.11E+04	1.90E+04	PI-2I-L-3-CI-6
	2	Loader	6	2.07	4.5	3.96E+04	1.05E+04	PI-2I-L-4-CI-6
	2	Loader	6	2.06	9.5	1.15E+04	3.12E+03	PI-2I-L-5-CI-6
	2	Loader	6	2.08	23.5	3.46E+03	9.34E+02	PI-2I-L-6-CI-6
	2	Loader	6	2.14	55.5	2.40E+03	6.33E+02	PI-2I-L-7-CI-6
	2	Loader	6	2.12	95.5	1.66E+03	4.42E+02	PI-2I-L-8-CI-6
	2	Loader	7	1.10	0.125	4.27E+05	8.32E+04	PI-2I-L-1-CI-7
	2	Loader	7	1.07	0.75	1.42E+05	2.87E+04	PI-2I-L-2-CI-7
	2	Loader	7	1.09	2	9.44E+04	1.89E+04	PI-2I-L-3-CI-7

Table A.7. *(continued)*

Phase	Shot	Position	CI stage	Median Dia (um)	MidTime (min)	U Conc, (ug/m3)	sd U Conc. (ug/m3)	SampleID
1	2	Loader	7	1.06	4.5	5.72E+04	1.14E+04	PI-2I-L-4-CI
1	2	Loader	7	1.06	9.5	1.68E+04	3.41E+03	PI-2I-L-5-CI
1	2	Loader	7	1.07	23.5	4.88E+03	9.84E+02	PI-2I-L-6-CI
1	2	Loader	7	1.10	55.5	3.19E+03	NA[(a)]	PI-2I-L-7-CI
1	2	Loader	7	1.09	95.5	2.50E+03	4.95E+02	PI-2I-L-8-CI
1	2	Loader	8	0.64	0.125	6.90E+05	9.72E+04	PI-2I-L-1-CI
1	2	Loader	8	0.63	0.75	1.71E+05	2.46E+04	PI-2I-L-2-CI
1	2	Loader	8	0.64	2	1.44E+05	2.05E+04	PI-2I-L-3-CI
1	2	Loader	8	0.63	4.5	2.03E+04	3.03E+03	PI-2I-L-4-CI
1	2	Loader	8	0.62	9.5	1.31E+04	1.92E+03	PI-2I-L-5-CI
1	2	Loader	8	0.63	23.5	6.28E+03	8.99E+02	PI-2I-L-6-CI
1	2	Loader	8	0.65	55.5	4.36E+03	NA[(a)]	PI-2I-L-7-CI
1	2	Loader	8	0.64	95.5	2.48E+03	3.54E+02	PI-2I-L-8-CI
1	2	Loader	9	0.22	0.125	1.25E+06	2.49E+05	PI-2I-L-1-CI
1	2	Loader	9	0.22	0.75	8.44E+05	1.70E+05	PI-2I-L-2-CI
1	2	Loader	9	0.22	2	1.30E+05	2.73E+04	PI-2I-L-3-CI
1	2	Loader	9	0.22	4.5	4.94E+04	1.05E+04	PI-2I-L-4-CI
1	2	Loader	9	0.22	9.5	1.91E+04	4.15E+03	PI-2I-L-5-CI
1	2	Loader	9	0.22	23.5	1.88E+04	3.81E+03	PI-2I-L-6-CI
1	2	Loader	9	0.22	55.5	1.56E+04	3.12E+03	PI-2I-L-7-CI
1	2	Loader	9	0.22	95.5	1.27E+04	2.55E+03	PI-2I-L-8-CI
1	3	Driver	1	31.36	0.083	2.41E+06	5.13E+05	PI-3/4I-D-1-C
1	3	Driver	1	31.53	3.25	1.32E+04	1.06E+05	PI-3/4I-D-2-C
1	3	Driver	1	30.90	9.5	2.31E+04	5.17E+04	PI-3/4I-D-3-C
1	3	Driver	2	16.42	0.083	1.37E+05	6.60E+04	PI-3/4I-D-1-C
1	3	Driver	2	16.60	3.25	5.41E+04	3.28E+04	PI-3/4I-D-2-C
1	3	Driver	2	15.94	9.5	2.12E+04	1.32E+04	PI-3/4I-D-3-C
1	3	Driver	3	10.99	0.083	1.01E+05	2.88E+04	PI-3/4I-D-1-C
1	3	Driver	3	11.11	3.25	7.41E+04	2.20E+04	PI-3/4I-D-2-C
1	3	Driver	3	10.67	9.5	3.02E+04	9.02E+03	PI-3/4I-D-3-C
1	3	Driver	4	7.01	0.083	1.08E+05	4.16E+04	PI-3/4I-D-1-C
1	3	Driver	4	7.09	3.25	2.91E+05	1.12E+05	PI-3/4I-D-2-C
1	3	Driver	4	6.80	9.5	7.13E+04	2.78E+04	PI-3/4I-D-3-C
1	3	Driver	5	4.21	0.083	2.74E+05	1.09E+05	PI-3/4I-D-1-C
1	3	Driver	5	4.26	3.25	2.97E+05	1.17E+05	PI-3/4I-D-2-C
1	3	Driver	5	4.09	9.5	1.24E+05	4.90E+04	PI-3/4I-D-3-C
1	3	Driver	6	2.11	0.083	1.87E+05	4.91E+04	PI-3/4I-D-1-C
1	3	Driver	6	2.13	3.25	2.07E+05	5.34E+04	PI-3/4I-D-2-C
1	3	Driver	6	2.05	9.5	6.67E+04	NA[(a)]	PI-3/4I-D-3-C
1	3	Driver	7	1.08	0.083	2.20E+05	4.42E+04	PI-3/4I-D-1-C
1	3	Driver	7	1.10	3.25	2.30E+05	4.48E+04	PI-3/4I-D-2-C
1	3	Driver	7	1.05	9.5	7.20E+04	1.40E+04	PI-3/4I-D-3-C

e A.7. *(continued)*

ase	Shot	Position	CI stage	Median Dia (um)	MidTime (min)	U Conc, (ug/m3)	sd U Conc. (ug/m3)	SampleID
	3	Driver	8	0.64	0.083	2.88E+05	4.21E+04	PI-3/4I-D-1-CI-8
	3	Driver	8	0.64	3.25	6.59E+04	1.01E+04	PI-3/4I-D-2-CI-8
	3	Driver	8	0.62	9.5	4.76E+04	6.96E+03	PI-3/4I-D-3-CI-8
	3	Driver	9	0.22	0.083	2.51E+06	NA[(a)]	PI-3/4I-D-1-CI-9
	3	Driver	9	0.22	3.25	2.79E+05	5.66E+04	PI-3/4I-D-2-CI-9
	3	Driver	9	0.22	9.5	7.60E+04	1.56E+04	PI-3/4I-D-3-CI-9
	3	Gunner	1	31.10	0.083	2.92E+06	4.97E+05	PI-3/4I-G-1-CI-1
	3	Gunner	1	31.15	3.25	2.55E+05	1.28E+05	PI-3/4I-G-2-CI-1
	3	Gunner	1	30.72	9.5	1.59E+05	6.66E+04	PI-3/4I-G-3-CI-1
	3	Gunner	2	16.14	0.083	4.21E+05	1.72E+05	PI-3/4I-G-1-CI-2
	3	Gunner	2	16.20	3.25	8.72E+04	4.45E+04	PI-3/4I-G-2-CI-2
	3	Gunner	2	15.76	9.5	1.90E+04	1.23E+04	PI-3/4I-G-3-CI-2
	3	Gunner	3	10.81	0.083	2.13E+05	4.62E+04	PI-3/4I-G-1-CI-3
	3	Gunner	3	10.84	3.25	1.04E+05	2.42E+04	PI-3/4I-G-2-CI-3
	3	Gunner	3	10.55	9.5	1.95E+04	5.27E+03	PI-3/4I-G-3-CI-3
	3	Gunner	4	6.89	0.083	2.94E+05	8.35E+04	PI-3/4I-G-1-CI-4
	3	Gunner	4	6.91	3.25	2.28E+05	6.57E+04	PI-3/4I-G-2-CI-4
	3	Gunner	4	6.73	9.5	4.47E+04	1.35E+04	PI-3/4I-G-3-CI-4
	3	Gunner	5	4.14	0.083	1.33E+06	5.29E+05	PI-3/4I-G-1-CI-5
	3	Gunner	5	4.15	3.25	3.32E+05	1.36E+05	PI-3/4I-G-2-CI-5
	3	Gunner	5	4.04	9.5	1.09E+05	4.51E+04	PI-3/4I-G-3-CI-5
	3	Gunner	6	2.07	0.083	8.17E+05	3.45E+05	PI-3/4I-G-1-CI-6
	3	Gunner	6	2.08	3.25	2.41E+05	1.03E+05	PI-3/4I-G-2-CI-6
	3	Gunner	6	2.03	9.5	7.23E+04	3.14E+04	PI-3/4I-G-3-CI-6
	3	Gunner	7	1.07	0.083	6.12E+05	1.55E+05	PI-3/4I-G-1-CI-7
	3	Gunner	7	1.07	3.25	2.25E+05	5.75E+04	PI-3/4I-G-2-CI-7
	3	Gunner	7	1.04	9.5	7.78E+04	2.02E+04	PI-3/4I-G-3-CI-7
	3	Gunner	8	0.63	0.083	9.21E+05	2.28E+05	PI-3/4I-G-1-CI-8
	3	Gunner	8	0.63	3.25	2.17E+05	5.83E+04	PI-3/4I-G-2-CI-8
	3	Gunner	8	0.61	9.5	8.62E+04	2.35E+04	PI-3/4I-G-3-CI-8
	3	Gunner	9	0.22	0.083	3.03E+06	9.98E+05	PI-3/4I-G-1-CI-9
	3	Gunner	9	0.22	3.25	3.28E+05	1.12E+05	PI-3/4I-G-2-CI-9
	3	Gunner	9	0.22	9.5	7.69E+04	2.72E+04	PI-3/4I-G-3-CI-9
	4	Commander	1	30.79	0.083	2.06E+06	3.50E+05	PI-3/4I-C-1-CI-1
	4	Commander	1	30.91	3.25	8.08E+04	2.87E+04	PI-3/4I-C-2-CI-1
	4	Commander	1	30.87	9.5	9.56E+04	2.34E+04	PI-3/4I-C-3-CI-1
	4	Commander	1	31.60	22	2.49E+04	8.24E+03	PI-3/4I-C-4-CI-1
	4	Commander	1	31.33	47	9.09E+03	3.53E+03	PI-3/4I-C-5-CI-1
	4	Commander	1	31.40	97	8.15E+03	2.36E+03	PI-3/4I-C-6-CI-1
	4	Commander	2	15.82	0.083	1.53E+06	6.21E+05	PI-3/4I-C-1-CI-2
	4	Commander	2	15.95	3.25	1.40E+05	5.79E+04	PI-3/4I-C-2-CI-2
	4	Commander	2	15.91	9.5	2.59E+04	1.13E+04	PI-3/4I-C-3-CI-2

Table A.7. *(continued)*

Phase	Shot	Position	CI stage	Median Dia (um)	MidTime (min)	U Conc, (ug/m3)	sd U Conc. (ug/m3)	SampleID
1	4	Commander	2	16.67	22	1.10E+03	9.82E+02	PI-3/4I-C-4-C
1	4	Commander	2	16.39	47	9.05E+03	3.84E+03	PI-3/4I-C-5-C
1	4	Commander	2	16.46	97	4.36E+02	2.99E+02	PI-3/4I-C-6-C
1	4	Commander	3	10.59	0.083	2.98E+05	6.31E+04	PI-3/4I-C-1-C
1	4	Commander	3	10.68	3.25	3.91E+04	8.77E+03	PI-3/4I-C-2-C
1	4	Commander	3	10.65	9.5	2.09E+04	4.74E+03	PI-3/4I-C-3-C
1	4	Commander	3	11.16	22	4.19E+03	1.05E+03	PI-3/4I-C-4-C
1	4	Commander	3	10.97	47	1.37E+03	3.63E+02	PI-3/4I-C-5-C
1	4	Commander	3	11.02	97	2.95E+02	1.26E+02	PI-3/4I-C-6-C
1	4	Commander	4	6.75	0.083	1.54E+05	4.37E+04	PI-3/4I-C-1-C
1	4	Commander	4	6.81	3.25	4.10E+04	1.23E+04	PI-3/4I-C-2-C
1	4	Commander	4	6.79	9.5	3.98E+04	1.16E+04	PI-3/4I-C-3-C
1	4	Commander	4	7.12	22	8.59E+03	2.66E+03	PI-3/4I-C-4-C
1	4	Commander	4	6.99	47	3.00E+03	9.45E+02	PI-3/4I-C-5-C
1	4	Commander	4	7.02	97	1.05E+03	3.47E+02	PI-3/4I-C-6-C
1	4	Commander	5	4.06	0.083	4.89E+05	1.96E+05	PI-3/4I-C-1-C
1	4	Commander	5	4.09	3.25	1.03E+05	4.26E+04	PI-3/4I-C-2-C
1	4	Commander	5	4.08	9.5	4.82E+04	1.99E+04	PI-3/4I-C-3-C
1	4	Commander	5	4.28	22	5.58E+03	2.53E+03	PI-3/4I-C-4-C
1	4	Commander	5	4.20	47	4.73E+03	2.04E+03	PI-3/4I-C-5-C
1	4	Commander	5	4.22	97	2.93E+03	1.26E+03	PI-3/4I-C-6-C
1	4	Commander	6	2.03	0.083	4.37E+05	1.93E+05	PI-3/4I-C-1-C
1	4	Commander	6	2.05	3.25	1.40E+05	6.40E+04	PI-3/4I-C-2-C
1	4	Commander	6	2.04	9.5	2.61E+04	1.30E+04	PI-3/4I-C-3-C
1	4	Commander	6	2.14	22	1.79E+04	8.47E+03	PI-3/4I-C-4-C
1	4	Commander	6	2.11	47	3.95E+03	2.15E+03	PI-3/4I-C-5-C
1	4	Commander	6	2.11	97	1.88E+03	9.79E+02	PI-3/4I-C-6-C
1	4	Commander	7	1.05	0.083	2.51E+05	6.86E+04	PI-3/4I-C-1-C
1	4	Commander	7	1.05	3.25	1.34E+05	3.56E+04	PI-3/4I-C-2-C
1	4	Commander	7	1.05	9.5	2.80E+04	8.22E+03	PI-3/4I-C-3-C
1	4	Commander	7	1.10	22	1.47E+04	4.39E+03	PI-3/4I-C-4-C
1	4	Commander	7	1.08	47	4.91E+03	1.61E+03	PI-3/4I-C-5-C
1	4	Commander	7	1.09	97	4.12E+03	1.16E+03	PI-3/4I-C-6-C
1	4	Commander	8	0.61	0.083	2.51E+05	7.51E+04	PI-3/4I-C-1-C
1	4	Commander	8	0.62	3.25	1.43E+05	4.03E+04	PI-3/4I-C-2-C
1	4	Commander	8	0.62	9.5	2.80E+04	9.84E+03	PI-3/4I-C-3-C
1	4	Commander	8	0.65	22	2.48E+04	7.47E+03	PI-3/4I-C-4-C
1	4	Commander	8	0.64	47	1.23E+04	3.74E+03	PI-3/4I-C-5-C
1	4	Commander	8	0.64	97	5.56E+03	1.68E+03	PI-3/4I-C-6-C
1	4	Commander	9	0.22	0.083	1.97E+06	6.77E+05	PI-3/4I-C-1-C
1	4	Commander	9	0.22	3.25	1.80E+05	6.96E+04	PI-3/4I-C-2-C
1	4	Commander	9	0.22	9.5	2.22E+04	1.19E+04	PI-3/4I-C-3-C

e A.7. *(continued)*

se	Shot	Position	CI stage	Median Dia (um)	MidTime (min)	U Conc, (ug/m3)	sd U Conc. (ug/m3)	SampleID
	4	Commander	9	0.22	22	1.77E+04	8.33E+03	PI-3/4I-C-4-CI-9
	4	Commander	9	0.22	47	4.55E+03	2.70E+03	PI-3/4I-C-5-CI-9
	4	Commander	9	0.22	97	2.50E+03	1.37E+03	PI-3/4I-C-6-CI-9
	4	Driver	1	30.87	0.083	1.51E+06	3.23E+05	PI-3/4I-D-4-CI-1
	4	Driver	1	31.15	3.25	-1.04E+04	9.74E+04	PI-3/4I-D-5-CI-1
	4	Driver	1	30.93	9.5	2.03E+05	8.51E+04	PI-3/4I-D-6-CI-1
	4	Driver	1	31.03	22	5.02E+04	3.31E+04	PI-3/4I-D-7-CI-1
	4	Driver	1	30.92	47	2.23E+02	1.20E+04	PI-3/4I-D-8-CI-1
	4	Driver	2	15.91	0.083	5.40E+05	2.61E+05	PI-3/4I-D-4-CI-2
	4	Driver	2	16.20	3.25	3.98E+04	2.55E+04	PI-3/4I-D-5-CI-2
	4	Driver	2	15.97	9.5	1.30E+04	9.40E+03	PI-3/4I-D-6-CI-2
	4	Driver	2	16.07	22	2.08E+03	2.71E+03	PI-3/4I-D-7-CI-2
	4	Driver	2	15.96	47	-8.55E+01	9.22E+02	PI-3/4I-D-8-CI-2
	4	Driver	3	10.65	0.083	3.60E+05	9.96E+04	PI-3/4I-D-4-CI-3
	4	Driver	3	10.84	3.25	9.62E+04	2.78E+04	PI-3/4I-D-5-CI-3
	4	Driver	3	10.69	9.5	1.10E+04	3.68E+03	PI-3/4I-D-6-CI-3
	4	Driver	3	10.76	22	2.79E+03	1.17E+03	PI-3/4I-D-7-CI-3
	4	Driver	3	10.68	47	8.80E+02	4.37E+02	PI-3/4I-D-8-CI-3
	4	Driver	4	6.79	0.083	2.72E+05	1.03E+05	PI-3/4I-D-4-CI-4
	4	Driver	4	6.92	3.25	6.50E+04	2.65E+04	PI-3/4I-D-5-CI-4
	4	Driver	4	6.82	9.5	3.16E+04	1.27E+04	PI-3/4I-D-6-CI-4
	4	Driver	4	6.86	22	4.75E+03	2.27E+03	PI-3/4I-D-7-CI-4
	4	Driver	4	6.81	47	2.34E+03	1.12E+03	PI-3/4I-D-8-CI-4
	4	Driver	5	4.08	0.083	3.47E+05	1.37E+05	PI-3/4I-D-4-CI-5
	4	Driver	5	4.16	3.25	1.26E+05	4.97E+04	PI-3/4I-D-5-CI-5
	4	Driver	5	4.10	9.5	4.48E+04	1.78E+04	PI-3/4I-D-6-CI-5
	4	Driver	5	4.12	22	8.46E+03	3.38E+03	PI-3/4I-D-7-CI-5
	4	Driver	5	4.09	47	6.41E+03	2.54E+03	PI-3/4I-D-8-CI-5
	4	Driver	6	2.04	0.083	1.42E+05	3.76E+04	PI-3/4I-D-4-CI-6
	4	Driver	6	2.08	3.25	9.97E+04	2.60E+04	PI-3/4I-D-5-CI-6
	4	Driver	6	2.05	9.5	2.58E+04	6.77E+03	PI-3/4I-D-6-CI-6
	4	Driver	6	2.07	22	7.56E+03	2.00E+03	PI-3/4I-D-7-CI-6
	4	Driver	6	2.05	47	2.21E+03	6.03E+02	PI-3/4I-D-8-CI-6
	4	Driver	7	1.05	0.083	1.26E+05	2.55E+04	PI-3/4I-D-4-CI-7
	4	Driver	7	1.07	3.25	6.82E+04	1.37E+04	PI-3/4I-D-5-CI-7
	4	Driver	7	1.06	9.5	2.91E+04	5.89E+03	PI-3/4I-D-6-CI-7
	4	Driver	7	1.06	22	9.30E+03	1.95E+03	PI-3/4I-D-7-CI-7
	4	Driver	7	1.05	47	6.87E+03	1.39E+03	PI-3/4I-D-8-CI-7
	4	Driver	8	0.62	0.083	2.30E+05	3.36E+04	PI-3/4I-D-4-CI-8
	4	Driver	8	0.63	3.25	8.30E+04	NA[a]	PI-3/4I-D-5-CI-8
	4	Driver	8	0.62	9.5	2.39E+04	3.74E+03	PI-3/4I-D-6-CI-8
	4	Driver	8	0.62	22	1.78E+04	2.65E+03	PI-3/4I-D-7-CI-8

Table A.7. *(continued)*

Phase	Shot	Position	CI stage	Median Dia (um)	MidTime (min)	U Conc, (ug/m3)	sd U Conc. (ug/m3)	SampleID
1	4	Driver	8	0.62	47	7.91E+03	1.19E+03	PI-3/4I-D-8-C
1	4	Driver	9	0.22	0.083	7.74E+05	1.54E+05	PI-3/4I-D-4-C
1	4	Driver	9	0.22	3.25	1.08E+05	2.27E+04	PI-3/4I-D-5-C
1	4	Driver	9	0.22	9.5	2.67E+04	5.87E+03	PI-3/4I-D-6-C
1	4	Driver	9	0.22	22	9.02E+03	2.10E+03	PI-3/4I-D-7-C
1	4	Driver	9	0.22	47	8.02E+03	1.73E+03	PI-3/4I-D-8-C
1	4	Gunner	1	31.04	0.083	3.32E+06	5.64E+05	PI-3/4I-G-4-C
1	4	Gunner	1	30.83	3.25	1.09E+05	1.03E+05	PI-3/4I-G-5-C
1	4	Gunner	1	30.73	9.5	2.35E+04	4.67E+04	PI-3/4I-G-6-C
1	4	Gunner	1	31.60	22	-7.58E+03	2.34E+04	PI-3/4I-G-7-C
1	4	Gunner	1	31.24	47	1.05E+04	1.30E+04	PI-3/4I-G-8-C
1	4	Gunner	2	16.09	0.083	1.30E+06	5.28E+05	PI-3/4I-G-4-C
1	4	Gunner	2	15.86	3.25	6.49E+04	3.54E+04	PI-3/4I-G-5-C
1	4	Gunner	2	15.76	9.5	8.51E+03	8.32E+03	PI-3/4I-G-6-C
1	4	Gunner	2	16.67	22	1.99E+03	3.70E+03	PI-3/4I-G-7-C
1	4	Gunner	2	16.29	47	1.11E+03	1.85E+03	PI-3/4I-G-8-C
1	4	Gunner	3	10.77	0.083	6.26E+05	NA[a]	PI-3/4I-G-4-C
1	4	Gunner	3	10.62	3.25	6.61E+04	1.61E+04	PI-3/4I-G-5-C
1	4	Gunner	3	10.55	9.5	1.86E+04	5.10E+03	PI-3/4I-G-6-C
1	4	Gunner	3	11.16	22	9.25E+02	9.36E+02	PI-3/4I-G-7-C
1	4	Gunner	3	10.91	47	5.42E+01	3.75E+02	PI-3/4I-G-8-C
1	4	Gunner	4	6.87	0.083	7.82E+05	2.21E+05	PI-3/4I-G-4-C
1	4	Gunner	4	6.77	3.25	5.12E+04	1.59E+04	PI-3/4I-G-5-C
1	4	Gunner	4	6.73	9.5	2.78E+04	8.53E+03	PI-3/4I-G-6-C
1	4	Gunner	4	7.11	22	7.01E+03	2.38E+03	PI-3/4I-G-7-C
1	4	Gunner	4	6.95	47	2.89E+03	1.01E+03	PI-3/4I-G-8-C
1	4	Gunner	5	4.13	0.083	4.94E+05	1.97E+05	PI-3/4I-G-4-C
1	4	Gunner	5	4.07	3.25	4.63E+04	2.22E+04	PI-3/4I-G-5-C
1	4	Gunner	5	4.04	9.5	2.68E+04	1.26E+04	PI-3/4I-G-6-C
1	4	Gunner	5	4.28	22	5.45E+03	3.42E+03	PI-3/4I-G-7-C
1	4	Gunner	5	4.18	47	6.21E+03	2.98E+03	PI-3/4I-G-8-C
1	4	Gunner	6	2.07	0.083	5.87E+05	2.49E+05	PI-3/4I-G-4-C
1	4	Gunner	6	2.04	3.25	8.21E+04	3.68E+04	PI-3/4I-G-5-C
1	4	Gunner	6	2.03	9.5	4.37E+04	1.93E+04	PI-3/4I-G-6-C
1	4	Gunner	6	2.14	22	1.19E+04	5.49E+03	PI-3/4I-G-7-C
1	4	Gunner	6	2.09	47	3.27E+03	1.60E+03	PI-3/4I-G-8-C
1	4	Gunner	7	1.06	0.083	3.57E+05	9.31E+04	PI-3/4I-G-4-C
1	4	Gunner	7	1.05	3.25	7.57E+04	2.11E+04	PI-3/4I-G-5-C
1	4	Gunner	7	1.04	9.5	5.02E+04	1.37E+04	PI-3/4I-G-6-C
1	4	Gunner	7	1.10	22	1.15E+04	3.55E+03	PI-3/4I-G-7-C
1	4	Gunner	7	1.08	47	5.20E+03	1.59E+03	PI-3/4I-G-8-C
1	4	Gunner	8	0.62	0.083	5.12E+05	1.31E+05	PI-3/4I-G-4-C

le A.7. *(continued)*

ase	Shot	Position	CI stage	Median Dia (um)	MidTime (min)	U Conc, (ug/m3)	sd U Conc. (ug/m3)	SampleID
	4	Gunner	8	0.62	3.25	9.59E+04	2.99E+04	PI-3/4I-G-5-CI-8
	4	Gunner	8	0.61	9.5	4.85E+04	1.49E+04	PI-3/4I-G-6-CI-8
	4	Gunner	8	0.65	22	2.22E+04	7.01E+03	PI-3/4I-G-7-CI-8
	4	Gunner	8	0.63	47	1.09E+04	3.33E+03	PI-3/4I-G-8-CI-8
	4	Gunner	9	0.22	0.083	1.52E+06	5.07E+05	PI-3/4I-G-4-CI-9
	4	Gunner	9	0.22	3.25	1.92E+05	6.70E+04	PI-3/4I-G-5-CI-9
	4	Gunner	9	0.22	9.5	6.00E+04	2.16E+04	PI-3/4I-G-6-CI-9
	4	Gunner	9	0.22	22	1.88E+04	7.33E+03	PI-3/4I-G-7-CI-9
	4	Gunner	9	0.22	47	1.45E+04	4.98E+03	PI-3/4I-G-8-CI-9
	4	Loader	1	31.74	0.083	6.98E+06	1.49E+06	PI-3/4I-L-1-CI-1
	4	Loader	1	31.36	3.25	6.31E+05	3.22E+05	PI-3/4I-L-2-CI-1
	4	Loader	1	31.71	9.5	4.41E+05	1.91E+05	PI-3/4I-L-3-CI-1
	4	Loader	1	31.41	22	1.93E+05	8.80E+04	PI-3/4I-L-4-CI-1
	4	Loader	1	31.28	47	2.04E+05	6.59E+04	PI-3/4I-L-5-CI-1
	4	Loader	1	31.26	97	4.97E+04	2.20E+04	PI-3/4I-L-6-CI-1
	4	Loader	2	16.82	0.083	1.01E+06	4.86E+05	PI-3/4I-L-1-CI-2
	4	Loader	2	16.42	3.25	7.82E+04	6.96E+04	PI-3/4I-L-2-CI-2
	4	Loader	2	16.79	9.5	6.47E+04	4.75E+04	PI-3/4I-L-3-CI-2
	4	Loader	2	16.47	22	4.40E+04	2.90E+04	PI-3/4I-L-4-CI-2
	4	Loader	2	16.34	47	2.23E+04	1.46E+04	PI-3/4I-L-5-CI-2
	4	Loader	2	16.31	97	6.48E+03	5.06E+03	PI-3/4I-L-6-CI-2
	4	Loader	3	11.26	0.083	8.99E+05	2.47E+05	PI-3/4I-L-1-CI-3
	4	Loader	3	10.99	3.25	7.32E+04	2.63E+04	PI-3/4I-L-2-CI-3
	4	Loader	3	11.24	9.5	9.20E+04	2.83E+04	PI-3/4I-L-3-CI-3
	4	Loader	3	11.02	22	1.44E+04	5.57E+03	PI-3/4I-L-4-CI-3
	4	Loader	3	10.93	47	8.53E+03	3.12E+03	PI-3/4I-L-5-CI-3
	4	Loader	3	10.92	97	6.94E+02	5.92E+02	PI-3/4I-L-6-CI-3
	4	Loader	4	7.18	0.083	3.07E+05	1.17E+05	PI-3/4I-L-1-CI-4
	4	Loader	4	7.01	3.25	8.52E+04	3.99E+04	PI-3/4I-L-2-CI-4
	4	Loader	4	7.17	9.5	5.61E+04	2.52E+04	PI-3/4I-L-3-CI-4
	4	Loader	4	7.03	22	2.05E+04	9.68E+03	PI-3/4I-L-4-CI-4
	4	Loader	4	6.97	47	6.22E+03	3.32E+03	PI-3/4I-L-5-CI-4
	4	Loader	4	6.96	97	1.80E+03	1.18E+03	PI-3/4I-L-6-CI-4
	4	Loader	5	4.31	0.083	3.12E+05	1.24E+05	PI-3/4I-L-1-CI-5
	4	Loader	5	4.21	3.25	1.21E+05	5.88E+04	PI-3/4I-L-2-CI-5
	4	Loader	5	4.31	9.5	6.25E+04	3.03E+04	PI-3/4I-L-3-CI-5
	4	Loader	5	4.22	22	2.92E+04	1.43E+04	PI-3/4I-L-4-CI-5
	4	Loader	5	4.19	47	8.06E+03	4.55E+03	PI-3/4I-L-5-CI-5
	4	Loader	5	4.18	97	3.75E+03	2.17E+03	PI-3/4I-L-6-CI-5
	4	Loader	6	2.16	0.083	3.01E+05	7.83E+04	PI-3/4I-L-1-CI-6
	4	Loader	6	2.11	3.25	7.99E+04	2.43E+04	PI-3/4I-L-2-CI-6
	4	Loader	6	2.16	9.5	2.84E+04	9.27E+03	PI-3/4I-L-3-CI-6

Table A.7. *(continued)*

Phase	Shot	Position	CI stage	Median Dia (um)	MidTime (min)	U Conc, (ug/m3)	sd U Conc. (ug/m3)	SampleID
1	4	Loader	6	2.12	22	9.15E+03	3.30E+03	PI-3/4I-L-4-C
1	4	Loader	6	2.10	47	2.92E+03	1.22E+03	PI-3/4I-L-5-C
1	4	Loader	6	2.10	97	2.61E+03	9.06E+02	PI-3/4I-L-6-C
1	4	Loader	7	1.11	0.083	2.39E+05	4.82E+04	PI-3/4I-L-1-C
1	4	Loader	7	1.09	3.25	9.13E+04	2.22E+04	PI-3/4I-L-2-C
1	4	Loader	7	1.11	9.5	3.71E+04	9.62E+03	PI-3/4I-L-3-C
1	4	Loader	7	1.09	22	1.30E+04	3.65E+03	PI-3/4I-L-4-C
1	4	Loader	7	1.08	47	4.81E+03	1.51E+03	PI-3/4I-L-5-C
1	4	Loader	7	1.08	97	3.13E+03	8.84E+02	PI-3/4I-L-6-C
1	4	Loader	8	0.65	0.083	5.81E+05	8.29E+04	PI-3/4I-L-1-C
1	4	Loader	8	0.64	3.25	6.68E+04	1.40E+04	PI-3/4I-L-2-C
1	4	Loader	8	0.65	9.5	3.06E+04	6.63E+03	PI-3/4I-L-3-C
1	4	Loader	8	0.64	22	1.14E+04	2.76E+03	PI-3/4I-L-4-C
1	4	Loader	8	0.63	47	5.29E+03	1.33E+03	PI-3/4I-L-5-C
1	4	Loader	8	0.63	97	3.30E+03	7.42E+02	PI-3/4I-L-6-C
1	4	Loader	9	0.22	0.083	1.20E+06	2.39E+05	PI-3/4I-L-1-C
1	4	Loader	9	0.22	3.25	1.75E+05	5.02E+04	PI-3/4I-L-2-C
1	4	Loader	9	0.22	9.5	3.25E+04	1.48E+04	PI-3/4I-L-3-C
1	4	Loader	9	0.22	22	1.48E+04	6.95E+03	PI-3/4I-L-4-C
1	4	Loader	9	0.22	47	1.02E+04	4.00E+03	PI-3/4I-L-5-C
1	4	Loader	9	0.22	97	6.67E+03	2.28E+03	PI-3/4I-L-6-C
1	5	Commander	1	30.58	0.083	1.25E+06	2.67E+05	PI-5I-C-1-CI
1	5	Commander	1	30.76	1.083	1.81E+05	8.02E+04	PI-5I-C-2-CI
1	5	Commander	1	30.74	3.083	2.10E+05	8.17E+04	PI-5I-C-3-CI
1	5	Commander	1	31.25	7.25	-1.83E+04	1.24E+04	PI-5I-C-4-CI
1	5	Commander	1	31.30	15.5	2.52E+04	1.22E+04	PI-5I-C-5-CI
1	5	Commander	1	31.25	32	1.25E+04	6.04E+03	PI-5I-C-6-CI
1	5	Commander	1	30.76	63	1.49E+03	2.07E+03	PI-5I-C-7-CI
1	5	Commander	1	31.02	125	-6.40E+02	8.43E+02	PI-5I-C-8-CI
1	5	Commander	2	15.62	0.083	8.01E+05	3.85E+05	PI-5I-C-1-CI
1	5	Commander	2	15.79	1.083	2.27E+05	1.24E+05	PI-5I-C-2-CI
1	5	Commander	2	15.78	3.083	2.19E+04	2.50E+04	PI-5I-C-3-CI
1	5	Commander	2	16.31	7.25	6.48E+03	8.42E+03	PI-5I-C-4-CI
1	5	Commander	2	16.36	15.5	3.26E+03	4.24E+03	PI-5I-C-5-CI
1	5	Commander	2	16.30	32	-3.64E+02	1.39E+03	PI-5I-C-6-CI
1	5	Commander	2	15.79	63	4.50E+02	8.79E+02	PI-5I-C-7-CI
1	5	Commander	2	16.06	125	2.33E+02	4.35E+02	PI-5I-C-8-CI
1	5	Commander	3	10.45	0.083	6.58E+05	1.80E+05	PI-5I-C-1-CI
1	5	Commander	3	10.57	1.083	3.35E+05	9.35E+04	PI-5I-C-2-CI
1	5	Commander	3	10.56	3.083	8.56E+04	2.54E+04	PI-5I-C-3-CI
1	5	Commander	3	10.91	7.25	8.73E+03	3.08E+03	PI-5I-C-4-CI
1	5	Commander	3	10.95	15.5	2.83E+03	1.26E+03	PI-5I-C-5-CI

A.7. *(continued)*

Shot	Position	CI stage	Median Dia (um)	MidTime (min)	U Conc, (ug/m3)	sd U Conc. (ug/m3)	SampleID
5	Commander	3	10.91	32	8.90E+02	4.71E+02	PI-5I-C-6-CI-3
5	Commander	3	10.57	63	7.82E+02	3.09E+02	PI-5I-C-7-CI-3
5	Commander	3	10.75	125	2.16E+02	1.03E+02	PI-5I-C-8-CI-3
5	Commander	4	6.66	0.083	6.64E+05	2.50E+05	PI-5I-C-1-CI-4
5	Commander	4	6.74	1.083	3.24E+05	1.24E+05	PI-5I-C-2-CI-4
5	Commander	4	6.73	3.083	1.59E+05	6.18E+04	PI-5I-C-3-CI-4
5	Commander	4	6.96	7.25	1.38E+04	6.00E+03	PI-5I-C-4-CI-4
5	Commander	4	6.98	15.5	1.93E+03	1.26E+03	PI-5I-C-5-CI-4
5	Commander	4	6.96	32	2.77E+03	1.26E+03	PI-5I-C-6-CI-4
5	Commander	4	6.74	63	2.68E+03	1.13E+03	PI-5I-C-7-CI-4
5	Commander	4	6.86	125	8.92E+02	3.97E+02	PI-5I-C-8-CI-4
5	Commander	5	4.00	0.083	4.65E+05	1.84E+05	PI-5I-C-1-CI-5
5	Commander	5	4.05	1.083	1.89E+05	7.86E+04	PI-5I-C-2-CI-5
5	Commander	5	4.05	3.083	2.03E+05	8.40E+04	PI-5I-C-3-CI-5
5	Commander	5	4.18	7.25	1.29E+04	6.71E+03	PI-5I-C-4-CI-5
5	Commander	5	4.19	15.5	5.22E+03	2.90E+03	PI-5I-C-5-CI-5
5	Commander	5	4.18	32	7.05E+03	3.14E+03	PI-5I-C-6-CI-5
5	Commander	5	4.05	63	3.41E+03	1.51E+03	PI-5I-C-7-CI-5
5	Commander	5	4.12	125	2.90E+03	1.24E+03	PI-5I-C-8-CI-5
5	Commander	6	2.01	0.083	4.14E+05	1.07E+05	PI-5I-C-1-CI-6
5	Commander	6	2.03	1.083	2.08E+05	5.57E+04	PI-5I-C-2-CI-6
5	Commander	6	2.03	3.083	9.47E+04	2.63E+04	PI-5I-C-3-CI-6
5	Commander	6	2.10	7.25	1.52E+04	4.75E+03	PI-5I-C-4-CI-6
5	Commander	6	2.10	15.5	8.50E+03	2.54E+03	PI-5I-C-5-CI-6
5	Commander	6	2.10	32	5.27E+03	1.53E+03	PI-5I-C-6-CI-6
5	Commander	6	2.03	63	2.57E+03	7.61E+02	PI-5I-C-7-CI-6
5	Commander	6	2.06	125	1.64E+03	4.66E+02	PI-5I-C-8-CI-6
5	Commander	7	1.03	0.083	6.44E+05	1.25E+05	PI-5I-C-1-CI-7
5	Commander	7	1.04	1.083	2.55E+05	5.22E+04	PI-5I-C-2-CI-7
5	Commander	7	1.04	3.083	1.43E+05	2.96E+04	PI-5I-C-3-CI-7
5	Commander	7	1.08	7.25	2.33E+04	5.32E+03	PI-5I-C-4-CI-7
5	Commander	7	1.08	15.5	5.89E+03	1.60E+03	PI-5I-C-5-CI-7
5	Commander	7	1.08	32	2.72E+03	7.67E+02	PI-5I-C-6-CI-7
5	Commander	7	1.04	63	2.25E+03	5.49E+02	PI-5I-C-7-CI-7
5	Commander	7	1.06	125	3.23E+03	6.69E+02	PI-5I-C-8-CI-7
5	Commander	8	0.61	0.083	4.11E+05	5.89E+04	PI-5I-C-1-CI-8
5	Commander	8	0.61	1.083	2.88E+05	4.22E+04	PI-5I-C-2-CI-8
5	Commander	8	0.61	3.083	1.10E+05	1.77E+04	PI-5I-C-3-CI-8
5	Commander	8	0.63	7.25	1.55E+04	2.95E+03	PI-5I-C-4-CI-8
5	Commander	8	0.64	15.5	9.44E+03	1.71E+03	PI-5I-C-5-CI-8
5	Commander	8	0.63	32	7.75E+03	1.29E+03	PI-5I-C-6-CI-8
5	Commander	8	0.61	63	4.02E+03	6.45E+02	PI-5I-C-7-CI-8

Table A.7. *(continued)*

Phase	Shot	Position	CI stage	Median Dia (um)	MidTime (min)	U Conc, (ug/m3)	sd U Conc. (ug/m3)	SampleI
1	5	Commander	8	0.62	125	3.02E+03	4.72E+02	PI-5I-C-8-C
1	5	Commander	9	0.21	0.083	1.47E+06	2.92E+05	PI-5I-C-1-C
1	5	Commander	9	0.22	1.083	3.41E+05	NA[(a)]	PI-5I-C-2-C
1	5	Commander	9	0.22	3.083	2.27E+04	1.05E+04	PI-5I-C-3-C
1	5	Commander	9	0.22	7.25	8.07E+02	2.43E+03	PI-5I-C-4-C
1	5	Commander	9	0.22	15.5	1.62E+03	1.42E+03	PI-5I-C-5-C
1	5	Commander	9	0.22	32	2.82E+03	1.07E+03	PI-5I-C-6-C
1	5	Commander	9	0.22	63	3.69E+03	9.46E+02	PI-5I-C-7-C
1	5	Commander	9	0.22	125	3.13E+03	7.33E+02	PI-5I-C-8-C
1	5	Driver	1	31.40	0.083	1.87E+06	3.16E+05	PI-5I-D-1-C
1	5	Driver	1	30.46	1.083	1.15E+06	2.76E+05	PI-5I-D-2-C
1	5	Driver	1	30.82	3.083	4.90E+05	1.64E+05	PI-5I-D-3-C
1	5	Driver	1	30.98	7.25	1.35E+05	5.10E+04	PI-5I-D-4-C
1	5	Driver	1	31.00	15.5	1.33E+05	3.59E+04	PI-5I-D-5-C
1	5	Driver	1	30.94	32	6.87E+04	1.83E+04	PI-5I-D-6-C
1	5	Driver	1	31.05	63	3.18E+04	8.78E+03	PI-5I-D-7-C
1	5	Driver	1	31.04	125	1.23E+04	3.80E+03	PI-5I-D-8-C
1	5	Driver	2	16.46	0.083	1.02E+06	4.18E+05	PI-5I-D-1-C
1	5	Driver	2	15.49	1.083	2.04E+05	8.73E+04	PI-5I-D-2-C
1	5	Driver	2	15.85	3.083	1.94E+04	1.37E+04	PI-5I-D-3-C
1	5	Driver	2	16.03	7.25	4.05E+03	3.36E+03	PI-5I-D-4-C
1	5	Driver	2	16.05	15.5	1.39E+03	1.50E+03	PI-5I-D-5-C
1	5	Driver	2	15.98	32	3.55E+03	1.87E+03	PI-5I-D-6-C
1	5	Driver	2	16.09	63	8.33E+02	5.60E+02	PI-5I-D-7-C
1	5	Driver	2	16.08	125	-6.68E+01	1.46E+02	PI-5I-D-8-C
1	5	Driver	3	11.02	0.083	1.07E+06	2.24E+05	PI-5I-D-1-C
1	5	Driver	3	10.37	1.083	3.55E+05	7.59E+04	PI-5I-D-2-C
1	5	Driver	3	10.61	3.083	7.99E+04	1.85E+04	PI-5I-D-3-C
1	5	Driver	3	10.73	7.25	8.22E+03	2.46E+03	PI-5I-D-4-C
1	5	Driver	3	10.74	15.5	2.62E+03	9.43E+02	PI-5I-D-5-C
1	5	Driver	3	10.70	32	1.55E+03	5.74E+02	PI-5I-D-6-C
1	5	Driver	3	10.77	63	4.06E+02	2.20E+02	PI-5I-D-7-C
1	5	Driver	3	10.76	125	7.68E+01	9.84E+01	PI-5I-D-8-C
1	5	Driver	4	7.03	0.083	6.16E+05	1.72E+05	PI-5I-D-1-C
1	5	Driver	4	6.61	1.083	3.63E+05	1.04E+05	PI-5I-D-2-C
1	5	Driver	4	6.77	3.083	1.34E+05	3.98E+04	PI-5I-D-3-C
1	5	Driver	4	6.84	7.25	1.25E+04	4.37E+03	PI-5I-D-4-C
1	5	Driver	4	6.85	15.5	4.05E+03	1.59E+03	PI-5I-D-5-C
1	5	Driver	4	6.82	32	4.19E+03	1.39E+03	PI-5I-D-6-C
1	5	Driver	4	6.87	63	2.01E+03	6.58E+02	PI-5I-D-7-C
1	5	Driver	4	6.86	125	8.39E+02	2.87E+02	PI-5I-D-8-C
1	5	Driver	5	4.22	0.083	9.51E+05	3.93E+05	PI-5I-D-1-C

A.7. *(continued)*

se	Shot	Position	CI stage	Median Dia (um)	MidTime (min)	U Conc, (ug/m3)	sd U Conc. (ug/m3)	SampleID
	5	Driver	5	3.97	1.083	6.21E+05	2.56E+05	PI-5I-D-2-CI-5
	5	Driver	5	4.07	3.083	1.49E+05	6.37E+04	PI-5I-D-3-CI-5
	5	Driver	5	4.11	7.25	2.55E+04	1.13E+04	PI-5I-D-4-CI-5
	5	Driver	5	4.12	15.5	8.23E+03	3.82E+03	PI-5I-D-5-CI-5
	5	Driver	5	4.10	32	7.14E+03	3.03E+03	PI-5I-D-6-CI-5
	5	Driver	5	4.13	63	6.10E+03	2.56E+03	PI-5I-D-7-CI-5
	5	Driver	5	4.12	125	3.20E+03	1.34E+03	PI-5I-D-8-CI-5
	5	Driver	6	2.12	0.083	7.38E+05	3.72E+05	PI-5I-D-1-CI-6
	5	Driver	6	1.99	1.083	5.33E+04	3.45E+04	PI-5I-D-2-CI-6
	5	Driver	6	2.04	3.083	6.22E+04	3.90E+04	PI-5I-D-3-CI-6
	5	Driver	6	2.06	7.25	1.71E+04	1.13E+04	PI-5I-D-4-CI-6
	5	Driver	6	2.06	15.5	1.51E+04	8.43E+03	PI-5I-D-5-CI-6
	5	Driver	6	2.05	32	6.48E+03	3.72E+03	PI-5I-D-6-CI-6
	5	Driver	6	2.07	63	1.34E+03	9.16E+02	PI-5I-D-7-CI-6
	5	Driver	6	2.07	125	3.02E+03	1.58E+03	PI-5I-D-8-CI-6
	5	Driver	7	1.09	0.083	3.05E+05	9.13E+04	PI-5I-D-1-CI-7
	5	Driver	7	1.02	1.083	1.23E+05	4.44E+04	PI-5I-D-2-CI-7
	5	Driver	7	1.05	3.083	4.95E+04	2.33E+04	PI-5I-D-3-CI-7
	5	Driver	7	1.06	7.25	1.61E+03	4.36E+03	PI-5I-D-4-CI-7
	5	Driver	7	1.06	15.5	5.24E+03	NA[a]	PI-5I-D-5-CI-7
	5	Driver	7	1.06	32	3.80E+03	1.54E+03	PI-5I-D-6-CI-7
	5	Driver	7	1.06	63	1.42E+03	7.50E+02	PI-5I-D-7-CI-7
	5	Driver	7	1.06	125	1.32E+03	5.25E+02	PI-5I-D-8-CI-7
	5	Driver	8	0.64	0.083	3.67E+05	1.31E+05	PI-5I-D-1-CI-8
	5	Driver	8	0.60	1.083	2.58E+05	1.20E+05	PI-5I-D-2-CI-8
	5	Driver	8	0.62	3.083	3.76E+04	4.61E+04	PI-5I-D-3-CI-8
	5	Driver	8	0.62	7.25	5.71E+03	1.38E+04	PI-5I-D-4-CI-8
	5	Driver	8	0.62	15.5	4.94E+02	6.49E+03	PI-5I-D-5-CI-8
	5	Driver	8	0.62	32	3.38E+03	NA[a]	PI-5I-D-6-CI-8
	5	Driver	8	0.62	63	1.32E+03	2.01E+03	PI-5I-D-7-CI-8
	5	Driver	8	0.62	125	2.69E+03	1.42E+03	PI-5I-D-8-CI-8
	5	Driver	9	0.22	0.083	1.03E+06	4.25E+05	PI-5I-D-1-CI-9
	5	Driver	9	0.21	1.083	1.07E+05	7.09E+04	PI-5I-D-2-CI-9
	5	Driver	9	0.22	3.083	-5.03E+04	2.52E+04	PI-5I-D-3-CI-9
	5	Driver	9	0.22	7.25	-5.46E+03	7.64E+03	PI-5I-D-4-CI-9
	5	Driver	9	0.22	15.5	6.25E+03	6.83E+03	PI-5I-D-5-CI-9
	5	Driver	9	0.22	32	9.69E+02	2.73E+03	PI-5I-D-6-CI-9
	5	Driver	9	0.22	63	2.26E+03	1.95E+03	PI-5I-D-7-CI-9
	5	Driver	9	0.22	125	2.31E+03	1.33E+03	PI-5I-D-8-CI-9
	5	Loader	1	31.24	0.083	2.12E+06	NA[a]	PI-5I-L-1-CI-1
	5	Loader	1	31.15	1.083	-1.43E+04	4.81E+04	PI-5I-L-2-CI-1
	5	Loader	1	30.92	3.083	2.06E+05	8.21E+04	PI-5I-L-3-CI-1

Table A.7. *(continued)*

Phase	Shot	Position	CI stage	Median Dia (um)	MidTime (min)	U Conc, (ug/m3)	sd U Conc. (ug/m3)	SampleII
1	5	Loader	1	31.01	7.25	3.64E+04	2.13E+04	PI-5I-L-4-C
1	5	Loader	1	30.94	15.5	1.21E+04	9.46E+03	PI-5I-L-5-C
1	5	Loader	1	31.06	32	-6.36E+03	3.18E+03	PI-5I-L-6-C
1	5	Loader	1	31.49	63	1.04E+04	3.89E+03	PI-5I-L-7-C
1	5	Loader	1	31.04	125	3.51E+03	1.56E+03	PI-5I-L-8-C
1	5	Loader	1	31.15	146	1.54E+04	4.84E+03	PI-5I-L-9-C
1	5	Loader	2	16.29	0.083	7.09E+05	3.41E+05	PI-5I-L-1-C
1	5	Loader	2	16.20	1.083	9.55E+04	NA[(a)]	PI-5I-L-2-C
1	5	Loader	2	15.96	3.083	1.86E+04	2.45E+04	PI-5I-L-3-C
1	5	Loader	2	16.05	7.25	3.72E+03	7.37E+03	PI-5I-L-4-C
1	5	Loader	2	15.98	15.5	5.66E+03	5.15E+03	PI-5I-L-5-C
1	5	Loader	2	16.11	32	-3.23E+01	1.44E+03	PI-5I-L-6-C
1	5	Loader	2	16.56	63	8.36E+02	1.08E+03	PI-5I-L-7-C
1	5	Loader	2	16.09	125	1.53E+02	4.10E+02	PI-5I-L-8-C
1	5	Loader	2	16.20	146	9.63E+02	1.13E+03	PI-5I-L-9-C
1	5	Loader	3	10.90	0.083	1.15E+06	3.14E+05	PI-5I-L-1-C
1	5	Loader	3	10.84	1.083	2.64E+05	7.44E+04	PI-5I-L-2-C
1	5	Loader	3	10.69	3.083	5.49E+04	1.71E+04	PI-5I-L-3-C
1	5	Loader	3	10.74	7.25	3.45E+03	2.00E+03	PI-5I-L-4-C
1	5	Loader	3	10.70	15.5	3.70E+03	1.42E+03	PI-5I-L-5-C
1	5	Loader	3	10.78	32	6.16E+02	4.53E+02	PI-5I-L-6-C
1	5	Loader	3	11.08	63	4.59E+02	2.55E+02	PI-5I-L-7-C
1	5	Loader	3	10.77	125	3.42E+02	1.47E+02	PI-5I-L-8-C
1	5	Loader	3	10.84	146	5.67E+02	2.48E+02	PI-5I-L-9-C
1	5	Loader	4	6.95	0.083	4.15E+05	1.56E+05	PI-5I-L-1-C
1	5	Loader	4	6.91	1.083	3.11E+05	1.20E+05	PI-5I-L-2-C
1	5	Loader	4	6.81	3.083	7.62E+04	3.12E+04	PI-5I-L-3-C
1	5	Loader	4	6.85	7.25	1.60E+04	6.98E+03	PI-5I-L-4-C
1	5	Loader	4	6.82	15.5	5.32E+03	2.56E+03	PI-5I-L-5-C
1	5	Loader	4	6.87	32	2.92E+03	1.32E+03	PI-5I-L-6-C
1	5	Loader	4	7.07	63	2.36E+03	1.00E+03	PI-5I-L-7-C
1	5	Loader	4	6.87	125	5.09E+02	2.62E+02	PI-5I-L-8-C
1	5	Loader	4	6.91	146	4.72E+02	3.29E+02	PI-5I-L-9-C
1	5	Loader	5	4.18	0.083	7.12E+05	2.81E+05	PI-5I-L-1-C
1	5	Loader	5	4.15	1.083	1.66E+05	7.00E+04	PI-5I-L-2-C
1	5	Loader	5	4.09	3.083	1.23E+05	5.30E+04	PI-5I-L-3-C
1	5	Loader	5	4.12	7.25	1.66E+04	8.18E+03	PI-5I-L-4-C
1	5	Loader	5	4.10	15.5	5.80E+03	3.06E+03	PI-5I-L-5-C
1	5	Loader	5	4.13	32	5.22E+03	2.42E+03	PI-5I-L-6-C
1	5	Loader	5	4.25	63	5.28E+03	2.25E+03	PI-5I-L-7-C
1	5	Loader	5	4.13	125	4.09E+03	1.69E+03	PI-5I-L-8-C
1	5	Loader	5	4.15	146	1.70E+03	8.91E+02	PI-5I-L-9-C

A.7. *(continued)*

e	Shot	Position	CI stage	Median Dia (um)	MidTime (min)	U Conc, (ug/m3)	sd U Conc. (ug/m3)	SampleID
	5	Loader	6	2.09	0.083	3.32E+05	8.60E+04	PI-5I-L-1-CI-6
	5	Loader	6	2.08	1.083	1.28E+05	3.51E+04	PI-5I-L-2-CI-6
	5	Loader	6	2.05	3.083	1.02E+05	2.80E+04	PI-5I-L-3-CI-6
	5	Loader	6	2.06	7.25	2.53E+04	7.27E+03	PI-5I-L-4-CI-6
	5	Loader	6	2.05	15.5	9.73E+03	2.86E+03	PI-5I-L-5-CI-6
	5	Loader	6	2.07	32	6.37E+03	1.79E+03	PI-5I-L-6-CI-6
	5	Loader	6	2.13	63	4.02E+03	1.12E+03	PI-5I-L-7-CI-6
	5	Loader	6	2.07	125	2.25E+03	6.20E+02	PI-5I-L-8-CI-6
	5	Loader	6	2.08	146	2.39E+03	7.11E+02	PI-5I-L-9-CI-6
	5	Loader	7	1.08	0.083	4.73E+05	9.27E+04	PI-5I-L-1-CI-7
	5	Loader	7	1.07	1.083	2.20E+05	4.48E+04	PI-5I-L-2-CI-7
	5	Loader	7	1.05	3.083	1.08E+05	2.34E+04	PI-5I-L-3-CI-7
	5	Loader	7	1.06	7.25	1.86E+04	4.68E+03	PI-5I-L-4-CI-7
	5	Loader	7	1.06	15.5	6.82E+03	1.75E+03	PI-5I-L-5-CI-7
	5	Loader	7	1.06	32	5.49E+03	1.22E+03	PI-5I-L-6-CI-7
	5	Loader	7	1.09	63	3.12E+03	7.19E+02	PI-5I-L-7-CI-7
	5	Loader	7	1.06	125	2.99E+03	6.28E+02	PI-5I-L-8-CI-7
	5	Loader	7	1.07	146	1.65E+03	4.37E+02	PI-5I-L-9-CI-7
	5	Loader	8	0.63	0.083	1.06E+06	1.49E+05	PI-5I-L-1-CI-8
	5	Loader	8	0.63	1.083	2.96E+05	4.33E+04	PI-5I-L-2-CI-8
	5	Loader	8	0.62	3.083	9.85E+04	1.60E+04	PI-5I-L-3-CI-8
	5	Loader	8	0.62	7.25	2.14E+04	3.69E+03	PI-5I-L-4-CI-8
	5	Loader	8	0.62	15.5	9.80E+03	1.75E+03	PI-5I-L-5-CI-8
	5	Loader	8	0.63	32	6.37E+03	1.10E+03	PI-5I-L-6-CI-8
	5	Loader	8	0.64	63	6.41E+03	9.88E+02	PI-5I-L-7-CI-8
	5	Loader	8	0.62	125	3.52E+03	5.34E+02	PI-5I-L-8-CI-8
	5	Loader	8	0.63	146	3.22E+03	5.49E+02	PI-5I-L-9-CI-8
	5	Loader	9	0.22	0.083	8.60E+05	1.72E+05	PI-5I-L-1-CI-9
	5	Loader	9	0.22	1.083	1.76E+05	4.10E+04	PI-5I-L-2-CI-9
	5	Loader	9	0.22	3.083	4.64E+04	1.47E+04	PI-5I-L-3-CI-9
	5	Loader	9	0.22	7.25	7.04E+03	3.33E+03	PI-5I-L-4-CI-9
	5	Loader	9	0.22	15.5	6.59E+03	2.27E+03	PI-5I-L-5-CI-9
	5	Loader	9	0.22	32	7.68E+03	1.97E+03	PI-5I-L-6-CI-9
	5	Loader	9	0.22	63	5.72E+03	1.38E+03	PI-5I-L-7-CI-9
	5	Loader	9	0.22	125	2.75E+03	6.68E+02	PI-5I-L-8-CI-9
	5	Loader	9	0.22	146	1.79E+03	5.88E+02	PI-5I-L-9-CI-9
	6	Commander	1	30.70	0.083	4.64E+06	7.85E+05	PI-6I-C-1-CI-1
	6	Commander	1	30.79	1.083	3.22E+05	1.99E+05	PI-6I-C-2-CI-1
	6	Commander	1	30.82	3.083	-4.74E+05	2.37E+05	PI-6I-C-3-CI-1
	6	Commander	1	31.33	7.25	-1.20E+05	6.02E+04	PI-6I-C-4-CI-1
	6	Commander	1	31.37	15.5	-1.56E+04	2.56E+04	PI-6I-C-5-CI-1
	6	Commander	1	31.43	32	8.15E+04	2.60E+04	PI-6I-C-6-CI-1

Table A.7. *(continued)*

Phase	Shot	Position	CI stage	Median Dia (um)	MidTime (min)	U Conc, (ug/m3)	sd U Conc. (ug/m3)	SampleI
1	6	Commander	1	30.87	63	5.49E+04	1.48E+04	PI-6I-C-7-(
1	6	Commander	1	31.19	125	2.37E+04	6.93E+03	PI-6I-C-8-(
1	6	Commander	2	15.74	0.083	7.77E+05	3.17E+05	PI-6I-C-1-(
1	6	Commander	2	15.83	1.083	2.77E+05	1.45E+05	PI-6I-C-2-(
1	6	Commander	2	15.85	3.083	-2.97E+04	3.29E+04	PI-6I-C-3-(
1	6	Commander	2	16.39	7.25	-1.33E+04	1.14E+04	PI-6I-C-4-(
1	6	Commander	2	16.43	15.5	7.32E+02	7.58E+03	PI-6I-C-5-(
1	6	Commander	2	16.49	32	-3.35E+03	2.88E+03	PI-6I-C-6-(
1	6	Commander	2	15.91	63	-7.73E+02	1.52E+03	PI-6I-C-7-(
1	6	Commander	2	16.24	125	-2.39E+02	8.25E+02	PI-6I-C-8-(
1	6	Commander	3	10.53	0.083	6.63E+05	1.40E+05	PI-6I-C-1-(
1	6	Commander	3	10.60	1.083	3.40E+05	7.71E+04	PI-6I-C-2-(
1	6	Commander	3	10.61	3.083	6.22E+04	1.95E+04	PI-6I-C-3-(
1	6	Commander	3	10.97	7.25	1.99E+03	3.17E+03	PI-6I-C-4-(
1	6	Commander	3	10.99	15.5	1.05E+02	1.37E+03	PI-6I-C-5-(
1	6	Commander	3	11.04	32	5.29E+01	6.92E+02	PI-6I-C-6-(
1	6	Commander	3	10.65	63	-4.68E+02	2.70E+02	PI-6I-C-7-(
1	6	Commander	3	10.87	125	-2.44E+02	1.34E+02	PI-6I-C-8-(
1	6	Commander	4	6.72	0.083	9.64E+05	2.71E+05	PI-6I-C-1-(
1	6	Commander	4	6.76	1.083	1.94E+05	5.97E+04	PI-6I-C-2-(
1	6	Commander	4	6.77	3.083	1.20E+05	3.94E+04	PI-6I-C-3-(
1	6	Commander	4	6.99	7.25	1.17E+04	5.53E+03	PI-6I-C-4-(
1	6	Commander	4	7.01	15.5	2.66E+03	1.91E+03	PI-6I-C-5-(
1	6	Commander	4	7.04	32	1.80E+03	1.06E+03	PI-6I-C-6-(
1	6	Commander	4	6.79	63	7.31E+02	4.88E+02	PI-6I-C-7-(
1	6	Commander	4	6.93	125	3.81E+02	2.49E+02	PI-6I-C-8-(
1	6	Commander	5	4.04	0.083	9.15E+05	3.67E+05	PI-6I-C-1-(
1	6	Commander	5	4.06	1.083	3.86E+05	1.70E+05	PI-6I-C-2-(
1	6	Commander	5	4.07	3.083	1.78E+05	8.79E+04	PI-6I-C-3-(
1	6	Commander	5	4.20	7.25	4.19E+04	2.28E+04	PI-6I-C-4-(
1	6	Commander	5	4.21	15.5	9.83E+03	7.20E+03	PI-6I-C-5-(
1	6	Commander	5	4.23	32	-9.13E+02	1.63E+03	PI-6I-C-6-(
1	6	Commander	5	4.08	63	3.11E+03	1.98E+03	PI-6I-C-7-(
1	6	Commander	5	4.17	125	1.62E+03	1.04E+03	PI-6I-C-8-(
1	6	Commander	6	2.02	0.083	4.87E+05	2.16E+05	PI-6I-C-1-(
1	6	Commander	6	2.03	1.083	1.72E+05	9.21E+04	PI-6I-C-2-(
1	6	Commander	6	2.04	3.083	1.56E+05	8.36E+04	PI-6I-C-3-(
1	6	Commander	6	2.11	7.25	3.61E+04	2.13E+04	PI-6I-C-4-(
1	6	Commander	6	2.11	15.5	1.22E+04	8.09E+03	PI-6I-C-5-(
1	6	Commander	6	2.12	32	9.64E+02	1.93E+03	PI-6I-C-6-(
1	6	Commander	6	2.04	63	5.55E+03	3.11E+03	PI-6I-C-7-(
1	6	Commander	6	2.09	125	2.79E+03	1.56E+03	PI-6I-C-8-(

A.7. *(continued)*

e	Shot	Position	CI stage	Median Dia (um)	MidTime (min)	U Conc, (ug/m3)	sd U Conc. (ug/m3)	SampleID
	6	Commander	7	1.04	0.083	4.05E+05	1.07E+05	PI-6I-C-1-CI-7
	6	Commander	7	1.05	1.083	2.36E+05	7.11E+04	PI-6I-C-2-CI-7
	6	Commander	7	1.05	3.083	9.99E+04	3.83E+04	PI-6I-C-3-CI-7
	6	Commander	7	1.08	7.25	4.49E+04	1.50E+04	PI-6I-C-4-CI-7
	6	Commander	7	1.09	15.5	3.53E+03	3.12E+03	PI-6I-C-5-CI-7
	6	Commander	7	1.09	32	6.04E+03	2.52E+03	PI-6I-C-6-CI-7
	6	Commander	7	1.05	63	1.13E+03	8.09E+02	PI-6I-C-7-CI-7
	6	Commander	7	1.07	125	9.48E+02	5.28E+02	PI-6I-C-8-CI-7
	6	Commander	8	0.61	0.083	6.48E+05	1.75E+05	PI-6I-C-1-CI-8
	6	Commander	8	0.61	1.083	3.26E+05	1.05E+05	PI-6I-C-2-CI-8
	6	Commander	8	0.62	3.083	2.34E+05	7.87E+04	PI-6I-C-3-CI-8
	6	Commander	8	0.64	7.25	3.07E+04	1.57E+04	PI-6I-C-4-CI-8
	6	Commander	8	0.64	15.5	1.58E+04	7.33E+03	PI-6I-C-5-CI-8
	6	Commander	8	0.64	32	2.13E+03	2.25E+03	PI-6I-C-6-CI-8
	6	Commander	8	0.62	63	3.32E+03	1.77E+03	PI-6I-C-7-CI-8
	6	Commander	8	0.63	125	4.76E+03	1.64E+03	PI-6I-C-8-CI-8
	6	Commander	9	0.22	0.083	1.10E+06	3.81E+05	PI-6I-C-1-CI-9
	6	Commander	9	0.22	1.083	1.37E+05	7.19E+04	PI-6I-C-2-CI-9
	6	Commander	9	0.22	3.083	5.26E+04	4.21E+04	PI-6I-C-3-CI-9
	6	Commander	9	0.22	7.25	1.71E+04	1.45E+04	PI-6I-C-4-CI-9
	6	Commander	9	0.22	15.5	-4.10E+02	4.72E+03	PI-6I-C-5-CI-9
	6	Commander	9	0.22	32	1.22E+04	6.11E+03	PI-6I-C-6-CI-9
	6	Commander	9	0.22	63	4.80E+03	2.75E+03	PI-6I-C-7-CI-9
	6	Commander	9	0.22	125	3.96E+03	1.91E+03	PI-6I-C-8-CI-9
	6	Driver	1	30.56	0.083	1.57E+05	3.51E+04	PI-6I-D-1-CI-1
	6	Driver	1	31.33	1.083	-2.35E+04	3.36E+04	PI-6I-D-2-CI-1
	6	Driver	1	30.65	3.083	1.81E+05	6.75E+04	PI-6I-D-3-CI-1
	6	Driver	1	30.14	7.25	-1.48E+04	8.43E+03	PI-6I-D-4-CI-1
	6	Driver	1	30.80	15.5	-1.24E+04	6.22E+03	PI-6I-D-5-CI-1
	6	Driver	1	30.93	32	7.82E+03	4.23E+03	PI-6I-D-6-CI-1
	6	Driver	1	31.25	63	5.63E+03	2.55E+03	PI-6I-D-7-CI-1
	6	Driver	1	30.57	125	2.40E+03	1.12E+03	PI-6I-D-8-CI-1
	6	Driver	2	15.59	0.083	1.81E+04	9.97E+03	PI-6I-D-1-CI-2
	6	Driver	2	16.39	1.083	7.93E+04	4.16E+04	PI-6I-D-2-CI-2
	6	Driver	2	15.68	3.083	2.49E+04	1.65E+04	PI-6I-D-3-CI-2
	6	Driver	2	15.17	7.25	1.35E+04	7.59E+03	PI-6I-D-4-CI-2
	6	Driver	2	15.83	15.5	3.61E+03	2.50E+03	PI-6I-D-5-CI-2
	6	Driver	2	15.97	32	5.65E+02	7.22E+02	PI-6I-D-6-CI-2
	6	Driver	2	16.31	63	1.46E+02	3.46E+02	PI-6I-D-7-CI-2
	6	Driver	2	15.60	125	4.38E+02	2.89E+02	PI-6I-D-8-CI-2
	6	Driver	3	10.43	0.083	1.11E+04	4.14E+03	PI-6I-D-1-CI-3
	6	Driver	3	10.97	1.083	2.54E+05	7.25E+04	PI-6I-D-2-CI-3

Table A.7. *(continued)*

Phase	Shot	Position	CI stage	Median Dia (um)	MidTime (min)	U Conc, (ug/m3)	sd U Conc. (ug/m3)	SampleI
1	6	Driver	3	10.50	3.083	6.58E+04	2.18E+04	PI-6I-D-3-C
1	6	Driver	3	10.15	7.25	2.58E+03	2.04E+03	PI-6I-D-4-C
1	6	Driver	3	10.60	15.5	9.17E+02	1.10E+03	PI-6I-D-5-C
1	6	Driver	3	10.69	32	-2.98E+01	4.28E+02	PI-6I-D-6-C
1	6	Driver	3	10.91	63	3.73E+02	3.05E+02	PI-6I-D-7-C
1	6	Driver	3	10.44	125	-7.11E+00	1.16E+02	PI-6I-D-8-C
1	6	Driver	4	6.65	0.083	5.22E+04	NA[(a)]	PI-6I-D-1-C
1	6	Driver	4	6.99	1.083	1.94E+05	7.87E+04	PI-6I-D-2-C
1	6	Driver	4	6.69	3.083	9.21E+04	3.99E+04	PI-6I-D-3-C
1	6	Driver	4	6.47	7.25	2.79E+03	3.22E+03	PI-6I-D-4-C
1	6	Driver	4	6.76	15.5	1.52E+03	1.72E+03	PI-6I-D-5-C
1	6	Driver	4	6.82	32	3.37E+02	7.60E+02	PI-6I-D-6-C
1	6	Driver	4	6.96	63	1.08E+03	6.99E+02	PI-6I-D-7-C
1	6	Driver	4	6.66	125	4.96E+02	3.01E+02	PI-6I-D-8-C
1	6	Driver	5	4.00	0.083	9.32E+03	5.21E+03	PI-6I-D-1-C
1	6	Driver	5	4.20	1.083	1.65E+05	6.73E+04	PI-6I-D-2-C
1	6	Driver	5	4.02	3.083	2.67E+05	NA[(a)]	PI-6I-D-3-C
1	6	Driver	5	3.89	7.25	6.69E+03	3.55E+03	PI-6I-D-4-C
1	6	Driver	5	4.06	15.5	3.25E+03	1.96E+03	PI-6I-D-5-C
1	6	Driver	5	4.10	32	2.67E+03	1.33E+03	PI-6I-D-6-C
1	6	Driver	5	4.18	63	3.30E+03	1.44E+03	PI-6I-D-7-C
1	6	Driver	5	4.00	125	1.56E+03	6.77E+02	PI-6I-D-8-C
1	6	Driver	6	2.00	0.083	6.91E+03	3.31E+03	PI-6I-D-1-C
1	6	Driver	6	2.11	1.083	1.81E+05	5.08E+04	PI-6I-D-2-C
1	6	Driver	6	2.02	3.083	7.28E+04	2.21E+04	PI-6I-D-3-C
1	6	Driver	6	1.95	7.25	6.67E+03	2.98E+03	PI-6I-D-4-C
1	6	Driver	6	2.03	15.5	2.84E+03	1.46E+03	PI-6I-D-5-C
1	6	Driver	6	2.05	32	3.06E+03	1.14E+03	PI-6I-D-6-C
1	6	Driver	6	2.10	63	2.01E+03	6.85E+02	PI-6I-D-7-C
1	6	Driver	6	2.01	125	1.64E+03	4.95E+02	PI-6I-D-8-C
1	6	Driver	7	1.03	0.083	9.12E+03	3.77E+03	PI-6I-D-1-C
1	6	Driver	7	1.08	1.083	1.79E+05	3.78E+04	PI-6I-D-2-C
1	6	Driver	7	1.04	3.083	5.79E+04	1.44E+04	PI-6I-D-3-C
1	6	Driver	7	1.00	7.25	1.08E+04	3.27E+03	PI-6I-D-4-C
1	6	Driver	7	1.05	15.5	4.34E+03	1.49E+03	PI-6I-D-5-C
1	6	Driver	7	1.06	32	5.00E+03	1.20E+03	PI-6I-D-6-C
1	6	Driver	7	1.08	63	2.09E+03	6.00E+02	PI-6I-D-7-C
1	6	Driver	7	1.03	125	2.81E+03	5.99E+02	PI-6I-D-8-C
1	6	Driver	8	0.61	0.083	2.46E+04	4.96E+03	PI-6I-D-1-C
1	6	Driver	8	0.64	1.083	2.37E+05	3.61E+04	PI-6I-D-2-C
1	6	Driver	8	0.61	3.083	1.38E+05	2.16E+04	PI-6I-D-3-C
1	6	Driver	8	0.59	7.25	7.08E+03	2.25E+03	PI-6I-D-4-C

A.7. *(continued)*

se	Shot	Position	CI stage	Median Dia (um)	MidTime (min)	U Conc, (ug/m3)	sd U Conc. (ug/m3)	SampleID
	6	Driver	8	0.61	15.5	8.47E+03	1.72E+03	PI-6I-D-5-CI-8
	6	Driver	8	0.62	32	4.11E+03	8.91E+02	PI-6I-D-6-CI-8
	6	Driver	8	0.63	63	4.80E+03	8.12E+02	PI-6I-D-7-CI-8
	6	Driver	8	0.61	125	3.92E+03	6.02E+02	PI-6I-D-8-CI-8
	6	Driver	9	0.21	0.083	8.85E+03	3.44E+03	PI-6I-D-1-CI-9
	6	Driver	9	0.22	1.083	1.76E+05	4.05E+04	PI-6I-D-2-CI-9
	6	Driver	9	0.21	3.083	1.79E+04	9.56E+03	PI-6I-D-3-CI-9
	6	Driver	9	0.21	7.25	7.69E+03	3.35E+03	PI-6I-D-4-CI-9
	6	Driver	9	0.22	15.5	9.52E+03	2.83E+03	PI-6I-D-5-CI-9
	6	Driver	9	0.22	32	5.81E+03	1.65E+03	PI-6I-D-6-CI-9
	6	Driver	9	0.22	63	4.34E+03	1.11E+03	PI-6I-D-7-CI-9
	6	Driver	9	0.21	125	3.19E+03	7.42E+02	PI-6I-D-8-CI-9
	6	Loader	1	31.34	0.17	8.79E+02	4.51E+02	PI-6I-L-1-CI-1
	6	Loader	1	31.19	11.5	3.04E+03	7.63E+02	PI-6I-L-2-CI-1
	6	Loader	1	30.93	29.5	1.70E+03	7.09E+02	PI-6I-L-3-CI-1
	6	Loader	1	31.23	43.5	1.86E+03	7.10E+02	PI-6I-L-4-CI-1
	6	Loader	1	31.06	62	1.69E+03	5.47E+02	PI-6I-L-5-CI-1
	6	Loader	1	31.13	75	2.45E+03	3.33E+03	PI-6I-L-6-CI-1
	6	Loader	1	31.73	78	3.59E+03	2.17E+03	PI-6I-L-7-CI-1
	6	Loader	1	31.17	84	8.06E+02	8.83E+02	PI-6I-L-8-CI-1
	6	Loader	1	31.28	96	4.55E+03	1.13E+03	PI-6I-L-9-CI-1
	6	Loader	2	16.39	0.17	1.80E+02	2.36E+02	PI-6I-L-1-CI-2
	6	Loader	2	16.24	11.5	3.59E+02	2.92E+02	PI-6I-L-2-CI-2
	6	Loader	2	15.97	29.5	4.96E+02	4.42E+02	PI-6I-L-3-CI-2
	6	Loader	2	16.29	43.5	1.73E+01	2.49E+02	PI-6I-L-4-CI-2
	6	Loader	2	16.11	62	6.97E+02	4.26E+02	PI-6I-L-5-CI-2
	6	Loader	2	16.18	75	-1.50E+03	1.25E+03	PI-6I-L-6-CI-2
	6	Loader	2	16.81	78	-4.60E+02	7.67E+02	PI-6I-L-7-CI-2
	6	Loader	2	16.21	84	6.86E+02	6.72E+02	PI-6I-L-8-CI-2
	6	Loader	2	16.34	96	7.64E+02	5.14E+02	PI-6I-L-9-CI-2
	6	Loader	3	10.97	0.17	3.59E+02	1.01E+02	PI-6I-L-1-CI-3
	6	Loader	3	10.87	11.5	4.24E+02	1.13E+02	PI-6I-L-2-CI-3
	6	Loader	3	10.69	29.5	3.26E+02	1.12E+02	PI-6I-L-3-CI-3
	6	Loader	3	10.90	43.5	2.71E+02	9.15E+01	PI-6I-L-4-CI-3
	6	Loader	3	10.78	62	4.13E+02	1.11E+02	PI-6I-L-5-CI-3
	6	Loader	3	10.83	75	1.71E+03	6.54E+02	PI-6I-L-6-CI-3
	6	Loader	3	11.25	78	6.50E+02	3.05E+02	PI-6I-L-7-CI-3
	6	Loader	3	10.85	84	8.14E+02	2.32E+02	PI-6I-L-8-CI-3
	6	Loader	3	10.94	96	6.41E+02	1.66E+02	PI-6I-L-9-CI-3
	6	Loader	4	7.00	0.17	3.16E+02	1.20E+02	PI-6I-L-1-CI-4
	6	Loader	4	6.93	11.5	5.52E+02	1.77E+02	PI-6I-L-2-CI-4
	6	Loader	4	6.82	29.5	5.17E+02	1.80E+02	PI-6I-L-3-CI-4

Table A.7. *(continued)*

Phase	Shot	Position	CI stage	Median Dia (um)	MidTime (min)	U Conc, (ug/m3)	sd U Conc. (ug/m3)	SampleI
1	6	Loader	4	6.95	43.5	3.32E+02	1.32E+02	PI-6I-L-4-C
1	6	Loader	4	6.88	62	4.04E+02	1.36E+02	PI-6I-L-5-C
1	6	Loader	4	6.91	75	2.86E+03	1.04E+03	PI-6I-L-6-C
1	6	Loader	4	7.18	78	9.40E+02	4.05E+02	PI-6I-L-7-C
1	6	Loader	4	6.92	84	6.05E+02	2.36E+02	PI-6I-L-8-C
1	6	Loader	4	6.97	96	3.64E+02	1.38E+02	PI-6I-L-9-C
1	6	Loader	5	4.20	0.17	7.09E+02	3.72E+02	PI-6I-L-1-C
1	6	Loader	5	4.17	11.5	4.84E+02	2.67E+02	PI-6I-L-2-C
1	6	Loader	5	4.10	29.5	8.58E+02	4.67E+02	PI-6I-L-3-C
1	6	Loader	5	4.18	43.5	7.49E+02	4.12E+02	PI-6I-L-4-C
1	6	Loader	5	4.13	62	4.02E+02	2.41E+02	PI-6I-L-5-C
1	6	Loader	5	4.15	75	2.93E+03	1.97E+03	PI-6I-L-6-C
1	6	Loader	5	4.31	78	2.49E+03	1.44E+03	PI-6I-L-7-C
1	6	Loader	5	4.16	84	1.10E+03	6.55E+02	PI-6I-L-8-C
1	6	Loader	5	4.19	96	6.14E+02	3.44E+02	PI-6I-L-9-C
1	6	Loader	6	2.11	0.17	7.72E+02	4.90E+02	PI-6I-L-1-C
1	6	Loader	6	2.09	11.5	5.40E+02	3.52E+02	PI-6I-L-2-C
1	6	Loader	6	2.05	29.5	2.86E+03	1.56E+03	PI-6I-L-3-C
1	6	Loader	6	2.09	43.5	1.38E+03	8.06E+02	PI-6I-L-4-C
1	6	Loader	6	2.07	62	6.69E+02	4.16E+02	PI-6I-L-5-C
1	6	Loader	6	2.08	75	6.07E+03	3.90E+03	PI-6I-L-6-C
1	6	Loader	6	2.16	78	1.71E+03	1.31E+03	PI-6I-L-7-C
1	6	Loader	6	2.08	84	1.00E+03	7.12E+02	PI-6I-L-8-C
1	6	Loader	6	2.10	96	1.09E+03	6.32E+02	PI-6I-L-9-C
1	6	Loader	7	1.08	0.17	3.72E+02	1.54E+02	PI-6I-L-1-C
1	6	Loader	7	1.07	11.5	3.23E+02	1.35E+02	PI-6I-L-2-C
1	6	Loader	7	1.06	29.5	9.81E+02	3.28E+02	PI-6I-L-3-C
1	6	Loader	7	1.08	43.5	9.78E+02	3.15E+02	PI-6I-L-4-C
1	6	Loader	7	1.06	62	3.32E+02	1.28E+02	PI-6I-L-5-C
1	6	Loader	7	1.07	75	2.81E+03	1.19E+03	PI-6I-L-6-C
1	6	Loader	7	1.11	78	1.52E+03	6.30E+02	PI-6I-L-7-C
1	6	Loader	7	1.07	84	7.05E+02	2.93E+02	PI-6I-L-8-C
1	6	Loader	7	1.08	96	4.89E+02	1.87E+02	PI-6I-L-9-C
1	6	Loader	8	0.64	0.17	2.19E+03	8.29E+02	PI-6I-L-1-C
1	6	Loader	8	0.63	11.5	7.59E+02	3.74E+02	PI-6I-L-2-C
1	6	Loader	8	0.62	29.5	2.93E+03	1.09E+03	PI-6I-L-3-C
1	6	Loader	8	0.63	43.5	2.47E+03	9.13E+02	PI-6I-L-4-C
1	6	Loader	8	0.63	62	3.21E+02	1.97E+02	PI-6I-L-5-C
1	6	Loader	8	0.63	75	7.93E+03	3.64E+03	PI-6I-L-6-C
1	6	Loader	8	0.65	78	3.63E+03	1.63E+03	PI-6I-L-7-C
1	6	Loader	8	0.63	84	2.50E+03	9.86E+02	PI-6I-L-8-C
1	6	Loader	8	0.63	96	9.09E+02	4.17E+02	PI-6I-L-9-C

A.7. *(continued)*

e	Shot	Position	CI stage	Median Dia (um)	MidTime (min)	U Conc, (ug/m3)	sd U Conc. (ug/m3)	SampleID
	6	Loader	9	0.22	0.17	5.51E+02	3.57E+02	PI-6I-L-1-CI-9
	6	Loader	9	0.22	11.5	5.48E+02	3.19E+02	PI-6I-L-2-CI-9
	6	Loader	9	0.22	29.5	1.83E+03	8.78E+02	PI-6I-L-3-CI-9
	6	Loader	9	0.22	43.5	1.39E+03	6.95E+02	PI-6I-L-4-CI-9
	6	Loader	9	0.22	62	3.09E+02	2.28E+02	PI-6I-L-5-CI-9
	6	Loader	9	0.22	75	4.57E+03	3.06E+03	PI-6I-L-6-CI-9
	6	Loader	9	0.22	78	5.69E+03	2.65E+03	PI-6I-L-7-CI-9
	6	Loader	9	0.22	84	2.49E+03	1.17E+03	PI-6I-L-8-CI-9
	6	Loader	9	0.22	96	1.04E+03	5.20E+02	PI-6I-L-9-CI-9
	7	Commander	1	30.67	0.083	4.23E+06	7.15E+05	PI-7I-C-1-CI-1
	7	Commander	1	30.80	1.083	8.79E+05	3.53E+05	PI-7I-C-2-CI-1
	7	Commander	1	30.76	3.125	8.54E+04	1.62E+05	PI-7I-C-3-CI-1
	7	Commander	1	31.03	7.5	5.18E+04	4.73E+04	PI-7I-C-4-CI-1
	7	Commander	1	31.10	16	5.11E+04	2.68E+04	PI-7I-C-5-CI-1
	7	Commander	1	31.27	33	1.09E+04	1.15E+04	PI-7I-C-6-CI-1
	7	Commander	1	30.72	65	3.88E+02	4.72E+03	PI-7I-C-7-CI-1
	7	Commander	1	31.18	125	-1.65E+02	4.93E+03	PI-7I-C-8-CI-1
	7	Commander	2	15.70	0.083	1.01E+06	4.11E+05	PI-7I-C-1-CI-2
	7	Commander	2	15.84	1.083	1.85E+05	1.26E+05	PI-7I-C-2-CI-2
	7	Commander	2	15.80	3.125	4.31E+04	5.49E+04	PI-7I-C-3-CI-2
	7	Commander	2	16.07	7.5	1.16E+04	1.48E+04	PI-7I-C-4-CI-2
	7	Commander	2	16.15	16	1.41E+04	1.02E+04	PI-7I-C-5-CI-2
	7	Commander	2	16.32	33	-3.76E+03	1.96E+03	PI-7I-C-6-CI-2
	7	Commander	2	15.75	65	-5.15E+02	1.14E+03	PI-7I-C-7-CI-2
	7	Commander	2	16.23	125	-1.04E+03	1.09E+03	PI-7I-C-8-CI-2
	7	Commander	3	10.51	0.083	1.05E+06	2.20E+05	PI-7I-C-1-CI-3
	7	Commander	3	10.60	1.083	2.89E+05	6.53E+04	PI-7I-C-2-CI-3
	7	Commander	3	10.58	3.125	2.99E+05	6.59E+04	PI-7I-C-3-CI-3
	7	Commander	3	10.76	7.5	5.25E+04	1.20E+04	PI-7I-C-4-CI-3
	7	Commander	3	10.81	16	3.24E+03	1.15E+03	PI-7I-C-5-CI-3
	7	Commander	3	10.92	33	4.87E+02	3.53E+02	PI-7I-C-6-CI-3
	7	Commander	3	10.54	65	5.89E+02	2.35E+02	PI-7I-C-7-CI-3
	7	Commander	3	10.86	125	3.69E+02	1.95E+02	PI-7I-C-8-CI-3
	7	Commander	4	6.70	0.083	6.64E+05	1.87E+05	PI-7I-C-1-CI-4
	7	Commander	4	6.76	1.083	2.96E+04	1.30E+04	PI-7I-C-2-CI-4
	7	Commander	4	6.74	3.125	5.72E+04	1.88E+04	PI-7I-C-3-CI-4
	7	Commander	4	6.86	7.5	3.00E+04	9.20E+03	PI-7I-C-4-CI-4
	7	Commander	4	6.89	16	6.57E+03	2.24E+03	PI-7I-C-5-CI-4
	7	Commander	4	6.97	33	1.65E+03	6.77E+02	PI-7I-C-6-CI-4
	7	Commander	4	6.72	65	1.78E+03	5.85E+02	PI-7I-C-7-CI-4
	7	Commander	4	6.93	125	1.04E+03	3.83E+02	PI-7I-C-8-CI-4
	7	Commander	5	4.03	0.083	1.00E+06	4.02E+05	PI-7I-C-1-CI-5

Table A.7. *(continued)*

Phase	Shot	Position	CI stage	Median Dia (um)	MidTime (min)	U Conc, (ug/m3)	sd U Conc. (ug/m3)	SampleI
1	7	Commander	5	4.06	1.083	2.73E+04	1.92E+04	PI-7I-C-2-(
1	7	Commander	5	4.05	3.125	2.19E+05	9.29E+04	PI-7I-C-3-(
1	7	Commander	5	4.12	7.5	2.37E+04	1.10E+04	PI-7I-C-4-(
1	7	Commander	5	4.14	16	1.15E+04	5.31E+03	PI-7I-C-5-(
1	7	Commander	5	4.19	33	7.06E+03	3.16E+03	PI-7I-C-6-(
1	7	Commander	5	4.04	65	2.89E+03	1.32E+03	PI-7I-C-7-(
1	7	Commander	5	4.16	125	3.86E+03	1.71E+03	PI-7I-C-8-(
1	7	Commander	6	2.02	0.083	1.01E+06	4.42E+05	PI-7I-C-1-(
1	7	Commander	6	2.04	1.083	-1.24E+03	1.92E+04	PI-7I-C-2-(
1	7	Commander	6	2.03	3.125	8.61E+04	4.84E+04	PI-7I-C-3-(
1	7	Commander	6	2.07	7.5	3.88E+04	1.97E+04	PI-7I-C-4-(
1	7	Commander	6	2.08	16	8.35E+03	5.05E+03	PI-7I-C-5-(
1	7	Commander	6	2.10	33	6.16E+03	3.37E+03	PI-7I-C-6-(
1	7	Commander	6	2.02	65	2.82E+03	1.49E+03	PI-7I-C-7-(
1	7	Commander	6	2.09	125	3.62E+03	1.84E+03	PI-7I-C-8-(
1	7	Commander	7	1.04	0.083	6.53E+05	1.69E+05	PI-7I-C-1-(
1	7	Commander	7	1.05	1.083	-1.51E+03	1.24E+04	PI-7I-C-2-(
1	7	Commander	7	1.04	3.125	1.65E+05	4.99E+04	PI-7I-C-3-(
1	7	Commander	7	1.06	7.5	3.79E+04	1.17E+04	PI-7I-C-4-(
1	7	Commander	7	1.07	16	9.67E+03	3.48E+03	PI-7I-C-5-(
1	7	Commander	7	1.08	33	3.79E+03	1.47E+03	PI-7I-C-6-(
1	7	Commander	7	1.04	65	2.64E+03	9.06E+02	PI-7I-C-7-(
1	7	Commander	7	1.07	125	3.68E+03	1.17E+03	PI-7I-C-8-(
1	7	Commander	8	0.61	0.083	7.31E+05	1.94E+05	PI-7I-C-1-(
1	7	Commander	8	0.62	1.083	3.72E+04	2.67E+04	PI-7I-C-2-(
1	7	Commander	8	0.61	3.125	2.04E+05	6.10E+04	PI-7I-C-3-(
1	7	Commander	8	0.62	7.5	4.40E+04	1.39E+04	PI-7I-C-4-(
1	7	Commander	8	0.63	16	1.16E+04	4.32E+03	PI-7I-C-5-(
1	7	Commander	8	0.63	33	1.33E+04	3.91E+03	PI-7I-C-6-(
1	7	Commander	8	0.61	65	4.43E+03	1.42E+03	PI-7I-C-7-(
1	7	Commander	8	0.63	125	4.25E+03	1.35E+03	PI-7I-C-8-(
1	7	Commander	9	0.22	0.083	1.92E+06	6.61E+05	PI-7I-C-1-(
1	7	Commander	9	0.22	1.083	1.14E+04	1.96E+04	PI-7I-C-2-(
1	7	Commander	9	0.22	3.125	5.46E+04	2.91E+04	PI-7I-C-3-(
1	7	Commander	9	0.22	7.5	8.95E+03	6.35E+03	PI-7I-C-4-(
1	7	Commander	9	0.22	16	2.97E+03	2.29E+03	PI-7I-C-5-(
1	7	Commander	9	0.22	33	3.54E+03	2.01E+03	PI-7I-C-6-(
1	7	Commander	9	0.22	65	5.84E+03	2.31E+03	PI-7I-C-7-(
1	7	Commander	9	0.22	125	5.45E+03	2.12E+03	PI-7I-C-8-(
1	7	Driver	1	31.15	0.083	1.62E+06	2.78E+05	PI-7I-D-1-(
1	7	Driver	1	31.30	1.083	1.45E+06	3.19E+05	PI-7I-D-2-(
1	7	Driver	1	30.64	3.125	1.27E+05	7.14E+04	PI-7I-D-3-(

A.7. *(continued)*

e	Shot	Position	CI stage	Median Dia (um)	MidTime (min)	U Conc, (ug/m3)	sd U Conc. (ug/m3)	SampleID
	7	Driver	1	30.62	7.5	-4.12E+03	1.28E+04	PI-7I-D-4-CI-1
	7	Driver	1	30.85	16	5.98E+03	7.65E+03	PI-7I-D-5-CI-1
	7	Driver	1	30.78	33	1.48E+04	5.56E+03	PI-7I-D-6-CI-1
	7	Driver	1	30.94	65	8.50E+03	2.99E+03	PI-7I-D-7-CI-1
	7	Driver	1	30.86	125	4.63E+03	2.37E+03	PI-7I-D-8-CI-1
	7	Driver	2	16.20	0.083	3.56E+05	1.46E+05	PI-7I-D-1-CI-2
	7	Driver	2	16.36	1.083	1.84E+05	9.55E+04	PI-7I-D-2-CI-2
	7	Driver	2	15.67	3.125	7.10E+04	4.17E+04	PI-7I-D-3-CI-2
	7	Driver	2	15.66	7.5	1.04E+04	7.62E+03	PI-7I-D-4-CI-2
	7	Driver	2	15.89	16	-2.51E+03	1.57E+03	PI-7I-D-5-CI-2
	7	Driver	2	15.82	33	-6.20E+02	8.77E+02	PI-7I-D-6-CI-2
	7	Driver	2	15.98	65	-3.96E+02	4.33E+02	PI-7I-D-7-CI-2
	7	Driver	2	15.89	125	-3.13E+02	4.38E+02	PI-7I-D-8-CI-2
	7	Driver	3	10.84	0.083	9.24E+05	1.96E+05	PI-7I-D-1-CI-3
	7	Driver	3	10.95	1.083	4.81E+05	1.04E+05	PI-7I-D-2-CI-3
	7	Driver	3	10.49	3.125	1.97E+05	4.36E+04	PI-7I-D-3-CI-3
	7	Driver	3	10.48	7.5	2.00E+04	4.82E+03	PI-7I-D-4-CI-3
	7	Driver	3	10.63	16	3.41E+03	1.04E+03	PI-7I-D-5-CI-3
	7	Driver	3	10.59	33	9.59E+02	3.59E+02	PI-7I-D-6-CI-3
	7	Driver	3	10.69	65	2.41E+02	1.48E+02	PI-7I-D-7-CI-3
	7	Driver	3	10.64	125	1.77E+02	1.35E+02	PI-7I-D-8-CI-3
	7	Driver	4	6.91	0.083	7.80E+05	2.20E+05	PI-7I-D-1-CI-4
	7	Driver	4	6.98	1.083	4.45E+05	1.29E+05	PI-7I-D-2-CI-4
	7	Driver	4	6.69	3.125	1.68E+05	4.96E+04	PI-7I-D-3-CI-4
	7	Driver	4	6.68	7.5	2.15E+04	6.81E+03	PI-7I-D-4-CI-4
	7	Driver	4	6.78	16	1.39E+03	7.71E+02	PI-7I-D-5-CI-4
	7	Driver	4	6.75	33	2.39E+03	8.21E+02	PI-7I-D-6-CI-4
	7	Driver	4	6.82	65	1.11E+03	3.95E+02	PI-7I-D-7-CI-4
	7	Driver	4	6.78	125	5.09E+02	2.42E+02	PI-7I-D-8-CI-4
	7	Driver	5	4.15	0.083	8.52E+05	3.36E+05	PI-7I-D-1-CI-5
	7	Driver	5	4.20	1.083	6.68E+05	2.67E+05	PI-7I-D-2-CI-5
	7	Driver	5	4.02	3.125	1.43E+05	5.90E+04	PI-7I-D-3-CI-5
	7	Driver	5	4.02	7.5	4.20E+04	1.73E+04	PI-7I-D-4-CI-5
	7	Driver	5	4.07	16	1.37E+04	5.83E+03	PI-7I-D-5-CI-5
	7	Driver	5	4.06	33	2.71E+03	1.25E+03	PI-7I-D-6-CI-5
	7	Driver	5	4.10	65	2.81E+03	1.21E+03	PI-7I-D-7-CI-5
	7	Driver	5	4.08	125	1.62E+03	7.36E+02	PI-7I-D-8-CI-5
	7	Driver	6	2.08	0.083	8.01E+05	3.28E+05	PI-7I-D-1-CI-6
	7	Driver	6	2.10	1.083	6.99E+05	2.86E+05	PI-7I-D-2-CI-6
	7	Driver	6	2.01	3.125	2.88E+05	1.19E+05	PI-7I-D-3-CI-6
	7	Driver	6	2.01	7.5	8.65E+04	3.56E+04	PI-7I-D-4-CI-6
	7	Driver	6	2.04	16	2.81E+04	1.17E+04	PI-7I-D-5-CI-6

Table A.7. *(continued)*

Phase	Shot	Position	CI stage	Median Dia (um)	MidTime (min)	U Conc, (ug/m3)	sd U Conc. (ug/m3)	SampleI
1	7	Driver	6	2.03	33	9.30E+03	3.86E+03	PI-7I-D-6-(
1	7	Driver	6	2.05	65	8.43E+03	3.46E+03	PI-7I-D-7-(
1	7	Driver	6	2.04	125	4.99E+03	2.09E+03	PI-7I-D-8-(
1	7	Driver	7	1.07	0.083	5.04E+05	1.26E+05	PI-7I-D-1-(
1	7	Driver	7	1.08	1.083	4.33E+05	1.12E+05	PI-7I-D-2-(
1	7	Driver	7	1.04	3.125	2.93E+05	7.53E+04	PI-7I-D-3-(
1	7	Driver	7	1.03	7.5	7.72E+04	1.98E+04	PI-7I-D-4-(
1	7	Driver	7	1.05	16	1.31E+04	3.92E+03	PI-7I-D-5-(
1	7	Driver	7	1.04	33	5.99E+03	1.76E+03	PI-7I-D-6-(
1	7	Driver	7	1.06	65	4.70E+03	1.29E+03	PI-7I-D-7-(
1	7	Driver	7	1.05	125	2.97E+03	9.03E+02	PI-7I-D-8-(
1	7	Driver	8	0.63	0.083	5.21E+05	1.28E+05	PI-7I-D-1-(
1	7	Driver	8	0.64	1.083	6.63E+05	1.60E+05	PI-7I-D-2-(
1	7	Driver	8	0.61	3.125	3.09E+05	7.49E+04	PI-7I-D-3-(
1	7	Driver	8	0.61	7.5	8.12E+04	1.97E+04	PI-7I-D-4-(
1	7	Driver	8	0.62	16	1.41E+04	3.99E+03	PI-7I-D-5-(
1	7	Driver	8	0.61	33	5.66E+03	1.77E+03	PI-7I-D-6-(
1	7	Driver	8	0.62	65	6.85E+03	1.72E+03	PI-7I-D-7-(
1	7	Driver	8	0.62	125	6.20E+03	1.55E+03	PI-7I-D-8-(
1	7	Driver	9	0.22	0.083	1.68E+06	5.35E+05	PI-7I-D-1-(
1	7	Driver	9	0.22	1.083	9.63E+05	3.14E+05	PI-7I-D-2-(
1	7	Driver	9	0.21	3.125	1.88E+05	6.57E+04	PI-7I-D-3-(
1	7	Driver	9	0.21	7.5	5.62E+04	1.96E+04	PI-7I-D-4-(
1	7	Driver	9	0.22	16	1.63E+04	6.19E+03	PI-7I-D-5-(
1	7	Driver	9	0.22	33	1.47E+04	5.19E+03	PI-7I-D-6-(
1	7	Driver	9	0.22	65	1.19E+04	4.03E+03	PI-7I-D-7-(
1	7	Driver	9	0.22	125	6.52E+03	2.30E+03	PI-7I-D-8-(
1	7	Gunner	1	30.61	0.083	1.73E+06	3.67E+05	PI-7I-G-1-(
1	7	Gunner	1	30.65	1.083	1.42E+05	9.29E+04	PI-7I-G-3-(
1	7	Gunner	1	30.73	3.125	-6.13E+04	3.73E+04	PI-7I-G-4-(
1	7	Gunner	1	30.61	7.5	2.78E+04	1.62E+04	PI-7I-G-5-(
1	7	Gunner	1	30.51	16	2.78E+03	5.99E+03	PI-7I-G-6-(
1	7	Gunner	1	31.44	33	-2.21E+03	2.80E+03	PI-7I-G-7-(
1	7	Gunner	1	31.37	65	6.78E+02	1.66E+03	PI-7I-G-8-(
1	7	Gunner	1	31.42	125	1.38E+03	1.79E+03	PI-7I-G-9-(
1	7	Gunner	2	15.64	0.083	5.64E+05	2.70E+05	PI-7I-G-1-(
1	7	Gunner	2	15.69	1.083	1.73E+05	9.82E+04	PI-7I-G-3-(
1	7	Gunner	2	15.76	3.125	5.95E+04	3.99E+04	PI-7I-G-4-(
1	7	Gunner	2	15.64	7.5	1.77E+04	1.11E+04	PI-7I-G-5-(
1	7	Gunner	2	15.54	16	6.22E+02	1.94E+03	PI-7I-G-6-(
1	7	Gunner	2	16.50	33	-1.58E+02	8.93E+02	PI-7I-G-7-(
1	7	Gunner	2	16.43	65	-2.46E+02	4.20E+02	PI-7I-G-8-(

A.7. (continued)

e	Shot	Position	CI stage	Median Dia (um)	MidTime (min)	U Conc, (ug/m3)	sd U Conc. (ug/m3)	SampleID
	7	Gunner	2	16.48	125	5.91E+00	4.87E+02	PI-7I-G-9-CI-2
	7	Gunner	3	10.47	0.083	5.09E+05	1.39E+05	PI-7I-G-1-CI-3
	7	Gunner	3	10.50	1.083	2.80E+05	7.78E+04	PI-7I-G-3-CI-3
	7	Gunner	3	10.55	3.125	1.77E+05	4.94E+04	PI-7I-G-4-CI-3
	7	Gunner	3	10.47	7.5	8.11E+04	2.23E+04	PI-7I-G-5-CI-3
	7	Gunner	3	10.40	16	7.16E+03	2.11E+03	PI-7I-G-6-CI-3
	7	Gunner	3	11.04	33	8.53E+02	NA[a]	PI-7I-G-7-CI-3
	7	Gunner	3	11.00	65	4.89E+02	2.03E+02	PI-7I-G-8-CI-3
	7	Gunner	3	11.03	125	2.27E+02	1.33E+02	PI-7I-G-9-CI-3
	7	Gunner	4	6.68	0.083	4.74E+05	1.78E+05	PI-7I-G-1-CI-4
	7	Gunner	4	6.69	1.083	3.03E+05	1.15E+05	PI-7I-G-3-CI-4
	7	Gunner	4	6.73	3.125	2.07E+05	7.83E+04	PI-7I-G-4-CI-4
	7	Gunner	4	6.67	7.5	7.60E+04	2.87E+04	PI-7I-G-5-CI-4
	7	Gunner	4	6.63	16	1.22E+04	4.70E+03	PI-7I-G-6-CI-4
	7	Gunner	4	7.04	33	2.80E+03	1.14E+03	PI-7I-G-7-CI-4
	7	Gunner	4	7.01	65	2.08E+03	8.21E+02	PI-7I-G-8-CI-4
	7	Gunner	4	7.03	125	1.05E+03	4.36E+02	PI-7I-G-9-CI-4
	7	Gunner	5	4.01	0.083	6.13E+05	2.41E+05	PI-7I-G-1-CI-5
	7	Gunner	5	4.02	1.083	5.57E+05	2.22E+05	PI-7I-G-3-CI-5
	7	Gunner	5	4.04	3.125	2.97E+05	1.19E+05	PI-7I-G-4-CI-5
	7	Gunner	5	4.01	7.5	7.08E+04	2.83E+04	PI-7I-G-5-CI-5
	7	Gunner	5	3.99	16	1.53E+04	6.27E+03	PI-7I-G-6-CI-5
	7	Gunner	5	4.23	33	5.02E+03	2.12E+03	PI-7I-G-7-CI-5
	7	Gunner	5	4.21	65	3.62E+03	1.49E+03	PI-7I-G-8-CI-5
	7	Gunner	5	4.23	125	2.29E+03	9.77E+02	PI-7I-G-9-CI-5
	7	Gunner	6	2.01	0.083	4.70E+05	1.22E+05	PI-7I-G-1-CI-6
	7	Gunner	6	2.02	1.083	1.80E+05	4.87E+04	PI-7I-G-3-CI-6
	7	Gunner	6	2.03	3.125	1.01E+05	2.76E+04	PI-7I-G-4-CI-6
	7	Gunner	6	2.01	7.5	5.07E+04	1.34E+04	PI-7I-G-5-CI-6
	7	Gunner	6	2.00	16	9.18E+03	2.53E+03	PI-7I-G-6-CI-6
	7	Gunner	6	2.12	33	3.56E+03	1.00E+03	PI-7I-G-7-CI-6
	7	Gunner	6	2.11	65	3.21E+03	8.79E+02	PI-7I-G-8-CI-6
	7	Gunner	6	2.12	125	2.42E+03	6.74E+02	PI-7I-G-9-CI-6
	7	Gunner	7	1.03	0.083	5.02E+05	9.80E+04	PI-7I-G-1-CI-7
	7	Gunner	7	1.04	1.083	3.42E+05	6.86E+04	PI-7I-G-3-CI-7
	7	Gunner	7	1.04	3.125	1.96E+05	3.95E+04	PI-7I-G-4-CI-7
	7	Gunner	7	1.03	7.5	6.63E+04	1.32E+04	PI-7I-G-5-CI-7
	7	Gunner	7	1.03	16	1.23E+04	2.63E+03	PI-7I-G-6-CI-7
	7	Gunner	7	1.09	33	2.25E+03	5.94E+02	PI-7I-G-7-CI-7
	7	Gunner	7	1.09	65	3.44E+03	7.30E+02	PI-7I-G-8-CI-7
	7	Gunner	7	1.09	125	2.93E+03	6.38E+02	PI-7I-G-9-CI-7
	7	Gunner	8	0.61	0.083	1.01E+06	1.42E+05	PI-7I-G-1-CI-8

Table A.7. *(continued)*

Phase	Shot	Position	CI stage	Median Dia (um)	MidTime (min)	U Conc, (ug/m3)	sd U Conc. (ug/m3)	SampleI
1	7	Gunner	8	0.61	1.083	4.73E+05	6.93E+04	PI-7I-G-3-C
1	7	Gunner	8	0.61	3.125	1.83E+05	2.78E+04	PI-7I-G-4-C
1	7	Gunner	8	0.61	7.5	3.42E+04	5.35E+03	PI-7I-G-5-C
1	7	Gunner	8	0.60	16	1.11E+04	1.83E+03	PI-7I-G-6-C
1	7	Gunner	8	0.64	33	1.02E+04	1.59E+03	PI-7I-G-7-C
1	7	Gunner	8	0.64	65	9.89E+03	1.45E+03	PI-7I-G-8-C
1	7	Gunner	8	0.64	125	6.56E+03	9.88E+02	PI-7I-G-9-C
1	7	Gunner	9	0.21	0.083	1.28E+06	2.55E+05	PI-7I-G-1-C
1	7	Gunner	9	0.21	1.083	1.23E+05	3.59E+04	PI-7I-G-3-C
1	7	Gunner	9	0.22	3.125	6.94E+04	2.14E+04	PI-7I-G-4-C
1	7	Gunner	9	0.21	7.5	2.41E+04	6.67E+03	PI-7I-G-5-C
1	7	Gunner	9	0.21	16	4.22E+03	1.85E+03	PI-7I-G-6-C
1	7	Gunner	9	0.22	33	7.23E+03	1.96E+03	PI-7I-G-7-C
1	7	Gunner	9	0.22	65	5.94E+03	1.43E+03	PI-7I-G-8-C
1	7	Gunner	9	0.22	125	3.66E+03	9.89E+02	PI-7I-G-9-C
1	7	Loader	1	31.29	0.17	2.59E+03	7.27E+02	PI-7I-L-1-C
1	7	Loader	1	31.14	10	6.10E+03	1.31E+03	PI-7I-L-2-C
1	7	Loader	1	30.89	30	2.76E+03	7.38E+02	PI-7I-L-3-C
1	7	Loader	1	31.12	45	7.94E+02	7.23E+02	PI-7I-L-4-C
1	7	Loader	2	16.35	0.17	1.28E+01	1.90E+02	PI-7I-L-1-C
1	7	Loader	2	16.19	10	-2.00E+01	1.83E+02	PI-7I-L-2-C
1	7	Loader	2	15.93	30	-1.77E+02	1.16E+02	PI-7I-L-3-C
1	7	Loader	2	16.17	45	-2.35E+02	2.83E+02	PI-7I-L-4-C
1	7	Loader	3	10.94	0.17	2.79E+02	8.53E+01	PI-7I-L-1-C
1	7	Loader	3	10.84	10	2.48E+02	7.79E+01	PI-7I-L-2-C
1	7	Loader	3	10.66	30	1.66E+02	6.12E+01	PI-7I-L-3-C
1	7	Loader	3	10.82	45	-1.40E+01	6.81E+01	PI-7I-L-4-C
1	7	Loader	4	6.98	0.17	3.83E+02	1.35E+02	PI-7I-L-1-C
1	7	Loader	4	6.91	10	3.82E+03	1.10E+03	PI-7I-L-2-C
1	7	Loader	4	6.80	30	1.25E+02	6.12E+01	PI-7I-L-3-C
1	7	Loader	4	6.90	45	1.25E+02	9.74E+01	PI-7I-L-4-C
1	7	Loader	5	4.19	0.17	5.99E+02	3.20E+02	PI-7I-L-1-C
1	7	Loader	5	4.15	10	3.31E+03	1.41E+03	PI-7I-L-2-C
1	7	Loader	5	4.09	30	1.42E+02	1.40E+02	PI-7I-L-3-C
1	7	Loader	5	4.15	45	2.10E+02	2.43E+02	PI-7I-L-4-C
1	7	Loader	6	2.10	0.17	1.68E+03	8.42E+02	PI-7I-L-1-C
1	7	Loader	6	2.08	10	2.56E+03	1.26E+03	PI-7I-L-2-C
1	7	Loader	6	2.05	30	2.65E+01	1.11E+02	PI-7I-L-3-C
1	7	Loader	6	2.08	45	1.79E+02	2.68E+02	PI-7I-L-4-C
1	7	Loader	7	1.08	0.17	1.04E+03	3.10E+02	PI-7I-L-1-C
1	7	Loader	7	1.07	10	2.50E+03	7.11E+02	PI-7I-L-2-C
1	7	Loader	7	1.05	30	3.12E+02	1.14E+02	PI-7I-L-3-C

A.7. *(continued)*

Shot	Position	CI stage	Median Dia (um)	MidTime (min)	U Conc, (ug/m3)	sd U Conc. (ug/m3)	SampleID
7	Loader	7	1.07	45	2.87E+01	1.40E+02	PI-7I-L-4-CI-7
7	Loader	8	0.63	0.17	4.61E+02	2.50E+02	PI-7I-L-1-CI-8
7	Loader	8	0.63	10	3.43E+03	1.05E+03	PI-7I-L-2-CI-8
7	Loader	8	0.62	30	2.93E+01	1.26E+02	PI-7I-L-3-CI-8
7	Loader	8	0.63	45	-2.20E+02	2.27E+02	PI-7I-L-4-CI-8
7	Loader	9	0.22	0.17	4.26E+02	2.60E+02	PI-7I-L-1-CI-9
7	Loader	9	0.22	10	2.85E+03	1.13E+03	PI-7I-L-2-CI-9
7	Loader	9	0.22	30	3.85E+01	1.31E+02	PI-7I-L-3-CI-9
7	Loader	9	0.22	45	1.98E+02	2.77E+02	PI-7I-L-4-CI-9
1	Commander	1	30.60	0.083	4.41E+05	7.59E+04	PII-1/2I-C-1-CI-1
1	Commander	1	31.01	3.5	4.15E+04	9.02E+03	PII-1/2I-C-2-CI-1
1	Commander	1	30.74	10	4.07E+04	7.78E+03	PII-1/2I-C-3-CI-1
1	Commander	2	15.63	0.083	9.86E+04	4.08E+04	PII-1/2I-C-1-CI-2
1	Commander	2	16.06	3.5	1.54E+04	7.90E+03	PII-1/2I-C-2-CI-2
1	Commander	2	15.77	10	8.02E+03	3.98E+03	PII-1/2I-C-3-CI-2
1	Commander	3	10.46	0.083	6.00E+04	1.35E+04	PII-1/2I-C-1-CI-3
1	Commander	3	10.75	3.5	2.06E+04	4.89E+03	PII-1/2I-C-2-CI-3
1	Commander	3	10.56	10	1.43E+04	3.25E+03	PII-1/2I-C-3-CI-3
1	Commander	4	6.67	0.083	1.10E+05	3.12E+04	PII-1/2I-C-1-CI-4
1	Commander	4	6.85	3.5	8.45E+04	2.41E+04	PII-1/2I-C-2-CI-4
1	Commander	4	6.73	10	1.42E+04	4.23E+03	PII-1/2I-C-3-CI-4
1	Commander	5	4.01	0.083	3.82E+05	1.56E+05	PII-1/2I-C-1-CI-5
1	Commander	5	4.12	3.5	1.08E+05	4.42E+04	PII-1/2I-C-2-CI-5
1	Commander	5	4.04	10	4.99E+04	NA[a]	PII-1/2I-C-3-CI-5
1	Commander	6	2.01	0.083	3.90E+05	1.87E+05	PII-1/2I-C-1-CI-6
1	Commander	6	2.06	3.5	1.76E+05	8.44E+04	PII-1/2I-C-2-CI-6
1	Commander	6	2.03	10	1.52E+04	7.67E+03	PII-1/2I-C-3-CI-6
1	Commander	7	1.03	0.083	2.30E+05	6.28E+04	PII-1/2I-C-1-CI-7
1	Commander	7	1.06	3.5	7.42E+04	2.05E+04	PII-1/2I-C-2-CI-7
1	Commander	7	1.04	10	1.26E+04	3.75E+03	PII-1/2I-C-3-CI-7
1	Commander	8	0.61	0.083	2.09E+05	6.92E+04	PII-1/2I-C-1-CI-8
1	Commander	8	0.62	3.5	1.27E+05	3.93E+04	PII-1/2I-C-2-CI-8
1	Commander	8	0.61	10	1.15E+04	4.49E+03	PII-1/2I-C-3-CI-8
1	Commander	9	0.21	0.083	3.25E+05	1.31E+05	PII-1/2I-C-1-CI-9
1	Commander	9	0.22	3.5	1.14E+05	4.42E+04	PII-1/2I-C-2-CI-9
1	Commander	9	0.22	10	2.08E+04	8.86E+03	PII-1/2I-C-3-CI-9
1	Driver	1	31.00	0.083	3.57E+05	6.24E+04	PII-1/2I-D-1-CI-1
1	Driver	1	31.31	3.5	5.64E+04	1.16E+04	PII-1/2I-D-2-CI-1
1	Driver	1	30.70	10	1.09E+04	2.86E+03	PII-1/2I-D-3-CI-1
1	Driver	2	16.05	0.083	7.31E+04	3.02E+04	PII-1/2I-D-1-CI-2
1	Driver	2	16.37	3.5	8.61E+03	5.23E+03	PII-1/2I-D-2-CI-2
1	Driver	2	15.74	10	3.67E+03	2.27E+03	PII-1/2I-D-3-CI-2

Table A.7. *(continued)*

Phase	Shot	Position	CI stage	Median Dia (um)	MidTime (min)	U Conc, (ug/m3)	sd U Conc. (ug/m3)	SampleI
2	1	Driver	3	10.74	0.083	8.13E+04	1.81E+04	PII-1/2I-D-1
2	1	Driver	3	10.95	3.5	1.78E+04	4.35E+03	PII-1/2I-D-2
2	1	Driver	3	10.53	10	1.28E+04	3.00E+03	PII-1/2I-D-3
2	1	Driver	4	6.85	0.083	1.00E+05	2.86E+04	PII-1/2I-D-1
2	1	Driver	4	6.99	3.5	8.14E+04	2.33E+04	PII-1/2I-D-2
2	1	Driver	4	6.72	10	3.74E+04	1.07E+04	PII-1/2I-D-3
2	1	Driver	5	4.12	0.083	4.15E+05	1.71E+05	PII-1/2I-D-1
2	1	Driver	5	4.20	3.5	1.24E+05	5.13E+04	PII-1/2I-D-2
2	1	Driver	5	4.04	10	4.34E+04	1.80E+04	PII-1/2I-D-3
2	1	Driver	6	2.06	0.083	4.25E+05	2.14E+05	PII-1/2I-D-1
2	1	Driver	6	2.10	3.5	1.54E+05	7.78E+04	PII-1/2I-D-2
2	1	Driver	6	2.02	10	5.96E+04	3.03E+04	PII-1/2I-D-3
2	1	Driver	7	1.06	0.083	2.27E+05	6.43E+04	PII-1/2I-D-1
2	1	Driver	7	1.08	3.5	4.69E+04	1.36E+04	PII-1/2I-D-2
2	1	Driver	7	1.04	10	5.54E+04	1.55E+04	PII-1/2I-D-3
2	1	Driver	8	0.62	0.083	2.77E+05	9.69E+04	PII-1/2I-D-1
2	1	Driver	8	0.64	3.5	8.48E+04	2.93E+04	PII-1/2I-D-2
2	1	Driver	8	0.61	10	5.29E+04	1.80E+04	PII-1/2I-D-3
2	1	Driver	9	0.22	0.083	4.44E+05	1.91E+05	PII-1/2I-D-1
2	1	Driver	9	0.22	3.5	1.40E+05	5.72E+04	PII-1/2I-D-2
2	1	Driver	9	0.22	10	4.31E+04	1.76E+04	PII-1/2I-D-3
2	1	Loader	1	31.36	0.083	5.10E+05	1.08E+05	PII-1/2I-L-1
2	1	Loader	1	31.44	3.5	3.11E+04	8.95E+03	PII-1/2I-L-2
2	1	Loader	1	30.96	10	3.45E+04	8.27E+03	PII-1/2I-L-3
2	1	Loader	2	16.42	0.083	7.25E+04	3.59E+04	PII-1/2I-L-1
2	1	Loader	2	16.50	3.5	1.55E+04	9.11E+03	PII-1/2I-L-2
2	1	Loader	2	16.01	10	8.58E+03	4.90E+03	PII-1/2I-L-3
2	1	Loader	3	10.99	0.083	4.73E+04	1.35E+04	PII-1/2I-L-1
2	1	Loader	3	11.04	3.5	1.33E+04	4.31E+03	PII-1/2I-L-2
2	1	Loader	3	10.71	10	1.27E+04	3.75E+03	PII-1/2I-L-3
2	1	Loader	4	7.01	0.083	4.97E+04	1.95E+04	PII-1/2I-L-1
2	1	Loader	4	7.04	3.5	4.33E+04	1.67E+04	PII-1/2I-L-2
2	1	Loader	4	6.83	10	4.85E+04	1.82E+04	PII-1/2I-L-3
2	1	Loader	5	4.21	0.083	9.83E+04	3.94E+04	PII-1/2I-L-1
2	1	Loader	5	4.23	3.5	6.18E+04	2.45E+04	PII-1/2I-L-2
2	1	Loader	5	4.10	10	4.36E+04	1.72E+04	PII-1/2I-L-3
2	1	Loader	6	2.11	0.083	9.21E+04	2.52E+04	PII-1/2I-L-1
2	1	Loader	6	2.12	3.5	5.83E+04	1.56E+04	PII-1/2I-L-2
2	1	Loader	6	2.06	10	2.72E+04	7.27E+03	PII-1/2I-L-3
2	1	Loader	7	1.08	0.083	9.88E+04	2.12E+04	PII-1/2I-L-1
2	1	Loader	7	1.09	3.5	3.27E+04	6.84E+03	PII-1/2I-L-2
2	1	Loader	7	1.06	10	2.26E+04	4.62E+03	PII-1/2I-L-3

A.7. *(continued)*

Shot	Position	CI stage	Median Dia (um)	MidTime (min)	U Conc, (ug/m3)	sd U Conc. (ug/m3)	SampleID
1	Loader	8	0.64	0.083	2.58E+05	3.80E+04	PII-1/2I-L-1-CI-8
1	Loader	8	0.64	3.5	4.48E+04	6.94E+03	PII-1/2I-L-2-CI-8
1	Loader	8	0.62	10	2.19E+04	3.42E+03	PII-1/2I-L-3-CI-8
1	Loader	9	0.22	0.083	3.98E+05	8.10E+04	PII-1/2I-L-1-CI-9
1	Loader	9	0.22	3.5	1.98E+05	3.98E+04	PII-1/2I-L-2-CI-9
1	Loader	9	0.22	10	4.05E+04	8.21E+03	PII-1/2I-L-3-CI-9
2	Commander	1	31.24	0.083	3.02E+05	5.39E+04	PII-1/2I-C-4-CI-1
2	Commander	1	31.76	3.5	7.30E+04	1.45E+04	PII-1/2I-C-5-CI-1
2	Commander	1	31.30	10	2.35E+04	5.05E+03	PII-1/2I-C-6-CI-1
2	Commander	1	30.73	23	1.00E+04	2.20E+03	PII-1/2I-C-7-CI-1
2	Commander	1	30.99	49	5.83E+03	1.24E+03	PII-1/2I-C-8-CI-1
2	Commander	1	31.18	101	1.16E+03	3.36E+02	PII-1/2I-C-9-CI-1
2	Commander	2	16.29	0.083	1.03E+05	4.41E+04	PII-1/2I-C-4-CI-2
2	Commander	2	16.84	3.5	5.59E+03	4.20E+03	PII-1/2I-C-5-CI-2
2	Commander	2	16.36	10	3.64E+03	2.44E+03	PII-1/2I-C-6-CI-2
2	Commander	2	15.76	23	7.60E+02	7.83E+02	PII-1/2I-C-7-CI-2
2	Commander	2	16.04	49	-8.67E+01	2.51E+02	PII-1/2I-C-8-CI-2
2	Commander	2	16.23	101	1.20E+02	1.80E+02	PII-1/2I-C-9-CI-2
2	Commander	3	10.91	0.083	1.98E+05	4.25E+04	PII-1/2I-C-4-CI-3
2	Commander	3	11.27	3.5	1.55E+04	3.87E+03	PII-1/2I-C-5-CI-3
2	Commander	3	10.95	10	1.23E+04	2.91E+03	PII-1/2I-C-6-CI-3
2	Commander	3	10.55	23	2.91E+03	7.71E+02	PII-1/2I-C-7-CI-3
2	Commander	3	10.73	49	6.26E+00	1.19E+02	PII-1/2I-C-8-CI-3
2	Commander	3	10.87	101	9.94E+01	7.44E+01	PII-1/2I-C-9-CI-3
2	Commander	4	6.95	0.083	4.64E+05	1.30E+05	PII-1/2I-C-4-CI-4
2	Commander	4	7.19	3.5	1.31E+04	4.34E+03	PII-1/2I-C-5-CI-4
2	Commander	4	6.98	10	1.74E+04	5.21E+03	PII-1/2I-C-6-CI-4
2	Commander	4	6.73	23	4.15E+03	1.32E+03	PII-1/2I-C-7-CI-4
2	Commander	4	6.84	49	2.20E+03	7.12E+02	PII-1/2I-C-8-CI-4
2	Commander	4	6.93	101	1.07E+03	3.42E+02	PII-1/2I-C-9-CI-4
2	Commander	5	4.18	0.083	8.90E+05	3.63E+05	PII-1/2I-C-4-CI-5
2	Commander	5	4.32	3.5	5.08E+04	2.12E+04	PII-1/2I-C-5-CI-5
2	Commander	5	4.20	10	3.83E+04	1.59E+04	PII-1/2I-C-6-CI-5
2	Commander	5	4.04	23	8.45E+03	3.58E+03	PII-1/2I-C-7-CI-5
2	Commander	5	4.11	49	2.63E+03	1.14E+03	PII-1/2I-C-8-CI-5
2	Commander	5	4.16	101	1.90E+03	7.96E+02	PII-1/2I-C-9-CI-5
2	Commander	6	2.09	0.083	8.83E+05	4.22E+05	PII-1/2I-C-4-CI-6
2	Commander	6	2.16	3.5	9.13E+04	4.43E+04	PII-1/2I-C-5-CI-6
2	Commander	6	2.10	10	3.27E+04	1.60E+04	PII-1/2I-C-6-CI-6
2	Commander	6	2.03	23	1.18E+04	5.88E+03	PII-1/2I-C-7-CI-6
2	Commander	6	2.06	49	3.78E+03	1.92E+03	PII-1/2I-C-8-CI-6
2	Commander	6	2.09	101	3.92E+03	1.84E+03	PII-1/2I-C-9-CI-6

Table A.7. *(continued)*

Phase	Shot	Position	CI stage	Median Dia (um)	MidTime (min)	U Conc, (ug/m3)	sd U Conc. (ug/m3)	SampleI
2	2	Commander	7	1.08	0.083	7.47E+05	2.00E+05	PII-1/2I-C-4
2	2	Commander	7	1.11	3.5	5.89E+04	1.67E+04	PII-1/2I-C-5
2	2	Commander	7	1.08	10	4.54E+04	1.25E+04	PII-1/2I-C-6
2	2	Commander	7	1.04	23	1.61E+04	4.56E+03	PII-1/2I-C-7
2	2	Commander	7	1.06	49	3.81E+03	1.15E+03	PII-1/2I-C-8
2	2	Commander	7	1.07	101	1.93E+03	5.98E+02	PII-1/2I-C-9
2	2	Commander	8	0.63	0.083	7.91E+05	2.38E+05	PII-1/2I-C-4
2	2	Commander	8	0.65	3.5	1.20E+05	3.73E+04	PII-1/2I-C-5
2	2	Commander	8	0.64	10	7.19E+04	2.19E+04	PII-1/2I-C-6
2	2	Commander	8	0.61	23	2.83E+04	8.76E+03	PII-1/2I-C-7
2	2	Commander	8	0.62	49	5.59E+03	1.84E+03	PII-1/2I-C-8
2	2	Commander	8	0.63	101	3.27E+03	1.03E+03	PII-1/2I-C-9
2	2	Commander	9	0.22	0.083	1.16E+06	4.39E+05	PII-1/2I-C-4
2	2	Commander	9	0.22	3.5	2.50E+04	1.11E+04	PII-1/2I-C-5
2	2	Commander	9	0.22	10	3.03E+04	1.19E+04	PII-1/2I-C-6
2	2	Commander	9	0.22	23	2.02E+04	7.84E+03	PII-1/2I-C-7
2	2	Commander	9	0.22	49	4.05E+03	1.65E+03	PII-1/2I-C-8
2	2	Commander	9	0.22	101	7.58E+03	2.80E+03	PII-1/2I-C-9
2	2	Driver	1	30.60	0.083	2.74E+05	4.80E+04	PII-1/2I-D-4
2	2	Driver	1	30.84	3.5	1.41E+04	4.53E+03	PII-1/2I-D-5
2	2	Driver	1	30.82	10	1.11E+04	2.84E+03	PII-1/2I-D-6
2	2	Driver	1	30.91	23	2.63E+03	9.71E+02	PII-1/2I-D-7
2	2	Driver	1	30.61	49	2.16E+03	6.03E+02	PII-1/2I-D-8
2	2	Driver	1	31.22	101	7.33E+02	2.63E+02	PII-1/2I-D-9
2	2	Driver	2	15.63	0.083	2.12E+05	8.68E+04	PII-1/2I-D-4
2	2	Driver	2	15.88	3.5	6.22E+03	4.17E+03	PII-1/2I-D-5
2	2	Driver	2	15.85	10	1.23E+03	1.44E+03	PII-1/2I-D-6
2	2	Driver	2	15.95	23	7.79E+02	7.88E+02	PII-1/2I-D-7
2	2	Driver	2	15.64	49	-8.25E+01	2.33E+02	PII-1/2I-D-8
2	2	Driver	2	16.27	101	-4.46E+01	1.23E+02	PII-1/2I-D-9
2	2	Driver	3	10.46	0.083	1.40E+05	2.98E+04	PII-1/2I-D-4
2	2	Driver	3	10.63	3.5	1.82E+04	4.34E+03	PII-1/2I-D-5
2	2	Driver	3	10.61	10	5.65E+03	1.48E+03	PII-1/2I-D-6
2	2	Driver	3	10.68	23	3.23E+03	8.77E+02	PII-1/2I-D-7
2	2	Driver	3	10.47	49	6.55E+01	1.30E+02	PII-1/2I-D-8
2	2	Driver	3	10.89	101	3.54E+01	6.46E+01	PII-1/2I-D-9
2	2	Driver	4	6.67	0.083	2.13E+05	6.01E+04	PII-1/2I-D-4
2	2	Driver	4	6.78	3.5	1.21E+04	3.95E+03	PII-1/2I-D-5
2	2	Driver	4	6.77	10	2.08E+04	6.08E+03	PII-1/2I-D-6
2	2	Driver	4	6.81	23	5.66E+03	1.74E+03	PII-1/2I-D-7
2	2	Driver	4	6.68	49	1.94E+03	6.28E+02	PII-1/2I-D-8
2	2	Driver	4	6.94	101	8.79E+02	2.83E+02	PII-1/2I-D-9

A.7. *(continued)*

Shot	Position	CI stage	Median Dia (um)	MidTime (min)	U Conc, (ug/m3)	sd U Conc. (ug/m3)	SampleID
2	Driver	5	4.01	0.083	2.93E+05	1.22E+05	PII-1/2I-D-4-CI-5
2	Driver	5	4.07	3.5	5.57E+04	2.34E+04	PII-1/2I-D-5-CI-5
2	Driver	5	4.07	10	2.79E+04	1.18E+04	PII-1/2I-D-6-CI-5
2	Driver	5	4.09	23	6.62E+03	2.90E+03	PII-1/2I-D-7-CI-5
2	Driver	5	4.01	49	4.60E+03	1.96E+03	PII-1/2I-D-8-CI-5
2	Driver	5	4.17	101	3.02E+03	1.26E+03	PII-1/2I-D-9-CI-5
2	Driver	6	2.01	0.083	2.25E+05	1.15E+05	PII-1/2I-D-4-CI-6
2	Driver	6	2.04	3.5	7.40E+04	3.78E+04	PII-1/2I-D-5-CI-6
2	Driver	6	2.04	10	2.57E+04	1.34E+04	PII-1/2I-D-6-CI-6
2	Driver	6	2.05	23	9.90E+03	5.20E+03	PII-1/2I-D-7-CI-6
2	Driver	6	2.01	49	2.58E+03	1.38E+03	PII-1/2I-D-8-CI-6
2	Driver	6	2.09	101	3.49E+03	1.72E+03	PII-1/2I-D-9-CI-6
2	Driver	7	1.03	0.083	1.26E+05	3.77E+04	PII-1/2I-D-4-CI-7
2	Driver	7	1.05	3.5	5.08E+04	1.49E+04	PII-1/2I-D-5-CI-7
2	Driver	7	1.05	10	3.54E+04	1.01E+04	PII-1/2I-D-6-CI-7
2	Driver	7	1.05	23	1.42E+04	4.14E+03	PII-1/2I-D-7-CI-7
2	Driver	7	1.03	49	4.97E+03	1.47E+03	PII-1/2I-D-8-CI-7
2	Driver	7	1.07	101	1.99E+03	6.12E+02	PII-1/2I-D-9-CI-7
2	Driver	8	0.61	0.083	1.84E+05	7.00E+04	PII-1/2I-D-4-CI-8
2	Driver	8	0.62	3.5	1.28E+05	4.30E+04	PII-1/2I-D-5-CI-8
2	Driver	8	0.62	10	9.68E+04	3.20E+04	PII-1/2I-D-6-CI-8
2	Driver	8	0.62	23	3.05E+04	1.03E+04	PII-1/2I-D-7-CI-8
2	Driver	8	0.61	49	6.96E+03	2.34E+03	PII-1/2I-D-8-CI-8
2	Driver	8	0.63	101	4.93E+03	1.62E+03	PII-1/2I-D-9-CI-8
2	Driver	9	0.21	0.083	2.52E+05	1.09E+05	PII-1/2I-D-4-CI-9
2	Driver	9	0.22	3.5	7.96E+04	3.29E+04	PII-1/2I-D-5-CI-9
2	Driver	9	0.22	10	5.44E+04	2.20E+04	PII-1/2I-D-6-CI-9
2	Driver	9	0.22	23	1.80E+04	7.45E+03	PII-1/2I-D-7-CI-9
2	Driver	9	0.21	49	3.95E+03	1.68E+03	PII-1/2I-D-8-CI-9
2	Driver	9	0.22	101	6.30E+03	2.47E+03	PII-1/2I-D-9-CI-9
2	Gunner	1	30.50	0.083	3.42E+05	7.28E+04	PII-1/2I-G-1-CI-1
2	Gunner	1	30.83	3.5	-1.34E+03	NA[a]	PII-1/2I-G-3-CI-1
2	Gunner	1	30.82	10	8.14E+03	2.90E+03	PII-1/2I-G-4-CI-1
2	Gunner	1	30.71	23	4.56E+03	1.52E+03	PII-1/2I-G-5-CI-1
2	Gunner	1	30.59	49	9.06E+02	4.85E+02	PII-1/2I-G-6-CI-1
2	Gunner	1	31.35	101	3.52E+02	2.32E+02	PII-1/2I-G-7-CI-1
2	Gunner	2	15.53	0.083	1.66E+05	7.94E+04	PII-1/2I-G-1-CI-2
2	Gunner	2	15.87	3.5	5.59E+03	4.45E+03	PII-1/2I-G-3-CI-2
2	Gunner	2	15.86	10	-2.61E+01	1.14E+03	PII-1/2I-G-4-CI-2
2	Gunner	2	15.75	23	1.22E+03	1.00E+03	PII-1/2I-G-5-CI-2
2	Gunner	2	15.62	49	1.51E+03	9.20E+02	PII-1/2I-G-6-CI-2
2	Gunner	2	16.41	101	2.06E+02	2.27E+02	PII-1/2I-G-7-CI-2

Table A.7. *(continued)*

Phase	Shot	Position	CI stage	Median Dia (um)	MidTime (min)	U Conc, (ug/m3)	sd U Conc. (ug/m3)	SampleI
2	2	Gunner	3	10.40	0.083	1.07E+05	2.93E+04	PII-1/2I-G-1
2	2	Gunner	3	10.62	3.5	1.48E+04	4.58E+03	PII-1/2I-G-3
2	2	Gunner	3	10.62	10	6.39E+03	2.07E+03	PII-1/2I-G-4
2	2	Gunner	3	10.54	23	3.27E+03	1.03E+03	PII-1/2I-G-5
2	2	Gunner	3	10.46	49	2.43E+02	1.76E+02	PII-1/2I-G-6
2	2	Gunner	3	10.98	101	2.00E+02	9.94E+01	PII-1/2I-G-7
2	2	Gunner	4	6.63	0.083	1.19E+05	4.40E+04	PII-1/2I-G-1
2	2	Gunner	4	6.77	3.5	1.64E+04	6.64E+03	PII-1/2I-G-3
2	2	Gunner	4	6.77	10	2.06E+04	7.90E+03	PII-1/2I-G-4
2	2	Gunner	4	6.72	23	3.40E+03	1.41E+03	PII-1/2I-G-5
2	2	Gunner	4	6.67	49	1.20E+03	5.29E+02	PII-1/2I-G-6
2	2	Gunner	4	7.00	101	7.22E+02	3.18E+02	PII-1/2I-G-7
2	2	Gunner	5	3.98	0.083	1.34E+05	5.30E+04	PII-1/2I-G-1
2	2	Gunner	5	4.07	3.5	2.66E+04	1.09E+04	PII-1/2I-G-3
2	2	Gunner	5	4.07	10	3.10E+04	1.23E+04	PII-1/2I-G-4
2	2	Gunner	5	4.04	23	8.63E+03	3.48E+03	PII-1/2I-G-5
2	2	Gunner	5	4.01	49	4.88E+03	1.95E+03	PII-1/2I-G-6
2	2	Gunner	5	4.21	101	3.85E+03	1.53E+03	PII-1/2I-G-7
2	2	Gunner	6	2.00	0.083	9.62E+04	2.63E+04	PII-1/2I-G-1
2	2	Gunner	6	2.04	3.5	2.33E+04	6.49E+03	PII-1/2I-G-3
2	2	Gunner	6	2.04	10	1.42E+04	3.92E+03	PII-1/2I-G-4
2	2	Gunner	6	2.02	23	6.32E+03	1.77E+03	PII-1/2I-G-5
2	2	Gunner	6	2.01	49	2.92E+03	8.08E+02	PII-1/2I-G-6
2	2	Gunner	6	2.11	101	1.74E+03	4.86E+02	PII-1/2I-G-7
2	2	Gunner	7	1.03	0.083	1.18E+05	2.48E+04	PII-1/2I-G-1
2	2	Gunner	7	1.05	3.5	3.89E+04	8.08E+03	PII-1/2I-G-3
2	2	Gunner	7	1.05	10	3.48E+04	7.00E+03	PII-1/2I-G-4
2	2	Gunner	7	1.04	23	1.51E+04	3.06E+03	PII-1/2I-G-5
2	2	Gunner	7	1.03	49	4.67E+03	9.80E+02	PII-1/2I-G-6
2	2	Gunner	7	1.08	101	2.86E+03	5.94E+02	PII-1/2I-G-7
2	2	Gunner	8	0.60	0.083	1.51E+05	2.28E+04	PII-1/2I-G-1
2	2	Gunner	8	0.62	3.5	2.18E+04	3.86E+03	PII-1/2I-G-3
2	2	Gunner	8	0.62	10	5.18E+04	7.57E+03	PII-1/2I-G-4
2	2	Gunner	8	0.61	23	1.35E+04	2.07E+03	PII-1/2I-G-5
2	2	Gunner	8	0.61	49	5.96E+03	9.22E+02	PII-1/2I-G-6
2	2	Gunner	8	0.64	101	4.73E+03	7.10E+02	PII-1/2I-G-7
2	2	Gunner	9	0.21	0.083	4.81E+05	9.88E+04	PII-1/2I-G-1
2	2	Gunner	9	0.22	3.5	9.33E+04	1.90E+04	PII-1/2I-G-3
2	2	Gunner	9	0.22	10	4.13E+04	8.48E+03	PII-1/2I-G-4
2	2	Gunner	9	0.22	23	2.28E+04	4.66E+03	PII-1/2I-G-5
2	2	Gunner	9	0.21	49	1.32E+04	2.70E+03	PII-1/2I-G-6
2	2	Gunner	9	0.22	101	1.07E+04	2.17E+03	PII-1/2I-G-7

A.7. *(continued)*

Shot	Position	CI stage	Median Dia (um)	MidTime (min)	U Conc, (ug/m3)	sd U Conc. (ug/m3)	SampleID
2	Loader	1	31.00	0.083	4.73E+05	1.00E+05	PII-1/2I-L-4-CI-1
2	Loader	1	31.00	3.5	3.09E+04	8.73E+03	PII-1/2I-L-5-CI-1
2	Loader	1	31.11	10	1.34E+04	4.02E+03	PII-1/2I-L-6-CI-1
2	Loader	1	31.29	23	9.79E+03	2.63E+03	PII-1/2I-L-7-CI-1
2	Loader	1	31.15	49	9.31E+03	2.23E+03	PII-1/2I-L-8-CI-1
2	Loader	1	31.10	101	3.10E+03	7.88E+02	PII-1/2I-L-9-CI-1
2	Loader	2	16.04	0.083	1.19E+05	5.74E+04	PII-1/2I-L-4-CI-2
2	Loader	2	16.04	3.5	1.08E+04	6.86E+03	PII-1/2I-L-5-CI-2
2	Loader	2	16.16	10	7.12E+03	4.30E+03	PII-1/2I-L-6-CI-2
2	Loader	2	16.35	23	1.15E+03	1.00E+03	PII-1/2I-L-7-CI-2
2	Loader	2	16.20	49	3.20E+02	3.97E+02	PII-1/2I-L-8-CI-2
2	Loader	2	16.15	101	6.87E+02	4.42E+02	PII-1/2I-L-9-CI-2
2	Loader	3	10.74	0.083	1.38E+05	3.77E+04	PII-1/2I-L-4-CI-3
2	Loader	3	10.74	3.5	1.66E+04	5.01E+03	PII-1/2I-L-5-CI-3
2	Loader	3	10.81	10	4.85E+03	1.63E+03	PII-1/2I-L-6-CI-3
2	Loader	3	10.94	23	1.18E+03	5.14E+02	PII-1/2I-L-7-CI-3
2	Loader	3	10.84	49	8.37E+02	3.02E+02	PII-1/2I-L-8-CI-3
2	Loader	3	10.81	101	2.57E+02	1.10E+02	PII-1/2I-L-9-CI-3
2	Loader	4	6.85	0.083	1.79E+05	6.61E+04	PII-1/2I-L-4-CI-4
2	Loader	4	6.85	3.5	2.11E+04	8.43E+03	PII-1/2I-L-5-CI-4
2	Loader	4	6.90	10	2.19E+04	8.38E+03	PII-1/2I-L-6-CI-4
2	Loader	4	6.98	23	5.72E+03	2.28E+03	PII-1/2I-L-7-CI-4
2	Loader	4	6.91	49	1.35E+03	5.91E+02	PII-1/2I-L-8-CI-4
2	Loader	4	6.89	101	6.44E+02	2.88E+02	PII-1/2I-L-9-CI-4
2	Loader	5	4.11	0.083	1.87E+05	7.41E+04	PII-1/2I-L-4-CI-5
2	Loader	5	4.11	3.5	3.83E+04	1.54E+04	PII-1/2I-L-5-CI-5
2	Loader	5	4.14	10	1.21E+04	4.96E+03	PII-1/2I-L-6-CI-5
2	Loader	5	4.19	23	5.67E+03	2.31E+03	PII-1/2I-L-7-CI-5
2	Loader	5	4.15	49	3.68E+03	1.49E+03	PII-1/2I-L-8-CI-5
2	Loader	5	4.14	101	1.70E+03	6.90E+02	PII-1/2I-L-9-CI-5
2	Loader	6	2.06	0.083	1.07E+05	2.89E+04	PII-1/2I-L-4-CI-6
2	Loader	6	2.06	3.5	3.96E+04	1.07E+04	PII-1/2I-L-5-CI-6
2	Loader	6	2.08	10	1.58E+04	4.30E+03	PII-1/2I-L-6-CI-6
2	Loader	6	2.10	23	4.91E+03	1.39E+03	PII-1/2I-L-7-CI-6
2	Loader	6	2.08	49	3.34E+03	9.25E+02	PII-1/2I-L-8-CI-6
2	Loader	6	2.08	101	1.61E+03	4.46E+02	PII-1/2I-L-9-CI-6
2	Loader	7	1.06	0.083	1.59E+05	3.17E+04	PII-1/2I-L-4-CI-7
2	Loader	7	1.06	3.5	5.99E+04	1.21E+04	PII-1/2I-L-5-CI-7
2	Loader	7	1.07	10	3.33E+04	6.62E+03	PII-1/2I-L-6-CI-7
2	Loader	7	1.08	23	1.17E+04	2.38E+03	PII-1/2I-L-7-CI-7
2	Loader	7	1.07	49	4.56E+03	9.50E+02	PII-1/2I-L-8-CI-7
2	Loader	7	1.07	101	2.70E+03	5.48E+02	PII-1/2I-L-9-CI-7

Table A.7. *(continued)*

Phase	Shot	Position	CI stage	Median Dia (um)	MidTime (min)	U Conc, (ug/m3)	sd U Conc. (ug/m3)	SampleI
2	2	Loader	8	0.62	0.083	1.56E+05	2.36E+04	PII-1/2I-L-4
2	2	Loader	8	0.62	3.5	5.46E+04	8.32E+03	PII-1/2I-L-5
2	2	Loader	8	0.63	10	3.27E+04	4.89E+03	PII-1/2I-L-6
2	2	Loader	8	0.63	23	8.24E+03	1.32E+03	PII-1/2I-L-7
2	2	Loader	8	0.63	49	4.45E+03	7.09E+02	PII-1/2I-L-8
2	2	Loader	8	0.63	101	2.71E+03	4.23E+02	PII-1/2I-L-9
2	2	Loader	9	0.22	0.083	4.15E+05	8.45E+04	PII-1/2I-L-4
2	2	Loader	9	0.22	3.5	6.80E+04	1.38E+04	PII-1/2I-L-5
2	2	Loader	9	0.22	10	4.54E+04	9.22E+03	PII-1/2I-L-6
2	2	Loader	9	0.22	23	3.26E+04	6.58E+03	PII-1/2I-L-7
2	2	Loader	9	0.22	49	1.57E+04	3.18E+03	PII-1/2I-L-8
2	2	Loader	9	0.22	101	4.16E+03	8.44E+02	PII-1/2I-L-9
2	3	Commander	1	30.98	0.083	2.52E+05	5.44E+04	PII-3I-C-2-
2	3	Commander	1	30.79	1.083	6.64E+04	2.49E+04	PII-3I-C-3-
2	3	Commander	1	31.34	3.125	5.68E+04	2.08E+04	PII-3I-C-4-
2	3	Commander	1	31.38	7.5	2.53E+04	7.36E+03	PII-3I-C-5-
2	3	Commander	1	31.30	16	6.69E+03	2.44E+03	PII-3I-C-6-
2	3	Commander	1	30.88	33	2.98E+03	1.15E+03	PII-3I-C-7-
2	3	Commander	1	31.22	65	1.27E+03	5.47E+02	PII-3I-C-8-
2	3	Commander	1	31.31	125	1.97E+03	6.74E+02	PII-3I-C-9-
2	3	Commander	2	16.02	0.083	9.96E+04	4.71E+04	PII-3I-C-2-
2	3	Commander	2	15.83	1.083	-1.20E+03	1.84E+04	PII-3I-C-3-
2	3	Commander	2	16.39	3.125	-1.42E+04	1.00E+04	PII-3I-C-4-
2	3	Commander	2	16.44	7.5	1.80E+03	4.04E+03	PII-3I-C-5-
2	3	Commander	2	16.36	16	3.56E+03	3.09E+03	PII-3I-C-6-
2	3	Commander	2	15.92	33	4.23E+02	9.48E+02	PII-3I-C-7-
2	3	Commander	2	16.27	65	-1.91E+02	3.65E+02	PII-3I-C-8-
2	3	Commander	2	16.37	125	7.23E+02	7.09E+02	PII-3I-C-9-
2	3	Commander	3	10.72	0.083	6.89E+04	1.94E+04	PII-3I-C-2-
2	3	Commander	3	10.60	1.083	8.08E+03	9.62E+03	PII-3I-C-3-
2	3	Commander	3	10.97	3.125	-6.78E+03	4.45E+03	PII-3I-C-4-
2	3	Commander	3	11.00	7.5	4.00E+02	1.54E+03	PII-3I-C-5-
2	3	Commander	3	10.95	16	-1.10E+03	5.52E+02	PII-3I-C-6-
2	3	Commander	3	10.66	33	5.87E+02	4.42E+02	PII-3I-C-7-
2	3	Commander	3	10.89	65	1.78E+02	2.07E+02	PII-3I-C-8-
2	3	Commander	3	10.95	125	-3.41E+02	1.71E+02	PII-3I-C-9-
2	3	Commander	4	6.84	0.083	4.73E+04	1.83E+04	PII-3I-C-2-
2	3	Commander	4	6.76	1.083	9.11E+03	9.95E+03	PII-3I-C-3-
2	3	Commander	4	7.00	3.125	2.84E+03	5.83E+03	PII-3I-C-4-
2	3	Commander	4	7.02	7.5	7.14E+02	1.54E+03	PII-3I-C-5-
2	3	Commander	4	6.98	16	1.27E+03	1.01E+03	PII-3I-C-6-
2	3	Commander	4	6.80	33	2.44E+03	1.11E+03	PII-3I-C-7-

A.7. *(continued)*

e	Shot	Position	CI stage	Median Dia (um)	MidTime (min)	U Conc, (ug/m3)	sd U Conc. (ug/m3)	SampleID
	3	Commander	4	6.95	65	9.91E+02	4.86E+02	PII-3I-C-8-CI-4
	3	Commander	4	6.99	125	2.60E+02	2.52E+02	PII-3I-C-9-CI-4
	3	Commander	5	4.11	0.083	6.89E+04	2.75E+04	PII-3I-C-2-CI-5
	3	Commander	5	4.06	1.083	3.42E+04	1.97E+04	PII-3I-C-3-CI-5
	3	Commander	5	4.20	3.125	2.58E+02	5.30E+03	PII-3I-C-4-CI-5
	3	Commander	5	4.22	7.5	6.58E+03	3.55E+03	PII-3I-C-5-CI-5
	3	Commander	5	4.20	16	5.19E+03	2.52E+03	PII-3I-C-6-CI-5
	3	Commander	5	4.08	33	4.28E+03	1.90E+03	PII-3I-C-7-CI-5
	3	Commander	5	4.17	65	3.50E+03	1.49E+03	PII-3I-C-8-CI-5
	3	Commander	5	4.20	125	1.35E+03	6.51E+02	PII-3I-C-9-CI-5
	3	Commander	6	2.06	0.083	4.14E+04	1.20E+04	PII-3I-C-2-CI-6
	3	Commander	6	2.03	1.083	1.01E+04	8.07E+03	PII-3I-C-3-CI-6
	3	Commander	6	2.11	3.125	-2.99E+03	4.06E+03	PII-3I-C-4-CI-6
	3	Commander	6	2.11	7.5	4.38E+03	2.16E+03	PII-3I-C-5-CI-6
	3	Commander	6	2.10	16	5.55E+03	1.88E+03	PII-3I-C-6-CI-6
	3	Commander	6	2.05	33	7.74E+03	2.24E+03	PII-3I-C-7-CI-6
	3	Commander	6	2.09	65	5.67E+03	1.58E+03	PII-3I-C-8-CI-6
	3	Commander	6	2.10	125	3.56E+03	1.03E+03	PII-3I-C-9-CI-6
	3	Commander	7	1.06	0.083	4.34E+04	1.02E+04	PII-3I-C-2-CI-7
	3	Commander	7	1.05	1.083	1.45E+04	7.96E+03	PII-3I-C-3-CI-7
	3	Commander	7	1.08	3.125	-4.74E+03	3.44E+03	PII-3I-C-4-CI-7
	3	Commander	7	1.09	7.5	3.46E+03	1.64E+03	PII-3I-C-5-CI-7
	3	Commander	7	1.08	16	4.86E+03	1.36E+03	PII-3I-C-6-CI-7
	3	Commander	7	1.05	33	4.09E+03	9.91E+02	PII-3I-C-7-CI-7
	3	Commander	7	1.08	65	2.29E+03	5.47E+02	PII-3I-C-8-CI-7
	3	Commander	7	1.08	125	2.05E+03	5.13E+02	PII-3I-C-9-CI-7
	3	Commander	8	0.62	0.083	7.79E+04	1.25E+04	PII-3I-C-2-CI-8
	3	Commander	8	0.61	1.083	-2.12E+03	6.39E+03	PII-3I-C-3-CI-8
	3	Commander	8	0.64	3.125	-3.16E+03	4.41E+03	PII-3I-C-4-CI-8
	3	Commander	8	0.64	7.5	1.54E+04	3.10E+03	PII-3I-C-5-CI-8
	3	Commander	8	0.64	16	1.34E+04	2.30E+03	PII-3I-C-6-CI-8
	3	Commander	8	0.62	33	6.03E+03	1.07E+03	PII-3I-C-7-CI-8
	3	Commander	8	0.63	65	4.22E+03	7.07E+02	PII-3I-C-8-CI-8
	3	Commander	8	0.64	125	4.06E+03	6.82E+02	PII-3I-C-9-CI-8
	3	Commander	9	0.22	0.083	2.01E+05	4.23E+04	PII-3I-C-2-CI-9
	3	Commander	9	0.22	1.083	1.83E+04	1.46E+04	PII-3I-C-3-CI-9
	3	Commander	9	0.22	3.125	8.16E+02	8.64E+03	PII-3I-C-4-CI-9
	3	Commander	9	0.22	7.5	1.19E+04	4.22E+03	PII-3I-C-5-CI-9
	3	Commander	9	0.22	16	1.58E+04	4.03E+03	PII-3I-C-6-CI-9
	3	Commander	9	0.22	33	1.87E+04	4.15E+03	PII-3I-C-7-CI-9
	3	Commander	9	0.22	65	1.17E+04	2.57E+03	PII-3I-C-8-CI-9
	3	Commander	9	0.22	125	7.01E+03	1.61E+03	PII-3I-C-9-CI-9

Table A.7. *(continued)*

Phase	Shot	Position	CI stage	Median Dia (um)	MidTime (min)	U Conc, (ug/m3)	sd U Conc. (ug/m3)	SampleID
2	3	Driver	1	31.12	0.083	1.56E+05	3.39E+04	PII-3I-D-1-C
2	3	Driver	1	31.14	1.083	7.42E+04	2.75E+04	PII-3I-D-2-C
2	3	Driver	1	30.60	3.125	3.31E+03	9.06E+03	PII-3I-D-3-C
2	3	Driver	1	30.63	7.5	2.98E+03	2.51E+03	PII-3I-D-4-C
2	3	Driver	1	30.80	16	1.16E+03	1.30E+03	PII-3I-D-5-C
2	3	Driver	1	30.72	33	2.56E+03	1.01E+03	PII-3I-D-6-C
2	3	Driver	1	30.91	65	1.04E+03	4.49E+02	PII-3I-D-7-C
2	3	Driver	1	30.62	125	1.80E+03	6.05E+02	PII-3I-D-8-C
2	3	Driver	2	16.17	0.083	5.07E+04	2.55E+04	PII-3I-D-1-C
2	3	Driver	2	16.19	1.083	1.09E+04	1.05E+04	PII-3I-D-2-C
2	3	Driver	2	15.63	3.125	-5.35E+02	3.49E+03	PII-3I-D-3-C
2	3	Driver	2	15.66	7.5	3.54E+02	9.28E+02	PII-3I-D-4-C
2	3	Driver	2	15.84	16	8.68E+02	7.27E+02	PII-3I-D-5-C
2	3	Driver	2	15.75	33	1.82E+02	2.74E+02	PII-3I-D-6-C
2	3	Driver	2	15.95	65	1.41E+02	NA[a]	PII-3I-D-7-C
2	3	Driver	2	15.65	125	3.64E+02	2.29E+02	PII-3I-D-8-C
2	3	Driver	3	10.82	0.083	7.33E+04	2.01E+04	PII-3I-D-1-C
2	3	Driver	3	10.83	1.083	1.56E+04	9.26E+03	PII-3I-D-2-C
2	3	Driver	3	10.46	3.125	2.66E+03	4.14E+03	PII-3I-D-3-C
2	3	Driver	3	10.48	7.5	1.00E+03	1.12E+03	PII-3I-D-4-C
2	3	Driver	3	10.60	16	-4.64E+02	3.94E+02	PII-3I-D-5-C
2	3	Driver	3	10.54	33	2.53E+02	2.84E+02	PII-3I-D-6-C
2	3	Driver	3	10.67	65	6.19E+00	1.23E+02	PII-3I-D-7-C
2	3	Driver	3	10.48	125	5.96E+00	1.19E+02	PII-3I-D-8-C
2	3	Driver	4	6.90	0.083	5.89E+04	2.25E+04	PII-3I-D-1-C
2	3	Driver	4	6.91	1.083	-3.92E+03	1.68E+04	PII-3I-D-2-C
2	3	Driver	4	6.67	3.125	-4.10E+03	9.92E+03	PII-3I-D-3-C
2	3	Driver	4	6.68	7.5	-6.11E+02	2.63E+03	PII-3I-D-4-C
2	3	Driver	4	6.76	16	5.44E+02	1.60E+03	PII-3I-D-5-C
2	3	Driver	4	6.72	33	3.74E+02	8.03E+02	PII-3I-D-6-C
2	3	Driver	4	6.81	65	6.26E+02	5.60E+02	PII-3I-D-7-C
2	3	Driver	4	6.68	125	-1.29E+02	3.24E+02	PII-3I-D-8-C
2	3	Driver	5	4.15	0.083	4.79E+04	1.95E+04	PII-3I-D-1-C
2	3	Driver	5	4.15	1.083	-1.77E+04	2.06E+04	PII-3I-D-2-C
2	3	Driver	5	4.01	3.125	-3.20E+03	1.48E+04	PII-3I-D-3-C
2	3	Driver	5	4.02	7.5	2.33E+03	4.62E+03	PII-3I-D-4-C
2	3	Driver	5	4.06	16	2.91E+02	2.07E+03	PII-3I-D-5-C
2	3	Driver	5	4.04	33	1.78E+03	1.59E+03	PII-3I-D-6-C
2	3	Driver	5	4.09	65	4.52E+03	2.17E+03	PII-3I-D-7-C
2	3	Driver	5	4.01	125	2.35E+03	1.32E+03	PII-3I-D-8-C
2	3	Driver	6	2.08	0.083	4.51E+04	1.28E+04	PII-3I-D-1-C
2	3	Driver	6	2.08	1.083	-5.67E+03	1.54E+04	PII-3I-D-2-C

A.7. *(continued)*

Shot	Position	CI stage	Median Dia (um)	MidTime (min)	U Conc, (ug/m3)	sd U Conc. (ug/m3)	SampleID
3	Driver	6	2.01	3.125	-1.74E+04	8.72E+03	PII-3I-D-3-CI-6
3	Driver	6	2.01	7.5	4.93E+03	3.57E+03	PII-3I-D-4-CI-6
3	Driver	6	2.04	16	2.52E+03	1.86E+03	PII-3I-D-5-CI-6
3	Driver	6	2.02	33	5.07E+03	1.86E+03	PII-3I-D-6-CI-6
3	Driver	6	2.05	65	4.26E+03	1.38E+03	PII-3I-D-7-CI-6
3	Driver	6	2.01	125	2.65E+03	9.53E+02	PII-3I-D-8-CI-6
3	Driver	7	1.07	0.083	4.17E+04	9.85E+03	PII-3I-D-1-CI-7
3	Driver	7	1.07	1.083	1.51E+04	7.27E+03	PII-3I-D-2-CI-7
3	Driver	7	1.03	3.125	6.30E+03	4.54E+03	PII-3I-D-3-CI-7
3	Driver	7	1.03	7.5	2.73E+03	1.33E+03	PII-3I-D-4-CI-7
3	Driver	7	1.05	16	2.77E+03	9.01E+02	PII-3I-D-5-CI-7
3	Driver	7	1.04	33	4.19E+03	9.63E+02	PII-3I-D-6-CI-7
3	Driver	7	1.05	65	2.49E+03	5.63E+02	PII-3I-D-7-CI-7
3	Driver	7	1.03	125	1.54E+03	3.79E+02	PII-3I-D-8-CI-7
3	Driver	8	0.63	0.083	6.25E+04	1.03E+04	PII-3I-D-1-CI-8
3	Driver	8	0.63	1.083	1.46E+04	6.15E+03	PII-3I-D-2-CI-8
3	Driver	8	0.61	3.125	1.81E+04	4.95E+03	PII-3I-D-3-CI-8
3	Driver	8	0.61	7.5	8.30E+03	1.84E+03	PII-3I-D-4-CI-8
3	Driver	8	0.62	16	7.51E+03	1.37E+03	PII-3I-D-5-CI-8
3	Driver	8	0.61	33	8.18E+03	1.29E+03	PII-3I-D-6-CI-8
3	Driver	8	0.62	65	4.59E+03	7.16E+02	PII-3I-D-7-CI-8
3	Driver	8	0.61	125	4.23E+03	6.63E+02	PII-3I-D-8-CI-8
3	Driver	9	0.22	0.083	1.10E+05	2.35E+04	PII-3I-D-1-CI-9
3	Driver	9	0.22	1.083	3.34E+04	8.39E+03	PII-3I-D-2-CI-9
3	Driver	9	0.21	3.125	1.78E+04	4.83E+03	PII-3I-D-3-CI-9
3	Driver	9	0.21	7.5	1.64E+04	3.52E+03	PII-3I-D-4-CI-9
3	Driver	9	0.22	16	1.75E+04	NA[(a)]	PII-3I-D-5-CI-9
3	Driver	9	0.22	33	8.66E+03	1.80E+03	PII-3I-D-6-CI-9
3	Driver	9	0.22	65	5.65E+03	1.17E+03	PII-3I-D-7-CI-9
3	Driver	9	0.21	125	2.09E+04	NA[(a)]	PII-3I-D-8-CI-9
3	Gunner	1	30.52	0.083	7.62E+04	1.43E+04	PII-3I-G-1-CI-1
3	Gunner	1	30.65	1.083	9.46E+03	1.11E+04	PII-3I-G-3-CI-1
3	Gunner	1	30.63	3.125	2.92E+04	1.03E+04	PII-3I-G-4-CI-1
3	Gunner	1	30.53	7.5	1.56E+03	1.83E+03	PII-3I-G-5-CI-1
3	Gunner	1	30.56	16	4.26E+02	9.33E+02	PII-3I-G-6-CI-1
3	Gunner	1	31.24	33	4.27E+02	5.17E+02	PII-3I-G-7-CI-1
3	Gunner	1	30.33	65	7.92E+02	3.11E+02	PII-3I-G-8-CI-1
3	Gunner	1	30.38	125	7.10E+02	3.02E+02	PII-3I-G-9-CI-1
3	Gunner	2	15.55	0.083	2.53E+04	1.19E+04	PII-3I-G-1-CI-2
3	Gunner	2	15.68	1.083	-1.26E+04	6.87E+03	PII-3I-G-3-CI-2
3	Gunner	2	15.66	3.125	1.37E+03	5.73E+03	PII-3I-G-4-CI-2
3	Gunner	2	15.56	7.5	-2.07E+03	1.09E+03	PII-3I-G-5-CI-2

Table A.7. *(continued)*

Phase	Shot	Position	CI stage	Median Dia (um)	MidTime (min)	U Conc, (ug/m3)	sd U Conc. (ug/m3)	SampleII
2	3	Gunner	2	15.60	16	-1.04E+03	5.89E+02	PII-3I-G-6-C
2	3	Gunner	2	16.30	33	1.58E+03	8.80E+02	PII-3I-G-7-C
2	3	Gunner	2	15.36	65	-3.23E+01	1.48E+02	PII-3I-G-8-C
2	3	Gunner	2	15.41	125	7.80E+02	4.09E+02	PII-3I-G-9-C
2	3	Gunner	3	10.41	0.083	4.80E+04	1.11E+04	PII-3I-G-1-C
2	3	Gunner	3	10.50	1.083	1.72E+04	5.24E+03	PII-3I-G-3-C
2	3	Gunner	3	10.48	3.125	1.91E+03	2.26E+03	PII-3I-G-4-C
2	3	Gunner	3	10.42	7.5	-4.71E+01	4.42E+02	PII-3I-G-5-C
2	3	Gunner	3	10.44	16	4.73E+02	3.24E+02	PII-3I-G-6-C
2	3	Gunner	3	10.91	33	1.42E+03	3.65E+02	PII-3I-G-7-C
2	3	Gunner	3	10.28	65	2.87E+02	9.29E+01	PII-3I-G-8-C
2	3	Gunner	3	10.31	125	-1.73E+01	5.86E+01	PII-3I-G-9-C
2	3	Gunner	4	6.64	0.083	5.44E+04	1.59E+04	PII-3I-G-1-C
2	3	Gunner	4	6.69	1.083	1.26E+04	5.85E+03	PII-3I-G-3-C
2	3	Gunner	4	6.68	3.125	1.67E+04	5.61E+03	PII-3I-G-4-C
2	3	Gunner	4	6.64	7.5	1.65E+03	8.47E+02	PII-3I-G-5-C
2	3	Gunner	4	6.66	16	1.87E+03	6.33E+02	PII-3I-G-6-C
2	3	Gunner	4	6.96	33	2.83E+03	8.65E+02	PII-3I-G-7-C
2	3	Gunner	4	6.56	65	1.51E+02	9.40E+01	PII-3I-G-8-C
2	3	Gunner	4	6.58	125	-1.01E+02	7.83E+01	PII-3I-G-9-C
2	3	Gunner	5	3.99	0.083	5.80E+04	2.44E+04	PII-3I-G-1-C
2	3	Gunner	5	4.02	1.083	4.47E+04	1.93E+04	PII-3I-G-3-C
2	3	Gunner	5	4.02	3.125	1.57E+04	7.63E+03	PII-3I-G-4-C
2	3	Gunner	5	3.99	7.5	8.51E+03	3.68E+03	PII-3I-G-5-C
2	3	Gunner	5	4.00	16	8.36E+03	3.39E+03	PII-3I-G-6-C
2	3	Gunner	5	4.18	33	5.31E+03	2.16E+03	PII-3I-G-7-C
2	3	Gunner	5	3.94	65	3.77E+02	1.88E+02	PII-3I-G-8-C
2	3	Gunner	5	3.95	125	-2.37E+02	1.33E+02	PII-3I-G-9-C
2	3	Gunner	6	2.00	0.083	5.28E+04	2.43E+04	PII-3I-G-1-C
2	3	Gunner	6	2.01	1.083	9.69E+03	1.33E+04	PII-3I-G-3-C
2	3	Gunner	6	2.01	3.125	2.35E+04	1.43E+04	PII-3I-G-4-C
2	3	Gunner	6	2.00	7.5	8.49E+03	4.92E+03	PII-3I-G-5-C
2	3	Gunner	6	2.00	16	1.41E+04	6.39E+03	PII-3I-G-6-C
2	3	Gunner	6	2.09	33	9.68E+03	4.37E+03	PII-3I-G-7-C
2	3	Gunner	6	1.97	65	5.41E+01	2.12E+02	PII-3I-G-8-C
2	3	Gunner	6	1.98	125	-2.27E+02	1.71E+02	PII-3I-G-9-C
2	3	Gunner	7	1.03	0.083	1.82E+04	1.11E+04	PII-3I-G-1-C
2	3	Gunner	7	1.04	1.083	9.23E+02	8.58E+03	PII-3I-G-3-C
2	3	Gunner	7	1.03	3.125	2.00E+04	8.06E+03	PII-3I-G-4-C
2	3	Gunner	7	1.03	7.5	3.41E+03	2.19E+03	PII-3I-G-5-C
2	3	Gunner	7	1.03	16	9.33E+03	2.64E+03	PII-3I-G-6-C
2	3	Gunner	7	1.08	33	5.51E+03	1.59E+03	PII-3I-G-7-C

A.7. *(continued)*

se	Shot	Position	CI stage	Median Dia (um)	MidTime (min)	U Conc, (ug/m3)	sd U Conc. (ug/m3)	SampleID
	3	Gunner	7	1.01	65	7.85E+02	2.67E+02	PII-3I-G-8-CI-7
	3	Gunner	7	1.02	125	-3.25E+02	1.76E+02	PII-3I-G-9-CI-7
	3	Gunner	8	0.60	0.083	3.56E+04	1.73E+04	PII-3I-G-1-CI-8
	3	Gunner	8	0.61	1.083	1.13E+04	1.66E+04	PII-3I-G-3-CI-8
	3	Gunner	8	0.61	3.125	1.51E+04	1.21E+04	PII-3I-G-4-CI-8
	3	Gunner	8	0.60	7.5	7.43E+03	4.29E+03	PII-3I-G-5-CI-8
	3	Gunner	8	0.61	16	1.21E+04	3.70E+03	PII-3I-G-6-CI-8
	3	Gunner	8	0.63	33	1.62E+04	4.25E+03	PII-3I-G-7-CI-8
	3	Gunner	8	0.60	65	1.09E+03	3.86E+02	PII-3I-G-8-CI-8
	3	Gunner	8	0.60	125	1.32E+03	4.00E+02	PII-3I-G-9-CI-8
	3	Gunner	9	0.21	0.083	4.63E+04	2.10E+04	PII-3I-G-1-CI-9
	3	Gunner	9	0.21	1.083	5.37E+04	2.23E+04	PII-3I-G-3-CI-9
	3	Gunner	9	0.21	3.125	7.45E+03	7.84E+03	PII-3I-G-4-CI-9
	3	Gunner	9	0.21	7.5	3.09E+04	1.09E+04	PII-3I-G-5-CI-9
	3	Gunner	9	0.21	16	1.92E+04	6.76E+03	PII-3I-G-6-CI-9
	3	Gunner	9	0.22	33	8.07E+03	3.02E+03	PII-3I-G-7-CI-9
	3	Gunner	9	0.21	65	1.12E+03	4.01E+02	PII-3I-G-8-CI-9
	3	Gunner	9	0.21	125	4.51E+01	1.59E+02	PII-3I-G-9-CI-9
	3	Loader	1	31.38	0.083	4.26E+05	7.48E+04	PII-3I-L-1-CI-1
	3	Loader	1	31.13	1.083	1.61E+05	3.69E+04	PII-3I-L-2-CI-1
	3	Loader	1	30.96	3.125	4.53E+04	1.36E+04	PII-3I-L-3-CI-1
	3	Loader	1	31.03	7.5	1.22E+04	3.66E+03	PII-3I-L-4-CI-1
	3	Loader	1	30.96	16	4.55E+03	1.58E+03	PII-3I-L-5-CI-1
	3	Loader	1	31.06	33	2.87E+03	8.78E+02	PII-3I-L-6-CI-1
	3	Loader	1	31.25	65	2.44E+03	6.07E+02	PII-3I-L-7-CI-1
	3	Loader	1	31.03	125	3.98E+03	8.62E+02	PII-3I-L-8-CI-1
	3	Loader	2	16.44	0.083	1.01E+05	4.21E+04	PII-3I-L-1-CI-2
	3	Loader	2	16.17	1.083	3.28E+03	1.01E+04	PII-3I-L-2-CI-2
	3	Loader	2	16.00	3.125	3.42E+03	6.61E+03	PII-3I-L-3-CI-2
	3	Loader	2	16.07	7.5	3.43E+03	2.70E+03	PII-3I-L-4-CI-2
	3	Loader	2	16.00	16	4.27E+02	8.81E+02	PII-3I-L-5-CI-2
	3	Loader	2	16.11	33	3.78E+02	4.78E+02	PII-3I-L-6-CI-2
	3	Loader	2	16.30	65	-1.37E+02	1.64E+02	PII-3I-L-7-CI-2
	3	Loader	2	16.08	125	1.88E+02	2.38E+02	PII-3I-L-8-CI-2
	3	Loader	3	11.00	0.083	8.21E+04	1.85E+04	PII-3I-L-1-CI-3
	3	Loader	3	10.83	1.083	3.32E+04	1.01E+04	PII-3I-L-2-CI-3
	3	Loader	3	10.71	3.125	1.17E+04	4.47E+03	PII-3I-L-3-CI-3
	3	Loader	3	10.76	7.5	2.96E+03	1.26E+03	PII-3I-L-4-CI-3
	3	Loader	3	10.71	16	1.47E+03	6.23E+02	PII-3I-L-5-CI-3
	3	Loader	3	10.78	33	2.01E+03	5.70E+02	PII-3I-L-6-CI-3
	3	Loader	3	10.91	65	3.16E+02	1.54E+02	PII-3I-L-7-CI-3
	3	Loader	3	10.76	125	-7.04E+01	8.50E+01	PII-3I-L-8-CI-3

Table A.7. *(continued)*

Phase	Shot	Position	CI stage	Median Dia (um)	MidTime (min)	U Conc, (ug/m3)	sd U Conc. (ug/m3)	SampleII
2	3	Loader	4	7.02	0.083	5.26E+04	1.62E+04	PII-3I-L-1-C
2	3	Loader	4	6.90	1.083	1.33E+04	1.01E+04	PII-3I-L-2-C
2	3	Loader	4	6.83	3.125	5.17E+03	5.75E+03	PII-3I-L-3-C
2	3	Loader	4	6.86	7.5	6.15E+03	2.60E+03	PII-3I-L-4-C
2	3	Loader	4	6.83	16	4.14E+03	1.64E+03	PII-3I-L-5-C
2	3	Loader	4	6.87	33	3.20E+03	1.13E+03	PII-3I-L-6-C
2	3	Loader	4	6.96	65	1.41E+03	5.27E+02	PII-3I-L-7-C
2	3	Loader	4	6.86	125	4.39E+02	2.46E+02	PII-3I-L-8-C
2	3	Loader	5	4.22	0.083	8.30E+04	3.25E+04	PII-3I-L-1-C
2	3	Loader	5	4.15	1.083	1.51E+04	1.48E+04	PII-3I-L-2-C
2	3	Loader	5	4.10	3.125	3.77E+04	1.96E+04	PII-3I-L-3-C
2	3	Loader	5	4.12	7.5	1.28E+04	6.28E+03	PII-3I-L-4-C
2	3	Loader	5	4.10	16	1.09E+04	4.80E+03	PII-3I-L-5-C
2	3	Loader	5	4.13	33	8.30E+03	3.52E+03	PII-3I-L-6-C
2	3	Loader	5	4.18	65	7.65E+03	3.12E+03	PII-3I-L-7-C
2	3	Loader	5	4.12	125	5.73E+03	2.36E+03	PII-3I-L-8-C
2	3	Loader	6	2.11	0.083	4.62E+04	1.91E+04	PII-3I-L-1-C
2	3	Loader	6	2.08	1.083	2.24E+04	1.85E+04	PII-3I-L-2-C
2	3	Loader	6	2.06	3.125	2.11E+04	1.44E+04	PII-3I-L-3-C
2	3	Loader	6	2.07	7.5	5.74E+03	3.95E+03	PII-3I-L-4-C
2	3	Loader	6	2.06	16	6.08E+03	3.15E+03	PII-3I-L-5-C
2	3	Loader	6	2.07	33	7.70E+03	3.25E+03	PII-3I-L-6-C
2	3	Loader	6	2.10	65	5.99E+03	2.43E+03	PII-3I-L-7-C
2	3	Loader	6	2.07	125	1.69E+03	8.47E+02	PII-3I-L-8-C
2	3	Loader	7	1.09	0.083	3.22E+04	1.05E+04	PII-3I-L-1-C
2	3	Loader	7	1.07	1.083	4.35E+04	1.51E+04	PII-3I-L-2-C
2	3	Loader	7	1.06	3.125	2.52E+04	8.13E+03	PII-3I-L-3-C
2	3	Loader	7	1.06	7.5	4.75E+03	2.11E+03	PII-3I-L-4-C
2	3	Loader	7	1.06	16	2.13E+03	9.82E+02	PII-3I-L-5-C
2	3	Loader	7	1.06	33	1.70E+03	6.98E+02	PII-3I-L-6-C
2	3	Loader	7	1.08	65	2.27E+03	6.45E+02	PII-3I-L-7-C
2	3	Loader	7	1.06	125	1.40E+03	4.61E+02	PII-3I-L-8-C
2	3	Loader	8	0.64	0.083	7.96E+04	1.97E+04	PII-3I-L-1-C
2	3	Loader	8	0.63	1.083	3.46E+04	1.51E+04	PII-3I-L-2-C
2	3	Loader	8	0.62	3.125	3.67E+04	1.25E+04	PII-3I-L-3-C
2	3	Loader	8	0.62	7.5	1.32E+04	4.01E+03	PII-3I-L-4-C
2	3	Loader	8	0.62	16	4.39E+03	1.85E+03	PII-3I-L-5-C
2	3	Loader	8	0.63	33	5.51E+03	1.59E+03	PII-3I-L-6-C
2	3	Loader	8	0.63	65	6.54E+03	1.45E+03	PII-3I-L-7-C
2	3	Loader	8	0.62	125	3.54E+03	9.14E+02	PII-3I-L-8-C
2	3	Loader	9	0.22	0.083	4.68E+04	1.70E+04	PII-3I-L-1-C
2	3	Loader	9	0.22	1.083	4.06E+04	1.76E+04	PII-3I-L-2-C

A.7. *(continued)*

se	Shot	Position	CI stage	Median Dia (um)	MidTime (min)	U Conc, (ug/m3)	sd U Conc. (ug/m3)	SampleID
	3	Loader	9	0.22	3.125	2.44E+04	1.22E+04	PII-3I-L-3-CI-9
	3	Loader	9	0.22	7.5	1.36E+04	5.16E+03	PII-3I-L-4-CI-9
	3	Loader	9	0.22	16	1.49E+04	4.87E+03	PII-3I-L-5-CI-9
	3	Loader	9	0.22	33	1.41E+04	4.38E+03	PII-3I-L-6-CI-9
	3	Loader	9	0.22	65	8.51E+03	2.55E+03	PII-3I-L-7-CI-9
	3	Loader	9	0.22	125	4.45E+03	1.40E+03	PII-3I-L-8-CI-9
	1	Commander	1	30.55	0.083	8.04E+05	1.39E+05	PIII-1I-C-1-CI-1
	1	Commander	1	30.73	1.083	5.18E+05	1.19E+05	PIII-1I-C-2-CI-1
	1	Commander	1	30.71	3.125	8.95E+05	1.72E+05	PIII-1I-C-3-CI-1
	1	Commander	1	31.17	7.5	5.00E+04	1.41E+04	PIII-1I-C-4-CI-1
	1	Commander	1	31.18	16	3.85E+04	9.34E+03	PIII-1I-C-5-CI-1
	1	Commander	1	31.20	33	1.50E+04	3.97E+03	PIII-1I-C-6-CI-1
	1	Commander	1	30.68	65	2.08E+03	1.06E+03	PIII-1I-C-7-CI-1
	1	Commander	1	31.02	125	4.71E+03	1.51E+03	PIII-1I-C-8-CI-1
	1	Commander	2	15.58	0.083	3.81E+05	1.55E+05	PIII-1I-C-1-CI-2
	1	Commander	2	15.76	1.083	1.54E+05	8.27E+04	PIII-1I-C-2-CI-2
	1	Commander	2	15.74	3.125	5.78E+04	3.80E+04	PIII-1I-C-3-CI-2
	1	Commander	2	16.22	7.5	9.44E+03	7.69E+03	PIII-1I-C-4-CI-2
	1	Commander	2	16.23	16	7.02E+03	4.73E+03	PIII-1I-C-5-CI-2
	1	Commander	2	16.25	33	-4.26E+02	1.01E+03	PIII-1I-C-6-CI-2
	1	Commander	2	15.71	65	-5.83E+02	3.98E+02	PIII-1I-C-7-CI-2
	1	Commander	2	16.06	125	-6.90E+02	4.10E+02	PIII-1I-C-8-CI-2
	1	Commander	3	10.43	0.083	5.22E+05	1.11E+05	PIII-1I-C-1-CI-3
	1	Commander	3	10.55	1.083	2.11E+05	4.77E+04	PIII-1I-C-2-CI-3
	1	Commander	3	10.54	3.125	1.00E+05	2.31E+04	PIII-1I-C-3-CI-3
	1	Commander	3	10.86	7.5	1.12E+04	2.96E+03	PIII-1I-C-4-CI-3
	1	Commander	3	10.87	16	3.30E+03	9.94E+02	PIII-1I-C-5-CI-3
	1	Commander	3	10.88	33	-1.54E+01	1.83E+02	PIII-1I-C-6-CI-3
	1	Commander	3	10.51	65	1.13E+02	9.36E+01	PIII-1I-C-7-CI-3
	1	Commander	3	10.75	125	1.18E+02	9.78E+01	PIII-1I-C-8-CI-3
	1	Commander	4	6.65	0.083	4.08E+05	1.16E+05	PIII-1I-C-1-CI-4
	1	Commander	4	6.73	1.083	2.07E+05	6.24E+04	PIII-1I-C-2-CI-4
	1	Commander	4	6.72	3.125	1.31E+05	3.97E+04	PIII-1I-C-3-CI-4
	1	Commander	4	6.92	7.5	5.14E+04	1.52E+04	PIII-1I-C-4-CI-4
	1	Commander	4	6.93	16	6.87E+03	2.35E+03	PIII-1I-C-5-CI-4
	1	Commander	4	6.94	33	1.75E+02	2.94E+02	PIII-1I-C-6-CI-4
	1	Commander	4	6.70	65	1.87E+02	1.62E+02	PIII-1I-C-7-CI-4
	1	Commander	4	6.86	125	1.40E+02	1.49E+02	PIII-1I-C-8-CI-4
	1	Commander	5	4.00	0.083	4.17E+05	1.65E+05	PIII-1I-C-1-CI-5
	1	Commander	5	4.04	1.083	3.82E+05	1.57E+05	PIII-1I-C-2-CI-5
	1	Commander	5	4.04	3.125	1.89E+05	7.91E+04	PIII-1I-C-3-CI-5
	1	Commander	5	4.16	7.5	8.76E+04	3.58E+04	PIII-1I-C-4-CI-5

Table A.7. *(continued)*

Phase	Shot	Position	CI stage	Median Dia (um)	MidTime (min)	U Conc, (ug/m3)	sd U Conc. (ug/m3)	SampleID
3	1	Commander	5	4.16	16	1.52E+04	6.63E+03	PIII-1I-C-5-C
3	1	Commander	5	4.17	33	5.10E+03	2.30E+03	PIII-1I-C-6-C
3	1	Commander	5	4.03	65	1.49E+03	7.39E+02	PIII-1I-C-7-C
3	1	Commander	5	4.12	125	4.28E+02	3.40E+02	PIII-1I-C-8-C
3	1	Commander	6	2.00	0.083	3.71E+05	1.53E+05	PIII-1I-C-1-C
3	1	Commander	6	2.03	1.083	2.43E+05	1.05E+05	PIII-1I-C-2-C
3	1	Commander	6	2.02	3.125	2.32E+05	9.85E+04	PIII-1I-C-3-C
3	1	Commander	6	2.09	7.5	5.09E+04	2.18E+04	PIII-1I-C-4-C
3	1	Commander	6	2.09	16	1.96E+04	8.52E+03	PIII-1I-C-5-C
3	1	Commander	6	2.09	33	6.70E+03	2.98E+03	PIII-1I-C-6-C
3	1	Commander	6	2.02	65	2.06E+03	9.65E+02	PIII-1I-C-7-C
3	1	Commander	6	2.06	125	1.39E+03	7.06E+02	PIII-1I-C-8-C
3	1	Commander	7	1.03	0.083	2.12E+05	5.44E+04	PIII-1I-C-1-C
3	1	Commander	7	1.04	1.083	5.70E+04	2.69E+04	PIII-1I-C-2-C
3	1	Commander	7	1.04	3.125	1.19E+05	3.67E+04	PIII-1I-C-3-C
3	1	Commander	7	1.07	7.5	4.26E+04	1.27E+04	PIII-1I-C-4-C
3	1	Commander	7	1.07	16	1.02E+04	3.65E+03	PIII-1I-C-5-C
3	1	Commander	7	1.07	33	3.35E+03	1.49E+03	PIII-1I-C-6-C
3	1	Commander	7	1.04	65	6.49E+02	4.55E+02	PIII-1I-C-7-C
3	1	Commander	7	1.06	125	-2.78E+01	3.14E+02	PIII-1I-C-8-C
3	1	Commander	8	0.60	0.083	2.84E+05	6.94E+04	PIII-1I-C-1-C
3	1	Commander	8	0.61	1.083	1.33E+05	5.21E+04	PIII-1I-C-2-C
3	1	Commander	8	0.61	3.125	1.02E+05	3.65E+04	PIII-1I-C-3-C
3	1	Commander	8	0.63	7.5	8.41E+04	2.19E+04	PIII-1I-C-4-C
3	1	Commander	8	0.63	16	2.05E+04	6.11E+03	PIII-1I-C-5-C
3	1	Commander	8	0.63	33	5.08E+03	2.01E+03	PIII-1I-C-6-C
3	1	Commander	8	0.61	65	1.19E+03	7.06E+02	PIII-1I-C-7-C
3	1	Commander	8	0.62	125	3.07E+02	NA[(a)]	PIII-1I-C-8-C
3	1	Commander	9	0.21	0.083	1.55E+05	5.28E+04	PIII-1I-C-1-C
3	1	Commander	9	0.22	1.083	1.67E+05	6.64E+04	PIII-1I-C-2-C
3	1	Commander	9	0.22	3.125	6.02E+04	2.80E+04	PIII-1I-C-3-C
3	1	Commander	9	0.22	7.5	1.72E+04	7.91E+03	PIII-1I-C-4-C
3	1	Commander	9	0.22	16	4.60E+03	2.63E+03	PIII-1I-C-5-C
3	1	Commander	9	0.22	33	-3.01E+02	6.74E+02	PIII-1I-C-6-C
3	1	Commander	9	0.22	65	1.87E+02	4.06E+02	PIII-1I-C-7-C
3	1	Commander	9	0.22	125	5.88E+02	5.08E+02	PIII-1I-C-8-C
3	1	Driver	1	31.07	0.083	1.38E+06	2.93E+05	PIII-1I-D-1-C
3	1	Driver	1	31.21	1.083	1.95E+06	4.62E+05	PIII-1I-D-2-C
3	1	Driver	1	30.59	3.125	6.02E+04	5.02E+04	PIII-1I-D-3-C
3	1	Driver	1	30.63	7.5	3.45E+04	1.61E+04	PIII-1I-D-4-C
3	1	Driver	1	30.83	16	-8.04E+03	4.02E+03	PIII-1I-D-5-C
3	1	Driver	1	30.70	33	-5.94E+03	2.97E+03	PIII-1I-D-6-C

e A.7. *(continued)*

se	Shot	Position	CI stage	Median Dia (um)	MidTime (min)	U Conc, (ug/m3)	sd U Conc. (ug/m3)	SampleID
	1	Driver	1	30.91	65	-8.17E+01	NA[(a)]	PIII-1I-D-7-CI-1
	1	Driver	1	30.68	125	2.80E+03	1.77E+03	PIII-1I-D-8-CI-1
	1	Driver	2	16.12	0.083	1.28E+05	6.17E+04	PIII-1I-D-1-CI-2
	1	Driver	2	16.26	1.083	5.93E+04	5.43E+04	PIII-1I-D-2-CI-2
	1	Driver	2	15.62	3.125	1.46E+04	2.39E+04	PIII-1I-D-3-CI-2
	1	Driver	2	15.66	7.5	-1.21E+03	4.17E+03	PIII-1I-D-4-CI-2
	1	Driver	2	15.86	16	3.19E+02	2.47E+03	PIII-1I-D-5-CI-2
	1	Driver	2	15.74	33	1.24E+03	1.64E+03	PIII-1I-D-6-CI-2
	1	Driver	2	15.95	65	3.18E+02	7.18E+02	PIII-1I-D-7-CI-2
	1	Driver	2	15.71	125	-7.52E+01	5.55E+02	PIII-1I-D-8-CI-2
	1	Driver	3	10.79	0.083	1.76E+05	4.91E+04	PIII-1I-D-1-CI-3
	1	Driver	3	10.88	1.083	1.66E+05	4.84E+04	PIII-1I-D-2-CI-3
	1	Driver	3	10.45	3.125	7.72E+04	2.32E+04	PIII-1I-D-3-CI-3
	1	Driver	3	10.48	7.5	3.13E+04	NA[(a)]	PIII-1I-D-4-CI-3
	1	Driver	3	10.62	16	4.08E+03	1.34E+03	PIII-1I-D-5-CI-3
	1	Driver	3	10.53	33	2.00E+02	2.29E+02	PIII-1I-D-6-CI-3
	1	Driver	3	10.68	65	5.94E+01	1.03E+02	PIII-1I-D-7-CI-3
	1	Driver	3	10.51	125	5.20E+02	2.09E+02	PIII-1I-D-8-CI-3
	1	Driver	4	6.88	0.083	1.33E+05	5.07E+04	PIII-1I-D-1-CI-4
	1	Driver	4	6.94	1.083	2.45E+05	9.61E+04	PIII-1I-D-2-CI-4
	1	Driver	4	6.67	3.125	1.16E+05	4.58E+04	PIII-1I-D-3-CI-4
	1	Driver	4	6.68	7.5	3.45E+04	1.37E+04	PIII-1I-D-4-CI-4
	1	Driver	4	6.77	16	5.25E+03	2.30E+03	PIII-1I-D-5-CI-4
	1	Driver	4	6.72	33	3.63E+02	3.40E+02	PIII-1I-D-6-CI-4
	1	Driver	4	6.81	65	1.32E+02	1.64E+02	PIII-1I-D-7-CI-4
	1	Driver	4	6.70	125	4.97E+02	2.80E+02	PIII-1I-D-8-CI-4
	1	Driver	5	4.13	0.083	2.39E+05	9.51E+04	PIII-1I-D-1-CI-5
	1	Driver	5	4.17	1.083	2.59E+05	1.02E+05	PIII-1I-D-2-CI-5
	1	Driver	5	4.01	3.125	2.56E+05	1.01E+05	PIII-1I-D-3-CI-5
	1	Driver	5	4.02	7.5	3.21E+04	1.28E+04	PIII-1I-D-4-CI-5
	1	Driver	5	4.07	16	2.07E+04	NA[(a)]	PIII-1I-D-5-CI-5
	1	Driver	5	4.04	33	4.75E+03	1.91E+03	PIII-1I-D-6-CI-5
	1	Driver	5	4.09	65	7.62E+02	3.20E+02	PIII-1I-D-7-CI-5
	1	Driver	5	4.03	125	9.37E+02	3.77E+02	PIII-1I-D-8-CI-5
	1	Driver	6	2.07	0.083	1.09E+05	2.94E+04	PIII-1I-D-1-CI-6
	1	Driver	6	2.09	1.083	1.89E+05	5.17E+04	PIII-1I-D-2-CI-6
	1	Driver	6	2.01	3.125	1.10E+05	3.03E+04	PIII-1I-D-3-CI-6
	1	Driver	6	2.01	7.5	3.92E+04	1.06E+04	PIII-1I-D-4-CI-6
	1	Driver	6	2.04	16	1.16E+04	3.23E+03	PIII-1I-D-5-CI-6
	1	Driver	6	2.02	33	6.18E+03	1.72E+03	PIII-1I-D-6-CI-6
	1	Driver	6	2.05	65	8.61E+02	2.92E+02	PIII-1I-D-7-CI-6
	1	Driver	6	2.02	125	8.84E+02	2.94E+02	PIII-1I-D-8-CI-6

Table A.7. *(continued)*

Phase	Shot	Position	CI stage	Median Dia (um)	MidTime (min)	U Conc, (ug/m3)	sd U Conc. (ug/m3)	SampleID
3	1	Driver	7	1.06	0.083	1.05E+05	2.16E+04	PIII-1I-D-1-C
3	1	Driver	7	1.07	1.083	1.22E+05	2.54E+04	PIII-1I-D-2-C
3	1	Driver	7	1.03	3.125	7.03E+04	1.50E+04	PIII-1I-D-3-C
3	1	Driver	7	1.03	7.5	4.64E+04	9.22E+03	PIII-1I-D-4-C
3	1	Driver	7	1.05	16	2.09E+04	4.17E+03	PIII-1I-D-5-C
3	1	Driver	7	1.04	33	7.56E+03	1.53E+03	PIII-1I-D-6-C
3	1	Driver	7	1.05	65	1.54E+03	3.39E+02	PIII-1I-D-7-C
3	1	Driver	7	1.04	125	9.66E+02	2.28E+02	PIII-1I-D-8-C
3	1	Driver	8	0.63	0.083	1.39E+05	2.13E+04	PIII-1I-D-1-C
3	1	Driver	8	0.63	1.083	1.29E+05	2.35E+04	PIII-1I-D-2-C
3	1	Driver	8	0.61	3.125	7.50E+04	1.41E+04	PIII-1I-D-3-C
3	1	Driver	8	0.61	7.5	5.38E+04	8.27E+03	PIII-1I-D-4-C
3	1	Driver	8	0.62	16	2.45E+04	3.80E+03	PIII-1I-D-5-C
3	1	Driver	8	0.61	33	4.38E+03	8.18E+02	PIII-1I-D-6-C
3	1	Driver	8	0.62	65	1.61E+03	3.28E+02	PIII-1I-D-7-C
3	1	Driver	8	0.61	125	1.57E+03	3.25E+02	PIII-1I-D-8-C
3	1	Driver	9	0.22	0.083	2.51E+06	NA[(a)]	PIII-1I-D-1-C
3	1	Driver	9	0.22	1.083	2.00E+05	4.75E+04	PIII-1I-D-2-C
3	1	Driver	9	0.21	3.125	3.70E+04	1.26E+04	PIII-1I-D-3-C
3	1	Driver	9	0.21	7.5	2.20E+04	5.64E+03	PIII-1I-D-4-C
3	1	Driver	9	0.22	16	8.60E+03	2.36E+03	PIII-1I-D-5-C
3	1	Driver	9	0.22	33	1.22E+03	5.95E+02	PIII-1I-D-6-C
3	1	Driver	9	0.22	65	7.24E+02	3.21E+02	PIII-1I-D-7-C
3	1	Driver	9	0.22	125	1.92E+03	5.33E+02	PIII-1I-D-8-C
3	1	Gunner	1	30.36	0.083	1.92E+06	3.27E+05	PIII-1I-G-1-C
3	1	Gunner	1	30.59	1.083	5.01E+04	5.23E+04	PIII-1I-G-3-C
3	1	Gunner	1	30.56	3.125	1.34E+04	3.13E+04	PIII-1I-G-4-C
3	1	Gunner	1	30.49	7.5	-1.22E+04	6.08E+03	PIII-1I-G-5-C
3	1	Gunner	1	30.37	16	-9.10E+03	4.55E+03	PIII-1I-G-6-C
3	1	Gunner	1	31.34	33	-2.41E+03	1.76E+03	PIII-1I-G-7-C
3	1	Gunner	1	31.10	65	-1.26E+03	8.48E+02	PIII-1I-G-8-C
3	1	Gunner	1	31.42	125	1.46E+03	1.22E+03	PIII-1I-G-9-C
3	1	Gunner	2	15.38	0.083	5.23E+05	2.13E+05	PIII-1I-G-1-C
3	1	Gunner	2	15.62	1.083	1.93E+05	9.74E+04	PIII-1I-G-3-C
3	1	Gunner	2	15.59	3.125	1.21E+04	1.97E+04	PIII-1I-G-4-C
3	1	Gunner	2	15.52	7.5	1.44E+04	9.08E+03	PIII-1I-G-5-C
3	1	Gunner	2	15.40	16	-2.94E+03	1.47E+03	PIII-1I-G-6-C
3	1	Gunner	2	16.40	33	-1.67E+03	8.35E+02	PIII-1I-G-7-C
3	1	Gunner	2	16.15	65	-6.48E+02	3.96E+02	PIII-1I-G-8-C
3	1	Gunner	2	16.48	125	-8.43E+02	4.22E+02	PIII-1I-G-9-C
3	1	Gunner	3	10.30	0.083	6.49E+05	1.37E+05	PIII-1I-G-1-C
3	1	Gunner	3	10.45	1.083	1.40E+05	3.10E+04	PIII-1I-G-3-C

A.7. *(continued)*

e	Shot	Position	CI stage	Median Dia (um)	MidTime (min)	U Conc, (ug/m3)	sd U Conc. (ug/m3)	SampleID
	1	Gunner	3	10.43	3.125	1.04E+05	2.30E+04	PIII-1I-G-4-CI-3
	1	Gunner	3	10.39	7.5	1.65E+04	3.80E+03	PIII-1I-G-5-CI-3
	1	Gunner	3	10.31	16	2.36E+03	6.96E+02	PIII-1I-G-6-CI-3
	1	Gunner	3	10.98	33	4.21E+02	2.43E+02	PIII-1I-G-7-CI-3
	1	Gunner	3	10.81	65	-1.13E+02	9.52E+01	PIII-1I-G-8-CI-3
	1	Gunner	3	11.03	125	-5.15E+01	NA[(a)]	PIII-1I-G-9-CI-3
	1	Gunner	4	6.57	0.083	4.14E+05	1.17E+05	PIII-1I-G-1-CI-4
	1	Gunner	4	6.67	1.083	2.73E+05	7.95E+04	PIII-1I-G-3-CI-4
	1	Gunner	4	6.65	3.125	1.43E+05	4.21E+04	PIII-1I-G-4-CI-4
	1	Gunner	4	6.63	7.5	2.81E+04	8.33E+03	PIII-1I-G-5-CI-4
	1	Gunner	4	6.57	16	1.16E+04	3.49E+03	PIII-1I-G-6-CI-4
	1	Gunner	4	7.00	33	2.67E+02	3.16E+02	PIII-1I-G-7-CI-4
	1	Gunner	4	6.89	65	1.84E+01	NA[(a)]	PIII-1I-G-8-CI-4
	1	Gunner	4	7.03	125	-9.67E+01	1.20E+02	PIII-1I-G-9-CI-4
	1	Gunner	5	3.95	0.083	5.56E+05	2.23E+05	PIII-1I-G-1-CI-5
	1	Gunner	5	4.01	1.083	4.27E+05	1.78E+05	PIII-1I-G-3-CI-5
	1	Gunner	5	4.00	3.125	1.71E+05	7.33E+04	PIII-1I-G-4-CI-5
	1	Gunner	5	3.98	7.5	5.79E+04	2.44E+04	PIII-1I-G-5-CI-5
	1	Gunner	5	3.95	16	2.51E+04	1.06E+04	PIII-1I-G-6-CI-5
	1	Gunner	5	4.21	33	1.86E+03	1.20E+03	PIII-1I-G-7-CI-5
	1	Gunner	5	4.14	65	3.31E+02	3.62E+02	PIII-1I-G-8-CI-5
	1	Gunner	5	4.23	125	-1.44E+02	2.48E+02	PIII-1I-G-9-CI-5
	1	Gunner	6	1.98	0.083	5.63E+05	2.49E+05	PIII-1I-G-1-CI-6
	1	Gunner	6	2.01	1.083	3.53E+05	1.69E+05	PIII-1I-G-3-CI-6
	1	Gunner	6	2.00	3.125	1.47E+05	7.40E+04	PIII-1I-G-4-CI-6
	1	Gunner	6	1.99	7.5	2.91E+04	1.53E+04	PIII-1I-G-5-CI-6
	1	Gunner	6	1.98	16	1.04E+04	5.70E+03	PIII-1I-G-6-CI-6
	1	Gunner	6	2.11	33	1.22E+03	1.36E+03	PIII-1I-G-7-CI-6
	1	Gunner	6	2.08	65	8.50E+02	7.56E+02	PIII-1I-G-8-CI-6
	1	Gunner	6	2.12	125	5.09E+02	6.26E+02	PIII-1I-G-9-CI-6
	1	Gunner	7	1.02	0.083	2.76E+05	7.40E+04	PIII-1I-G-1-CI-7
	1	Gunner	7	1.03	1.083	3.21E+05	8.70E+04	PIII-1I-G-3-CI-7
	1	Gunner	7	1.03	3.125	1.04E+05	3.05E+04	PIII-1I-G-4-CI-7
	1	Gunner	7	1.03	7.5	3.78E+04	1.08E+04	PIII-1I-G-5-CI-7
	1	Gunner	7	1.02	16	2.01E+04	5.48E+03	PIII-1I-G-6-CI-7
	1	Gunner	7	1.08	33	4.01E+03	1.42E+03	PIII-1I-G-7-CI-7
	1	Gunner	7	1.07	65	7.19E+02	NA[(a)]	PIII-1I-G-8-CI-7
	1	Gunner	7	1.09	125	-3.14E+02	2.74E+02	PIII-1I-G-9-CI-7
	1	Gunner	8	0.60	0.083	2.55E+05	7.25E+04	PIII-1I-G-1-CI-8
	1	Gunner	8	0.61	1.083	2.16E+05	7.35E+04	PIII-1I-G-3-CI-8
	1	Gunner	8	0.61	3.125	7.77E+04	3.61E+04	PIII-1I-G-4-CI-8
	1	Gunner	8	0.60	7.5	6.73E+04	2.02E+04	PIII-1I-G-5-CI-8

Table A.7. *(continued)*

Phase	Shot	Position	CI stage	Median Dia (um)	MidTime (min)	U Conc, (ug/m3)	sd U Conc. (ug/m3)	SampleID
3	1	Gunner	8	0.60	16	2.35E+04	6.95E+03	PIII-1I-G-6-C
3	1	Gunner	8	0.64	33	4.86E+03	2.01E+03	PIII-1I-G-7-C
3	1	Gunner	8	0.63	65	1.01E+03	7.03E+02	PIII-1I-G-8-C
3	1	Gunner	8	0.64	125	1.93E+03	8.90E+02	PIII-1I-G-9-C
3	1	Gunner	9	0.21	0.083	1.92E+05	6.89E+04	PIII-1I-G-1-C
3	1	Gunner	9	0.21	1.083	5.11E+04	5.50E+04	PIII-1I-G-3-C
3	1	Gunner	9	0.21	3.125	-1.18E+04	2.55E+04	PIII-1I-G-4-C
3	1	Gunner	9	0.21	7.5	-7.32E+02	7.00E+03	PIII-1I-G-5-C
3	1	Gunner	9	0.21	16	9.00E+02	3.65E+03	PIII-1I-G-6-C
3	1	Gunner	9	0.22	33	-3.27E+03	1.63E+03	PIII-1I-G-7-C
3	1	Gunner	9	0.22	65	-1.49E+03	7.43E+02	PIII-1I-G-8-C
3	1	Gunner	9	0.22	125	-1.44E+03	7.22E+02	PIII-1I-G-9-C
3	1	Loader	1	31.29	0.083	4.59E+05	8.00E+04	PIII-1I-L-1-C
3	1	Loader	1	31.15	1.083	3.64E+04	3.50E+04	PIII-1I-L-2-C
3	1	Loader	1	31.02	3.125	-1.83E+04	1.74E+04	PIII-1I-L-3-C
3	1	Loader	1	31.12	7.5	-6.93E+03	4.15E+03	PIII-1I-L-4-C
3	1	Loader	1	31.15	16	-2.71E+03	2.18E+03	PIII-1I-L-5-C
3	1	Loader	1	31.13	33	-5.89E+02	1.15E+03	PIII-1I-L-6-C
3	1	Loader	1	31.31	65	4.81E+02	6.88E+02	PIII-1I-L-7-C
3	1	Loader	1	31.11	125	-1.44E+03	7.18E+02	PIII-1I-L-8-C
3	1	Loader	2	16.35	0.083	4.11E+05	1.68E+05	PIII-1I-L-1-C
3	1	Loader	2	16.20	1.083	1.40E+05	6.05E+04	PIII-1I-L-2-C
3	1	Loader	2	16.06	3.125	6.37E+04	2.78E+04	PIII-1I-L-3-C
3	1	Loader	2	16.17	7.5	1.16E+04	5.28E+03	PIII-1I-L-4-C
3	1	Loader	2	16.20	16	1.20E+04	5.14E+03	PIII-1I-L-5-C
3	1	Loader	2	16.18	33	6.16E+02	4.67E+02	PIII-1I-L-6-C
3	1	Loader	2	16.37	65	2.32E+02	2.08E+02	PIII-1I-L-7-C
3	1	Loader	2	16.15	125	2.26E+02	2.02E+02	PIII-1I-L-8-C
3	1	Loader	3	10.94	0.083	2.50E+05	5.33E+04	PIII-1I-L-1-C
3	1	Loader	3	10.84	1.083	1.72E+05	3.79E+04	PIII-1I-L-2-C
3	1	Loader	3	10.75	3.125	1.01E+05	2.21E+04	PIII-1I-L-3-C
3	1	Loader	3	10.82	7.5	1.44E+04	3.39E+03	PIII-1I-L-4-C
3	1	Loader	3	10.84	16	1.59E+03	4.98E+02	PIII-1I-L-5-C
3	1	Loader	3	10.83	33	9.17E+01	1.51E+02	PIII-1I-L-6-C
3	1	Loader	3	10.95	65	2.75E+02	1.05E+02	PIII-1I-L-7-C
3	1	Loader	3	10.81	125	7.75E+01	6.57E+01	PIII-1I-L-8-C
3	1	Loader	4	6.98	0.083	2.08E+05	5.96E+04	PIII-1I-L-1-C
3	1	Loader	4	6.91	1.083	2.34E+05	6.67E+04	PIII-1I-L-2-C
3	1	Loader	4	6.86	3.125	9.33E+04	2.69E+04	PIII-1I-L-3-C
3	1	Loader	4	6.90	7.5	4.77E+04	1.36E+04	PIII-1I-L-4-C
3	1	Loader	4	6.91	16	1.05E+04	3.09E+03	PIII-1I-L-5-C
3	1	Loader	4	6.91	33	1.12E+03	3.64E+02	PIII-1I-L-6-C

e A.7. *(continued)*

se	Shot	Position	CI stage	Median Dia (um)	MidTime (min)	U Conc, (ug/m3)	sd U Conc. (ug/m3)	SampleID
	1	Loader	4	6.99	65	1.14E+02	8.06E+01	PIII-1I-L-7-CI-4
	1	Loader	4	6.89	125	5.56E+01	7.79E+01	PIII-1I-L-8-CI-4
	1	Loader	5	4.19	0.083	3.12E+05	1.25E+05	PIII-1I-L-1-CI-5
	1	Loader	5	4.15	1.083	2.39E+05	9.68E+04	PIII-1I-L-2-CI-5
	1	Loader	5	4.12	3.125	1.87E+05	7.56E+04	PIII-1I-L-3-CI-5
	1	Loader	5	4.15	7.5	3.97E+04	1.61E+04	PIII-1I-L-4-CI-5
	1	Loader	5	4.15	16	2.29E+04	9.30E+03	PIII-1I-L-5-CI-5
	1	Loader	5	4.15	33	2.93E+03	1.31E+03	PIII-1I-L-6-CI-5
	1	Loader	5	4.20	65	6.98E+02	3.38E+02	PIII-1I-L-7-CI-5
	1	Loader	5	4.14	125	6.28E+02	3.13E+02	PIII-1I-L-8-CI-5
	1	Loader	6	2.10	0.083	4.38E+05	1.87E+05	PIII-1I-L-1-CI-6
	1	Loader	6	2.08	1.083	2.29E+05	9.85E+04	PIII-1I-L-2-CI-6
	1	Loader	6	2.06	3.125	2.37E+05	1.02E+05	PIII-1I-L-3-CI-6
	1	Loader	6	2.08	7.5	5.21E+04	2.25E+04	PIII-1I-L-4-CI-6
	1	Loader	6	2.08	16	1.93E+04	8.39E+03	PIII-1I-L-5-CI-6
	1	Loader	6	2.08	33	4.04E+03	1.84E+03	PIII-1I-L-6-CI-6
	1	Loader	6	2.10	65	2.38E+03	1.06E+03	PIII-1I-L-7-CI-6
	1	Loader	6	2.08	125	6.19E+02	3.60E+02	PIII-1I-L-8-CI-6
	1	Loader	7	1.08	0.083	1.76E+05	4.87E+04	PIII-1I-L-1-CI-7
	1	Loader	7	1.07	1.083	4.72E+04	2.12E+04	PIII-1I-L-2-CI-7
	1	Loader	7	1.06	3.125	1.60E+05	4.39E+04	PIII-1I-L-3-CI-7
	1	Loader	7	1.07	7.5	3.52E+04	1.02E+04	PIII-1I-L-4-CI-7
	1	Loader	7	1.07	16	1.56E+04	4.56E+03	PIII-1I-L-5-CI-7
	1	Loader	7	1.07	33	4.42E+03	1.49E+03	PIII-1I-L-6-CI-7
	1	Loader	7	1.08	65	2.68E+02	3.13E+02	PIII-1I-L-7-CI-7
	1	Loader	7	1.07	125	1.59E+02	2.61E+02	PIII-1I-L-8-CI-7
	1	Loader	8	0.63	0.083	1.92E+05	5.54E+04	PIII-1I-L-1-CI-8
	1	Loader	8	0.63	1.083	1.07E+05	4.46E+04	PIII-1I-L-2-CI-8
	1	Loader	8	0.62	3.125	1.05E+05	3.68E+04	PIII-1I-L-3-CI-8
	1	Loader	8	0.63	7.5	5.70E+04	1.61E+04	PIII-1I-L-4-CI-8
	1	Loader	8	0.63	16	2.14E+04	6.46E+03	PIII-1I-L-5-CI-8
	1	Loader	8	0.63	33	3.93E+03	1.75E+03	PIII-1I-L-6-CI-8
	1	Loader	8	0.64	65	8.79E+02	5.83E+02	PIII-1I-L-7-CI-8
	1	Loader	8	0.63	125	8.56E+02	6.37E+02	PIII-1I-L-8-CI-8
	1	Loader	9	0.22	0.083	1.80E+05	6.30E+04	PIII-1I-L-1-CI-9
	1	Loader	9	0.22	1.083	1.05E+05	4.52E+04	PIII-1I-L-2-CI-9
	1	Loader	9	0.22	3.125	7.21E+04	3.00E+04	PIII-1I-L-3-CI-9
	1	Loader	9	0.22	7.5	1.63E+04	7.46E+03	PIII-1I-L-4-CI-9
	1	Loader	9	0.22	16	6.18E+03	2.87E+03	PIII-1I-L-5-CI-9
	1	Loader	9	0.22	33	4.97E+02	7.63E+02	PIII-1I-L-6-CI-9
	1	Loader	9	0.22	65	4.58E+02	4.24E+02	PIII-1I-L-7-CI-9
	1	Loader	9	0.22	125	6.44E+02	4.92E+02	PIII-1I-L-8-CI-9

Table A.7. *(continued)*

Phase	Shot	Position	CI stage	Median Dia (um)	MidTime (min)	U Conc, (ug/m3)	sd U Conc. (ug/m3)	SampleID
3	2	Commander	1	30.47	0.083	2.31E+06	3.95E+05	PIII-2I-C-1-C
3	2	Commander	1	30.63	1.083	3.91E+06	7.91E+05	PIII-2I-C-2-C
3	2	Commander	1	30.58	3.125	1.53E+06	3.44E+05	PIII-2I-C-3-C
3	2	Commander	1	31.15	7.5	-2.74E+04	2.20E+04	PIII-2I-C-4-C
3	2	Commander	1	31.14	16	1.63E+03	1.31E+04	PIII-2I-C-5-C
3	2	Commander	1	31.13	33	1.78E+04	9.09E+03	PIII-2I-C-6-C
3	2	Commander	1	30.47	65	2.86E+03	3.38E+03	PIII-2I-C-7-C
3	2	Commander	1	30.95	125	4.12E+04	9.76E+03	PIII-2I-C-8-C
3	2	Commander	2	15.50	0.083	7.53E+05	3.07E+05	PIII-2I-C-1-C
3	2	Commander	2	15.66	1.083	5.21E+05	2.23E+05	PIII-2I-C-2-C
3	2	Commander	2	15.61	3.125	1.22E+05	5.82E+04	PIII-2I-C-3-C
3	2	Commander	2	16.20	7.5	1.52E+04	8.36E+03	PIII-2I-C-4-C
3	2	Commander	2	16.19	16	1.72E+03	1.99E+03	PIII-2I-C-5-C
3	2	Commander	2	16.17	33	2.00E+03	1.44E+03	PIII-2I-C-6-C
3	2	Commander	2	15.50	65	2.06E+01	3.40E+02	PIII-2I-C-7-C
3	2	Commander	2	15.99	125	2.60E+02	4.31E+02	PIII-2I-C-8-C
3	2	Commander	3	10.38	0.083	7.15E+05	1.53E+05	PIII-2I-C-1-C
3	2	Commander	3	10.48	1.083	2.38E+05	5.36E+04	PIII-2I-C-2-C
3	2	Commander	3	10.45	3.125	1.64E+05	3.69E+04	PIII-2I-C-3-C
3	2	Commander	3	10.84	7.5	6.69E+03	2.12E+03	PIII-2I-C-4-C
3	2	Commander	3	10.83	16	1.53E+03	6.86E+02	PIII-2I-C-5-C
3	2	Commander	3	10.83	33	3.95E+02	2.72E+02	PIII-2I-C-6-C
3	2	Commander	3	10.38	65	1.83E+02	1.21E+02	PIII-2I-C-7-C
3	2	Commander	3	10.71	125	3.17E+02	1.43E+02	PIII-2I-C-8-C
3	2	Commander	4	6.62	0.083	1.01E+06	2.86E+05	PIII-2I-C-1-C
3	2	Commander	4	6.68	1.083	1.73E+05	5.70E+04	PIII-2I-C-2-C
3	2	Commander	4	6.66	3.125	1.15E+05	3.82E+04	PIII-2I-C-3-C
3	2	Commander	4	6.91	7.5	8.91E+03	4.18E+03	PIII-2I-C-4-C
3	2	Commander	4	6.91	16	1.32E+03	1.25E+03	PIII-2I-C-5-C
3	2	Commander	4	6.90	33	1.55E+03	8.47E+02	PIII-2I-C-6-C
3	2	Commander	4	6.62	65	8.75E+02	4.40E+02	PIII-2I-C-7-C
3	2	Commander	4	6.83	125	1.04E+02	2.43E+02	PIII-2I-C-8-C
3	2	Commander	5	3.98	0.083	8.98E+05	3.52E+05	PIII-2I-C-1-C
3	2	Commander	5	4.02	1.083	2.34E+05	9.85E+04	PIII-2I-C-2-C
3	2	Commander	5	4.00	3.125	1.82E+05	7.59E+04	PIII-2I-C-3-C
3	2	Commander	5	4.15	7.5	2.88E+04	1.25E+04	PIII-2I-C-4-C
3	2	Commander	5	4.15	16	8.09E+03	3.83E+03	PIII-2I-C-5-C
3	2	Commander	5	4.15	33	4.77E+03	2.20E+03	PIII-2I-C-6-C
3	2	Commander	5	3.98	65	3.52E+03	1.53E+03	PIII-2I-C-7-C
3	2	Commander	5	4.10	125	2.23E+03	1.04E+03	PIII-2I-C-8-C
3	2	Commander	6	1.99	0.083	6.31E+05	2.52E+05	PIII-2I-C-1-C
3	2	Commander	6	2.01	1.083	1.68E+05	7.99E+04	PIII-2I-C-2-C

e A.7. *(continued)*

se	Shot	Position	CI stage	Median Dia (um)	MidTime (min)	U Conc, (ug/m3)	sd U Conc. (ug/m3)	SampleID
	2	Commander	6	2.01	3.125	1.19E+05	5.70E+04	PIII-2I-C-3-CI-6
	2	Commander	6	2.08	7.5	2.41E+04	1.23E+04	PIII-2I-C-4-CI-6
	2	Commander	6	2.08	16	6.26E+03	3.77E+03	PIII-2I-C-5-CI-6
	2	Commander	6	2.08	33	2.30E+03	1.59E+03	PIII-2I-C-6-CI-6
	2	Commander	6	1.99	65	1.83E+03	1.04E+03	PIII-2I-C-7-CI-6
	2	Commander	6	2.06	125	3.04E+03	1.51E+03	PIII-2I-C-8-CI-6
	2	Commander	7	1.02	0.083	3.52E+05	8.89E+04	PIII-2I-C-1-CI-7
	2	Commander	7	1.03	1.083	1.65E+05	4.74E+04	PIII-2I-C-2-CI-7
	2	Commander	7	1.03	3.125	1.20E+05	3.49E+04	PIII-2I-C-3-CI-7
	2	Commander	7	1.07	7.5	2.81E+04	8.32E+03	PIII-2I-C-4-CI-7
	2	Commander	7	1.07	16	1.76E+03	1.36E+03	PIII-2I-C-5-CI-7
	2	Commander	7	1.07	33	1.08E+03	6.85E+02	PIII-2I-C-6-CI-7
	2	Commander	7	1.02	65	1.21E+03	5.08E+02	PIII-2I-C-7-CI-7
	2	Commander	7	1.06	125	1.78E+03	6.18E+02	PIII-2I-C-8-CI-7
	2	Commander	8	0.60	0.083	5.66E+05	1.31E+05	PIII-2I-C-1-CI-8
	2	Commander	8	0.61	1.083	2.94E+05	7.80E+04	PIII-2I-C-2-CI-8
	2	Commander	8	0.61	3.125	1.85E+05	4.88E+04	PIII-2I-C-3-CI-8
	2	Commander	8	0.63	7.5	2.39E+04	7.47E+03	PIII-2I-C-4-CI-8
	2	Commander	8	0.63	16	9.73E+03	3.26E+03	PIII-2I-C-5-CI-8
	2	Commander	8	0.63	33	3.86E+03	1.55E+03	PIII-2I-C-6-CI-8
	2	Commander	8	0.60	65	3.42E+03	1.03E+03	PIII-2I-C-7-CI-8
	2	Commander	8	0.62	125	3.80E+03	1.12E+03	PIII-2I-C-8-CI-8
	2	Commander	9	0.21	0.083	7.77E+05	2.44E+05	PIII-2I-C-1-CI-9
	2	Commander	9	0.21	1.083	2.59E+05	9.26E+04	PIII-2I-C-2-CI-9
	2	Commander	9	0.21	3.125	5.19E+04	2.44E+04	PIII-2I-C-3-CI-9
	2	Commander	9	0.22	7.5	2.79E+03	3.23E+03	PIII-2I-C-4-CI-9
	2	Commander	9	0.22	16	9.94E+02	1.46E+03	PIII-2I-C-5-CI-9
	2	Commander	9	0.22	33	1.19E+03	9.37E+02	PIII-2I-C-6-CI-9
	2	Commander	9	0.21	65	9.67E+02	6.49E+02	PIII-2I-C-7-CI-9
	2	Commander	9	0.22	125	1.50E+03	7.52E+02	PIII-2I-C-8-CI-9
	2	Driver	1	31.14	0.083	6.93E+05	1.47E+05	PIII-2I-D-1-CI-1
	2	Driver	1	31.22	1.083	6.31E+05	2.30E+05	PIII-2I-D-2-CI-1
	2	Driver	1	30.72	3.125	5.60E+05	1.77E+05	PIII-2I-D-3-CI-1
	2	Driver	1	30.65	7.5	2.45E+04	2.18E+04	PIII-2I-D-4-CI-1
	2	Driver	1	30.89	16	-1.40E+04	7.16E+03	PIII-2I-D-5-CI-1
	2	Driver	1	30.72	33	-4.14E+03	3.84E+03	PIII-2I-D-6-CI-1
	2	Driver	1	30.83	65	1.48E+04	5.02E+03	PIII-2I-D-7-CI-1
	2	Driver	1	30.66	125	1.46E+04	4.93E+03	PIII-2I-D-8-CI-1
	2	Driver	2	16.19	0.083	4.03E+05	1.92E+05	PIII-2I-D-1-CI-2
	2	Driver	2	16.27	1.083	1.19E+05	6.16E+04	PIII-2I-D-2-CI-2
	2	Driver	2	15.75	3.125	3.26E+04	1.84E+04	PIII-2I-D-3-CI-2
	2	Driver	2	15.68	7.5	2.55E+03	1.98E+03	PIII-2I-D-4-CI-2

Table A.7. *(continued)*

Phase	Shot	Position	CI stage	Median Dia (um)	MidTime (min)	U Conc, (ug/m3)	sd U Conc. (ug/m3)	SampleID
3	2	Driver	2	15.93	16	4.01E+02	6.91E+02	PIII-2I-D-5-C
3	2	Driver	2	15.75	33	2.12E+02	3.12E+02	PIII-2I-D-6-C
3	2	Driver	2	15.87	65	2.48E+02	2.23E+02	PIII-2I-D-7-C
3	2	Driver	2	15.69	125	7.79E+02	4.54E+02	PIII-2I-D-8-C
3	2	Driver	3	10.83	0.083	5.91E+05	1.60E+05	PIII-2I-D-1-C
3	2	Driver	3	10.89	1.083	1.09E+05	3.18E+04	PIII-2I-D-2-C
3	2	Driver	3	10.54	3.125	1.28E+05	3.62E+04	PIII-2I-D-3-C
3	2	Driver	3	10.50	7.5	6.43E+03	2.22E+03	PIII-2I-D-4-C
3	2	Driver	3	10.67	16	1.33E+02	3.18E+02	PIII-2I-D-5-C
3	2	Driver	3	10.54	33	7.78E+02	3.21E+02	PIII-2I-D-6-C
3	2	Driver	3	10.62	65	1.50E+02	1.07E+02	PIII-2I-D-7-C
3	2	Driver	3	10.50	125	8.63E+01	9.26E+01	PIII-2I-D-8-C
3	2	Driver	4	6.91	0.083	3.92E+05	1.46E+05	PIII-2I-D-1-C
3	2	Driver	4	6.94	1.083	4.30E+04	3.54E+04	PIII-2I-D-2-C
3	2	Driver	4	6.72	3.125	1.96E+05	8.31E+04	PIII-2I-D-3-C
3	2	Driver	4	6.69	7.5	1.84E+04	9.66E+03	PIII-2I-D-4-C
3	2	Driver	4	6.80	16	-1.11E+03	1.48E+03	PIII-2I-D-5-C
3	2	Driver	4	6.72	33	3.04E+02	9.44E+02	PIII-2I-D-6-C
3	2	Driver	4	6.77	65	4.76E+02	5.75E+02	PIII-2I-D-7-C
3	2	Driver	4	6.70	125	1.03E+02	4.60E+02	PIII-2I-D-8-C
3	2	Driver	5	4.15	0.083	5.53E+05	2.18E+05	PIII-2I-D-1-C
3	2	Driver	5	4.17	1.083	2.93E+04	2.50E+04	PIII-2I-D-2-C
3	2	Driver	5	4.04	3.125	3.03E+05	1.26E+05	PIII-2I-D-3-C
3	2	Driver	5	4.02	7.5	3.64E+04	1.61E+04	PIII-2I-D-4-C
3	2	Driver	5	4.09	16	3.76E+03	2.52E+03	PIII-2I-D-5-C
3	2	Driver	5	4.04	33	2.93E+03	1.65E+03	PIII-2I-D-6-C
3	2	Driver	5	4.07	65	3.40E+03	1.57E+03	PIII-2I-D-7-C
3	2	Driver	5	4.02	125	2.14E+03	1.08E+03	PIII-2I-D-8-C
3	2	Driver	6	2.08	0.083	1.67E+05	4.41E+04	PIII-2I-D-1-C
3	2	Driver	6	2.09	1.083	1.02E+05	4.07E+04	PIII-2I-D-2-C
3	2	Driver	6	2.02	3.125	8.60E+04	3.11E+04	PIII-2I-D-3-C
3	2	Driver	6	2.01	7.5	2.09E+04	7.58E+03	PIII-2I-D-4-C
3	2	Driver	6	2.05	16	1.76E+03	1.73E+03	PIII-2I-D-5-C
3	2	Driver	6	2.02	33	2.33E+03	1.18E+03	PIII-2I-D-6-C
3	2	Driver	6	2.04	65	1.23E+03	6.16E+02	PIII-2I-D-7-C
3	2	Driver	6	2.02	125	1.25E+03	6.12E+02	PIII-2I-D-8-C
3	2	Driver	7	1.07	0.083	1.82E+05	3.68E+04	PIII-2I-D-1-C
3	2	Driver	7	1.07	1.083	2.31E+05	4.94E+04	PIII-2I-D-2-C
3	2	Driver	7	1.04	3.125	1.42E+05	3.02E+04	PIII-2I-D-3-C
3	2	Driver	7	1.04	7.5	4.70E+04	9.82E+03	PIII-2I-D-4-C
3	2	Driver	7	1.05	16	2.62E+03	9.40E+02	PIII-2I-D-5-C
3	2	Driver	7	1.04	33	2.35E+03	6.38E+02	PIII-2I-D-6-C

e A.7. *(continued)*

se	Shot	Position	CI stage	Median Dia (um)	MidTime (min)	U Conc, (ug/m3)	sd U Conc. (ug/m3)	SampleID
	2	Driver	7	1.05	65	2.28E+03	5.41E+02	PIII-2I-D-7-CI-7
	2	Driver	7	1.04	125	2.27E+03	5.32E+02	PIII-2I-D-8-CI-7
	2	Driver	8	0.63	0.083	2.70E+05	3.97E+04	PIII-2I-D-1-CI-8
	2	Driver	8	0.63	1.083	2.70E+05	4.27E+04	PIII-2I-D-2-CI-8
	2	Driver	8	0.61	3.125	1.37E+05	2.18E+04	PIII-2I-D-3-CI-8
	2	Driver	8	0.61	7.5	3.09E+04	4.96E+03	PIII-2I-D-4-CI-8
	2	Driver	8	0.62	16	2.55E+03	7.75E+02	PIII-2I-D-5-CI-8
	2	Driver	8	0.61	33	4.00E+03	7.35E+02	PIII-2I-D-6-CI-8
	2	Driver	8	0.62	65	4.59E+03	7.26E+02	PIII-2I-D-7-CI-8
	2	Driver	8	0.61	125	3.59E+03	5.84E+02	PIII-2I-D-8-CI-8
	2	Driver	9	0.22	0.083	1.28E+06	2.57E+05	PIII-2I-D-1-CI-9
	2	Driver	9	0.22	1.083	2.55E+05	5.51E+04	PIII-2I-D-2-CI-9
	2	Driver	9	0.22	3.125	3.62E+04	9.85E+03	PIII-2I-D-3-CI-9
	2	Driver	9	0.21	7.5	7.09E+03	2.10E+03	PIII-2I-D-4-CI-9
	2	Driver	9	0.22	16	5.40E+03	1.44E+03	PIII-2I-D-5-CI-9
	2	Driver	9	0.22	33	4.62E+03	1.08E+03	PIII-2I-D-6-CI-9
	2	Driver	9	0.22	65	2.77E+03	6.40E+02	PIII-2I-D-7-CI-9
	2	Driver	9	0.21	125	1.12E+03	3.17E+02	PIII-2I-D-8-CI-9
	2	Gunner	1	30.52	0.083	6.84E+06	1.16E+06	PIII-2I-G-1-CI-1
	2	Gunner	1	30.67	1.083	5.76E+06	1.27E+06	PIII-2I-G-3-CI-1
	2	Gunner	1	30.74	3.125	8.21E+06	1.58E+06	PIII-2I-G-4-CI-1
	2	Gunner	1	30.57	7.5	1.31E+06	2.69E+05	PIII-2I-G-5-CI-1
	2	Gunner	1	30.54	16	2.72E+05	7.11E+04	PIII-2I-G-6-CI-1
	2	Gunner	1	31.41	33	4.59E+04	2.29E+04	PIII-2I-G-7-CI-1
	2	Gunner	1	31.28	65	1.28E+05	2.84E+04	PIII-2I-G-8-CI-1
	2	Gunner	1	31.33	125	1.10E+05	2.53E+04	PIII-2I-G-9-CI-1
	2	Gunner	2	15.55	0.083	1.54E+06	6.25E+05	PIII-2I-G-1-CI-2
	2	Gunner	2	15.70	1.083	3.58E+05	1.66E+05	PIII-2I-G-3-CI-2
	2	Gunner	2	15.78	3.125	1.69E+05	8.26E+04	PIII-2I-G-4-CI-2
	2	Gunner	2	15.60	7.5	1.59E+04	1.02E+04	PIII-2I-G-5-CI-2
	2	Gunner	2	15.57	16	1.87E+03	NA[(a)]	PIII-2I-G-6-CI-2
	2	Gunner	2	16.47	33	1.72E+03	1.83E+03	PIII-2I-G-7-CI-2
	2	Gunner	2	16.34	65	5.97E+02	8.30E+02	PIII-2I-G-8-CI-2
	2	Gunner	2	16.39	125	1.69E+03	1.21E+03	PIII-2I-G-9-CI-2
	2	Gunner	3	10.41	0.083	1.65E+06	3.47E+05	PIII-2I-G-1-CI-3
	2	Gunner	3	10.51	1.083	4.44E+05	9.92E+04	PIII-2I-G-3-CI-3
	2	Gunner	3	10.56	3.125	1.17E+05	2.87E+04	PIII-2I-G-4-CI-3
	2	Gunner	3	10.44	7.5	1.10E+04	3.27E+03	PIII-2I-G-5-CI-3
	2	Gunner	3	10.42	16	1.93E+03	9.11E+02	PIII-2I-G-6-CI-3
	2	Gunner	3	11.03	33	-2.38E+02	3.13E+02	PIII-2I-G-7-CI-3
	2	Gunner	3	10.94	65	9.22E+02	3.32E+02	PIII-2I-G-8-CI-3
	2	Gunner	3	10.97	125	2.74E+02	2.06E+02	PIII-2I-G-9-CI-3

Table A.7. *(continued)*

Phase	Shot	Position	CI stage	Median Dia (um)	MidTime (min)	U Conc, (ug/m3)	sd U Conc. (ug/m3)	SampleID
3	2	Gunner	4	6.64	0.083	1.19E+06	3.34E+05	PIII-2I-G-1-C
3	2	Gunner	4	6.70	1.083	2.98E+05	9.55E+04	PIII-2I-G-3-C
3	2	Gunner	4	6.73	3.125	2.09E+05	6.70E+04	PIII-2I-G-4-C
3	2	Gunner	4	6.66	7.5	1.03E+04	5.21E+03	PIII-2I-G-5-C
3	2	Gunner	4	6.65	16	-2.34E+02	1.23E+03	PIII-2I-G-6-C
3	2	Gunner	4	7.03	33	1.01E+02	7.43E+02	PIII-2I-G-7-C
3	2	Gunner	4	6.97	65	2.77E+02	4.14E+02	PIII-2I-G-8-C
3	2	Gunner	4	6.99	125	4.98E+01	3.59E+02	PIII-2I-G-9-C
3	2	Gunner	5	3.99	0.083	1.01E+06	4.01E+05	PIII-2I-G-1-C
3	2	Gunner	5	4.03	1.083	8.64E+05	3.52E+05	PIII-2I-G-3-C
3	2	Gunner	5	4.05	3.125	1.14E+05	5.19E+04	PIII-2I-G-4-C
3	2	Gunner	5	4.00	7.5	4.72E+04	2.04E+04	PIII-2I-G-5-C
3	2	Gunner	5	3.99	16	8.42E+03	4.13E+03	PIII-2I-G-6-C
3	2	Gunner	5	4.22	33	6.34E+03	2.97E+03	PIII-2I-G-7-C
3	2	Gunner	5	4.19	65	3.60E+03	1.66E+03	PIII-2I-G-8-C
3	2	Gunner	5	4.20	125	2.06E+03	1.03E+03	PIII-2I-G-9-C
3	2	Gunner	6	2.00	0.083	7.69E+05	3.27E+05	PIII-2I-G-1-C
3	2	Gunner	6	2.02	1.083	3.06E+05	1.53E+05	PIII-2I-G-3-C
3	2	Gunner	6	2.03	3.125	8.01E+04	5.02E+04	PIII-2I-G-4-C
3	2	Gunner	6	2.00	7.5	4.57E+04	2.31E+04	PIII-2I-G-5-C
3	2	Gunner	6	2.00	16	1.51E+04	8.20E+03	PIII-2I-G-6-C
3	2	Gunner	6	2.12	33	7.27E+03	4.13E+03	PIII-2I-G-7-C
3	2	Gunner	6	2.10	65	3.37E+03	1.95E+03	PIII-2I-G-8-C
3	2	Gunner	6	2.11	125	2.86E+03	1.72E+03	PIII-2I-G-9-C
3	2	Gunner	7	1.03	0.083	2.88E+05	7.74E+04	PIII-2I-G-1-C
3	2	Gunner	7	1.04	1.083	2.84E+05	8.61E+04	PIII-2I-G-3-C
3	2	Gunner	7	1.04	3.125	1.69E+05	5.24E+04	PIII-2I-G-4-C
3	2	Gunner	7	1.03	7.5	4.40E+04	1.34E+04	PIII-2I-G-5-C
3	2	Gunner	7	1.03	16	1.04E+03	1.72E+03	PIII-2I-G-6-C
3	2	Gunner	7	1.09	33	8.28E+01	8.25E+02	PIII-2I-G-7-C
3	2	Gunner	7	1.08	65	2.01E+03	8.28E+02	PIII-2I-G-8-C
3	2	Gunner	7	1.08	125	1.65E+03	7.68E+02	PIII-2I-G-9-C
3	2	Gunner	8	0.60	0.083	1.19E+06	2.94E+05	PIII-2I-G-1-C
3	2	Gunner	8	0.61	1.083	5.93E+05	1.59E+05	PIII-2I-G-3-C
3	2	Gunner	8	0.61	3.125	3.30E+05	8.97E+04	PIII-2I-G-4-C
3	2	Gunner	8	0.61	7.5	4.19E+04	1.29E+04	PIII-2I-G-5-C
3	2	Gunner	8	0.60	16	8.40E+03	3.52E+03	PIII-2I-G-6-C
3	2	Gunner	8	0.64	33	4.18E+03	1.87E+03	PIII-2I-G-7-C
3	2	Gunner	8	0.63	65	5.79E+03	1.78E+03	PIII-2I-G-8-C
3	2	Gunner	8	0.64	125	3.92E+03	1.41E+03	PIII-2I-G-9-C
3	2	Gunner	9	0.21	0.083	9.74E+05	3.26E+05	PIII-2I-G-1-C
3	2	Gunner	9	0.22	1.083	2.67E+05	1.08E+05	PIII-2I-G-3-C

e A.7. *(continued)*

ase	Shot	Position	CI stage	Median Dia (um)	MidTime (min)	U Conc, (ug/m3)	sd U Conc. (ug/m3)	SampleID
	2	Gunner	9	0.22	3.125	6.51E+04	3.54E+04	PIII-2I-G-4-CI-9
	2	Gunner	9	0.21	7.5	1.07E+04	6.93E+03	PIII-2I-G-5-CI-9
	2	Gunner	9	0.21	16	2.76E+03	2.83E+03	PIII-2I-G-6-CI-9
	2	Gunner	9	0.22	33	6.90E+03	3.08E+03	PIII-2I-G-7-CI-9
	2	Gunner	9	0.22	65	1.98E+03	1.13E+03	PIII-2I-G-8-CI-9
	2	Gunner	9	0.22	125	1.63E+03	1.03E+03	PIII-2I-G-9-CI-9
	2	Loader	1	31.22	0.083	1.17E+06	2.48E+05	PIII-2I-L-1-CI-1
	2	Loader	1	31.09	1.083	2.05E+05	7.56E+04	PIII-2I-L-2-CI-1
	2	Loader	1	30.83	3.125	6.45E+04	3.56E+04	PIII-2I-L-3-CI-1
	2	Loader	1	30.85	7.5	1.93E+04	9.47E+03	PIII-2I-L-4-CI-1
	2	Loader	1	30.84	16	5.49E+03	3.91E+03	PIII-2I-L-5-CI-1
	2	Loader	1	31.05	33	5.28E+03	2.47E+03	PIII-2I-L-6-CI-1
	2	Loader	1	31.15	65	3.92E+03	NA[(a)]	PIII-2I-L-7-CI-1
	2	Loader	1	30.97	125	5.70E+03	1.83E+03	PIII-2I-L-8-CI-1
	2	Loader	2	16.27	0.083	3.00E+05	1.43E+05	PIII-2I-L-1-CI-2
	2	Loader	2	16.14	1.083	2.52E+04	3.28E+04	PIII-2I-L-2-CI-2
	2	Loader	2	15.87	3.125	6.17E+03	1.75E+04	PIII-2I-L-3-CI-2
	2	Loader	2	15.89	7.5	-1.60E+03	3.34E+03	PIII-2I-L-4-CI-2
	2	Loader	2	15.88	16	1.44E+02	2.01E+03	PIII-2I-L-5-CI-2
	2	Loader	2	16.09	33	1.69E+03	1.62E+03	PIII-2I-L-6-CI-2
	2	Loader	2	16.20	65	4.29E+02	6.58E+02	PIII-2I-L-7-CI-2
	2	Loader	2	16.01	125	1.16E+02	5.31E+02	PIII-2I-L-8-CI-2
	2	Loader	3	10.89	0.083	5.50E+05	NA[(a)]	PIII-2I-L-1-CI-3
	2	Loader	3	10.80	1.083	1.30E+05	3.79E+04	PIII-2I-L-2-CI-3
	2	Loader	3	10.62	3.125	7.97E+04	2.37E+04	PIII-2I-L-3-CI-3
	2	Loader	3	10.64	7.5	1.80E+03	1.08E+03	PIII-2I-L-4-CI-3
	2	Loader	3	10.63	16	2.37E+03	9.17E+02	PIII-2I-L-5-CI-3
	2	Loader	3	10.77	33	4.61E+02	2.78E+02	PIII-2I-L-6-CI-3
	2	Loader	3	10.84	65	1.70E+02	1.28E+02	PIII-2I-L-7-CI-3
	2	Loader	3	10.72	125	1.66E+02	1.25E+02	PIII-2I-L-8-CI-3
	2	Loader	4	6.94	0.083	5.20E+05	NA[(a)]	PIII-2I-L-1-CI-4
	2	Loader	4	6.89	1.083	1.76E+05	6.64E+04	PIII-2I-L-2-CI-4
	2	Loader	4	6.77	3.125	1.39E+05	5.27E+04	PIII-2I-L-3-CI-4
	2	Loader	4	6.78	7.5	9.48E+03	3.68E+03	PIII-2I-L-4-CI-4
	2	Loader	4	6.78	16	4.52E+03	1.75E+03	PIII-2I-L-5-CI-4
	2	Loader	4	6.87	33	2.21E+03	8.54E+02	PIII-2I-L-6-CI-4
	2	Loader	4	6.91	65	1.12E+03	4.38E+02	PIII-2I-L-7-CI-4
	2	Loader	4	6.83	125	6.56E+02	2.70E+02	PIII-2I-L-8-CI-4
	2	Loader	5	4.17	0.083	5.23E+05	NA[(a)]	PIII-2I-L-1-CI-5
	2	Loader	5	4.14	1.083	1.63E+05	6.48E+04	PIII-2I-L-2-CI-5
	2	Loader	5	4.07	3.125	1.37E+05	5.44E+04	PIII-2I-L-3-CI-5
	2	Loader	5	4.08	7.5	2.02E+04	8.08E+03	PIII-2I-L-4-CI-5

Table A.7. *(continued)*

Phase	Shot	Position	CI stage	Median Dia (um)	MidTime (min)	U Conc, (ug/m3)	sd U Conc. (ug/m3)	SampleID
3	2	Loader	5	4.07	16	5.64E+03	2.30E+03	PIII-2I-L-5-C
3	2	Loader	5	4.13	33	3.73E+03	1.52E+03	PIII-2I-L-6-C
3	2	Loader	5	4.15	65	3.67E+03	1.45E+03	PIII-2I-L-7-C
3	2	Loader	5	4.11	125	1.95E+03	7.82E+02	PIII-2I-L-8-C
3	2	Loader	6	2.09	0.083	2.94E+05	7.69E+04	PIII-2I-L-1-C
3	2	Loader	6	2.07	1.083	1.58E+05	4.31E+04	PIII-2I-L-2-C
3	2	Loader	6	2.04	3.125	1.13E+05	3.05E+04	PIII-2I-L-3-C
3	2	Loader	6	2.04	7.5	2.64E+04	7.09E+03	PIII-2I-L-4-C
3	2	Loader	6	2.04	16	4.39E+03	1.30E+03	PIII-2I-L-5-C
3	2	Loader	6	2.07	33	3.59E+03	9.85E+02	PIII-2I-L-6-C
3	2	Loader	6	2.08	65	3.32E+03	8.93E+02	PIII-2I-L-7-C
3	2	Loader	6	2.06	125	1.67E+03	4.74E+02	PIII-2I-L-8-C
3	2	Loader	7	1.07	0.083	1.99E+05	4.05E+04	PIII-2I-L-1-C
3	2	Loader	7	1.07	1.083	1.10E+05	2.43E+04	PIII-2I-L-2-C
3	2	Loader	7	1.05	3.125	1.42E+05	2.91E+04	PIII-2I-L-3-C
3	2	Loader	7	1.05	7.5	1.54E+04	3.39E+03	PIII-2I-L-4-C
3	2	Loader	7	1.05	16	3.36E+03	8.16E+02	PIII-2I-L-5-C
3	2	Loader	7	1.06	33	3.04E+03	6.83E+02	PIII-2I-L-6-C
3	2	Loader	7	1.07	65	2.67E+03	5.58E+02	PIII-2I-L-7-C
3	2	Loader	7	1.06	125	2.96E+03	6.20E+02	PIII-2I-L-8-C
3	2	Loader	8	0.63	0.083	5.38E+05	7.74E+04	PIII-2I-L-1-C
3	2	Loader	8	0.63	1.083	1.54E+05	2.82E+04	PIII-2I-L-2-C
3	2	Loader	8	0.62	3.125	1.41E+05	2.34E+04	PIII-2I-L-3-C
3	2	Loader	8	0.62	7.5	6.24E+03	2.02E+03	PIII-2I-L-4-C
3	2	Loader	8	0.62	16	3.12E+03	9.66E+02	PIII-2I-L-5-C
3	2	Loader	8	0.62	33	3.09E+03	7.14E+02	PIII-2I-L-6-C
3	2	Loader	8	0.63	65	4.28E+03	7.18E+02	PIII-2I-L-7-C
3	2	Loader	8	0.62	125	3.20E+03	5.75E+02	PIII-2I-L-8-C
3	2	Loader	9	0.22	0.083	4.24E+05	8.66E+04	PIII-2I-L-1-C
3	2	Loader	9	0.22	1.083	1.47E+05	3.75E+04	PIII-2I-L-2-C
3	2	Loader	9	0.22	3.125	1.83E+04	9.58E+03	PIII-2I-L-3-C
3	2	Loader	9	0.22	7.5	2.68E+03	2.00E+03	PIII-2I-L-4-C
3	2	Loader	9	0.22	16	4.40E+03	1.56E+03	PIII-2I-L-5-C
3	2	Loader	9	0.22	33	6.49E+03	1.64E+03	PIII-2I-L-6-C
3	2	Loader	9	0.22	65	4.68E+03	1.10E+03	PIII-2I-L-7-C
3	2	Loader	9	0.22	125	2.38E+03	6.38E+02	PIII-2I-L-8-C
4	1	Driver	1	32.00	0.5	1.71E+03	6.23E+02	PIV-1I-D-6-C
4	1	Driver	1	32.01	2	-2.13E+02	3.00E+02	PIV-1I-D-7-C
4	1	Driver	1	31.13	7	1.91E+01	1.34E+02	PIV-1I-D-8-C
4	1	Driver	1	31.39	29	9.88E+01	8.95E+01	PIV-1I-D-9-C
4	1	Driver	2	17.09	0.5	7.27E+02	8.04E+02	PIV-1I-D-6-C
4	1	Driver	2	17.10	2	-1.09E+02	2.55E+02	PIV-1I-D-7-C

e A.7. *(continued)*

se	Shot	Position	CI stage	Median Dia (um)	MidTime (min)	U Conc, (ug/m3)	sd U Conc. (ug/m3)	SampleID
	1	Driver	2	16.18	7	6.52E+02	2.69E+02	PIV-1I-D-8-CI-2
	1	Driver	2	16.45	29	4.21E+02	1.90E+02	PIV-1I-D-9-CI-2
	1	Driver	3	11.44	0.5	-5.69E+02	3.46E+02	PIV-1I-D-6-CI-3
	1	Driver	3	11.45	2	1.14E+01	1.14E+02	PIV-1I-D-7-CI-3
	1	Driver	3	10.83	7	2.55E+02	1.05E+02	PIV-1I-D-8-CI-3
	1	Driver	3	11.01	29	6.59E+00	3.29E+01	PIV-1I-D-9-CI-3
	1	Driver	4	7.30	0.5	1.50E+02	3.49E+02	PIV-1I-D-6-CI-4
	1	Driver	4	7.30	2	7.48E+02	2.61E+02	PIV-1I-D-7-CI-4
	1	Driver	4	6.91	7	-1.12E+01	7.82E+01	PIV-1I-D-8-CI-4
	1	Driver	4	7.02	29	1.15E+01	2.89E+01	PIV-1I-D-9-CI-4
	1	Driver	5	4.38	0.5	4.67E+02	6.09E+02	PIV-1I-D-6-CI-5
	1	Driver	5	4.39	2	2.34E+02	2.82E+02	PIV-1I-D-7-CI-5
	1	Driver	5	4.15	7	2.09E+02	1.27E+02	PIV-1I-D-8-CI-5
	1	Driver	5	4.22	29	-5.41E+01	5.97E+01	PIV-1I-D-9-CI-5
	1	Driver	6	2.20	0.5	4.62E+02	7.87E+02	PIV-1I-D-6-CI-6
	1	Driver	6	2.20	2	-2.31E+02	3.94E+02	PIV-1I-D-7-CI-6
	1	Driver	6	2.08	7	-1.04E+02	1.35E+02	PIV-1I-D-8-CI-6
	1	Driver	6	2.11	29	5.35E+01	1.07E+02	PIV-1I-D-9-CI-6
	1	Driver	7	1.13	0.5	9.15E+02	7.83E+02	PIV-1I-D-6-CI-7
	1	Driver	7	1.13	2	2.29E+02	3.90E+02	PIV-1I-D-7-CI-7
	1	Driver	7	1.07	7	6.15E+02	3.14E+02	PIV-1I-D-8-CI-7
	1	Driver	7	1.09	29	1.59E+02	8.63E+01	PIV-1I-D-9-CI-7
	1	Driver	8	0.66	0.5	4.48E+02	1.12E+03	PIV-1I-D-6-CI-8
	1	Driver	8	0.66	2	-1.12E+02	3.81E+02	PIV-1I-D-7-CI-8
	1	Driver	8	0.63	7	1.00E+02	2.11E+02	PIV-1I-D-8-CI-8
	1	Driver	8	0.64	29	-1.56E+02	8.97E+01	PIV-1I-D-9-CI-8
	1	Driver	9	0.22	0.5	1.77E+03	1.04E+03	PIV-1I-D-6-CI-9
	1	Driver	9	0.22	2	-2.22E+02	4.45E+02	PIV-1I-D-7-CI-9
	1	Driver	9	0.22	7	-2.98E+02	1.72E+02	PIV-1I-D-8-CI-9
	1	Driver	9	0.22	29	-1.03E+02	7.78E+01	PIV-1I-D-9-CI-9
	1	Loader	1	31.40	0.5	8.71E+03	1.80E+03	PIV-1I-L-1-CI-1
	1	Loader	1	30.95	2	7.47E+02	3.65E+02	PIV-1I-L-2-CI-1
	1	Loader	1	30.69	7	5.41E+01	1.72E+02	PIV-1I-L-3-CI-1
	1	Loader	1	30.60	29	4.46E+02	1.11E+02	PIV-1I-L-4-CI-1
	1	Loader	2	16.46	0.5	-4.45E+02	1.61E+03	PIV-1I-L-1-CI-2
	1	Loader	2	16.00	2	-1.17E+03	NA[a]	PIV-1I-L-2-CI-2
	1	Loader	2	15.72	7	-1.02E+02	3.36E+02	PIV-1I-L-3-CI-2
	1	Loader	2	15.63	29	-1.26E+02	1.47E+02	PIV-1I-L-4-CI-2
	1	Loader	3	11.02	0.5	1.69E+02	8.87E+02	PIV-1I-L-1-CI-3
	1	Loader	3	10.71	2	-9.16E+02	4.58E+02	PIV-1I-L-2-CI-3
	1	Loader	3	10.52	7	-2.02E+02	1.81E+02	PIV-1I-L-3-CI-3
	1	Loader	3	10.46	29	-2.78E+02	NA[a]	PIV-1I-L-4-CI-3

Table A.7. *(continued)*

Phase	Shot	Position	CI stage	Median Dia (um)	MidTime (min)	U Conc, (ug/m3)	sd U Conc. (ug/m3)	SampleID
4	1	Loader	4	7.03	0.5	-3.24E+03	1.62E+03	PIV-1I-L-1-C
4	1	Loader	4	6.83	2	-1.53E+03	7.65E+02	PIV-1I-L-2-C
4	1	Loader	4	6.71	7	-8.44E+02	NA[(a)]	PIV-1I-L-3-C
4	1	Loader	4	6.67	29	-4.69E+02	2.35E+02	PIV-1I-L-4-C
4	1	Loader	5	4.22	0.5	6.50E+01	6.28E+02	PIV-1I-L-1-C
4	1	Loader	5	4.10	2	2.15E+02	3.44E+02	PIV-1I-L-2-C
4	1	Loader	5	4.03	7	1.39E+03	6.11E+02	PIV-1I-L-3-C
4	1	Loader	5	4.01	29	-4.64E+01	NA[(a)]	PIV-1I-L-4-C
4	1	Loader	6	2.12	0.5	8.92E+02	9.68E+02	PIV-1I-L-1-C
4	1	Loader	6	2.06	2	-1.86E+02	4.33E+02	PIV-1I-L-2-C
4	1	Loader	6	2.02	7	5.94E+02	3.69E+02	PIV-1I-L-3-C
4	1	Loader	6	2.01	29	-1.89E+02	9.98E+01	PIV-1I-L-4-C
4	1	Loader	7	1.09	0.5	8.49E+02	8.63E+02	PIV-1I-L-1-C
4	1	Loader	7	1.06	2	1.00E+03	5.97E+02	PIV-1I-L-2-C
4	1	Loader	7	1.04	7	1.26E+03	3.95E+02	PIV-1I-L-3-C
4	1	Loader	7	1.03	29	-1.43E+02	1.05E+02	PIV-1I-L-4-C
4	1	Loader	8	0.64	0.5	-2.45E+03	1.83E+03	PIV-1I-L-1-C
4	1	Loader	8	0.62	2	8.04E+01	9.30E+02	PIV-1I-L-2-C
4	1	Loader	8	0.61	7	-7.48E+02	3.74E+02	PIV-1I-L-3-C
4	1	Loader	8	0.61	29	-5.57E+02	2.78E+02	PIV-1I-L-4-C
4	1	Loader	9	0.22	0.5	4.12E+02	1.66E+03	PIV-1I-L-1-C
4	1	Loader	9	0.22	2	5.83E+02	8.08E+02	PIV-1I-L-2-C
4	1	Loader	9	0.22	7	9.38E+01	4.01E+02	PIV-1I-L-3-C
4	1	Loader	9	0.21	29	-9.27E+01	1.50E+02	PIV-1I-L-4-C
4	3	Driver	1	31.11	0.5	2.27E+03	5.35E+03	PIV-3I-D-1-C
4	3	Driver	1	30.86	2	7.99E+03	3.04E+03	PIV-3I-D-2-C
4	3	Driver	1	30.64	7	1.34E+04	2.62E+03	PIV-3I-D-3-C
4	3	Driver	1	30.43	29	-6.15E+02	3.08E+02	PIV-3I-D-4-C
4	3	Driver	2	16.15	0.5	-3.18E+04	1.59E+04	PIV-3I-D-1-C
4	3	Driver	2	15.90	2	-1.81E+04	9.03E+03	PIV-3I-D-2-C
4	3	Driver	2	15.67	7	-1.56E+04	7.80E+03	PIV-3I-D-3-C
4	3	Driver	2	15.46	29	-1.64E+03	8.20E+02	PIV-3I-D-4-C
4	3	Driver	3	10.81	0.5	-5.38E+03	3.14E+03	PIV-3I-D-1-C
4	3	Driver	3	10.65	2	-8.86E+03	4.43E+03	PIV-3I-D-2-C
4	3	Driver	3	10.49	7	-6.17E+03	3.08E+03	PIV-3I-D-3-C
4	3	Driver	3	10.35	29	-8.04E+02	4.02E+02	PIV-3I-D-4-C
4	3	Driver	4	6.89	0.5	-2.69E+03	3.04E+03	PIV-3I-D-1-C
4	3	Driver	4	6.79	2	-6.60E+03	3.30E+03	PIV-3I-D-2-C
4	3	Driver	4	6.69	7	-4.60E+03	2.30E+03	PIV-3I-D-3-C
4	3	Driver	4	6.60	29	-5.99E+02	3.00E+02	PIV-3I-D-4-C
4	3	Driver	5	4.14	0.5	5.17E+03	4.96E+03	PIV-3I-D-1-C
4	3	Driver	5	4.08	2	-1.30E+04	6.51E+03	PIV-3I-D-2-C

e A.7. *(continued)*

ase	Shot	Position	CI stage	Median Dia (um)	MidTime (min)	U Conc, (ug/m3)	sd U Conc. (ug/m3)	SampleID
	3	Driver	5	4.02	7	-6.05E+03	3.46E+03	PIV-3I-D-3-CI-5
	3	Driver	5	3.96	29	-1.18E+02	3.41E+02	PIV-3I-D-4-CI-5
	3	Driver	6	2.08	0.5	3.83E+04	1.71E+04	PIV-3I-D-1-CI-6
	3	Driver	6	2.04	2	-3.30E+04	1.65E+04	PIV-3I-D-2-CI-6
	3	Driver	6	2.01	7	-1.09E+04	6.02E+03	PIV-3I-D-3-CI-6
	3	Driver	6	1.99	29	-2.21E+03	1.10E+03	PIV-3I-D-4-CI-6
	3	Driver	7	1.07	0.5	4.07E+04	1.26E+04	PIV-3I-D-1-CI-7
	3	Driver	7	1.05	2	-4.86E+03	8.47E+03	PIV-3I-D-2-CI-7
	3	Driver	7	1.04	7	2.66E+03	6.73E+03	PIV-3I-D-3-CI-7
	3	Driver	7	1.02	29	-2.02E+03	1.01E+03	PIV-3I-D-4-CI-7
	3	Driver	8	0.63	0.5	1.03E+04	1.79E+04	PIV-3I-D-1-CI-8
	3	Driver	8	0.62	2	-1.30E+04	1.47E+04	PIV-3I-D-2-CI-8
	3	Driver	8	0.61	7	-6.05E+03	1.05E+04	PIV-3I-D-3-CI-8
	3	Driver	8	0.60	29	9.86E+03	2.12E+03	PIV-3I-D-4-CI-8
	3	Driver	9	0.22	0.5	-1.57E+04	9.75E+03	PIV-3I-D-1-CI-9
	3	Driver	9	0.22	2	-1.32E+04	8.01E+03	PIV-3I-D-2-CI-9
	3	Driver	9	0.21	7	-2.13E+04	1.06E+04	PIV-3I-D-3-CI-9
	3	Driver	9	0.21	29	2.35E+03	9.11E+02	PIV-3I-D-4-CI-9
	3	Loader	1	31.11	0.5	1.37E+04	6.24E+03	PIV-3I-L-1-CI-1
	3	Loader	1	30.86	2	-1.33E+03	5.11E+03	PIV-3I-L-2-CI-1
	3	Loader	1	30.64	7	-8.61E+03	4.31E+03	PIV-3I-L-3-CI-1
	3	Loader	1	30.43	29	-1.05E+03	5.39E+02	PIV-3I-L-4-CI-1
	3	Loader	2	16.15	0.5	-1.80E+04	8.98E+03	PIV-3I-L-1-CI-2
	3	Loader	2	15.90	2	-1.36E+04	6.85E+03	PIV-3I-L-2-CI-2
	3	Loader	2	15.67	7	-8.83E+03	4.66E+03	PIV-3I-L-3-CI-2
	3	Loader	2	15.46	29	-1.07E+03	5.40E+02	PIV-3I-L-4-CI-2
	3	Loader	3	10.81	0.5	-3.05E+03	3.11E+03	PIV-3I-L-1-CI-3
	3	Loader	3	10.65	2	2.95E+03	3.50E+03	PIV-3I-L-2-CI-3
	3	Loader	3	10.49	7	-1.91E+03	1.81E+03	PIV-3I-L-3-CI-3
	3	Loader	3	10.35	29	-2.33E+02	2.37E+02	PIV-3I-L-4-CI-3
	3	Loader	4	6.89	0.5	-5.34E+03	4.12E+03	PIV-3I-L-1-CI-4
	3	Loader	4	6.79	2	-5.18E+03	4.00E+03	PIV-3I-L-2-CI-4
	3	Loader	4	6.69	7	-5.03E+03	2.97E+03	PIV-3I-L-3-CI-4
	3	Loader	4	6.60	29	-4.08E+02	3.15E+02	PIV-3I-L-4-CI-4
	3	Loader	5	4.14	0.5	5.00E+02	6.54E+03	PIV-3I-L-1-CI-5
	3	Loader	5	4.08	2	-9.70E+03	7.49E+03	PIV-3I-L-2-CI-5
	3	Loader	5	4.02	7	-7.86E+03	5.13E+03	PIV-3I-L-3-CI-5
	3	Loader	5	3.96	29	-9.55E+02	6.08E+02	PIV-3I-L-4-CI-5
	3	Loader	6	2.08	0.5	7.42E+03	1.22E+04	PIV-3I-L-1-CI-6
	3	Loader	6	2.04	2	-1.20E+04	1.17E+04	PIV-3I-L-2-CI-6
	3	Loader	6	2.01	7	6.22E+03	8.30E+03	PIV-3I-L-3-CI-6
	3	Loader	6	1.99	29	-1.13E+03	9.12E+02	PIV-3I-L-4-CI-6

Table A.7. *(continued)*

Phase	Shot	Position	CI stage	Median Dia (um)	MidTime (min)	U Conc, (ug/m3)	sd U Conc. (ug/m3)	SampleID
4	3	Loader	7	1.07	0.5	-2.45E+03	9.90E+03	PIV-3I-L-1-C
4	3	Loader	7	1.05	2	-1.90E+04	1.04E+04	PIV-3I-L-2-C
4	3	Loader	7	1.04	7	-1.23E+04	6.99E+03	PIV-3I-L-3-C
4	3	Loader	7	1.02	29	-1.12E+03	8.43E+02	PIV-3I-L-4-C
4	3	Loader	8	0.63	0.5	-5.28E+04	2.64E+04	PIV-3I-L-1-C
4	3	Loader	8	0.62	2	-1.86E+04	1.59E+04	PIV-3I-L-2-C
4	3	Loader	8	0.61	7	-2.71E+04	1.36E+04	PIV-3I-L-3-C
4	3	Loader	8	0.60	29	-2.93E+03	1.48E+03	PIV-3I-L-4-C
4	3	Loader	9	0.22	0.5	-1.90E+04	1.48E+04	PIV-3I-L-1-C
4	3	Loader	9	0.22	2	-2.30E+04	1.55E+04	PIV-3I-L-2-C
4	3	Loader	9	0.21	7	-8.96E+03	8.68E+03	PIV-3I-L-3-C
4	3	Loader	9	0.21	29	-5.44E+02	1.02E+03	PIV-3I-L-4-C
4	4	Driver	1	31.18	0.083	6.00E+04	1.25E+04	PIV-4I-D-1-C
4	4	Driver	1	30.80	1.083	1.32E+04	5.90E+03	PIV-4I-D-2-C
4	4	Driver	1	31.17	5.125	3.07E+03	4.31E+03	PIV-4I-D-3-C
4	4	Driver	1	31.44	26	7.95E+02	5.27E+02	PIV-4I-D-4-C
4	4	Driver	2	16.23	0.083	7.86E+03	6.42E+03	PIV-4I-D-1-C
4	4	Driver	2	15.83	1.083	-7.49E+03	5.43E+03	PIV-4I-D-2-C
4	4	Driver	2	16.22	5.125	2.62E+03	3.54E+03	PIV-4I-D-3-C
4	4	Driver	2	16.50	26	6.78E+02	4.62E+02	PIV-4I-D-4-C
4	4	Driver	3	10.87	0.083	-3.07E+02	3.37E+03	PIV-4I-D-1-C
4	4	Driver	3	10.60	1.083	2.93E+03	3.22E+03	PIV-4I-D-2-C
4	4	Driver	3	10.86	5.125	-2.05E+03	2.09E+03	PIV-4I-D-3-C
4	4	Driver	3	11.05	26	2.12E+03	5.51E+02	PIV-4I-D-4-C
4	4	Driver	4	6.93	0.083	-2.69E+02	3.45E+03	PIV-4I-D-1-C
4	4	Driver	4	6.76	1.083	5.13E+03	4.17E+03	PIV-4I-D-2-C
4	4	Driver	4	6.92	5.125	3.59E+02	2.16E+03	PIV-4I-D-3-C
4	4	Driver	4	7.04	26	1.86E+03	6.40E+02	PIV-4I-D-4-C
4	4	Driver	5	4.16	0.083	-2.27E+03	6.09E+03	PIV-4I-D-1-C
4	4	Driver	5	4.06	1.083	-2.16E+03	5.32E+03	PIV-4I-D-2-C
4	4	Driver	5	4.16	5.125	-6.56E+03	3.96E+03	PIV-4I-D-3-C
4	4	Driver	5	4.23	26	3.72E+03	1.73E+03	PIV-4I-D-4-C
4	4	Driver	6	2.09	0.083	5.00E+03	8.49E+03	PIV-4I-D-1-C
4	4	Driver	6	2.03	1.083	3.57E+04	1.78E+04	PIV-4I-D-2-C
4	4	Driver	6	2.08	5.125	2.83E+04	1.29E+04	PIV-4I-D-3-C
4	4	Driver	6	2.12	26	1.72E+03	9.96E+02	PIV-4I-D-4-C
4	4	Driver	7	1.07	0.083	2.47E+03	8.82E+03	PIV-4I-D-1-C
4	4	Driver	7	1.05	1.083	5.65E+04	1.96E+04	PIV-4I-D-2-C
4	4	Driver	7	1.07	5.125	1.65E+03	5.75E+03	PIV-4I-D-3-C
4	4	Driver	7	1.09	26	1.49E+03	7.92E+02	PIV-4I-D-4-C
4	4	Driver	8	0.63	0.083	-2.42E+03	1.18E+04	PIV-4I-D-1-C
4	4	Driver	8	0.61	1.083	1.04E+05	2.97E+04	PIV-4I-D-2-C

e A.7. *(continued)*

se	Shot	Position	CI stage	Median Dia (um)	MidTime (min)	U Conc, (ug/m3)	sd U Conc. (ug/m3)	SampleID
	4	Driver	8	0.63	5.125	1.45E+04	8.42E+03	PIV-4I-D-3-CI-8
	4	Driver	8	0.64	26	2.09E+02	9.51E+02	PIV-4I-D-4-CI-8
	4	Driver	9	0.22	0.083	3.60E+04	1.51E+04	PIV-4I-D-1-CI-9
	4	Driver	9	0.22	1.083	2.74E+04	1.26E+04	PIV-4I-D-2-CI-9
	4	Driver	9	0.22	5.125	3.19E+02	5.20E+03	PIV-4I-D-3-CI-9
	4	Driver	9	0.22	26	2.07E+02	6.29E+02	PIV-4I-D-4-CI-9

J mass based on ICP-MS results; no analytical uncertainty was reported.

B APPENDIX Uranium Toxicity—Animal Studies

B.1 INTRODUCTION

This appendix provides an overview of the animal studies related to various forms of uranium. The focus is twofold: studies in which animals were exposed to oxides of uranium, as oxides are the principal forms of uranium likely to be produced when a target is struck by a DU munition, and studies in which animals were acutely exposed to uranium, as the scenarios evaluated in this report are for acute exposures only. More complete reviews of the toxicity of uranium or the toxicity of DU used in munitions have recently been published (ATSDR 1999; Harley et al. 1999; Fulco et al. 2000; WHO 2001; The Royal Society 2001, 2002).

B.2 RENAL TOXICITY

Uranium oxide exposures to the kidneys via inhalation, ingestion, dermal, and subcutaneous routes are discussed in the next sections. These discussions are followed by a summary of renal effects observed in animal studies and in humans after acute exposure to more soluble forms of uranium.

B.2.1 Exposure Via Inhalation

Following inhalation of uranium, uranium must dissolve and enter the bloodstream prior to affecting the kidneys or other target organs. This process depends on the size of the particle, the site of deposition within the lungs, and the solubility of the uranium. Smaller particles (<1-µm AMAD) are preferentially deposited in the alveoli; it is these particles that are most likely to be absorbed into the bloodstream. These particles may also be removed from the lungs via phagocytosis by alveolar macrophages. More soluble forms of uranium will enter the blood more quickly than less soluble forms, which may remain in the lungs and the regional lymph nodes for weeks to years. Regardless of particle size and solubility, once uranium is absorbed into the blood stream, the primary target organs are the kidneys.

B.2.2 Acute Exposure to Uranium Oxides

Overt effects of uranium toxicity that are indicative of renal toxicity include weight loss and a diminished frequency of urination. There are several animal studies in which acute exposure to oxides of uranium via the lungs resulted in overt effects suggestive of renal toxicity, as well as biochemical and histological changes indicative of renal damage. Biochemical changes typically noted include glucosuria, polyuria and proteinuria, increased blood non-protein nitrogen (NPN) and retention of phenolsulfophthalein (PSP) (Stradling et al. 1985; Thompson and Steadman 1953; Rubanovskaya 1970; Dygert et al. 1949). Histopathology examinations showed that damage was located in the proximal tubules of the kidneys (Stradling et al. 1985); however, it has also been found to extend into other regions of the kidneys, including the corticomedullary region and the glomerulus (Stradling et al. 1985; Rubanovskaya 1970). Reported kidney burdens of uranium in these studies ranged from 0.26 to 23.34 μg U/g kidney (Stradling et al. 1985); using basic assumptions about kidney weight, estimated kidney burdens in other studies ranged from 1.14 to 22.2 μg U/g kidney (Dygert et al. 1949; Rubanovskaya 1970). The biochemical and histopathological changes resulting from kidney injury were usually reversible. However, in the study where the highest concentration was noted (23.34 μg U/g kidney, Stradling et al. 1985); there were reductions in cortex size and increases in papillary space of the kidney, suggesting protracted damage to the kidney. Biochemical parameters were not assessed in this study. Mortalities were noted in only one of these studies (Rubanovskaya 1970); surviving animals showed transient overt and biochemical responses indicative of renal toxicity. Biochemical changes in these animals returned to normal after an unspecified period of time.

B.2.3 Sub-Chronic and Chronic Exposure to Uranium Oxides

Following sub-chronic (weeks to months), and chronic exposures (months to years) to oxides of uranium, changes similar to those following acute exposures have been observed. Overt symptoms of renal toxicity include mortality, weight changes, and changes in urinary output (Rothstein 1949; Dygert et al. 1949). Biochemical changes reported included proteinuria, and increased blood NPN and urine nitrogen (Rothstein 1949; Dygert et al. 1949; Wilson et al. 1952; Wilson and Sylvester 1955). Histological changes to the kidneys reported included mild to severe necrosis of the tubular epithelium and corticomedullary junction (Rothstein 1949); other researchers reported only "renal damage" (Dygert et al. 1949; Wilson et al. 1952; Wilson and Sylvester 1955). In one study, there was regeneration of the renal epithelium noted, suggesting that renal damage had occurred (Dygert et al. 1949). Toxic effects were not noted with particle sizes greater than 2.5-μm AMAD in any of these studies.

The lowest toxic concentration in rats following exposure to U_3O_8 was 10 mg/m^3, and data suggested that a concentration of 1 mg/m^3 approached a threshold for toxicity. Exposure times were 4 h/day for 4 months (Galibin 1975). Toxic effects included impaired liver function, changes in urine composition, and changes in leukocyte counts. Although these data are lacking in details, they bear mentioning because the NIOSH IDLH is based upon them.

B.2.4 Exposure to Other Forms of Uranium

Acute inhalation exposures to UF_6 (357 ± 89 mg/m^3 for 90 min) or UF_6 hydrolysis products (UO_2F_2 + HF) resulted in diuresis, proteinuria, enzymuria, glucosuria and aminoaciduria (Morrow et al. 1982). The estimated kidney concentration of uranium was between 19.26 μg U/g kidney and 42 μg U/g kidney. Similar effects were noted in rats and guinea pigs exposed to UF_6 (0.44 g U/m^3 to 277 g U/m^3) for up to 10 min (Leach et al. 1984). Sub-chronic exposure to UF_4 or uranyl nitrate [0.5 to 24 mg UF_4/m^3 or 0.3 to 20 mg UO_2 (NO_3)•$6H_2O$/m^3] resulted in severe renal tubular damage and decreased alkaline phosphatase activity. Adverse effects were noted at air concentrations at or below 4.0 mg UF_4/m^3 or 0.3 mg $UO_2(NO_3)$•$6H_2O$/m^3; this resulted in a mean kidney burden of 0.1 μg U/g kidney and 1.7 μg U/kidney for each of these compounds, respectively (Dygert et al. 1949).

B.3 EXPOSURE VIA ORAL INTAKE

Uranium is not efficiently absorbed from the intestine; the absorption in a 70-kg adult has been estimated to be 2% for water-soluble salts and less water-soluble oxides and 0.2% for relatively insoluble oxides (ICRP [69] 1995). This absorption is influenced by several factors. There is at least a tenfold difference in uptake depending on the solubility of the chemical form. Uptake is also influenced by the composition of the intestinal contents, and is generally greater when the stomach is empty. However, as with inhalation, if uranium is absorbed from the intestine, the primary target organs are the kidneys.

B.3.1 Exposure to Uranium Oxides

No studies describing acute oral exposures to oxides of uranium were identified. Sub-chronic and chronic studies showed effects only at very high doses or for prolonged exposure periods. In these studies, doses of >100 mg/kg resulted in overt effects including weight loss and lethality (Karsner and Reimann 1918; Tannenbaum and Silverstone 1951). Kidney burdens were between 13 and 28 μg U per kidney pair (Tannenbaum and Silverstone 1951); using basic assumptions, this burden is equivalent to kidney doses between 25 and 54 μgU/g kidney. The only biochemical effect reported in these studies was albuminuria in guinea pigs exposed to ~1-mg/kg U_3O_8 (Karsner and Reimann 1918). Histopathological changes were noted in several studies and included damage to the tubular epithelium (Maynard and Hodge 1949; Tolgskaya 1970). Rodent studies were also conducted in which no effects were observed. In these studies, rats or mice were exposed to up to 20% UO_2, UO_3, or U_3O_8 in their food, (Maynard and Hodge 1949; Tannenbaum and Silverstone 1951) or at a level of 80 mg/animal/day (Tannenbaum and Silverstone 1951). Kidney burdens of uranium were not reported for these studies.

B.3.2 Exposure to Other Forms of Uranium

Several drinking water studies have shown that sub-chronic exposure to soluble forms of uranium can result in both biochemical and morphological changes to the kidneys (McDonald-Taylor et al. 1997; Gilman et al. 1998a,b,c). Morphological changes included dose-dependent ultrastructural changes in proximal tubules, cytoplasmic vacuolation in the proximal tubule epithelium accompanied by vesiculation, and pyknosis of tubular nuclei. Kidney burdens of uranium in animals where these effects were noted were between 0.04 and 1.03 μg U/g kidney. More serious effects included sclerosis of the glomerular capsule; the kidney burden in this study was 1.6 μg U/g kidney (Gilman et al. 1998b).

B.4 EXPOSURE VIA OTHER ROUTES—URANIUM OXIDES

Other routes of exposure in which the renal toxicity of uranium oxides has been assessed include dermal and subcutaneous routes. In a study by Orcutt

(1949), neither symptoms of overt toxicity nor any changes in urinary protein levels following dermal exposure to UO_2 were identified. In studies of subcutaneous injection greater than 10 mg/kg, rabbits and rats showed evidence of nephrotoxicity (Rubanovskaya 1970; Tolgskaya 1970; deRey et al. 1984). Details of the portions of the kidney affected and the degree of damage were not provided.

B.5 EXPOSURE VIA OTHER ROUTES—SOLUBLE FORMS OF URANIUM

More recent subcutaneous studies with soluble forms of uranium have provided a more detailed description of effects to the kidneys. These effects include cellular and tubular necrosis of proximal tubule, less severe injury to the ascending limb of loop of the Henle and collecting tubule, and impaired tubular reabsorption, proteinuria, and enzymuria related to progression of renal injury (Diamond et al. 1989). These injuries were apparent at levels between 0.7 and 1.4 μg U/g wet kidney, and were most severe between 3.4 and 5.6 μg U/g kidney. Repair was rapid, with complete repair within 35 d of exposure. Studies examining the time-course of renal injury found diffuse focal brush border loss and increased vacuolization of the proximal tubule within 1 h of exposure to high doses of uranyl nitrate (Haley 1982). By 5 d post-exposure, the P2 and P3 segments were completely necrotic, and glomerular filtration rates were significantly reduced within 1 wk; SEM images showed abnormalities in glomerular epithelial cells similar to those seen in humans with chronic renal disease and in animal models characterized by proteinuria. Glomerular filtration rates returned to normal values by 4 wk, and tubular regeneration was noted within 2 wk (Haley et al. 1982).

Histopathological examinations following acute intravenous and intratracheal studies revealed persistent and severe acute tubular necrosis involving the loop of Henle and distal convoluted tubules. After 2 wk regeneration was evident; after 4 wk and more, the rate of regeneration had slowed; and at later periods, residual areas of damage remained, with evidence of inflammation (Morrow et al. 1982). Biochemical changes included glucosuria, proteinuria, enzymuria, amino aciduria, and reduced insulin clearance (a measurement of glomerular filtration). Estimated kidney burdens in these studies were between 3.5 and 35 μg U/ g kidney; effects at the higher kidney concentrations were more persistent than effects at higher kidney concentrations.

Effects to the kidneys are summarized in Appendix Tables B.1 and B.2. Data are limited to those studies where kidney burdens of uranium were reported or where sufficient information was provided to calculate kidney burdens of uranium.

Table B.1. Summary of Effects to the Kidney Following Acute Exposures to Uranium

Species	Duration	Dose (µg/g kidney)	Biochemical Effect[(a)]	Histopathological or Other Effect	Reference
Rat (albino)	acute (IT)	1.14	45% PSP retained; 40% Blood NPN	0% bw change	Dygert et al. (1949)
		2.86	54% PSP retained; 46% Blood NPN	-0.5% bw change	
		3.29	49% PSP retained; 52% Blood NPN	-1.2% bw change	
		5.29	52% PSP retained; 36% Blood NPN	-1.4% bw change	
		7.57	75% PSP retained; 58% Blood NPN	-1.8% bw change	
		8.86	92.5% PSP retained; 130% Blood NPN	-7.0% bw change	
		6.29	99.5% PSP retained; 120% Blood NPN	-10.1% bw change	
Rat	2 d	0.4	No changes noted at 14 or 30 days	Extremely slight change to kidney	Dygert et al. (1949)
	2 d	0.5	No changes noted at 14 or 30 days	Slight degeneration and necrosis of tubular epithelium	
Rat	1 h	23.34	No data	Foci of tubular degeneration, extending into corticomedullary region. Severity not noted	Stradling (1985)
	10 min	6.9	No data	Mild to severe foci of tubular degeneration. Tubular regeneration by 91 days, however reductions in size of cortex and increases in papillary space observed.	
	Single	0.26	No data	Mild tubular degeneration affecting PTs.	
Rat	Single IT	7.3	Transient proteinuria, glucosuria, PUN	Persistent renal damage including compression of glomeruli, small tubules, stroma infiltrated with inflammatory cells. Transient tubular necrosis	Morrow et al. (1982)
Rat	1.5 h	19.29	Persistent diuresis, proteinuria, aminiaciduria, NAG, transient glucosuria, ~85 % decrease in GFR	No data	
Rat	1 h	35.21[1]		No data	
Rat	Single IV	3.57[2]	Transient glucosuria, no change in PUN or creatinine clearance	No data	
Rat	Single IV	35.71[2]	Transient glucosuria, proteinuria, enzymuria; protracted urine alpha-amino nitrogen excretion, protracted decrease in creatinine clearance	No data	
Rat	2 min	1.29[3]	Increased urine vol, proteinuria, glucosuria	No data	Leach et al. (1984)
Rat	5 min	7.42[3]	Increased urine vol, proteinuria, glucosuria	No data	Leach et al. (1984)

ble B.1. *(continued)*

ecies	Duration	Dose (μg/g kidney)	Biochemical Effect [(a)]	Histopathological or Other Effect	Reference
Rat	10 min	1.13[3]	Increased urine vol, proteinuria, glucosuria	mild tubular damage	Leach et al. (1984)
abbit	Single IT	122	No data	"Protein dystrophy" of the tubule epithelium	Rubanovskaya (1970)
Dog	0.5 - 2 h	99.5	Transient proteinuria, protracted glucosuria, decreased GFR w/o recovery	No data	Morrow et al. (1982)
		51.4		No data	
		286		No data	
Dog	Single IT	3200	Increased blood NPN, albuminuria	Mortality in 50% of treated animals	Rubanovskaya (1970)

Effects acronyms: bw = body weight, IT = intratracheal, PSP = phenolsulfophthalein, NPN = non-protein nitrogen, PT = proximal tubule, PUN = plasma urea nitrogen; GFR = glomerular filtration rate

ts were exposed to 284 mg/m3 UF_6. Authors assumed absorbed dose of 2-3 mg. Assuming 200 g rat, 04 to mg absorbed. Based on similar experiments by Morrow et al. (1982), it was assumed that 9.86 ± 6.51% of orbed dose went to kidneys. This equals 28.7 to 42 μgU/g kidney; average is 35.2 μg/g kidney.

ts were given IV dose of 0.1 mg/kg or 1.0 mg/kg. Assuming 200 g rat, = 0.02 mg or 0.2 mg injected. Using the get and Pellmar (2003) model, it was assumed that 25% goes to kidneys, = 5 μg or 50 vg to kidney. It was umed that the kidney weight was 0.70 g/100 g body weight (Derelanko 2000). 1.4 g kidney, = 3.57 μg/g kidney mg/kg) or 35.71 μg U/g kidney (1.0 mg/kg).

eans for body weight, absorbed dose, and percent of absorbed dose by kidney were provided by authors. To culate μg U/g kidney, it was assumed that the kidney weight was 0.70 g/100 g body weight (Derelanko, 2000).

Table B.2. Summary of Effects to the Kidney Following Sub-Chronic or Chronic Exposures to Uranium

Species	Duration	Dose (µg/g kidney)	Biochemical Effect	Histopathological or Other Effect	Reference
Rat (S-D)	28 d	0.92	No significant changes	No change	Gilman et al. (1998)
	90 d	0.42		Sclerosis of glomerular capsules, reticulin sclerosis of tubular basement membranes and interstitial scarring	
	90 d	1.67			
	90 d	0.42		Dilation, apical displacement, and vesiculation of tubular nuclei, glomerular adhesions, focal tubular degranulation	
	90 d	2.12			
Rat	33 d	0.7	Significant and transient enzymuria, proteinuria, albuniuria, amino aciduria	Focally severe necrosis, apoptosis of tubular epithelium. Nearly complete recovery 3 weeks after exposure ended.	Diamond et al. (1989)
		1.4			
		3.45			
		5.6			
Rat	30 d	1.7	No changes noted in NPN	No data	Dygert et al. (1949)
	30 d	5.9	Decreased alkaline phosphatase activity	Damage to renal tubules	
Rat	30 d	0.1	No changes noted in NPN	No data	Dygert et al. (1949)
	30 d	0.3	Decreased alkaline phosphatase activity	Damage to renal tubules	
Rat	37 d	6.7		Slight but significant kidney damage	Wilson et al. (1955)
Rats	1 yr	1.1	No changes in blood NPN	No effects noted	Leach et al. (1970)
Rat	1 yr	1.6	No changes in blood NPN	No effects noted	Leach et al. (1973)
Rabbit	37 d	0.1	No significant changes	Minimal degree of regeneration of the renal epithelium	Wilson et al. (1955)
	37 d	1.4	Elevated urinary protein values at several time points.	Significant, severe renal damage	
Rabbit	91 d	0.04	No significant differences in biochemical parameters. Significant linear relationship between exposure level and rate of PSP secretion.	Dose-dependent changes in cytoplasmic vacuolation in proximal tubule epithelium accompanied by vesiculation and pyknosis of tubular nuclei.	Gilman et al. (1998)
		0.5			
		0.73			
		2.97			
		4.98			
		0.01			
		0.019			
		0.07			
		1.03			
Rabbit	91 d	0.18	Initial decrease in urinary output accompanied by increased excretion of glucose and protein. These changes were reversible.	No significant changes.	Gilman et al. (1998)
		3.48	No data	Renal tubular injury with degenerative nuclear changes, cytoplasmic vacuolation, and dubular dilation without evidence of recovery.	
Monkey	5-yr study	21	No changes in blood NPN	No effects noted	Leach et al. (1970)
Monkey	5-yr study	13	No changes in blood NPN	No effects noted	Leach et al. (1973)
Dog	5-yr study	6.3	No changes in blood NPN	No effects noted	Leach et al. (1973)
Dog	5-yr study	7.1	No changes in blood NPN	No effects noted	Leach et al. (1970)

B.6 EFFECTS TO THE LUNG FOLLOWING EXPOSURE TO URANIUM

B.6.1 Acute Inhalation Exposures to Uranium Oxides

Acute exposures to oxides of uranium have been shown to result in effects on the size, number, and viability of alveolar macrophage as well as Type-I and Type-II cells (Morris et al. 1989, 1990; Hengé-Napoli et al. 1996). Alveolar macrophages are cells of the lung involved in the phagocytosis and removal of particulates. Activation, damage, or death of these cells may result in the release of compounds involved in inflammation, the stimulation of fibroblasts, and lung fibrosis. Type-I alveolar cells line the alveolar spaces; they form the barrier between ambient air and the capillary endothelium and, while Type-II cells synthesize surfactant, act as stem cells for Type-I cells, and form part of the blood-gas barrier in the alveoli. In these studies, the lung burdens ranged from 26 to 500 μg of the oxide being tested; the particle size ranged from 2.7 to 3.2 μm AMAD in the Morris studies and was reported as ~4 μm 50% aerodynamic diameter in the study by Hengé-Napoli. It should be noted that some of the results from these studies are contradictory. Rats exposed for 100 min to 50 to 220 mg UO_2/m^3 (estimated lung burdens from 27 to 447 μg) had a significant increase in the size of alveolar macrophages and Type-II cells, and in the numbers of alveolar macrophages and Type-I cells (Morris et al. 1989, 1990). However, no cytotoxic effects were reported up to 2 yr post exposure. In contrast to this, rats instilled with 500 μg of UO_2 plus uranium metal (U_m) showed lysis of all observed cells by 3 wk post-exposure (Hengé-Napoli et al. 1996). Instillation of the same amount of U_3O_8 did not result in any adverse effects. A larger intratracheal dose of U_3O_8 (30 to 100 mg U_3O_8/kg) caused no overt symptoms of toxicity; however, microscopic examination of the lungs showed thickening of the interalveolar septa, and black U_3O_8 grains concentrated in the perivascular and peribronchial lung tissues (Rubanovskaya 1970; Tolgskaya 1970). Some animals displayed focal hyperplasia and metaplasia of the bronchial epithelium.

B.7 CHRONIC EXPOSURES OF URANIUM OXIDE

There were no indications of lung toxicity following chronic exposures in animals with U_3O_8 or UO_2 concentrations up 22 mg/m^3 (median particle size 0.4 to 2 μm) (Dygert et al. 1949; Leach et al. 1970, 1973). In one of these studies, several species were exposed for up to 5 yr (Leach et al. 1970) and followed for up to another 6-½ yr (Leach et al. 1973). Lung concentrations were estimated to be up to 3600 μg/g lung tissue (Leach et al. 1973). Higher exposures (80 mg/m^3) to these oxides did result in moderate to severe lung damage, as well as abnormalities in lung condition and weight (Wilson et al. 1952; Wilson and Sylvester 1955). Exposure to 19 mg/m^3 UO_3, a more soluble form of uranium, for an extended period of time resulted in high mortality; pathology examinations revealed lung hemorrhage (Rothstein 1949).

B.8 REFERENCES

ATSDR. 1999. *Toxicological Profile for Uranium*, Agency for Toxic Substances and Disease Registry, Report TP 90-29, Atlanta, Georgia.

deRey BM, HE Lanfranchi, and RL Cabrini. 1984. "Deposition Pattern and Toxicity of Subcutaneously Implanted Uranium Dioxide in Rats." *Health Physics* 46:688-692.

Derelanko MJ. 2000. *Toxicologist's Pocket Handbook.* CRC Press, Boca Raton.

Diamond GL, PE Morrow, BJ Panner, RM Gelein, and RB Baggs. 1989. "Reversible Uranyl Fluoride Nephrotoxicity in the Long Evans Rat." *Fundamental and Applied Toxicology* 13:65-78.

Dygert HP, CW LaBelle, S Laskin, UC Pozzani, E Roberts, JJ Rothermel, A Rothstein, CJ Spiegl, GF Sprague, and HE Stokinger. 1949. "Toxicity Following Inhalation." In: *Pharmacology and Toxicology of Uranium Compounds*, edited by C Voegtlin and HC Hodge, McGraw-Hill, New York, pp. 423-700.

Fulco CE, CT Liverman, and HC Sox, Eds. 2000. "Chapter 4 Depleted Uranium." In *Gulf War and Health. Vol 1 Depleted Uranium, Sarin, Pyridostigmine Bromide*, Vaccines. p. 88-167 National Academy Press, Washington DC.

Galibin GP. 1975. "Biological Action of Uranium Oxide-Protoxide in Chronic Inhalation Poisoning." *Gigiena Sanitariya* 2:41-46. Article in Russian, English translation.

Gilman AP, DC Villeneuve, VE Secours, AP Yagminas, BL Tracy, JM Quinn, VE Valli, RJ Willes, and MA Moss. 1998a. "Uranyl Nitrate: 28-day and 91-day Toxicity Studies in the Sprague-Dawley Rat." *Toxicological Sciences* 41:117-128.

Gilman AP, DC Villeneuve, VE Secours, AP Yagminas, BL Tracy, JM Quinn, VE Valli, and MA Moss. 1998b. "Uranyl nitrate: 91-day Toxicity Studies in the New Zealand White Rabbit. *Toxicological Sciences* 41:129-137.

Gilman AP, MA Moss, DC Villeneuve, VE Secours, AP Yagminas, BL Tracy, JM Quinn, G Long, and VE Valli. 1998c. "Uranyl Nitrate: 91-day Exposure and Recovery Studies in the Male New Zealand White Rabbit." *Toxicological Sciences* 41:128-151.

Haley DP. 1982. "Morphologic Changes in Uranyl Nitrate-Induced Acute Renal Failure in Saline-and Water-Drinking Rats." *Laboratory Investigation* 46:196-208.

Haley DP, RE Bulger, and DC Dobyan. 1982. "The Long-Term Effects of Uranyl Nitrate on the Structure and Function of the Rat Kidney." Virchows Arch B *Cell Pathology* 41:181-192.

Harley NH, FC Foulkes, LH Hilborne, A Hudson, and CR Anthony. 1999. *A Review of the Scientific Literature As It Pertains to Gulf War Illnesses*: Volume 7, Depleted Uranium. MR-1018/7-OSD, RAND.

Hengé-Napoli MH, E Ansoborlo, M Claraz, J-P Berry, and M-C Cheynet. 1996. "Role of Alveolar Macrophages in the Dissolution of Two Different Industrial Uranium Oxides." *Cellular and Molecular Biology* 42(3):413-420.

ICRP 69. 1995. International Commission on Radiological Protection. *Age-dependant Doses to Members of the Public from Intake of Radionuclides: Part 3 Ingestion Dose Coefficients,* ICRP Publication 69. Ann ICRP 25:57-74 Pergamon Press, Oxford.

Karsner HT and SP Reimann. 1918. "The Toxicity of Certain Water-Insoluble Salts of Uranium." *Journal of Medical Research* 39:157-161.

Leach LJ, EA Maynard, HC Hodge, JK Scott, CL Yuile, GE Sylvester, and HB Wilson. 1970. "A Five-Year Inhalation Study with Natural Uranium Dioxide (UO_2) Dust-I. Retention and Biologic Effect in the Monkey, Dog, and Rat." *Health Physics* 18:599-612.

Leach LJ, CL Yuile, HC Hodge, GE Sylvester, and HB Wilson. 1973. "A Five-Year Inhalation Study with Natural Uranium Dioxide (UO_2) Dust-II. Postexposure Retention and Biologic Effect in the Monkey, Dog, and Rat." *Health Physics* 25:239-258.

Leach LJ, RM Gelein, BJ Panner, CL Yuile, CC Cox, MM Malys, and PM Rolchigo. 1984. *The Acute Toxicity of the Hydrolysis Products of Uranium Hexafluoride (UF6) when Inhaled by the Rat and Guinea Pig*. NTIS DE84011539.

Legget RW and TC Pellmar. 2003. "The Biokinetics of Uranium Migrating from Embedded DU Fragments." *Journal of Environmental Radioactivity* 64:205-225.

Maynard EA and HC Hodge. 1949. "Studies of the Toxicity of Various Uranium Compounds when Fed to Experimental Animals." In: *Pharmacology and Toxicology of Uranium Compounds*, edited by C Voegtlin and HC Hodge, McGraw-Hill, New York. pp. 309-376.

McDonald-Taylor CK, A Singh, and A Gilman. 1997. "Uranyl Nitrate-Induced Proximal Tubule Alterations in Rabbits: a Quantitative Analysis." *Toxicologic Pathology* 25:381-389.

Morris KJ, KMS Townsend, and AL Batchelor. 1989. "Studies of Alveolar Cell Morphometry and Mass Clearance in the Rat Lung Following Inhalation of and Enriched Uranium Dioxide Aerosol." *Radiation and Environmental Biophysics* 28:141-154.

Morris KJ, P Khanna, and AL Batchelor. 1990. "Long-Term Clearance of Inhaled UO_2 Particles from the Pulmonary Region of the Rat." *Health Physics* 58*(4)*:477-485.

Morrow PE, LJ Leach, FA Smith, RM Gelein, JB Scott, HD Beiter, FJ Amato JJ Picano, CL Yuile, and TG Consler. 1982. *Metabolic Fate and Evaluation of Injury in Rats and Dogs Following Exposure to the Hydrolysis Products of Uranium*

Hexafluoride. Implications for a bioassay program related to potential releases of uranium hexafluroide. NUREG/CR-2268.

Orcutt JA. 1949. “The Toxicology of Compounds of Uranium Following Application to the Skin.” In: *Pharmacology and Toxicology of Uranium Compounds*, edited by C Voegtlin and HC Hodge, McGraw-Hill, New York, pp. 377-414.

Rothstein A. 1949. “Toxicity Following Inhalation-17. Uranium Trioxide.” In: *Pharmacology and Toxicology of Uranium Compounds*, edited by C Voegtlin and HC Hodge, McGraw-Hill, New York, pp. 527-645.

Rubanovskaya AA. 1970. “The Effect of Uranous-Uranic Oxide (U_3O_8) in Experimental Work.” In: *The Toxicology of Radioactive Substances*, edited by AA Letavet and EB Kurlyandskaya, translated by A Crozy, translation edited by G.W Dolphin, Pergamon Press, Oxford, United Kingdom, pp. 87-100.

Stradling GN, JW Stather, M Ellender, SA Sumner, JC Moody, CG Towndrow, A Hodgson, D Sedgwick, and N Cook. 1985. “Metabolism of an Industrial Uranium Trioxide Dust after Deposition in the Rat Lung.” *Human Toxicology* 4, 563-572.

Tannenbaum A and H Silverstone. 1951. “Some Aspects of the Toxicology of Uranium Compounds.” In: *Toxicology of Uranium*. Tannenbaum A Ed. Natl. Nuclear Energy Series, McGraw-Hill Book Co., New York, pp. 59-96.

The Royal Society. 2001. *The Health Hazards of Depleted Uranium Munitions Part I* Policy Document 6/01, London, UK. Online report available at URL: www.royalsoc.ac.uk in the Science Policy Section.

The Royal Society. 2002. *The Health Hazards of Depleted Uranium Munitions Part II* Policy Document 5/02, London, UK. Online report available at URL: www.royalsoc.ac.uk in the Science Policy Section.

Thompson HE and LT Steadman. 1953. “Insufflation Studies in Rabbits with Dusts of Uranium Compounds.” In: *Pharmacology and Toxicology of Uranium Compounds*, edited by C Voegtlin and HC Hodge, McGraw-Hill, New York, pp. 1779-1838.

Tolgskaya MS. 1970. “Pathomorphological Changes in Animals Following Acute and Chronic Poisoning with U_3O_8.” In: *The Toxicology of Radioactive Substances*, edited by AA Letavet and EB Kurlyandskaya, translated by A Crozy, translation edited by GW Dolphin, Pergamon Press, Oxford, United Kingdom, pp. 120-135.

Wilson HB, GE Sylvester, S Laskin, CW Labelle, JK Scott, and HE Stokinger. 1952. “Relation of Particle Size of Uranium Dioxide Dust to Toxicity Following Inhalation by Animals: II.” *Archives of Industrial Hygiene and Occupational Medicine* 6(2):93-104.

Wilson HB and GE Sylvester. 1955. “Relation of Particle Size of U_3O_8 Dust to Toxicity Following Inhalation by Animals.” *Archives of Industrial Health* 11:11-16.

World Health Organization (WHO). 2001. *Depleted Uranium: Sources, Exposure and Health Effects*. Department of the Protection of the Human Environment, World Health Organization, Geneva, Switzerland. Online report available at URL: www.who.int/environmental_information/radiation/depleted_uranium.htm.

Index

Pages containing tables are denoted by *t*. Pages containing figures are denoted by *f*.

50-yr committed effective dose. *see* E(50)
50-yr committed equivalent dose
see $H_T(50)$
50-yr committed equivalent lung dose
see $H_{Lung}(50)$

A
a one way analysis of variance (ANOVA), 148
abbreviations, 215–217
Aberdeen Proving Ground, 23
Abrams tanks. *see also* Bradley Fighting vehicles; vehicles
aerosol concentrations, 49–50, 146–147
armor, 143–147, 198
crew positions, 13 *f*
defined, xviii *f*, 219
doses, xxv, 114 *t*, 116 *f*, 117 *f*
exposure scenarios, 24–26, 165 *t*, 201 *t*
health risks, xxxi, 165–166, 165 *t*, 185, 190, 204–205
peak kidney uranium concentrations, xxiii, xxiii *t*, 118 *f*
perforations, 190–191
risk reduction, 209–210
shot tests, 24–26, 94–97, 95 *t*, 100 *t*
ventilation, xxvii, 45–48, 47 *t*, 102–103, 148, 189
absorbed dose, defined, 219
absorption models, 58–60
accidents, industrial, 129–134, 139
acronyms, 215–217
actions, follow-on, 211–212
activity median aerodynamic diameter (AMAD)
Bayesian analysis, 236–238 *t*, 239–246 *t*
collection efficiency, 67 *t*
defined, 219
deposition model, 57
dose coefficient, 106 *f*
monodisperse aerosol collection, 66
statistical analysis, 247–249 *t*
adult males, 20–21, 64
aerodynamic diameter (AD), 37 *f*, 78, 219
aerosol samplers
data mapping, 71
flow rates, 66
particle size distributions, 36, 70
positions in vehicles, xix, 24
aerosols
Capstone study, xv–xviii
collection and characterization, 26–27, 225–226
components, 66
concentration reduction with time, 23, 50 *t*, 52 *f*
aerosols *(cont.)*
cyclone samples, 29
defined, 219
deposition rates, 49 *t*
dose calculations, 63–64, 86–87
measurements, xviii–xix, xxx, 143
risk reduction, 193–194
solubility and dissolution, 44, 59–60
age based risk, 156
age structure, U.S. Army, 157
AIDE (computer program), 81–82
air, uranium concentrations in, 3
air filtration systems, 45–46
airborne release fraction, 19
alveolar-interstitial region (AI), 58, 102, 142
Andersen personal impactor, 37 *f*
animal studies
as follow-on actions, 211
kidney function, 186
lung disease, 140
reproductive system, 141–142
toxicity, 133–134, 329–338
armor, 1, 24–26, 143–148, 190
armor-piercing munitions. *see* munitions
Army. *see* U. S. Army
arrays. *see* sampler arrays
assumptions
breathing patterns, 64 *t*, 79
dose quantities, 67
IMBA uranium code, 82
Level I dose calculations, 63
modeling, 12–16, 72

B
ballistic hulls and turrets (BHT)
defined, xviii, 219
dose distributions inside, 114 *f*, 115 *f*
shot tests, 96, 197–198
test conditions, 44–48
test firing scenarios, 24–25
ventilation rates, 47 *t*, 50, 51 *t*
Battelle Eastern Science and Technology Center (BEST), xvi
Battelle (San Antonio, Texas), xvi
battle situations. *see* combat
Bayes Theorem, 67
Bayesian analysis
doses and concentrations, 68–70, 101, 239–246 *t*
E(50), 68 *f*, 236–238 *t*
posterior distributions, 112 *t*
uncertainty measurement, xxx–xxxi, 111–117
becquerel (Bq), defined, 219

best estimates of risk, 174–176, 180–181, 183–186, 183 *t*, 207–208
beta-radioactivity counting, 105
bioassays, 181, 191, 193, 211, 219
biochemical evaluation, 129–131, 137
Biokinetic Model (computer code) (BK)
  code comparisons, 83 *t*, 84 *t*, 85–87, 86 *t*
  description, 76
  validation and verification, 81–82
biokinetic models
  defined, 219
  dose estimates, 200, 202
  ICRP, 131, 142
  uranium systemic, 61–62, 61 *t*, 107
biomonitoring, xxxii, 206, 207–208. *see also* medical follow-up
bladder, 62
blank adjustments, 32–35
blast, initial, 32
blood, 55, 58, 60, 61–62
blood-brain barrier, 141
body, uranium retention in, 124 *f*
bone marrow
  cancer risks, 153, 161 *t*
  doses and concentrations, 94, 95 *t*, 239–246 *t*, 250–257 *t*
  E(50) and $H_T(50)$, xxv, 81 *t*, 159 *t*
  intakes, 103 *t*
  risk coefficients, 157–158
bone surfaces. *see also* skeleton
biokinetic model, 62
  cancer risks, 152–153, 161 *t*
  doses and concentrations, 94, 95 *t*, 239–246 *t*, 250–257 *t*
  E(50) and $H_T(50)$, xxv, 81 *t*, 111 *f*, 159 *t*
  intakes, 103 *t*
  risk coefficients, 157–158
bound state, 59–60
Bradley Fighting vehicles. *see also* Abrams tanks; vehicles
  air filtration systems, 46
  cancer mortality estimates, 165 *t*
  chemical and radiological risks, 189–190
  conventional armor, 144–145
  crew positions, 13 *f*, 24
  defined, xviii *f*, 219
  dose distribution, 115 *f*
  exposure scenarios, 24–25, 201 *t*
  risk reduction, 210
  shot tests, 25, 96, 100 *t*
  ventilation, xxxii, 46–48, 47 *t*, 50–51, 51 *t*
brain, 141
breathing patterns, 21–22, 64 *t*, 102, 103–105, 104 *f*
breathing rates
  assumptions, 14, 64 *t*, 79–81
  description, 21
  exposure scenarios, 14 *t*, 142 *t*
  lung concentration calculations, 101–102
  physiological model, 56–57
bronchial and bronchiolar regions, 56, 58–59, 102. *see also* lungs; respiratory tract
burn tests, 2
burns, 130–131, 139

C

calibration data, 37 *f*, 42–44
cancer
  bone, 152–153
  health risks, 150–152, 158–162, 164–166, 210
  lung, 152
  lymphatic, 153
  risk coefficients, xxvi, 154–158
cancer mortality risks
  comparisons, 210
  exposure scenarios, xxvi–xxvii, 160–162, 161 *t*, 162 *t*, 165 *t*, 204, 204 *t*
  health effects categories, 182 *t*, 206–208
  inhalation exposures, 162 *t*, 165 *t*
  lifetime risks, xxvii *t*, 161 *t*
  occupational, 151 *t*, 157 *t*
  Radiation Exposure Status (RES), 182 *t*, 184 *t*
  risk coefficients, 154, 155 *t*
Capstone study
  aerosol measurements, 199
  conclusions, xxix–xxxii
  data analysis, 17–53, 181, 209
  description, 1–2
  dissolution parameters, 74 *t*
  Level I dose calculations, 63–64
  peak uranium kidney concentrations, 142–143
  purpose, 173, 197
cartridges, large-caliber. *see* munitions
cascade impactors (CI)
  aerosol sampling, xviii, 26, 28–29, 64–65, 67 *t*, 85 *t*
  data adjustments and calibration, 32–35, 37 *f*, 42–44
  data mapping, 71 *t*
  defined, 222
  particle size, 36–44, 44 *t*, 70, 109 *t*, 199–200
  shot tests, xxi
  uncertainty analysis, 105–106
central nervous system, 141
cerebellum, 141
chemical risks, 136 *t*, 163–164, 207–208, 207 *t*
chronic exposures, 137–138
classification of exposures, 12 *t*
clearance, defined, 219
clearance model, 57–60
coal miners, 149–150, 149 *t*
coefficients
  cancer risk, xxvi, 154–158
  dose comparison, 82–85, 82 *t*, 84 *t*
  transfer, 199
cold war, 178–179

combat
best estimates of risk, 174–176
cancer risks, xxvii, 182 *t*
high-intensity, 176–177, 179
operational exposure guidance, 178–179
training, 193
veterans, 140–141
commanders, 174–176, 206
committed effective dose. *see* E(50)
committed equivalent dose. *see* H_T(50)
committed equivalent lung dose. *see* H_{Lung}(50)
comparisons
best estimates of risk, 183–186
cancer risks, 158–160
cascade impactors *vs.* IOM filter cassettes, 29–32
computer code, 81–87
intakes, 100 *t*
kidney dysfunction, 147–149
occupational standards, 180 *t*
vehicle ventilation rates, 51–52
computer code, 81–87, 82 *t*
concentrations, aerosol
Bayesian analysis, 68–70, 101, 236–238 *t,* 239–246 *t*
blank adjusted, 32–35
factors effecting, 22–23
Human Health Risk Assessment (HHRA), 226–229
multiple perforations, 187–188, 208
particle size, 36, 43, 64–65
reducing, 14–15, 48–52, 50 *f,* 193–194
shot test results, 100
statistical analysis, 30–32, 31 *t,* 247–249 *t,* 250–257 *t*
time analysis, 26, 28–29, 79 *t,* 84–85
in vehicles, 22–23, 198
concentrations, uranium. *see also* peak kidney uranium concentrations
chemical risks, 164 *t*
conventional armor perforations, 143–144
environmental exposures, 126, 126 *t*
exposure scenarios, xxiii *t,* 201–202, 202 *t*
occupational exposures, 126–127, 127 *t*
organs and tissues, 122
particle size, 259–327 *t*
renal dysfunction, 131
Renal Effects Groups, 135 *t,* 136–137
solubility parameters, 111 *f*
conclusions
crewmember risks, 189–191
health risks, xxxi–xxxii, 200–205, 207–208, 209–211
mathematical models, 199–200
multiple perforations, 208
U. S. Army policies, 208–209
ventilation and overpressure systems, 205–206
confounders
double shot tests, 94, 99
modeling considerations, 102, 119
occupational exposures, 139, 156–157
conventional armor, 24–25, 143–145, 190
conversion factors, 218
correction factors, wall losses, 34 *t*
counseling, xxxii, 193, 210
crew compartments
dose estimates, 93–94
perforations, xv
sampling arrays, 103, 143
shot tests, xxi–xxii
ventilation rates, 47 *t*
ventilation systems, 45–46, 102–103
crewmembers
biomonitoring, xxxii
cancer risks, 160–162, 164–166, 204–205
doses, 91, 94–97, 113–117, 115 *t*
exposure scenarios, 14 *t,* 15, 198 *t*
peak kidney uranium concentrations, xxiii *t*
positions in vehicles, xxi–xxii, 13 *f,* 24–25, 68–70, 105
renal dysfunction, 207
risk estimates, 180–181, 183 *t,* 189–191
risk reduction, 205–206
training, 193
crops, uranium content, 2–3
cyclone samplers, 29, 70–71, 71 *t,* 73, 74 *t*

D
data
agreement and calibration, 42–44, 99
field blank adjustments, 32–35
gaps, 1, 211
mapping, 71
reliability, 209
uncertainty measurements, 105–117
data collection
aerosol samples, 19, 26–29, 35–44, 63–65, 225–226
dissolution fit parameters, 75
efficiency, 37–41 *f*
equipment, 29–32
techniques, xviii–xix, xxix
test conditions, 2
uncertainty analysis, 105–107
Data Quality Objectives (DQO), xvi–xvii, 5, 220
deaths, worker, 139
decay, uranium, 3
depleted uranium, defined, 3–4, 220
deposition, 22–23, 56–57, 101–102, 101 *t,* 220
dermal exposures, 131–132
deterministic effects, defined, 221
detriment, defined, 221
dissolution rates, 26–27, 58–59, 75, 221
doctrine and policies, 208–209
dose and dose-rate effectiveness factor (DDREF), xxviii, 182
dose assessment
description, 18
medical follow-up, 191
quantities calculation, 67–70
techniques, xxx–xxxi, 199–200
time-sequence sampling, 29

dose models, 55–89
  absorption, 58–60
  aerosol concentrations, 64–70
  biokinetic, 61–62, 76
  breathing rate, 57
  dosimetric, 76
  E(50) and $H_T(50)$, 76–81
  gastrointestinal tract, 60
  respiratory tract, 55–60
  software quality assurance, 81–87
  thoracic tissues, 56
  in vitro dissolution, 70–75
dose regions, 113–117, 116 *t*
doses
  Bayesian analysis, 236–238 *t*, 239–246 *t*
  coefficients, 82–85, 82 *t*, 84 *t*, 106 *f*
  conversion factors, 218, 218 *t*
  distribution, 113–119, 114 *t*, 115 *f*, 117 *f*
  double shot tests, 94–97
  exposure scenarios, 92–95, 226–229
  factors effecting, 12–13
  Radiation Exposure Status (RES), xxviii–xxix, 180 *t*, 192 *t*
  radiological, 68–70, 104–105, 202–205, 207–208
  statistical analysis, 247–249 *t*, 250–257 *t*
  uncertainty analysis, 105–117
double shot tests, 94–97, 99–101, 118–119
drills. *see* training
drinking water, 3, 137, 138 *t*
DU armor. *see* heavy armor
duration estimates, 13
dust, 149–150, 152, 155

E

E(50)
  Bayesian analysis, 68 *f*, 236–238 *t*
  calculations, 76–81, 81 *t*, 111–117, 258 *t*
  computer code validation, 85–87
  defined, 77, 219
  effect of breathing patterns on, 104–105
  exposure scenarios, xxiv–xxv, xxiv *t*, 92–94, 93 *t*, 159 *t*
  peak kidney uranium concentrations, 103 *t*, 111 *t*
  risk analysis, 159–160, 160 *t*
  solubility, 110–111, 111 *f*
  uncertainty analysis, 200
  inside vehicles, 203 *t*
EC/NBC ventilation systems. *see* ventilation systems
effective dose, defined, 221
efficiency of data collection, 37–41 *f*
$E(50)G_{Org}$, 159–160, 160 *t*. *see also* summed organ risks
EPA. *see* U.S. Environmental Protection Agency
epidemiological studies, xxvi, 139, 150–157, 204, 211
equations
  aerosol components, 66
  aerosol removal, 48
  dose quantities, 67
  E(50) and $H_T(50)$, 77
  efficiency data, 42–44
  gastrointestinal tract clearance rates, 60
  intakes, 65, 77–78
  likelihood function, 65
  penetration functions, 37
  in vitro dissolution, 73
  wall loss correction, 33–35
equipment, data collection, 29–32
equivalent dose, defined, 221
estimates
  aerosol intakes, 78–81, 93 *t*
  best estimates of risk, 174–176, 180–181, 183–186, 183 *t*, 207–208
  cancer risk, 160–162, 165 *t*
  duration, 13
  particulates from initial blast, 32
  radiological doses, 91–94, 118, 181–183, 193, 202, 211
  uncertainty, 200
  upper bounds, xxiv, 181
evacuation tests, 13
excretion rates, 122–123
exercise, 79–80, 95 *t*, 104
experts, xvi, 5
exposure-control philosophy, xxviii, 173–174, 176–177
exposure levels, defined, 10–12
exposures. *see also* scenarios
  animal studies, 140, 141–142, 331–336
  duration, xx *t*, 15, 142 *t*
  limits, 178 *t*, 185, 192
  mass lung concentration, 101–102
  occupational, 137–140, 149–150, 149 *t*
  operational guidance for, 178–180
  radiation, xxviii, xxix *t*, 155–157, 180 *t*
  renal effects, 128–138
  special, 189, 190, 192, 203
  inside vehicles, 14 *t*, 193–194, 202
extrathoracic tissues
  cancer risks, 161 *t*
  doses, 94, 95 *t*, 102, 210, 239–246 *t*
  E(50) and $H_T(50)$, 81 *t*, 159 *t*
  Human Respiratory Tract model, 56
  radiation doses, xxv
  risk coefficients, 158
eyes, 203

F

fat, 126 *t*
fecal excretion, 123, 125 *f*
feeble poisons, 121
fibrotic lesions, 150
field blank adjustments, 32–35
field commanders, 174–176, 206
filtration systems, 45–46

findings, xxx–xxxii, 189–191, 199–211
fire suppression systems, xxix, 22
first responders
  defined, 221
  E(50), xxiv *t*
  exposure scenarios, 142 *t,* 198 *t*
  health risks, 162, 191, 210
  modeling assumptions, 15–16
  uranium concentrations, xxiii *t*
flow rates, 42–43, 46, 66, 80 *t,* 103
follow-on actions, 211–212
foods, uranium content, 2–3
fractional adsorption, defined, 221
fragments, xxxi, 19, 128, 138, 140–141
frontal cortex, 141

G
gastrointestinal tract, 60
Gaussian likelihood function, 65
gonads, 127 *t. see also* reproductive system
gray (Gy), 218, 221
guidelines
  exposure, 142
  occupational, 128, 148, 186–187, 201–202
  operational risk, 191–194
  radiological exposure, 179–180
Gulf War of 1991. *see* Operation Desert Storm (ODS)

H
Halon systems. *see* fire suppression systems
hatches, 14, 22, 46–48, 160–162, 185, 206
health risks
  characterization, 18, 163 *t,* 183–186
  chemical, 163–164, 189 *t*
  dust burdens, 149–150
  findings, xxxi–xxxii, 207–208, 209–211
  $H_{Lung}(50)$, 203–204
  $H_T(50)$, xxv–xxvii
  kidneys, 122–123, 128–138, 161 *t,* 186–188
  Level I exposures, 163, 200–205
  long-term exposures, 181
  lungs, 101–102, 138–140, 152, 154–157, 161 *t,* 166
  management, 174–175
  radiological, 150–162, 164–166, 165 *t,* 188–191, 189 *t*
heavy armor, 25–26, 145–147
heavy exercise, 79–80, 95 *t,* 101–102, 104
hexavalent uranium nitrate, 131
high-intensity combat, 177, 179
hippocampus, 141
$H_{Lung}(50)$, xxv *t,* 203–204, 204 *t*
$H_T(50)$
  calculations, 80–81, 258 *t*
  defined, 220
  exposure scenarios, 81 *t,* 159 *t*
  health risks, xxv–xxvii
  influence of solubility parameters on, 111 *f*
  models, 76–81
  peak kidney uranium concentrations, 103 *t,* 111 *t*
Human Health Risk Assessment (HHRA), 4–5, 5–6, 225
Human Respiratory Tract Model (HRTM), 55–60, 107–108, 199

I
illness, 190, 209
IMBA uranium code, 82–87, 84 *t,* 86 *t*
immune system, 142, 192, 207
impact tests, 2
in vitro solubility and dissolution
  aerosol samples, 44
  modeling, xxx, 200
  parameters, 70–75
  particle characteristics, xix, 59–60
  uncertainty analysis, 110
in vivo dissolution, 110–111
industrial accidents, 129–134, 139
ingestion exposures, 126, 129–130, 331–332
inhalability, defined, 222
inhalable fraction, defined, 222
inhalation exposures
  animal studies, 329, 331
  biokinetics, 123
  environmental, 126
  exposure scenarios, 198
  health risks, 131, 153–162, 164–166
  mortality risks, 162 *t,* 165 *t*
  radiation doses, 159
initial blast, 32
injection exposure, 131
injuries and wounds. *see* wounds and injuries
Institute of Medicine (IOM), 137–138, 152–153, 174–176, 181
instruments, sampling, 29–32
intakes
  assessment, 18
  calculations, 69–70 *t,* 72–73, 85
  dissolution parameters, 83–84, 84 *t*
  doses and concentrations, 91–97, 117–119, 124 *f*
  equations, 65, 77–78
  exposure estimates, 78–81, 93 *t*
  flow rates, 80 *t*
  health risks, xxxi–xxxii, 130 *t,* 134–142, 138 *t*
  peak kidney uranium concentrations, 98 *t,* 100 *t,* 103 *t*
  U-238, 156 *f*
  vehicle scenarios, xxii *t,* 12–13, 201, 201 *t*
International Atomic Energy Agency, 177
International Commission on Radiological Protection (ICRP)
  biokinetic models, xix, 76, 142
  dissolution parameters, 75 *t*
  Human Respiratory Tract model, 14, 55–60, 107–108, 199
  lung dissolution models, 72 *f,* 82–83, 83 *t*
  radiation standards, 177
  Reference Man, 20, 223
  risk coefficients, xxvi, 156–157, 182
intestines, gastrointestinal tract model, 60

IOM filter cassettes, 26, 28–35
isotopes
  animal studies, 331
  dose calculations, 63, 80–83, 83 *t,* 113
  dose models, 67
  health risks, 156 *f*
  munitions composition, 18–19, 19 *t*
  radioactivity, 3

K

kidney dysfunction, 129–131, 147–149, 194, 202
kidneys
  biokinetic model, 62
  chemical risks, 128, 163–164, 163 *t,* 201–202, 210
  doses and concentrations, 94, 95 *t,* 239–246 *t,* 250–257 *t*
  E(50) and $H_T(50)$, 81 *t,* 159 *t*
  health risks, 122–123, 128–134, 161 *t,* 186–188
  influence of solubility parameters on, 111 *f*
  peak kidney uranium concentrations, 97–101, 103 *t,* 144–149, 149 *t*
  risk coefficients, 158
  toxicity, 211
  uranium exposures, 124 *f,* 126 *t,* 127 *t,* 130 *t*
kinetic parameters, 72–75

L

large-caliber munitions. *see* munitions
larynx, 158
lens of the eye, 203
lesions, fibrotic, 150
leukemia, 153
Level I exposures
  in combat, 176
  defined, xvii, 10
  dose calculations, 63–64
  exposure scenarios, 142 *t,* 198, 198 *t*
  modeling assumptions, 12–16
  risk assessment, 210–211
  risk management, 180–181
  U. S. Army policies, 208–209
Level II exposures, defined, 10–11
Level III exposures, defined, 11
light exercise, 79–80, 95 *t*
likelihood function, 65, 111
limits
  dose, xxiv–xxvi
  exposure, xxxi, 178 *t,* 185, 192, 203
  occupational, 128
Live Fire Test Objectives, 103
liver
  biokinetic model, 62
  cancer risks, 161 *t*
  doses and concentrations, 94, 95 *t,* 239–246 *t,* 250–257 *t*
  E(50) and $H_T(50)$, xxv, 81 *t,* 159 *t*
  environmental exposures, 126 *t*
  occupational exposures, 127 *t*
  risk coefficients, 158
Livermore Solver for Ordinary Differential Equations (LSODE), 76
Los Alamos National Laboratory, xvi, 76
Lovelace Respiratory Research Institute, xvi
low-LET exposure, 191–193, 192 *t*
Lung Dose Uncertainty Calculation (LUDUC), 107–108
lungs. *see also* bronchial and bronchiolar regions; respiratory tract
  animal studies, 140, 337–338
  bound state model, 59–60
  chemical toxicity, 138–140
  dissolution models, 70–75, 72 *f,* 82–83
  doses and concentrations, xxv, 94, 95 *t,* 101–102, 203, 239–246 *t,* 250–257 *t*
  E(50) and $H_T(50)$, 111 *f,* 159 *t*
  environmental exposures, 126 *t,* 149 *t*
  fluid solubility and dissolution rates, 26–27
  health risks, 149–150, 152, 154–157, 161, 161 *t,* 166, 210
  intakes, 103 *t,* 124 *f*
  occupational exposures, 127 *t,* 149 *t*
  risk coefficients, 157–158
lymph nodes, xxv, 94, 127 *t,* 158, 210. *see also* thoracic lymph nodes (LN-TH)
lymphatic system, 55–56, 57–58, 142

M

males, adult, 20–21, 64
Markov chain Monte Carlo methods (MCMC), xxx, 67, 69–70 *t,* 111–112, 200
Marple cascade impactor (CI), 28, 222. *see also* cascade impactors (CI)
mathematical formulas
  aerosol components, 66
  aerosol removal, 48
  dose quantities, 67
  E(50) and $H_T(50)$, 77
  efficiency data, 42–44
  intakes, 65, 77–78
  likelihood function, 65
  particle transport, 60
  penetration functions, 37
  in vitro dissolution, 73
  wall loss corrections, 33–35
Matlab/Simulink, 76, 85
medical follow-up
  dose estimates, 174, 193
  individual doses, xxxii, 181, 206
  recommended actions, 191
  renal effects, 133, 133 *t*
medical personnel, 174–175, 187, 191, 193, 206
medical recommendations, 191–192, 192 *t,* 209
medical risks, 187
medical use of uranium, 131
military risk management, xxvii–xxix, 18, 174–176, 208–209

miners, coal, 149–150, 149 *t*
models
Abrams tanks, 49–50
adequacy of, 199–200
aerosols, 55–89
assessment criteria, 10
assumptions, 12–16, 72
combat, 23–26
confounder effects on, 102, 119
dose and risk, xvi–xx, xxx–xxxi, 134–137, 187
mathematical, xxi
particle size distribution, 36–44
physiology, 20–22
uncertainty analysis, 18
moderate absorption model, Type M, 58–59, 127–128
monitors, sampling, 29–32
monodisperse aerosols, 66, 67 *t,* 84, 106 *f*
mortality risks, cancer
comparisons, 210
exposure scenarios, xxvi–xxvii, 160–162, 161 *t,* 162 *t,* 165 *t,* 204, 204 *t*
health effects categories, 182 *t,* 206–208
inhalation exposures, 162 *t,* 165 *t*
lifetime risks, xxvii *t,* 161 *t*
occupational, 151 *t,* 157 *t*
Radiation Exposure Status (RES), 182 *t,* 184 *t*
risk coefficients, 154, 155 *t*
mouth, 158
multiple perforations, xxxii, 187–188, 208
munitions
defined, 222
depleted uranium use in, 1
heavy armor, 25–26
isotopic composition, 19 *t*
source term analysis, 18–19
tests, 17
muscle, 126 *t,* 141

N

nasal passages, 56, 57, 104, 158
National Academy of Sciences, Institute of Medicine, 137–138, 152–153, 174–176, 181
National Council on Radiation Protection and Measurements (NCRP), 156–157, 177, 218
National Radiological Protection Board, 76
nervous system, 141
neurocognitive function, 140, 141
nonradiological risks, 149–150
North Atlantic Treaty Organization (NATO), 178, 181–182
nose, 158
notional, defined, 222
NRC. *see* U.S. Nuclear Regulatory Commission

O

Oak Ridge National Laboratory, 76
occupational exposures, 126–127, 137–138, 138 *t,* 149 *t,* 150–153
occupational safety guidelines
combat operations, 175–176
kidneys, xxiii, 142, 186–187, 201–202, 210
public debate, 122
radiation protection standards, 177–178
Office of the Special Assistant for Gulf War Illnesses (OSAGWI), xvii, 1
one-way analysis of variance (ANOVA), 148
Operation Desert Storm (ODS)
combat models, 23–25
exposure classification, 12 *t*
exposures, 9–11
fragment wounds, 138
munitions and armor, xv–xvi, 1
veterans, xxxi, 9–10, 138, 140–141
Operational Exposure Guidance (OEG), 178–179
operational risks, 187, 193
ore, uranium, 137, 138 *t*
organs
doses, 92, 153, 199–200
limits to exposures, 203
risk coefficients, 157–158
risk estimates, 181
summed risks, xxvi, 158–160, 165, 181, 204
uranium concentrations in, 122, 126–127
overpressure systems, 45, 118, 166, 188, 205–206
oxides, 19–20, 130–131, 138, 142, 330–333

P

Pacific Northwest National Laboratory, xvi, 5
particle size
concentrations, 64–70, 199–200, 259–327 *t*
deposition, 109 *t*
distributions, 20, 26, 35–44, 57, 70, 222
increments and cutoff diameters, 44 *t*
sample mapping, 71
solubility and dissolution rates, 26–27
time analysis, 48
uncertainty analysis, 105–110
particle transport, 57–60, 222
particles
coagulation and aggregation, 23, 188
diameters, 79
initial and transformed states, 58, 82
oxide, 138, 142
sampling, 32–35
uncertainty analysis, 107–110
pathway analysis, 18
peacetime risk management, 176–178

peak kidney uranium concentrations
animal studies, 186
chemical risks, xxii–xxiv, xxiii *t,* 97–101
clearance times, 149 *t*
E(50) and $H_T(50)$, 103 *t,* 111 *t*
exposure scenarios, xxiii *t,* 142–149, 147 *f,* 202 *t*
occupational exposures, 126–127, 137–138, 138 *t*
percentiles, 98 *t,* 113, 163–164
perforations, 147 *t,* 194
Renal Effects Groups (REG), 135 *t,* 164 *t,* 201–202
uncertainty analysis, 110–111
peer review panel, 5
penetrated, defined, 222
penetration functions, 37
perforations
aerosol concentrations, 22, 78–79
through armor, 24–26, 143–147, 144 *t,* 145 *t,* 146 *t,* 147 *t,* 189–191
best estimates of risk, 183–186, 183 *t*
crew compartments, xv
crewmember risks, 189, 194
defined, 222
dose estimates, 91–92, 117–119
dust burdens, 150
multiple shots, xxxii, 187–188, 208
post-shot ventilation rates, 48
pharynx, 158
philosophy, radiation exposure-control, xxviii, 173–174, 176–177
physical activity, 21–22, 21 *t*
physiology, 20–22, 21 *t,* 56
plants, uranium content, 2–3
plutonium studies, 155–156, 158
poisons, feeble, 121
policies and doctrines, 208–209
polydisperse aerosols, 66, 84
posterior distributions, 112 *t,* 222
pressure pulse, 32
prior distribution, defined, 221
probability distributions, 112
protective equipment, 23, 45–46, 102, 173
proximal tubules, 128
pulmonary toxicity, 138–140

Q
quality assurance, 81–89, 222
quality control, defined, 222

R
rad, defined, 223
radiation
exposure-control philosophy, xxviii, 173–174, 176–177
occupational guidelines, 177–178, 179–180
risk estimates, 181, 211
toxicity, 150
radiation doses
Bayesian analysis, 68–70
double-shot tests, 96 *t,* 97 *f*
explained, 92
exposure scenarios, xxiv–xxv, 159 *t,* 189 *t,* 202–205
health risks, xxvi–xxvii, 101–102
multiple perforations, 187–188
Radiation Exposure Status (RES)
cancer risk, 182 *t,* 184 *t*
doses, xxviii–xxix
health effects categories, 206–209
medical implications, 192 *t*
peacetime standards comparison, 180 *t*
Renal Effects Groups (REG) comparison, xxix *t*
risk estimates, 178–179, 181–186
radiation weighting factor, 223
radiological risks
best estimate of, 185
cancer mortality, xx–xxi
epidemiology, 150–152
exposure scenarios, xx–xxi, 188–191, 207 *t*
findings, 188, 207–208, 211
health effects, 153–162, 164–166, 202–205
percentiles, 189 *t*
radionuclides, xxvi, 83 *t,* 202
radium, 152
radon, 155
rapid absorption model, Type F, 58, 127
rate constants, 61 *t*
recommendations, 191–194, 192 *t,* 209, 211–212
red marrow. *see* bone marrow
Reference Man, 20, 223
regions, dose, 113–117, 116 *t*
regression analysis, 31 *t*
rem, defined, 223
renal dysfunction, 188, 202, 207. *see also* kidney dysfunction
renal effects, xxiii–xxiv, xxix *t,* xxxi, 128–138, 329
Renal Effects Groups (REG)
chemical risks, 163–164
defined and described, 134–137, 223
exposure scenarios, 143–147
kidney toxicity distribution, 135 *f*
peak kidney uranium concentrations, 135 *t,* 164 *t,* 201–202
perforations, 144 *t,* 145 *t,* 146 *t*
risk assessment, 187, 206
reproductive system, 127 *t,* 140, 141–142
respirable fraction, defined, 223
respirators, 23
respiratory tract
aerosols deposition, 101 *t*
particle clearance, 108–110, 123
physiological models, 20, 21 *t*
radiation doses, 101–102
respirable fraction, 19

risk
coefficients, xxvi, 154–158, 182
defined, 223
levels, 178–179, 181–183, 209
management, xxvii–xxix, 18, 173–178, 180–181, 208–209
mitigation, 191–194
risk assessment
application of, 206
approach and assumptions, 179–181
cancer, xxvi–xxv
characterization, 136 *t,* 163 *t*
combat, 4–5
comparisons for, 160 *t*
interpretation factors, 27
Level I exposures, 210–211
purpose, 197
team, 5
The Royal Society (London), 129, 137, 150–151
Russian plutonium workers, 155–157

S
safety standards, 175–176
sample arrays
crew positions, 24, 143
field blanks, 32
scenarios, 230–235 *t*
schematic, 28 *f*
uncertainty calculations, 112
ventilation system tests, 103
sampling, 28–32, 32–35, 99
scenarios
cancer risk, 160–162, 165 *t,* 204, 204 *t*
categories and classifications, xxix *t,* 12 *t*
E(50) and $H_T(50)$, xxiv–xxv, xxiv *t,* 81 *t,* 203 *t*
explained, xvii–xviii, 9–16, 181, 197–198, 223
intakes and doses, xxv *t,* 92–95, 93 *t,* 103 *t,* 201 *t,* 204 *t*
Level I exposures, 63–76, 142 *t,* 198 *t*
peak kidney uranium concentrations, xxiii *t,* 99–101, 142–149, 147 *f,* 201–202, 202 *t*
perforations, 117–119, 144 t, 145 t, 146 t, 147 *t*
radiological and chemical risks, xxiv–xxv, 159 *t,* 188–191, 189 *t,* 202–205, 207 *t*
sample arrays, 230–235 *t*
shot tests, xxi *t,* xxii *t,* 23–26, 95 *t,* 117–119
stay-times, xx *t,* 14 *t,* 15, 198 *t*
SEECAL (computer program), 77
shape factor, defined, 223
shot lines, 143–146, 197, 223
shot tests, 23–26, 143–147
Abrams tanks, 49–50, 143–144, 145–147
Bradley Fighting vehicles, 50–51, 144–145
crew compartments, xxi–xxii
double shots, 94–101, 118–119
intakes and doses, 92–94, 93 *t*
shot tests *(cont.)*
test firing scenarios, 23–26
vehicle configurations, 143–147, 197–198
SI units, 218
sievert (Sv), 218, 223
SigmaStat (software), 148
skeleton, 62, 123, 126 *t,* 127, 127 *t. see also* bone surfaces
skin, 131–132, 158
slow absorption model, Type S, 59, 127–128
smoking, 156, 166
soft tissue, 62, 141
software quality assurance, 81–87
soils, uranium content, 2
soldiers, 140
solubility and dissolution
classes, 58–59
defined, 223
simulations, 82–85
uncertainty analysis, 110–111
in vitro, xix, 44, 59–60, 74 *t*
source term, xix, 17–20, 223
special exposures, 189, 190, 192, 203
specific effective energy (SEE), 80–81, 86
square root of Stokes number, 38–41 *t,* 43
stability and support operations, 176, 179
standards, occupational, 175–176, 177–178, 180 *t*
statistical analysis, 247–249 *t,* 250–257 *t*
stay-times
duration, xx *t*
exposure risks, 185–186, 186 *f*
exposure scenarios, 15, 142 *t,* 198 *t*
kidney toxicity, 211
stochastic risks, 177, 192 *t,* 223
Stokes number, 37–43
stomach, 60
stress and toxicity, 141
striatum, 141
subject matter experts, xvi, 5
summed organ risks
basing radiation risks on, 181
cancer mortality, xxvi, 165, 204
E(50) comparison, 160 *t*
whole body risk comparison, 158–160
System for Internal Dosimetry (SID), 76, 85, 86 *t,* 258 *t*

T
target organ, defined, 223
test firing scenarios, 23–26
tests
burn and impact, 2
design review, 5
double shot, 94–101, 118–119
evacuation, 13
shot, xxi–xxii, 23–26, 49–51, 143–147
ventilation, 46–48, 103
thoracic lymph nodes (LN-TH)
doses and concentrations, 239–246 *t,* 250–257 *t*

thoracic lymph nodes (LN-TH) *(cont.)*
  $H_{T(}50)$, 81 *t*
  insoluble uranium burdens, 153
  risk coefficients, 158
  uranium retention, 125 *f*
thoracic tissues, 58
thorium, 152
time functions
  aerosol concentrations, 28–31, 79 *t*
  concentration decay rates, 49–52, 50 *f,* 52 *f*
  particle distributions, 35–44
  particle size, 48
time-sequence sampling, 28–29
tissue weighting factor, 77, 86, 158, 224
tissues, xxv, 141
toxicity, 121–142
  animal studies, 329–338
  chemical, 121–122, 201–202
  health effects, xxii–xxiv
  immune system, 142
  kidneys, 128–138, 211
  lungs, 138–140
  nervous system, 141
  radiological, 121–122, 150
  reproductive system, 141–142
  risk model, 134–137
training, 193, 212
transfer compartment, defined, 224
transuranic elements, 155
turret, defined, 224
Type F absorption model, 58, 127
Type M absorption model, 58–59, 127–128
Type S absorption model, 59, 127–128

U
U.S. Army, 157, 178–179, 191, 193, 208–209
U.S. Army Center of Health Promotion and Preventive Medicine (USACHPPM), xv–xvi, 1
U.S. Army Heavy Metals Office, 1
U.S. Department of Defense, xv, 4, 208–209
U.S. Environmental Protection Agency, 154–156, 177, 182, 184
U.S. Nuclear Regulatory Commission, 177, 192, 203
U-238, 81 *t,* 86, 156 *f. see also* isotopes
uncertainty analysis
  approaches, 200
  dose calculations, 105–117
  likelihood function, 65–66
  Markov chain Monte Carlo method (MCMC), xxx
  peak kidney uranium concentrations, 110–111
  risk assessment parameters, 18
  variable parameters, 107
  wall-loss, 34–35
upper bounds estimates, xxiv, 181
uranium acetate, 129–130, 142
uranium hexafluoride, 132, 133, 138–139
uranium nitrate, 141
uranium oxide, 130–131
uranium systemic biokinetic model. *see* biokinetic models
uranium tetrafluoride, 131
uranyl acetate, 141
uranyl ions, 128
uranyl nitrate, 130–131, 132
urinary tract, 55
urine, 62, 123, 125 *f,* 140

V
vagal-mediated laryngeal reflex, 102
vehicles. *see also* Abrams tanks; Bradley Fighting vehicles
  armor, 1, 24–26, 143–148, 190
  crewmember health risks, 189–191
  crewmember positions in, xxi–xxii, 13 *f*
  E(50), xxiv–xxv
  exposure scenarios, xxix *t,* 12–16, 14 *t,* 117–119, 160–162
  factors effecting exposure, 22–23
  intakes, 201, 201 *t*
  perforations, 183–186, 187–188
  risk reduction, 210
  shot tests, xxii *t,* 24–26, 99–101, 197–198
  source term analysis, xix
ventilation, 45–46, 51–52, 202, 210
ventilation rates, 46–52
ventilation systems
  defined, 221
  dose and concentration effects, xxv, 27, 102–103, 147, 160–161
  findings, xxx
  modeling, 15, 102
  risk reduction, xxvii, 166, 185, 189, 193–194, 205–206
  test conditions, 44–45
  during vehicle perforations, 22
very slow dissolution, 60, 72–73, 199
veterans, xxxi, 9–10, 138, 140–141

W
wall losses, cascade impactors (CI), 33–35, 34 *t*
water, drinking, 3, 137, 138 *t*
whole body
  cancer risks, 154, 158–160, 161 *t*
  E(50), 159 *t*
  external radiation doses, 182
  low-LET exposure, 191–193, 192 *t*
worker
  deaths and injuries, 139
  exposure limits, 178 *t,* 192
  health risks, 152–157
  industrial exposures, 126–127, 137, 149, 150–152, 155–156
  lifetime exposures, 140
  mortality, 151 *t,* 157 *t*
wounds and injuries, xxxi, 128–129, 132, 139, 140–141

Y
yellowcake, 137, 138 *t*